COGNITIVE RADIO NETWORKS

Architectures, Protocols, and Standards

WIRELESS NETWORKS AND MOBILE COMMUNICATIONS

Dr. Yan Zhang, Series Editor
Simula Research Laboratory, Norway
E-mail: yanzhang@ieee.org

Broadband Mobile Multimedia: Techniques and Applications
Yan Zhang, Shiwen Mao, Laurence T. Yang, and Thomas M. Chen
ISBN: 978-1-4200-5184-1

Cognitive Radio Networks: Architectures, Protocols, and Standards
Yan Zhang, Jun Zheng, and Hsiao-Hwa Chen,
ISBN: 978-1-4200-7775-9

Cooperative Wireless Communications
Yan Zhang, Hsiao-Hwa Chen, and Mohsen Guizani
ISBN: 978-1-4200-6469-8

Distributed Antenna Systems: Open Architecture for Future Wireless Communications
Honglin Hu, Yan Zhang, and Jijun Luo
ISBN: 978-1-4200-4288-7

The Internet of Things: From RFID to the Next-Generation Pervasive Networked Systems
Lu Yan, Yan Zhang, Laurence T. Yang, and Huansheng Ning
ISBN: 978-1-4200-5281-7

Millimeter Wave Technology in Wireless PAN, LAN and MAN
Shao-Qiu Xiao, Ming-Tuo Zhou, and Yan Zhang
ISBN: 978-0-8493-8227-7

Mobile WiMAX: Toward Broadband Wireless Metropolitan Area Networks
Yan Zhang and Hsiao-Hwa Chen
ISBN: 978-0-8493-2624-0

Orthogonal Frequency Division Multiple Access Fundamentals and Applications
Tao Jiang, Lingyang Song, and Yan Zhang
ISBN: 978-1-4200-8824-3

Resource, Mobility, and Security Management in Wireless Networks and Mobile Communications
Yan Zhang, Honglin Hu, and Masayuki Fujise
ISBN: 978-0-8493-8036-5

RFID and Sensor Networks: Architectures, Protocols, Security and Integrations
Yan Zhang, Laurence T. Yang, and Jimlng Chen
ISBN: 978-1-4200-7777-3

Security in RFID and Sensor Networks
Yan Zhang and Paris Kitsos
ISBN: 978-1-4200-6839-9

Security in Wireless Mesh Networks
Yan Zhang, Jun Zheng, and Honglin Hu
ISBN: 978-0-8493-8250-5

Unlicensed Mobile Access Technology: Protocols, Architectures, Security, Standards, and Applications
Yan Zhang, Laurence T. Yang, and Jianhua Ma
ISBN: 978-1-4200-5537-5

WiMAX Network Planning and Optimization
Yan Zhang
ISBN: 978-1-4200-6662-3

Wireless Ad Hoc Networking: Personal-Area, Local-Area, and the Sensory-Area Networks
Shih-Lin Wu, Yu-Chee Tseng, and Hsin-Chu
ISBN: 978-0-8493-9254-2

Wireless Mesh Networking: Architectures, Protocols, and Standards
Yan Zhang, Jijun Luo, and Honglin Hu
ISBN: 978-0-8493-7399-2

Wireless Quality-of-Service: Techniques, Standards, and Applications
Maode Ma, Mieso K. Denko, and Yan Zhang
ISBN: 978-1-4200-5130-8

COGNITIVE RADIO NETWORKS

Architectures, Protocols, and Standards

Edited by

Yan Zhang • Jun Zheng • Hsiao-Hwa Chen

CRC Press is an imprint of the
Taylor & Francis Group, an **informa** business
AN AUERBACH BOOK

CRC Press
Taylor & Francis Group
6000 Broken Sound Parkway NW, Suite 300
Boca Raton, FL 33487-2742

First issued in paperback 2019

ISBN-13: 978-1-4200-7775-9 (hbk)
ISBN-13: 978-0-367-38398-5 (pbk)

Library of Congress Cataloging-in-Publication Data

Cognitive radio networks : architectures, protocols, and standards / editors, Yan Zhang, Jun Zheng, and Hsiao-Hwa Chen.
p. cm.
"A CRC title."
Includes bibliographical references and index.
ISBN 978-1-4200-7775-9 (alk. paper)
1. Cognitive radio networks. I. Zhang, Yan, 1977- II. Zheng, Jun, Ph.D. III. Chen, Hsiao-Hwa. IV. Title.

TK5103.4815.C646 2010
621.384--dc22 2009053747

Visit the Taylor & Francis Web site at
http://www.taylorandfrancis.com

and the CRC Press Web site at
http://www.crcpress.com

Contents

Preface

Spectrum is a scarce and precious resource in wireless communication systems and networks. Currently, wireless networks are regulated by a fixed spectrum assignment policy. This strategy partitions the spectrum into a large number of different ranges. Each piece is specified for a particular system. This leads to the undesirable situation that some systems may use only the allocated spectrum to a very limited extent while others have very serious spectrum insufficiency problems. In addition, future-generation broadband wireless networking promises to provide broadband multimedia services under heterogeneous networks coexistence. These challenges and requirements make the problem of scarce spectra even worse, and motivate new technologies to efficiently use spectra and combat the vulnerability of wireless channels.

Cognitive radio is believed to be a high-potential technology to address these issues. It refers to the potentiality that systems are aware of context and are capable of reconfiguring themselves based on the surrounding environments and their own properties with respect to spectrum, traffic load, congestion situation, network topology, and wireless channel propagation. This capability is particularly applicable to resolve heterogeneity, robustness, and openness. However, cognitive wireless networks are still in the very early stages of research and development. There are a number of technical, economical, and regulatory challenges to be addressed. In addition, there are unique complexities in aspects of spectrum sensing, spectrum management, spectrum sharing, and spectrum mobility.

This book systematically introduces and explains cognitive radio wireless networks. It provides a comprehensive technical guide covering introductory concepts, fundamental techniques, recent advances, and open issues in cognitive radio communications and networks. It also contains illustrative figures and allows for complete cross-referencing.

This book is organized into three parts:

- Part I: Physical Layer Issues
- Part II: Protocols and Economic Approaches
- Part III: Applications and Systems

Part I introduces the issues and solutions in the physical layer, including sensing, capacity, and power control. Part II introduces the issues and solutions in the protocol layers. This part also contributes to the applications of economic approaches in cognitive radio networks. Part III explores applications and practical cognitive radio systems.

This book has the following salient features:

- It serves as a comprehensive and essential reference on cognitive radio.
- It covers basics, a broad range of topics, and future development directions.
- It introduces architectures, protocols, security, and applications.
- It assists professionals, engineers, students, and researchers

This book can serve as an essential reference for students, educators, research strategists, scientists, researchers, and engineers in the field of wireless communications and networking. In particular, it will have an instant appeal to students, researchers, developers, and consultants in developing future-generation wireless systems and networks. The content in this book will enable readers to understand the necessary background, concepts, and principles in the framework of cognitive wireless systems. It will also provide readers with a comprehensive technical guidance on cognitive radio, cognitive wireless networks, and dynamic spectrum access. The issues covered include spectrum sensing, medium access control (MAC), cooperation schemes, resource management, mobility, game theoretical approach, and healthcare application.

We would like to acknowledge the time and effort invested by the contributors for their excellent work. All of them were extremely professional and cooperative. Special thanks go to Richard O'Hanley, Stephanie Morkert, and Joette Lynch of Taylor & Francis Group for their patience, support, and professionalism from the beginning until the final stage. We are very grateful to Sathyanarayanamoorthy Sridharan at SPi for his great efforts during the production process. Last but not least, a special thank you to our families and friends for their constant encouragement, patience, and understanding throughout this project.

Yan Zhang
Simula Research Laboratory, Norway

Jun Zheng
Southeast University, China

Hsiao-Hwa Chen
National Cheng Kung University, Taiwan

Editors

Yan Zhang received a BS in communication engineering from the Nanjing University of Post and Telecommunications, China; an MS in electrical engineering from the Beijing University of Aeronautics and Astronautics, China; and a PhD from the School of Electrical & Electronics Engineering, Nanyang Technological University, Singapore.

He is an associate editor or editorial board member of Wiley's *International Journal of Communication Systems* (*IJCS*); the *International Journal of Communication Networks and Distributed Systems* (*IJCNDS*); Springer's *International Journal of Ambient Intelligence and Humanized Computing* (*JAIHC*); the *International Journal of Adaptive, Resilient and Autonomic Systems* (*IJARAS*); Wiley's *Wireless Communications and Mobile Computing* (*WCMC*); Wiley's *Security and Communication Networks*; the *International Journal of Network Security;* the *International Journal of Ubiquitous Computing*; *Transactions on Internet and Information Systems* (*TIIS*); the *International Journal of Autonomous and Adaptive Communications Systems* (*IJAACS*); the *International Journal of Ultra Wideband Communications and Systems* (*IJUWBCS*); and the *International Journal of Smart Home* (*IJSH*).

He is currently serving as an editor for the book series Wireless Networks and Mobile Communications (Auerbach Publications, CRC Press, Taylor & Francis Group). He serves as a guest coeditor for Wiley's *Wireless Communications and Mobile Computing* (*WCMC*) special issue for best papers in the conference IWCMC 2009; ACM/Springer's *Multimedia Systems Journal* special issue on "wireless multimedia transmission technology and application"; Springer's *Journal of Wireless Personal Communications* special issue on "cognitive radio networks and communications"; Inderscience's *International Journal of Autonomous and Adaptive Communications Systems* (*IJAACS*) special issue on "ubiquitous/pervasive services and applications"; EURASIP's *Journal on Wireless Communications and Networking* (*JWCN*) special issue on "broadband wireless access"; *IEEE Intelligent Systems* special issue on "context-aware middleware and intelligent agents for smart environments"; Wiley's *Security and Communication Networks* special issue on "secure multimedia communication"; Elsevier's *Computer Communications* special issue on "adaptive multicarrier communications and networks"; Inderscience's *International Journal of*

Autonomous and Adaptive Communications Systems (*IJAACS*) special issue on "cognitive radio systems"; the *Journal of Universal Computer Science* (*JUCS*) special issue on "multimedia security in communication"; Springer's *Journal of Cluster Computing* special issue on "algorithm and distributed computing in wireless sensor networks"; EURASIP's *Journal on Wireless Communications and Networking* (*JWCN*) special issue on "OFDMA architectures, protocols, and applications"; and Springer's *Journal of Wireless Personal Communications* special issue on "security and multimodality in pervasive environments."

He is also serving as a coeditor for several books, including *Resource, Mobility and Security Management in Wireless Networks and Mobile Communications*; *Wireless Mesh Networking: Architectures, Protocols and Standards*; *Millimeter-Wave Technology in Wireless PAN, LAN and MAN*; *Distributed Antenna Systems: Open Architecture for Future Wireless Communications*; *Security in Wireless Mesh Networks*; *Mobile WiMAX: Toward Broadband Wireless Metropolitan Area Networks*; *Wireless Quality-of-Service: Techniques, Standards and Applications*; *Broadband Mobile Multimedia: Techniques and Applications*; *Internet of Things: From RFID to the Next-Generation Pervasive Networked Systems*; *Unlicensed Mobile Access Technology: Protocols, Architectures, Security, Standards and Applications*; *Cooperative Wireless Communications*; *WiMAX Network Planning and Optimization*; *RFID Security: Techniques, Protocols and System-on-Chip Design*; *Autonomic Computing and Networking*; *Security in RFID and Sensor Networks*; *Handbook of Research on Wireless Security*; *Handbook of Research on Secure Multimedia Distribution*; *RFID and Sensor Networks*; *Cognitive Radio Networks*; *Wireless Technologies for Intelligent Transportation Systems*; *Vehicular Networks: Techniques, Standards and Applications*; *Orthogonal Frequency Division Multiple Access* (*OFDMA*); *Game Theory for Wireless Communications and Networking*; and *Delay Tolerant Networks: Protocols and Applications.*

He serves or has served as industrial liaison cochair for UIC 2010, program cochair for WCNIS 2010, symposium vice chair for CMC 2010, program track chair for BodyNets 2010, program chair for IWCMC 2010, program cochair for WICON 2010, program vice chair for CloudCom 2009, publicity cochair for IEEE MASS 2009, publicity cochair for IEEE NSS 2009, publication chair for PSATS 2009, symposium cochair for ChinaCom 2009, program cochair for BROADNETS 2009, program cochair for IWCMC 2009, workshop cochair for ADHOCNETS 2009, general cochair for COGCOM 2009, program cochair for UC-Sec 2009, journal liasion chair for IEEE BWA 2009, track cochair for ITNG 2009, publicity cochair for SMPE 2009, publicity cochair for COMSWARE 2009, publicity cochair for ISA 2009, general cochair for WAMSNet 2008, publicity cochair for TrustCom 2008, general cochair for COGCOM 2008, workshop cochair for IEEE APSCC 2008, general cochair for WITS-08, program cochair for PCAC 2008, general cochair for CONET 2008, workshop chair for SecTech 2008, workshop chair for SEA 2008, workshop co-organizer for MUSIC'08, workshop co-organizer for 4G-WiMAX 2008, publicity cochair for SMPE-08, international journals coordinating cochair for FGCN-08, publicity cochair for ICCCAS 2008, workshop chair for ISA 2008, symposium cochair for ChinaCom 2008, industrial cochair for MobiHoc 2008, program cochair for UIC-08, general cochair for CoNET 2007, general cochair for WAMSNet 2007, workshop cochair for FGCN 2007, program vice cochair for IEEE ISM 2007, publicity cochair for UIC-07, publication chair for IEEE ISWCS 2007, program cochair for IEEE PCAC'07, special track cochair for Mobility and Resource Management in Wireless/Mobile Networks in ITNG 2007, special session co-organizer for Wireless Mesh Networks in PDCS 2006, a member of the Technical Program Committee for numerous international conferences, including ICC, GLOBECOM, WCNC, PIMRC, VTC, CCNC, AINA, ISWCS, etc. He received the Best Paper Award in the IEEE 21st International Conference on Advanced Information Networking and Applications (AINA-07).

Since August 2006, he has been working with Simula Research Laboratory, Lysaker, Norway (http://www.simula.no/). His research interests include resource, mobility, spectrum, data, energy, and security management in wireless networks and mobile computing. He is a member of IEEE and IEEE ComSoc.

Jun Zheng is a full professor with the National Mobile Communications Research Laboratory at Southeast University, Nanjing, China. He received a PhD in electrical and electronic engineering from the University of Hong Kong, China. Before joining Southeast University, he was with the School of Information Technology and Engineering of the University of Ottawa, Canada.

Dr. Zheng serves as a technical editor of *IEEE Communications Magazine* and *IEEE Communications Surveys & Tutorials*. He is also the founding editor in chief of *ICST Transactions on Mobile Communications and Applications*, and an editorial board member of several other refereed journals, including Wiley's *Wireless Communications and Mobile Computing*, Wiley's *Security and Communication Networks*, Inderscience's *International Journal of Communication Networks and Distributed Systems*, and Inderscience's *International Journal of Autonomous and Adaptive Communications Systems*. He has coedited eight special issues for different refereed journals and magazines, including *IEEE Journal on Selected Areas in Communications*, *IEEE Network*, Wiley's *Wireless Communications and Mobile Computing*, Wiley's *International Journal of Communication Systems*, and Springer's *Mobile Networks and Applications*, all as lead guest editor.

Dr. Zheng has served as general chair of AdHoctNets'09 and AccessNets'07, TPC cochair of AdHocNets'10 and AccessNets'08, and symposium cochair of IEEE GLOBECOM'08, ICC'09, GLOBECOM'10, and ICC'11. He is also serving on the steering committees of AdHocNets and AccessNets, and has served on the technical program committees of a number of international conferences and symposia, including IEEE ICC and GLOBECOM.

Dr. Zheng has conducted extensive research in the field of communication networks. The scope of his research includes design and analysis of network architecture and protocols for efficient and reliable communications, and their applications to different types of communication networks, covering wireless networks and wired networks. His current research interests are focused on mobile communications and wireless ad hoc networks. He has coauthored books published by Wiley–IEEE Press, and has published a number of technical papers in refereed journals and magazines as well as in peer-reviewed conference proceedings. He is a senior member of the IEEE.

Hsiao-Hwa Chen is currently a full professor in the Department of Engineering Science, National Cheng Kung University, Tainan, Taiwan. He received a BSc and MSc with the highest honor from Zhejiang University, Hangzhou, China, and a PhD from the University of Oulu, Finland, in 1982, 1985, and 1990, respectively, all in electrical engineering. He worked with the Academy of Finland as a research associate from 1991 to 1993, and with the National University of Singapore as a lecturer and then as a senior lecturer from 1992 to 1997. He joined the Department of Electrical Engineering, National Chung Hsing University, Taichung, Taiwan, as an associate professor in 1997 and was promoted to a full professor in 2000. In 2001, he joined National Sun Yat-Sen University, Kaohsiung, Taiwan, as the founding chair of the Institute of Communications Engineering of the university. Under his strong leadership, the institute was ranked second in the country in terms of SCI journal publications and National Science Council funding per faculty member in 2004. In particular, National Sun Yat-Sen University was ranked first in the world in terms of the number of SCI journal publications in wireless LAN research papers during 2004 to mid-2005, according to a research report released by The Office of Naval Research, United States. He was a visiting professor to the Department of Electrical Engineering, University of Kaiserslautern, Germany, in 1999; the Institute of Applied Physics, Tsukuba University, Japan, in 2000; the Institute of Experimental Mathematics, University of Essen, Germany, in 2002 (under DFG Fellowship); the Chinese University of Hong Kong in 2004; and the City University of Hong Kong in 2007.

His current research interests include wireless networking, MIMO systems, information security, and Beyond 3G wireless communications. He is the inventor of next-generation CDMA technologies. He is also a recipient of numerous research and teaching awards from the National Science Council, the Ministry of Education, and other professional groups in Taiwan. He has authored or coauthored over 200 technical papers in major international journals and conferences, and five books and several book chapters in the area of communications, including *Next Generation Wireless Systems and Networks* and *The Next Generation CDMA Technologies*, both of which were published by Wiley in 2005 and 2007, respectively.

He has been an active volunteer for IEEE for various technical activities for over 15 years. Currently, he is serving as the chair of IEEE Communications Society Radio Communications Committee, and the vice chair of IEEE Communications Society Communications & Information Security Technical Committee. He served or is serving as symposium chair/cochair of many major IEEE conferences, including IEEE VTC 2003 Fall, IEEE ICC 2004, IEEE Globecom 2004, IEEE ICC

2005, IEEE Globecom 2005, IEEE ICC 2006, IEEE Globecom 2006, IEEE ICC 2007, IEEE WCNC 2007, etc. He served or is serving as an editorial board member and/or guest editor of *IEEE Communications Letters, IEEE Communications Magazine, IEEE Wireless Communications Magazine, IEEE JSAC, IEEE Network Magazine, IEEE Transactions on Wireless Communications*, and *IEEE Vehicular Technology Magazine*. He is the editor in chief of Wiley's *Security and Communication Networks* journal (www.interscience.wiley.com/journal/security), and the special issue editor in chief of *Hindawi Journal of Computer Systems, Networks, and Communications* (http://www.hindawi.com/journals/jcsnc/). He is also serving as the chief editor (Asia and Pacific) for Wiley's *Wireless Communications and Mobile Computing* (*WCMC*) journal and *International Journal of Communication Systems*. His original work in CDMA wireless networks, digital communications, and radar systems has resulted in five U.S. patents, two Finnish patents, three Taiwanese patents, and two Chinese patents, some of which have been licensed to industry for commercial applications. He is an adjunct professor of Zhejiang University, China, and Shanghai Jiao Tong University, China. Professor Chen is the recipient of the Best Paper Award in IEEE WCNC 2008 and he is also a fellow of IEEE and IET.

Contributors

S. Anand
Department of Electrical and Computer Engineering
Stevens Institute of Technology
Hoboken, New Jersey

John Attia
Department of Electrical and Computer Engineering
Prairie View A&M University
Prairie View, Texas

Yong Bai
DOCOMO
Beijing Communications Labs
Beijing, China

Jack L. Burbank
Applied Physics Laboratory
Johns Hopkins University
Baltimore, Maryland

Sergio Camorlinga
Departments of Radiology and Computer Science
University of Manitoba

and

TRLabs
Winnipeg, Manitoba, Canada

Leonardo S. Cardoso
SUPELEC
Gif-sur-Yvette, France

R. Chandramouli
Department of Electrical and Computer Engineering
Stevens Institute of Technology
Hoboken, New Jersey

Jean-Marie Chaufray
Orange Labs
France Telecom R&D
Paris, France

Hsiao-Hwa Chen
Department of Engineering Science
National Cheng Kung University
Tainan, Taiwan

Lan Chen
DOCOMO
Beijing Communications Labs
Beijing, China

Mérouane Debbah
SUPELEC
Gif-sur-Yvette, France

Christian Doerr
Department of Computer Science
University of Colorado
Boulder, Colorado

and

Department of Telecommunications
Technische Universiteit Delft
Delft, the Netherlands

Dirk Grunwald
Department of Computer Science
University of Colorado
Boulder, Colorado

Zhu Han
Department of Electrical and Computer Engineering
University of Houston
Houston, Texas

Xuemin Hong
Joint Research Institute for Signal and Image Processing
School of Engineering and Physical Sciences
Heriot-Watt University
Edinburgh, United Kingdom

Ekram Hossain
Department of Electrical and Computer Engineering
University of Manitoba

and

TRLabs
Winnipeg, Manitoba, Canada

Jianwei Huang
Department of Information Engineering
The Chinese University of Hong Kong
Hong Kong, People's Republic of China

Deepak Kataria
LSI Corporation
Allentown, Pennsylvania

Mari Kobayashi
SUPELEC
Gif-sur-Yvette, France

Samson Lasaulce
Laboratoire des Signaux et Systèmes
Centre National de la Recherche Scientifique
SUPELEC
Gif-sur-Yvette, France

Victor C. M. Leung
Department of Electrical and Computer Engineering
The University of British Columbia
Vancouver, British Columbia, Canada

Haoming Li
Department of Electrical and Computer Engineering
The University of British Columbia
Vancouver, British Columbia, Canada

Xiangfang Li
Department of Electrical and Computer Engineering
Texas A&M University
College Station, Texas

Klaus Nolte
Alcatel-Lucent Deutschland AG
Bell Labs
Stuttgart, Germany

Jacques Palicot
Signal, Communication et
Electronique Embarquée
SUPELEC
Rennes, France

Qixiang Pang
General Dynamics Canada
Calgary, Alberta, Canada

Samir Medina Perlaza
Orange Labs
France Telecom R&D
Paris, France

Phond Phunchongharn
Department of Electrical and
Computer Engineering
University of Manitoba

and

TRLabs
Winnipeg, Manitoba, Canada

Lijun Qian
Department of Electrical and
Computer Engineering
Prairie View A&M University
Prairie View, Texas

S. Sengupta
Department of Mathematics and
Computer Science
City University of New York
New York, New York

Douglas C. Sicker
Department of Computer Science
University of Colorado
Boulder, Colorado

Mikhail Smirnov
Fraunhofer Institute for Open
Communication Systems
Berlin, Germany

John Thompson
Institute for Digital Communications
Joint Research Institute for Signal and
Image Processing
School of Engineering and
Electronics
The University of Edinburgh
Edinburgh, United Kingdom

Jens Tiemann
Fraunhofer Institute for Open
Communication Systems
Berlin, Germany

Cheng-Xiang Wang
Joint Research Institute for Signal and
Image Processing
School of Engineering and Physical
Sciences
Heriot-Watt University
Edinburgh, United Kingdom

Jie Xiang
Simula Research Laboratory
Lysaker, Norway

Yifan Yu
DOCOMO
Beijing Communications Labs
Beijing, China

Yan Zhang
Simula Research Laboratory
Lysaker, Norway

PHYSICAL LAYER ISSUES

Chapter 1

Spectrum Sensing in Cognitive Radio Networks

Leonardo S. Cardoso, Mérouane Debbah,
Samson Lasaulce, Mari Kobayashi, and
Jacques Palicot

Contents

Today, the creation of new radio access technologies is limited by the shortage of the available radio spectrum. These new technologies are becoming evermore bandwidth demanding due to their higher rate requirements. Cognitive radio networks and spectrum-sensing techniques are a natural way to allow these new technologies to be deployed.

In this chapter, we discuss spectrum sensing for cognitive radio networks. We begin by introducing the subject in Section 1.1, providing a brief background followed by a discussion of spectrum-sensing motivations and characteristics. Then we move on to the spectrum-sensing problem itself in Section 1.2, where we explain the issues that are inherent to spectrum sensing. In Section 1.3, we explore the classical noncooperative spectrum-sensing techniques that form the basis for the more elaborate, cooperative techniques presented in Section 1.4. Finally, we close this chapter with some conclusions and open issues.

1.1 Introduction

One of the most prominent features of cognitive radio networks will be the ability to switch between radio access technologies, transmitting in different portions of the radio spectrum as unused frequency band slots arise [1–3]. This dynamic spectrum access is one of the fundamental requirements for transmitters to adapt to varying channel qualities, network congestion, interference, and service requirements. Cognitive radio networks (from now on called secondary networks) will also need to coexist with legacy ones (hereafter called primary networks), which have the right to their spectrum slice and thus cannot accept interference.

Based on these facts, underutilization of the current spectrum and the need to increase the network capacity is pushing research toward new means of exploiting the wireless medium. In this direction, the Federal Communications Commission (FCC) Spectrum Policy Task Force has published a report [4] in 2002, in which it thoroughly investigates the underutilization of the radio spectrum. While the FCC is in charge of determining the spectrum usage and its policies, the Whitespace Coalition is studying ways to exploit the spectrum vacancies in the television band. Cognitive radio networks are envisioned to be able to opportunistically exploit those spectrum "leftovers," by means of knowledge of the environment and cognition capability, to adapt to their radio parameters accordingly. Spectrum sensing is the technique that will enable cognitive radio networks to achieve this goal.

1.1.1 Interference Management and Spectrum Sensing

To share the spectrum with legacy systems, cognitive radio networks will have to respect some set of policies defined by regulatory agencies [2,3]. These policies are based on the central idea where there are primary systems that have the right to the spectrum and secondary systems that are allowed to use the spectrum so long as they do not disturb the communications of the primary systems. Roughly speaking, these policies deal with controlling the amount of interference that the secondary systems can incur to primary ones. Thus, the problem is one of interference management [2,3]. We can address this problem from two different points of view: receiver centric or transmitter centric.

1.1.1.1 Receiver-Centric Interference Management

In the receiver-centric approach [2,3], an interference limit at the receiver is calculated and used to determine the restriction on the power of the transmitters around it. This interference limit, called the interference temperature, is chosen to be the worst interference level that can be accepted without disturbing the receiver operation beyond its operating point. Although very interesting, this approach requires knowledge of the interference limits of all receivers in a primary system. Such knowledge depends on many variables, including individual locations, fading situations, modulations, coding schemes, and services. Receiver-centric interference-management techniques are not addressed in this chapter as they have been recently ruled out by the IEEE SCC41 cognitive radio network standard.

1.1.1.2 Transmitter-Centric Interference Management

In the transmitter-centric approach, the focus is shifted to the source of interference [2,3]. The transmitter does not know the interference temperature, but by means of sensing, it tries to detect free bandwidth. The sensing procedure allows the transmitter to classify the channel status to decide whether it can transmit and with how much power. In actual systems, however, as the transmitter does not know the location of the receivers or their channel conditions, it is not able to infer how much interference these receivers can tolerate. Thus, spectrum sensing solves the problem for worst-case scenario, assuming strong interference channels, so that the secondary system transmits only when it senses an empty medium.

1.1.2 Characteristics of Spectrum Sensing

There are several techniques available for spectrum sensing, each with its own set of advantages and disadvantages that depend on the specific scenario. Some works in the literature [5–7] consider spectrum sensing as a method for distinguishing between two or more different types of signals or technologies in operation. Because this is not a question of detection (determining whether a given frequency band is being used), these types of signal identification issues [8] are not addressed in this chapter. Rather we focus on their detection.

Ultimately, a spectrum-sensing device must be able to give a general picture of the medium over the entire radio spectrum. This allows the cognitive radio network to analyze all degrees of freedom (time, frequency, and space) to predict the spectrum usage. Wideband spectrum-sensing works are also available in the literature [9–12]; however, an equipment able to perform wideband sensing all at once is prohibitively difficult to build with today's technology. Feasible spectrum-sensing devices can quickly sweep the radio spectrum, analyzing one narrowband segment at a time. This chapter focuses on narrowband-sensing techniques.

In this section, we have emphasized the importance of the spectrum-sensing technique for cognitive radio networks. In the next section, we aim at understanding the underlying characteristics of the spectrum-sensing problem, which will enable us to develop the approaches presented further in this chapter.

1.2 Problem Formulation

1.2.1 The General Spectrum-Sensing Problem

Spectrum sensing is based on a well-known technique called signal detection. In a nutshell, signal detection can be described as a method for identifying the presence of a signal in a noisy environment. Signal detection has been thoroughly studied for radar purposes since the 1950s [13]. Analytically, signal detection can be reduced to a simple identification problem, formalized as a hypothesis test [14–16]:

$$y(k) = \begin{cases} n(k): & \mathrm{H}_0 \\ s(k) + n(k): & \mathrm{H}_1 \end{cases}, \tag{1.1}$$

where

$y(k)$ is the sample to be analyzed at each instant k

$n(k)$ is the noise (not necessarily white Gaussian noise) of variance σ^2

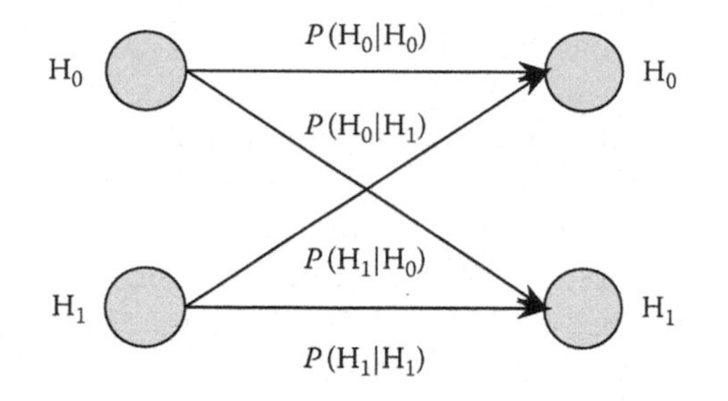

Figure 1.1 Hypothesis test and possible outcomes with their corresponding probabilities.

$s(k)$ is the signal the network wants to detect
H_0 and H_1 are the noise-only and signal-plus-noise hypotheses, respectively

H_0 and H_1 are the sensed states for the absence and presence of signal, respectively. Then, as shown in Figure 1.1 we can define four possible cases for the detected signal:

1. Declaring H_0 when H_0 is true $(H_0|H_0)$
2. Declaring H_1 when H_1 is true $(H_1|H_1)$
3. Declaring H_0 when H_1 is true $(H_0|H_1)$
4. Declaring H_1 when H_0 is true $(H_1|H_0)$

Case 2 is known as a correct detection, whereas cases 3 and 4 are known as a missed detection and a false alarm, respectively. Clearly, the aim of the signal detector is to achieve correct detection all of the time, but this can never be perfectly achieved in practice because of the statistical nature of the problem. Therefore, signal detectors are designed to operate within prescribed minimum error levels. Missed detections are the biggest issue for spectrum sensing, as it means possibly interfering with the primary system. Nevertheless, it is desirable to keep the false alarm rate as low as possible for spectrum sensing, so that the system can exploit all possible transmission opportunities.

The performance of the spectrum-sensing technique is usually influenced by the probability of false alarm $P_f = P(H_1|H_0)$, because this is the most influential metric. Usually, the performance is presented by receiver operation characteristic (ROC) curves, which plot the probability of detection $P_d = P(H_1|H_1)$ as a function of the probability of false alarm P_f.

Equation 1.1 shows that, to distinguish H_0 and H_1, a reliable way to differentiate signal from noise is required. This becomes very difficult in the case where the statistics of the noise are not well known or when the signal-to-noise ratio (SNR) is low, in which case the signal characteristics are buried under the noise, as shown by Tandra et al. in [17]. In fact, this work also shows that the lesser one knows about the statistics of the noise, the worse the performance of any signal detector is in the low-SNR regime.

Clearly, the noise characteristics are very important for the spectrum-sensing procedure. Most works on spectrum sensing consider noise to be additive white Gaussian noise (AWGN), because many independent sources of noise are added (central limit theory). Nevertheless, in realistic scenarios, this approximation may not be appropriate, because receivers modify the noise through processes such as filters, amplifier nonlinearities, and automatic gain controls [18,19].

Poor performance in a low-SNR regime means that all of the techniques available are negatively affected by poor channels. In the case of variable channel gains, Equation 1.1 is rewritten as

$$y(k) = \begin{cases} n(k): & \mathrm{H}_0 \\ h(k)s(k) + n(k): & \mathrm{H}_1 \end{cases}, \tag{1.2}$$

where $h(k)$ is the channel gain at each instant k. In a wireless radio network, as it is reasonable to assume that the spectrum-sensing device does not know the location of the transmitter, two options arise:

- A low $h(k)$ is solely due to the pathloss (distance) between the transmitter and the sensing device, meaning that the latter is out of range.
- A low $h(k)$ is due to shadowing or multipath, meaning that the sensing device might be within the range of the transmitter.

In the latter case, a critical issue arises. Therein, the fading plays an especially negative role in the well-known "hidden node" problem [20]. In this problem, the spectrum-sensing terminal is deeply faded with respect to the transmitting node while having a good channel to the receiving node. The spectrum-sensing node then senses a free medium and initiates its transmission, which produces interference on the primary transmission. Thus, fading here introduces uncertainty regarding the estimation problem. To solve this issue, cooperative sensing has been proposed. In this approach, several sensing terminals gather their information to make a joint decision about the medium availability. Cooperative spectrum sensing is further explored in Section 1.4.

1.2.2 Spectrum Sensing from the Cognitive Radio Network Perspective

In contrast to the general case, where only the signal detection aspect is considered, the problem of spectrum sensing as seen from a cognitive radio perspective has very stringent restrictions. These are mainly imposed by the policies these cognitive radio networks face to be able to operate alongside legacy networks. Some of these restrictions are summarized in Sections 1.2.2.1 through 1.2.2.3.

1.2.2.1 No Prior Knowledge on the Signal Structure

There are portions of the spectrum where multiple technologies (using different protocols) share the spectrum, such as the ones operating on the instrumentation scientific and medical (ISM) unlicensed band. Cognitive radio networks must be able to deal with existing multiple technologies, as well as new ones that may eventually appear across the span of the wireless radio spectrum. These networks should be able to discover the state of the medium irrespective of the technologies in use. Of course, if the technologies are known, then this information can be exploited to improve the accuracy of the spectrum sensing, for example, through the detection of known pilot sequences within the signal [17].

1.2.2.2 Sensing Time

Due to the primary importance of the legacy system, the secondary system must be designed to free the medium as soon as it senses that a legacy network has initiated a transmission. For efficient use of the spectrum, these secondary networks must also sense available spectrum as quickly as possible, in the least possible number of received samples. In general terms, spectrum-sensing techniques work through a compromise between the number of samples and accuracy. Cooperative spectrum sensing gives the opportunity to decrease the sensing time for the same level of accuracy.

1.2.2.3 Fading Channels

As discussed earlier, spectrum sensing is particularly sensitive to fading environments. Communication systems operate in diverse environments, including those prone to fading. Thus, in many situations, spectrum-sensing devices must be able to detect reliably even over heavily faded channels. Although several works have focused on sensing for the fading environment in the noncooperative setting [21,22], it is foreseen that cooperative sensing [23–31] is the best way to address this problem. Nevertheless, it creates other implications such as the distribution of metrics among the sensing terminals and the decision regarding which terminals are to be considered dependable or not.

1.3 Noncooperative Sensing Techniques

In a realistic spectrum-sensing scenario, there are situations in which only one sensing terminal is available or in which no cooperation is allowed due to the lack of communication between sensing terminals. In this section, we explore the main single-user sensing schemes, some of which will serve as a basis for the development of the cooperative ones investigated in Section 1.4.

Single-user spectrum-sensing approaches have been widely studied in the literature, in part because of the relationship to signal detection. There are several classical techniques for this purpose, including the energy detector (ED) [16,21,22], the matched filter (MF) [25,32], and the cyclostationary feature detection (CFD) [6,33–36].

1.3.1 Energy Detector

The most well-known spectrum-sensing technique is the ED. It is based on the principle that, at the reception, the energy of the signal to be detected is always higher than the energy of the noise. The ED is said to be a blind signal detector because it ignores the structure of the signal. It estimates the presence of a signal by comparing the energy received with a known threshold ν [16,21,22], derived from the statistics of the noise.

Let $y(k)$ be a sequence of received samples $k \in \{1, 2, \ldots, N\}$ at the signal detector, such as that in Equation 1.1. Then, the decision rule can be stated as

$$\text{decide for} \begin{cases} \mathrm{H}_0, & \text{if } \mathcal{E} < \nu \\ \mathrm{H}_1, & \text{if } \mathcal{E} \geq \nu \end{cases},$$

where

$\mathcal{E} = E[\mid y(k) \mid^2]$ is the estimated energy of the received signal

ν is chosen to be the noise variance σ^2

In practice, one does not dispose of the actual received energy power $\mathcal{E}$. The ED uses instead the approximation $\hat{\mathcal{E}}$, where

$$\hat{\mathcal{E}} \triangleq \frac{1}{N} \sum_{k=1}^{N} \mid y(k) \mid^2 .$$

As the number of samples N becomes large, by the law of the large numbers, $\hat{\mathcal{E}}$ converges to $\mathcal{E}$.

The ED is one of the simplest signal detectors. Its operation is very straightforward, and it has a very easy implementation, because it depends only on simple and readily available information.

Nevertheless, in spite of its simplicity, the ED is not a perfect solution. The approximation of signal energy $\mathcal{E}$ gets better as N increases. Thus, the performance of the ED is directly linked to the number of samples. Furthermore, the ED relies completely on the variance of the noise σ^2, which is taken as a fixed value. This is generally not true in practice, where the noise floor varies. Essentially, this means

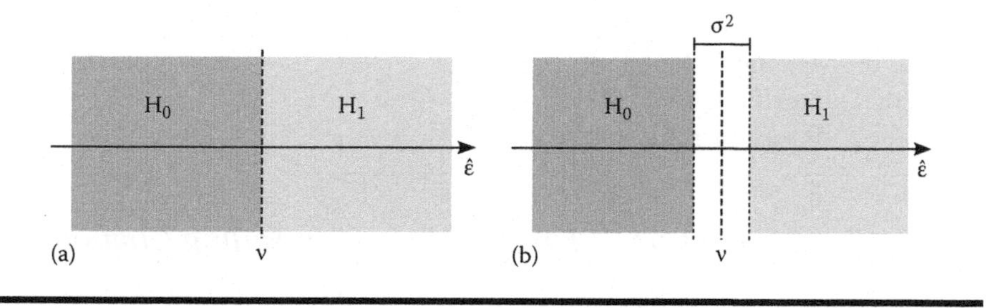

Figure 1.2 (a) Ideal ED scheme. (b) Detection uncertainty for the ED.

that the ED will generate errors during those variations, especially when the SNR is very low, as shown in Figure 1.2b, where we see an area of uncertainty surrounding the threshold ν in contrast to the case portrayed in Figure 1.2a, in which perfect noise knowledge is considered.

1.3.1.1 Characterization of Energy Detector in AWGN Channels

This case has been studied in the work of Urkowitz in 1967 [16]. It is known that the energy detection is the optimal signal detector in AWGN considering no prior information on the signal structure [17]. To understand the inner workings of the ED in this scenario, we need to understand how the probability of detection $P_\mathrm{d} = \mathrm{Prob}\{\hat{\mathcal{E}} > \nu | \mathrm{H}_1\}$ and false alarm $P_\mathrm{f} = \mathrm{Prob}\{\hat{\mathcal{E}} > \nu | \mathrm{H}_0\}$ behave with the measured received signal energy.

Take $n(k) \sim \mathcal{NC}\left(0, \sigma^2\right)$ as the AWGN noise sample. Then, we know that for the noise-only case, the distribution of the energy of n over T samples can be approximated by a zero mean chi-square distribution χ^2_{2TW} [16], where W is the total bandwidth. Similarly, the energy over T samples of the sum of a signal plus noise can be represented by a noncentral chi-square distribution $\chi^2_{2TW}(\lambda)$ [16], where λ is the noncentrality parameter. Briefly:

$$\hat{\mathcal{E}} \sim \begin{cases} \chi^2_{2TW}, & \mathrm{H}_0 \\ \chi^2_{2TW}(\lambda), & \mathrm{H}_1 \end{cases}.$$

With these considerations in mind

$$P_\mathrm{f} = Q_\mathrm{m}(\sqrt{\lambda\,\xi}, \sqrt{\nu}) \tag{1.3}$$

and

$$P_\mathrm{d} = \frac{\Gamma(TW, \nu/2)}{\Gamma(TW)}, \tag{1.4}$$

where

Q_m is the Marcum Q-function
Γ is the gamma function
ξ is the SNR seen by the signal detector

1.3.1.2 Characterization of Energy Detector in Fading Channels

In 2002, Kostylev [21] studied the performance of the ED in fading channels. He derived analytical expressions for the ED over the Rayleigh fading channel case (also analyzed the Rice and Nakagami cases numerically). In 2003, the problem was revisited by Digham et al. [22], who provided an alternative analytical development for these three kinds of fading channels. In this section, however, we will restrict the analysis to the more commonly adopted Rayleigh fading.

Let us begin by recalling that, in this case, the model of interest is the one shown in Equation 1.14. As such, similar to what Urkowitz did in [16], Kostylev characterized the statistics of the energy of the signal for both the H_0 and H_1 cases, under the assumption that $h(k)$ is Rayleigh distributed:

$$\hat{\mathcal{E}} \sim \begin{cases} \chi^2_{2(TW+1)}, & H_0 \\ e_{2(\xi^2+1)} + \chi^2_{2TW}(\lambda), & H_1 \end{cases},$$

where

$e_{2(d^2+1)}$ is the exponential distribution with parameter $\alpha = 2(\xi^2 + 1)$ with probability density function $f(x, \alpha) = \alpha e^{-\alpha x}$
ξ is the SNR

It is clear that, under the hypothesis H_0, the statistics are the same as for the AWGN channel case, so the probability of false alarm is the same as in Equation 1.3.

$$P_f = Q_m(\sqrt{\lambda}\,\xi, \sqrt{\nu}). \tag{1.5}$$

The H_1 case behaves differently and has the probability of detection given by [22]

$$P_d = e^{\frac{\hat{\mathcal{E}}}{2}} \sum_{m=0}^{TW-2} \frac{1}{m!}\left(\frac{\hat{\mathcal{E}}}{2}\right) + \left(\frac{1+\xi}{\xi}\right)^{TW-1}\left[e^{\frac{\hat{\mathcal{E}}}{2(1+\xi)}} - e^{\frac{\hat{\mathcal{E}}}{2}} \sum_{m=0}^{TW-2} \frac{1}{m!}\frac{\hat{\mathcal{E}}\xi}{2(1+\xi)}\right]. \tag{1.6}$$

1.3.2 Matched Filter Detector

We have seen previously in Section 1.3.1 that the best sensing technique in an AWGN environment, and without any knowledge of the signal structure, is the

ED. If we do assume some knowledge of the signal structure, then we can achieve a better performance.

Most of the wireless technologies in operation include the transmission of some sort of pilot sequence to allow channel estimation, to beacon its presence to other terminals, and to give a synchronization reference for subsequent messages. Secondary systems can exploit pilot signals to detect the presence of transmissions of primary systems in their vicinity.

If a pilot signal is known, then the MF signal detector achieves the optimal detection performance in AWGN channel, since it maximizes the SNR, as shown by Tandra and Sahai in [17].

Let us assume that

- The signal detector knows the pilot sequence $x(k)$, the bandwidth, and the center frequency in which it will be transmitted.
- The pilot sequence is always appended to the transmission of each primary system (uplink or downlink).
- The signal detector can always receive coherently.

Then, if $y(k)$ is a sequence of received samples at instant $k \in \{1, 2, \ldots, N\}$ at the signal detector, the decision rule can be stated as [25]

$$\text{decide for} \begin{cases} \mathrm{H}_0, & \text{if } \hat{\mathcal{S}} < \nu \\ \mathrm{H}_1, & \text{if } \hat{\mathcal{S}} \geq \nu \end{cases},$$

where

$$\hat{\mathcal{S}} = \sum_{k=1}^{N} y(k)x(k)^* \tag{1.7}$$

is the decision criterion

ν is the threshold to be compared

$x(k)^*$ is the transpose conjugate of the pilot sequence

Here the threshold ν is not the noise variance as it was for the ED. The hypothesis decision is simplified as the MF maximizes the power of $\hat{\mathcal{S}}$ as shown in Equation 1.7. This means that it performs well even in a low-SNR regime.

The MF has some drawbacks. First, a cognitive spectrum sensor might not know which networks are in operation in the environment at a given moment. Therefore, it may not know which sets of pilots to look for. One must remember that if it tries to match an incorrect pilot, it will sense an empty medium and will incorrectly conclude that the medium is free. Second, the MF requires that every medium access be "signed" by a pilot transmission, but this is not the case in general. Furthermore, pilot sequences are only transmitted in the downlink direction. This leaves the uplink transmissions uncovered. Third, the MF requires coherent reception, which is generally hard to achieve in practice.

1.3.2.1 Characterization of the Matched Filter

Signal detection using the MF was studied in 2006 by Cabric et al. in [25]. They showed that $\hat{\mathcal{S}}$ is Gaussian:

$$\hat{\mathcal{S}} \sim \begin{cases} \mathcal{N}\left(0, \sigma_{\mathrm{n}}^2 \varepsilon\right), & \mathrm{H}_0 \\ \mathcal{N}\left(\varepsilon, \sigma_{\mathrm{n}}^2 \varepsilon\right), & \mathrm{H}_1 \end{cases},$$

where σ_{n}^2 is the variance of the noise and

$$\varepsilon = \sum_{k=1}^{N} x(k)^2.$$

Based on this information, the probabilities of false alarm P_{f} and detection P_{d} are

$$P_{\mathrm{f}} = Q\left(\frac{\hat{\mathcal{S}}}{\sqrt{\varepsilon \sigma_{\mathrm{n}}^2}}\right) \tag{1.8}$$

and

$$P_{\mathrm{d}} = Q\left(\frac{\hat{\mathcal{S}} - \varepsilon}{\sqrt{\varepsilon \sigma_{\mathrm{n}}^2}}\right). \tag{1.9}$$

1.3.3 Cyclostationary Feature Detection

As we have seen, although it performs well, even in the low-SNR regime, the MF requires a good knowledge of the signal structure, which secondary terminals may not have. The natural question to ask is whether we can still be able to perform spectrum sensing with a limited knowledge of the signal structure, perhaps based on a characteristic that is common to most known transmitted signals. In the following text, we show that it is indeed possible.

The cyclostationary feature detector relies on the fact that most signals exhibit periodic features, present in pilots, cyclic prefixes, modulations, carriers, and other repetitive characteristics [6,33–37]. Because the noise is not periodic, the signal can be successfully detected.

The works by Gardner [33] in 1991 and Enserink et al. [34] in 1995 have studied this signal detection scheme in detail. The work of Enserink et al. follows the same line of the one by Gardner, in which the cyclostationary feature detector is based on the squared magnitude of the spectral coherence, which for any random process X is given by

$$|\rho_X^{\alpha}(f)| = \frac{|S_X^{\alpha}(f)|^2}{\left[S_X\left(f + \frac{\alpha}{2}\right) S_X\left(f - \frac{\alpha}{2}\right)\right]^{\frac{1}{2}}}, \tag{1.10}$$

where

S_X is the spectral correlation density function
α is the cyclic frequency
f is the spectral frequency

In the specific case of the cyclostationary feature detector, substituting $\rho_X^{\alpha}(f)$ by $\hat{\rho}_X^{\alpha}(f)$ and S_X by $\hat{S}_X$, which are the estimated versions of the same quantities, we have the decision metric:

$$\hat{\mathcal{M}} = |\hat{\rho}_X^{\alpha}(f)| = \frac{|S_X^{\alpha}(f)|^2}{\hat{S}_X\left(f + \frac{\alpha}{2}\right) \hat{S}_X\left(f - \frac{\alpha}{2}\right)}, \tag{1.11}$$

which goes into the decision statistic, given by

$$\text{decide for} \begin{cases} \text{H}_0, & \text{if } \hat{\mathcal{M}} < \nu \\ \text{H}_1, & \text{if } \hat{\mathcal{M}} \geq \nu \end{cases},$$

A recent work focuses on a cyclostationary feature detector for cognitive radio networks [37], called multi-cycles detector. In this work, a cyclostationarity detector scheme is employed on nonfiltered signals, such as OFDM, to detect the cyclic frequency and its harmonics. Finally, it is thought that the cyclostationary feature detector is the most promising signal detection technique as it combines good performance with low requirements on the knowledge of the signal structure [35].

1.4 Cooperative Sensing Techniques

Although for simple AWGN channels most classical approaches perform well, as we have seen, in the case of fading these techniques are not able to provide satisfactory results due to their inherent limitations and to the hidden node problem. To this end, several works [23–31] have looked into the case in which cooperation is employed in sensing the spectrum.

Consider the scenario depicted in Figure 1.3, in which primary users (in white) communicate with their dedicated (primary) base station. Secondary receivers $\{\text{RX}_1, \text{RX}_2, \text{RX}_3, \ldots, \text{RX}_K\}$ cooperatively sense the channel to identify a white space and exploit the medium. The main idea of the cooperative sensing techniques is that each receiver RX_i can individually measure the channel and interact on their

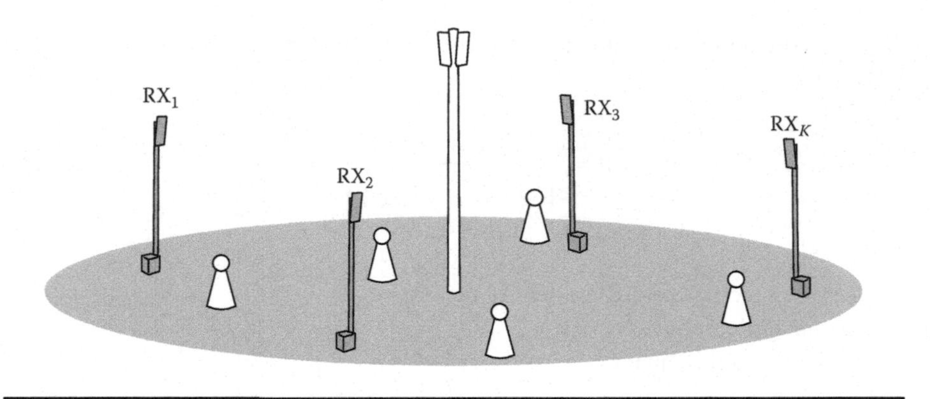

Figure 1.3 Cooperative sensing scenario.

findings to decide if the medium is available. The main drive behind this idea is that each secondary receiver will have a different perception of the spectrum, as its channel to the receiver will be different from the other secondary receivers, thus decreasing the chances of interfering with hidden nodes.

We will concentrate on the scenario depicted in Figure 1.3, although all sensing techniques presented herein can be also applied to alternative scenarios available in the literature, that is, [38].

The cooperative spectrum sensing can be [31]

- Centralized, in which a central entity gathers all information from all secondary receivers to make a decision about the medium status, which is then transmitted back to the receivers
- Distributed, in which the receivers share their information to make their own decision

In both these situations, the cooperative spectrum sensing is plagued with one problem: how to report or distribute the measures in a resource-constrained network. In fact, if these measurements are the basis for deciding whether a transmission can be made or not, then it does not make any sense to propagate the measurements before the decision is made. To overcome the problem, one could create a dedicated channel for signaling (such as that in [39]) or use an unregulated band (such as ISM). Other works [23–27,30,31] try to restrict the reporting to the minimum possible (often one bit) to ease the process of distributing this information. Finally, [28] considers a hierarchically structured secondary network, in which the secondary spectrum sensors are the secondary base stations, distributed over the sensing area. These base stations would make use of a backbone with enough bandwidth to distribute the measurements among themselves, irrespective of being a single bit or the actual acquired data. Then, during a white space, the terminals are allowed to transmit. Nevertheless, secondary base stations, as opposed to secondary terminals,

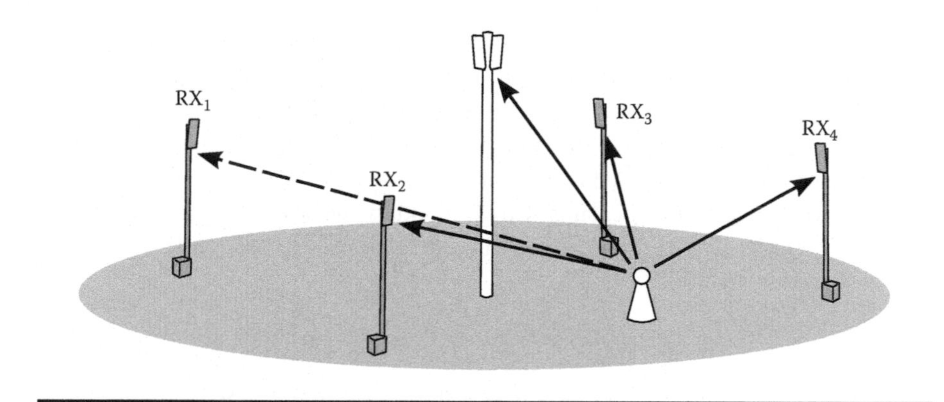

Figure 1.4 Cooperative sensing scenario.

have more processing power and fewer power constraints so that they can perform the spectrum-sensing task better. It should be noted that both of these approaches have their own target applications; neither can be considered the best approach in every case.

Another problem of cooperative spectrum sensing is identifying which secondary receivers offer reliable estimations. Let us consider the situation depicted in Figure 1.4, in which one primary terminal is transmitting data in the uplink channel (with low power) toward its primary base station. Several spectrum sensors $\{RX_1, RX_2, RX_3, RX_4\}$ are monitoring the medium detect its state. In this example, $\{RX_2, RX_3, RX_4\}$ are in range of the transmitter and can correctly sense its ongoing transmission, but RX_1 is not.* Thus, when the measures of all of the sensors are gathered, how does one select the individual receivers that are performing a reliable measurement? Without knowing the position of the primary transmitter and the channels between secondary receivers and the primary transmitter, this is a complicated task. The work by Mishra et al. [30] looks further into the performance impacts of the lack of reliability. Some works [35,40] discuss about a weighting scheme to give different scales to different secondary receivers based on their channel. Other works [23–27] propose a voting scheme to make a trustworthy decision, even with the presence of doubtful measurements.

In the remainder of this section, we explore some of the state-of-the-art cooperative sensing techniques.

1.4.1 Voting-Based Sensing

We saw in Section 1.3.1 that, in the low-SNR regime, the ED is highly vulnerable to fading and fluctuations in the level of the noise power. What if, instead of employing

* This would also apply to the case where RX_1 is shadowed or is in a deep multipath fading.

the ED at one location, we could do the same thing in other locations as well? It is expected that among several secondary receivers, even though some will suffer from fading or imprecisions due to the choice of the threshold, some will be able to correctly sense the medium. This is the main idea behind the collaborative spectrum sensing based on voting, studied in a number of works [23–27].

In the voting spectrum sensing, each secondary receiver RX_i uses spectrum sensing to form its own decision, as presented in Section 1.3.1. Consider the vector of all responses $\mathbf{r}$ such that

$$\mathbf{r} = [r_1\ r_2\ r_3\ \ldots\ r_K],$$

where $r_i \in \{1, 0\}$ is the binary response for each sensor i. After all measurements are gathered, the voting procedure takes place [23–25]:

$$\text{decide for} \begin{cases} \mathrm{H}_0, & \text{if } \mathcal{V} = 0 \\ \mathrm{H}_1, & \text{if } \mathcal{V} \geq 0 \end{cases}$$

where

$$\mathcal{V} = \sum_{k=1}^{K} r_k.$$

Briefly, the voting schemes select H_1 if at least one of the secondary receivers decides for H_1, which is known as the OR rule. Although this may seem too pessimistic, as it will favor false alarms, according to [23–25], this already gives improvements over the simple energy detection case even for two users. This is reasonable if we remark that with a high number of sensors, higher the probability of reliable spectrum sensing among secondary receivers will be. The probabilities of detection and false alarm for the cooperative approach are

$$Q_{\mathrm{f}} = 1 - (1 - P_{\mathrm{f}})^K \tag{1.12}$$

and

$$Q_{\mathrm{d}} = 1 - (1 - P_{\mathrm{d}})^K, \tag{1.13}$$

respectively.

The work by Sun et al. [26] revisits this scheme to estimate the reliability of each node. In this scheme, only the nodes with reliable sensing are allowed to report their detection. The reliability measure is based on how close the energy of $y(k)$ is to γ, as shown in Figure 1.5.

This work defines two new thresholds, γ_1 and γ_2, that are used to define a "no decision" region. Thus the decision rule can be stated as

$$\text{decide for} \begin{cases} \mathrm{H}_0, & \text{if } 0 \leq \mathcal{E} \leq \gamma_1 \\ \mathrm{H}_1, & \text{if } \mathcal{E} \geq \gamma_2 \end{cases}.$$

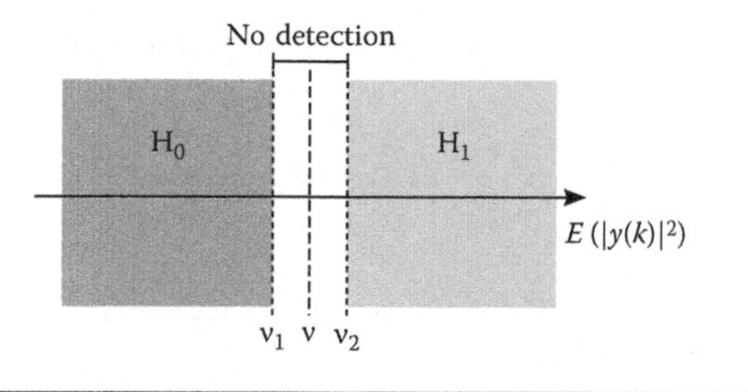

Figure 1.5 Reliability decision scheme.

If $\mathcal{E}$ falls in (v_1, v_2), then the secondary receiver decides not to report. This way the overall decision, based on the OR rule, concentrates on the reports of M users with a reliable detection out of K total users. The results from this work suggest an increased performance over the conventional case, where no reliability information is used.

Another work by Sun et al. [27] proposes a cluster-based spectrum sensing. In this work, a cluster is a grouping of secondary receivers that are spatially close. In each cluster, one receiver, called the cluster head, is elected to do the local decision and the reporting to the central decision entity. There the final decision takes place.

1.4.2 Correlator-Based Sensing

Another possibility is to gather all received samples at a central entity that will take the decision instead of leaving the decision of the medium availability to the secondary receivers. With an overall view of the situation, the central entity can decide how to manage the measurements for the decision-taking task better. The schemes presented in this and the following sections all involve such a central entity.

Let us, for simplicity sake, suppose that all secondary receivers $\{RX_1, RX_2, RX_3, \ldots, RX_K\}$, shown in Figure 1.3 are within the range of a certain primary transmitter. Then, considering a flat-faded environment, we have

$$y_i(k) = \begin{cases} n_i(k): & \mathrm{H}_0 \\ h_i(k)s(k) + n_i(k): & \mathrm{H}_1 \end{cases},$$

where the subscript i means that each value is to be taken for each user i.

We can see that for the H_0 hypothesis, all $y_i(k)$ are independent because they are only composed of AWGN noise. On the other hand, in the H_1 hypothesis, all $y_i(k)$ are composed of not only the noise but also the signal component $s(k)$ modulated by the channel $h_i(k)$. As we know, the signal is common for all users, because it is broadcast by the primary transmitter. We can exploit this fact to detect the presence of transmitted signals by focusing on the correlation between received signals from secondary receivers.

This correlation is calculated via the cyclic convolution, defined as

$$R_{(ij)}(k) = \sum_{k=1}^{N} y_i(a)y_j((k-a) \bmod N)$$

where i and j are the indices of any two secondary receivers.

In this scheme, the decision rule is given by

$$\text{decide for} \begin{cases} H_0, & \text{if } \mathcal{L} < \nu \\ H_1, & \text{if } \mathcal{L} \geq \nu \end{cases},$$

where $\mathcal{L}$ is the decision statistic calculated as

$$\mathcal{L} = \max_{(i,j)\in\mathcal{B}} \max_{k}(R_{(ij)}), \tag{1.14}$$

where

$R_{(ij)}$ is the pairwise cyclic convolution for all permutation of secondary receivers $\mathcal{B} = \{(x,y) \in \mathcal{A} \times \mathcal{A} | y \geq x+1\}$, $\mathcal{A} = \{1,2,\ldots,N\}$. Note that unlike the MF, this scheme does not require coherent reception, as it looks for the highest correlation between any two pairs of sensors. Nevertheless, in the case of coherent reception, we could rewrite Equation 1.14 as

$$\mathcal{L} = \sum_{k=1}^{N}(R_{(ij)}), \quad \forall\, (i,j) \mid i \neq j,$$

which would effectively maximize the SNR.

As far as the authors know, this spectrum sensing scheme has not yet been studied in the literature and thus its performance is not known. It would likely suffer from the same problem as the MF, namely, the challenge of correctly choosing ν. The main limitation of this scheme would be its necessity to report all the measurements, which would require an infrastructure with a very high bandwidth dedicated for the task.

1.4.3 Eigenvalue-Based Sensing

Eigenvalue-based sensing is another technique for cooperative sensing, introduced by Cardoso et al. [28] and Zeng et al. [29], based on evaluating the eigenvalues of a matrix formed by the samples collected by multiple sensors in relation to the Marchenko–Pastur law. Herein, we explore the approach as was presented in [28] because the approach in [29] is very similar.

To better understand how this spectrum-sensing procedure works, we start with the following assumption:

- The K base stations in the secondary system share information between them. This can be performed by transmission over a wired high-speed backbone.
- The base stations are analyzing the same portion of the spectrum.

Let us consider the following $K \times N$ matrix consisting of the samples received by all the K secondary receivers RX_i:

$$\mathbf{Y} = \begin{bmatrix} y_1(1) & y_1(2) & \cdots & y_1(N) \\ y_2(1) & y_2(2) & \cdots & y_2(N) \\ y_3(1) & y_3(2) & \cdots & y_3(N) \\ \vdots & \vdots & & \vdots \\ y_K(1) & y_K(2) & \cdots & y_K(N) \end{bmatrix}.$$

Then, the objective of the eigenvalue-based approach is to perform a test of independence of the signals received at RX_i. As said before, in the H_1 case, all the received samples are expected to be correlated, whereas in the H_0 case, the samples are decorrelated. Hence, in this case, for a fixed K and $N \rightarrow \infty$, under the H_0 assumption the sample covariance matrix $\frac{1}{N}\mathbf{Y}\mathbf{Y}^H$ converges to $\sigma^2\mathbf{I}$. However, in practice, N can be of the same order of magnitude as K and therefore one cannot infer directly $\frac{1}{N}\mathbf{Y}\mathbf{Y}^H$ independence of the samples. This can be formalized using tools from random matrix theory [41]. In the case where the entries of $\mathbf{Y}$ are independent (irrespective of the specific probability distribution, which corresponds to H_0), we can use the following result from asymptotic random matrix theory [41]:

THEOREM 1.1 Consider a $K \times N$ matrix $\mathbf{W}$ whose entries are independent zero-mean complex (or real) random variables with variance $\frac{\sigma^2}{N}$ and fourth moments of order $O\left(\frac{1}{N^2}\right)$. As $K, N \rightarrow \infty$ with $\frac{K}{N} \rightarrow \alpha$, the empirical distribution of $\mathbf{W}\mathbf{W}^H$ converges almost surely to a nonrandom limiting distribution with density

$$f(x) = \left(1 - \frac{1}{\alpha}\right)^+ \delta(x) + \frac{\sqrt{(x-a)^+(b-x)^+}}{2\pi\alpha x}$$

where

$$a = \sigma^2(1 - \sqrt{\alpha})^2 \quad \text{and} \quad b = \sigma^2(1 + \sqrt{\alpha})^2,$$

which is known as the Marchenko–Pastur law.

Interestingly, under the H_0 hypothesis, the support of the eigenvalues of the sample covariance matrix (in Figure 1.6, denoted by M̌P) is finite. The Marchenko–Pastur law thus serves as a theoretical prediction under the assumption that the

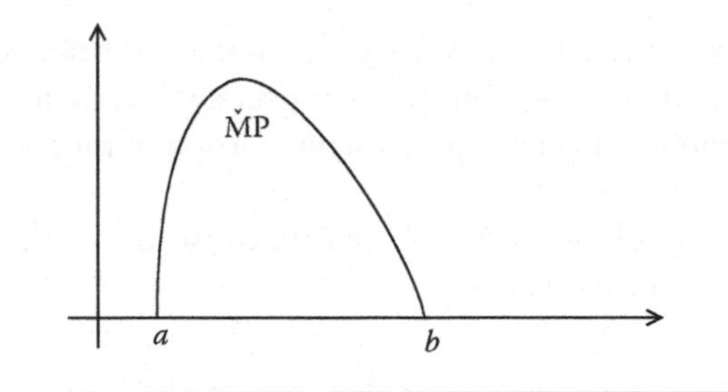

Figure 1.6 The Marchenko–Pastur support (H_0 hypothesis).

matrix is "all noise." Deviations from this theoretical limit in the eigenvalue distribution should indicate nonnoisy components.

In the case in which a signal is present (H_1), $\mathbf{Y}$ can be rewritten as

$$\mathbf{Y} = \begin{bmatrix} h_1 & \sigma & & 0 \\ \vdots & & \ddots & \\ h_K & 0 & & \sigma \end{bmatrix} \begin{bmatrix} s(1) & \cdots & s(N) \\ z_1(1) & \cdots & z_1(N) \\ \vdots & & \vdots \\ z_K(1) & \cdots & z_K(N) \end{bmatrix},$$

where $s(k)$ and $z_i(k) = \sigma n_i(k)$ are, respectively, the independent signal and noise with unit variance at instant k and secondary receiver i. Let us denote by $\mathbf{T}$ the matrix

$$\mathbf{T} = \begin{bmatrix} h_1 & \sigma & & 0 \\ \vdots & & \ddots & \\ h_K & 0 & & \sigma \end{bmatrix}.$$

$\mathbf{TT}^H$ clearly has one eigenvalue equal to $\lambda_1 = \sum |h_i|^2 + \sigma^2$ and all the rest equal to σ^2. The behavior of the eigenvalues of $\frac{1}{N}\mathbf{YY}^H$ is related to the study of the eigenvalues of large sample covariance matrices of spiked population models [42]. Here, the SNR ξ is defined as

$$\xi = \frac{\sum |h_i|^2}{\sigma^2}.$$

The works by Baik et al. [42,43] have shown that when

$$\frac{K}{N} < 1 \quad \text{and} \quad \xi > \sqrt{\frac{K}{N}} \tag{1.15}$$

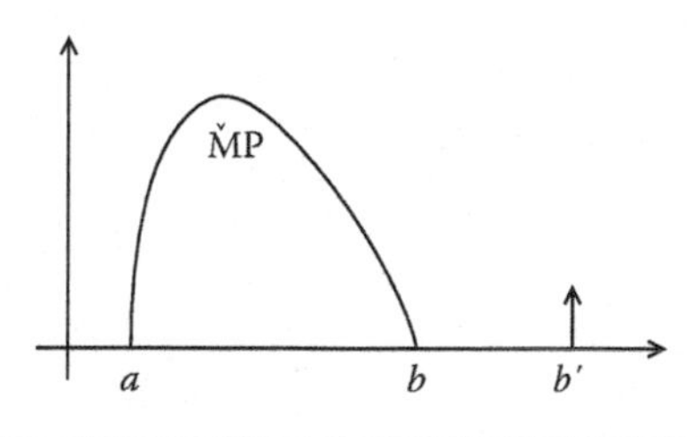

Figure 1.7 The Marchenko–Pastur support plus a signal component.

(which are assumptions that are clearly met when the number of samples N are sufficiently high), the maximum eigenvalue of $\frac{1}{N}\mathbf{YY}^H$ converges almost surely to

$$b' = \left(\sum |h_i|^2 + \sigma^2\right)\left(1 + \frac{\alpha}{\xi}\right),$$

which is greater than the value of $b = \sigma^2(1 + \sqrt{\alpha})^2$ seen in the H_0 case.

Therefore, whenever the distribution of the eigenvalues of the matrix $\frac{1}{N}\mathbf{YY}^H$ departs from the Marchenko–Pastur law, as shown in Figure 1.7, the detector decides that the signal is present. Hence, we apply this feature from a spectrum-sensing point of view.

Considering λ_i as the eigenvalues of $\frac{1}{N}\mathbf{YY}^H$ and $G = [a, b]$, the cooperative sensing scheme works in two possible ways.

1.4.3.1 Noise Distribution Unknown, Variance Known

In this case, the decision criteria used is

$$\text{decide for} \begin{cases} H_0: & \text{if } \lambda_i \in G \\ H_1: & \text{otherwise.} \end{cases} \tag{1.16}$$

1.4.3.2 Both Noise Distribution and Variance Unknown

The ratio of the maximum and the minimum eigenvalues in the H_0 hypothesis case does not depend on the noise variance and thus serves well as a criteria independent of the noise

$$\text{decide for} \begin{cases} H_0: & \text{if } \frac{\lambda_{\max}}{\lambda_{\min}} \le \frac{(1+\sqrt{\alpha})^2}{(1-\sqrt{\alpha})^2} \\ H_1: & \text{otherwise.} \end{cases} \tag{1.17}$$

It should be noted that, in this case, one still needs to take a sufficiently high number of samples N such that the conditions in Equation 1.15 are met. In other words, the number of samples scales quadratically with the inverse of the SNR.

Note, moreover, that the test under H_1 hypothesis also provides a good estimator of the SNR ρ. Indeed, the ratio of largest eigenvalue (b') and smallest (a) of $\frac{1}{N}\mathbf{Y}\mathbf{Y}^H$ is related solely to ρ and α:

$$\frac{b'}{a} = \frac{(\rho + 1)\left(1 + \frac{\alpha}{\rho}\right)}{(1 - \sqrt{\alpha})^2}.$$

1.5 Conclusions and Open Issues

In this chapter, the state of the art of spectrum-sensing techniques for cognitive radio networks were covered. We presented not only the classical techniques, inspired by the signal detection approaches developed for radar systems, but also some newly developed ones, carefully tailored for the cognitive radio network scenario. Furthermore, we presented their operation, characteristics, advantages, and limitations.

In spite of the popularity of spectrum sensing as a study subject for cognitive radio networks, there are still some open issues in this area. Generally, the study has tackled the sensing techniques themselves but little work has considered the systemic point of view, implementation issues, and the complexity of techniques concerning spectrum sensing. Some open issues can be highlighted:

- *Adaptive spectrum sensing.* The techniques for spectrum sensing studied so far consider well-behaved scenarios. For some of these techniques, it is quite clear that time-varying environments would greatly compromise their performance. Because cognitive radio networks will most likely operate in such environments, it is important that adaptive spectrum-sensing techniques be devised.
- *Cooperation between primary and secondary systems.* Is spectrum sensing the best way to find out the medium availability? In some scenarios, maybe not. It is possible that by sharing some information to spectrum brokers, primary systems may benefit from less, or even zero, interference from secondary systems.
- *Cooperative sensing.* It is clear that cooperative sensing may be the best option for spectrum sensing in many faded environments. However, there are still some open issues in this field, such as the impact of imperfect information exchange between secondary receivers.
- *Complexity and implementation issues.* One of the main limitations of the cognitive radios, and hence of spectrum sensing, is the physical limitation of the hardware and radio frequency (RF) components required. Today, no one knows how to create these cognitive radio transceivers in production scale, with a small package and consuming low power. Another open question is how to find the right bandwidth size for spectrum sensing. Although wideband

sensing would give a faster and clearer overall picture of the spectrum, it would provide a very rough estimate, because the sensing energy is distributed over a large spectrum. Sweeping the spectrum with narrowband sensing concentrates the sensing energy, but might be too slow in relation to the fast-changing environments. Furthermore, because it is envisioned that sensing will be done by terminals, how do all sensing techniques compare in terms of implementation complexity, energy usage, and processing power?

- *Cognitive pilot channel.* The CPC is a specific frequency channel reserved for the diffusion of cognitive radio-related information, such as current frequency band allocation. This interesting new concept could alleviate the requirements of spectrum sensing and provide better performances. It requires further studies to evaluate its gains over the traditional approach.

References

1. J. Mitola. Cognitive radio an integrated agent architecture for software defined radio, PhD thesis. Royal Institute of Technology (KTH), Stockholm, Sweden, May 2000.
2. S. Haykin. Cognitive radio: Brain-empowered wireless communications. *IEEE Journal on Selected Areas in Communications*, 23(2):201–220, 2005.
3. I.F. Akyildiz, W.Y. Lee, M.C. Vuran, and S. Mohanty. Next generation/dynamic spectrum access/cognitive radio wireless networks: A survey. *Computer Networks*, 50(13): 2127–2159, 2006.
4. Spectrum Efficiency Working Group. Report of the spectrum efficiency working group. Technical report, FCC, Washington, DC, November 2002.
5. J. Palicot and C. Roland. A new concept for wireless reconfigurable receivers. *IEEE Communications Magazine*, 41(7):124–132, 2003.
6. A. Fehske, J. Gaeddert, and J.H. Reed. A new approach to signal classification using spectral correlation and neural networks. In *Proceedings of the IEEE International Symposium on New Frontiers Dynamic Spectrum Access Networks*, Vol. 1, Baltimore, MD, 2005, pp. 144–150.
7. R. Hachemani, J. Palicot, and C. Moy. A new standard recognition sensor for cognitive radio terminal. In *USIPCO 2007*, Poznan, Poland, September 3–7, 2007.
8. A. Bouzegzi, P. Jallon, and P. Ciblat. A second order statistics based algorithm for blind recognition of OFDM based systems. In *Proceedings of Globecom 2008*, New Orleans, LA, 2008.
9. A. Sahai and D. Cabric. Spectrum sensing: Fundamental limits and practical challenges. In *Tutorial Presented at the 1st IEEE Conference on Dynamic Spectrum Management (DySPAN'05)*, Baltimore, MD, 2005.
10. Z. Tian and G.B. Giannakis. A wavelet approach to wideband spectrum sensing for cognitive radios. In *1st International Conference on Cognitive Radio Oriented Wireless Networks and Communications 2006*, Mykonos Island, Greece, 2006, pp. 1–5.
11. Y. Hur, J. Park, W. Woo, K. Lim, C.H. Lee, H.S. Kim, and J. Laskar. A wideband analog multi-resolution spectrum sensing (MRSS) technique for cognitive radio (CR)

systems. In *Proceedings of the IEEE International Symposium on Circuits and Systems 2006* (*ISCAS'06*), Island of Kos, Greece, 2006, p. 4.

12. Z. Quan, S. Cui, A.H. Sayed, and V. Poor. Wideband spectrum sensing in cognitive radio networks. arXiv:0802.4130, 2008.
13. M.I. Skolnik. *Introduction to Radar Systems*. McGraw-Hill, Singapore, 1980.
14. H.L. Vantrees. *Detection, Estimation and Modulation Theory*, Vol. 1. Wiley, New York, 1968.
15. V. Poor. *An Introduction to Signal Detection and Estimation*. Springer, New York, 1994.
16. H. Urkowitz. Energy detection of unknown deterministic signals. *Proceedings of the IEEE*, 55:523–531, 1967.
17. R. Tandra and A. Sahai. Fundamental limits on detection in low SNR under noise uncertainty. In *International Conference on Wireless Networks*, Maui, HI, 2005.
18. J.G. Proakis and M. Salehi. *Digital Communications*. McGraw-Hill, New York, 1995.
19. S.S. Haykin. *Communication Systems*. Wiley, New York, 2001.
20. C.L. Fullmer and J.J. Garcia-Luna-Aceves. Solutions to hidden terminal problems in wireless networks. In *Proceedings of the ACM SIGCOMM'97 Conference on Applications, Technologies, Architectures, and Protocols for Computer Communication*, Cannes, France, 1997, pp. 39–49.
21. V.I. Kostylev. Energy detection of a signal with random amplitude. In *Proceedings of the IEEE International Conference on Communications* (*ICC'02*), Vol. 3, New York, April–May 2002.
22. F.F. Digham, M.S. Alouini, and M.K. Simon. On the energy detection of unknown signals over fading channels. In *Proceedings of the IEEE International Conference on Communications* (*ICC'03*), Vol. 5, Anchorage, AK, May 11–15, 2003.
23. A. Ghasemi and E.S. Sousa. Collaborative spectrum sensing for opportunistic access in fading environments. In *First IEEE International Symposium on New Frontiers in Dynamic Spectrum Access Networks 2005* (*DySPAN'05*), Piscataway, NJ, 2005, pp. 131–136.
24. A. Ghasemi and E.S. Sousa. Spectrum sensing in cognitive radio networks: The cooperation-processing tradeoff. *Wireless Communications and Mobile Computing*, 7:1049–1060, 2007.
25. D. Cabric, A. Tkachenko, and R.W. Brodersen. Spectrum sensing measurements of pilot, energy, and collaborative detection. In *Proceedings of IEEE Military Communications Conference*, Washington, DC, October 2006.
26. C. Sun, W. Zhang, and K.B. Letaief. Cooperative spectrum sensing for cognitive radios under bandwidth constraints. In *IEEE Wireless Communications and Networking Conference 2007* (*WCNC'07*), Hong Kong, China, 2007, pp. 1–5.
27. C. Sun, W. Zhang, and K.B. Letaief. Cluster-based cooperative spectrum sensing in cognitive radio systems. In *IEEE International Conference on Communications 2007* (*ICC'07*), Glasgow, U.K., 2007, pp. 2511–2515.
28. L.S. Cardoso, M. Debbah, P. Bianchi, and J. Najim. Cooperative spectrum sensing using random matrix theory. In *International Symposium on Wireless Pervasive Computing* (*ISWPC'08*), Santorini, Greece, 2008.

29. Y. Zeng and Y.C. Liang. Eigenvalue base spectrum sensing algorithms for cognitive radio. arXiV:0804.2960, 2008.
30. S.M. Mishra, A. Sahai, and R.W. Brodersen. Cooperative sensing among cognitive radios. In *IEEE International Conference on Communications 2006* (*ICC'06*), Vol. 4, San Francisco, CA, 2006, pp. 1658–1663.
31. G. Ganesan and Y. Li. Cooperative spectrum sensing in cognitive radio networks. In *First IEEE International Symposium on New Frontiers in Dynamic Spectrum Access Networks 2005* (*DySPAN'05*), Baltimore, MD, 2005, pp. 137–143.
32. A. Sahai, N. Hoven, and R. Tandra. Some fundamental limits on cognitive radio. In *Allerton Conference on Communication, Control, and Computing*, Monticello, IL, 2003.
33. W.A. Gardner. Exploitation of spectral redundancy in cyclostationary signals. *IEEE Signal Processing Magazine*, 8(2):14–36, 1991.
34. S. Enserink and D. Cochran. A cyclostationary feature detector. In *Conference Record of the Twenty-Eighth Asilomar Conference on Signals, Systems and Computers 1994*, Vol. 2, Pacific Grove, CA, 1995.
35. D. Cabric, S.M. Mishra, and R.W. Brodersen. Implementation issues in spectrum sensing for cognitive radios. In *Conference Record of the Thirty-Eighth Asilomar Conference on Signals, Systems and Computers 2004*, Vol. 1, Pacific Grove, CA, 2004, pp. 772–776.
36. H. Tang. Some physical layer issues of wide-band cognitive radio systems. In *Proceedings of IEEE International Symposium on New Frontiers Dynamic Spectrum Access Networks*, Vol. 1, Baltimore, MD, 2005, pp. 151–159.
37. M. Ghozzi, M. Dohler, F. Marx, and J. Palicot. Cognitive radio: Methods for the detection of free bands. *Comptes Rendus Physique*, 7:794–804, September 2006.
38. S.N. Shankar, C. Cordeiro, and K. Challapali. SpectrumAgile radios: Utilization and sensing architectures. In *First IEEE International Symposium on New Frontiers in Dynamic Spectrum Access Networks*, Baltimore, MD, 2005, pp. 160–169.
39. P. Codier, D. Bourse, D. Grandblaise, K. Moessner, J. Luo, C. Kloeck et al. Cognitive pilot channel. In *Proceedings of WWRF15*, Vol. 8, Paris, France, 2005.
40. X. Huang, N. Han, G. Zheng, S. Sohn, and J. Kim. Weighted-collaborative spectrum sensing in cognitive radio. In *Second International Conference on Communications and Networking in China 2007* (*CHINACOM'07*), Shanghai, China, 2007, pp. 110–114.
41. V.A. Marchenko and L.A. Pastur. Distributions of eigenvalues for some sets of random matrices. *Math USSR-Sbornik*, 1:457–483, 1967.
42. J. Baik, G.B. Arous, and S. Peche. Phase transition of the largest eigenvalue for nonnull complex sample covariance matrices. *Annals of Probability*, 33(5):1643–1697, 2005.
43. J. Baik and J.W. Silverstein. Eigenvalues of large sample covariance matrices of spiked population models. *Journal of Multivariate Analysis*, 97(6):1382–1408, 2006.

Chapter 2

Capacity Analysis of Cognitive Radio Networks

Xuemin Hong, Cheng-Xiang Wang, John Thompson, and Hsiao-Hwa Chen

Contents

Current static and rigid spectrum licensing policy has resulted in very inefficient spectrum utilization [1–3]. Cognitive radio (CR) [4–8] has been extensively researched in recent years as a promising technology to improve spectrum utilization. The ultimate goal of CR research is to establish a CR network that is either self-sufficient in delivering a multitude of wireless services or capable of assisting existing wireless networks to enhance their performance. The performance of a CR network is inevitably affected by the coexisting primary systems. Most importantly, the CR transmissions should be carefully controlled to guarantee that the primary services are not jeopardized. To better understand the ultimate performance limits and potential applications of CR networks, it is crucial to study the CR network capacity to provide theoretical insights into the CR network design.

In this chapter, we first introduce the classifications of CR networks. We then analyze the capacities of two promising CR networks under average interference power constraints. The first one is a central access CR network, which aims to provide broadband access to CR devices with central base stations (BSs). The second one is a cooperative CR network, where multiple dual-mode CR-cellular users collaborate in the CR band to improve the access performance in the cellular band. Under a simple power control framework, the uplink channel capacities of both CR networks are analyzed and compared, taking into account various system-level factors such as the densities and locations of primary/CR users and path loss in radio propagation channels. Finally, the chapter concludes with some open research issues.

2.1 Classification of Cognitive Radio Networks

The core of a CR network is a coexistence mechanism that controls the spectrum sharing in such a way that the operations of the primary system are not compromised. Based on different coexistence methods, CR networks can be classified into noninterfering CR networks [9–14] and interference-tolerant CR networks [14–17]. On the other hand, based on different radio access types [6], CR networks can be classified as central access/infrastructure-based CR networks [9,12,13] and ad hoc CR networks [18]. In what follows, we briefly explain these four types of CR networks.

2.1.1 Noninterfering CR Networks

Noninterfering CR networks exploit the existence of underutilized spectrum, which refers to the frequency segments that have been licensed to a particular primary service, but are completely unused or partly utilized at a given location or a given time. The unused frequency segments are also called frequency voids, spectrum holes, or white spaces [7], while the partly used spectra are often referred to as *grey spaces*. A noninterfering CR network seeks to collect these underutilized spectra and reuse them on an opportunistic basis. With careful design, a noninterfering CR network can coexist well with the primary system because it essentially seeks to operate in a signal space orthogonal to the primary signals. A number of measurement campaigns have shown that a large amount of white space exists in two frequency bands: 400–800 MHz and 3–10 GHz. Therefore, noninterfering CR might start to operate first in these two bands in the near future.

The concept of noninterfering CR networks has been widely accepted and studied, for example, in [9–14], due to its two obvious advantages. First, the "noninterfering" philosophy means that the primary networks can be well protected. Second, the implementation of a noninterfering CR is relatively simple. Typically, a noninterfering CR is an intelligent wireless device that can dynamically sense the radio spectrum, locate unused or underutilized spectrum segments (or wireless channels) in a target spectrum pool, and automatically adjust its transceiver parameters to communicate in the discovered free channels. Such a sensing-based approach allows minimum changes to the primary system to tolerate CR networks.

The IEEE 802.22 working group is currently developing the first wireless standard [9,10] based on the noninterfering CR networks. The aim is to construct a fixed point-to-multipoint wireless regional area network (WRAN) utilizing white spaces in the TV frequency band between 54 and 862 MHz.

2.1.2 Interference-Tolerant CR Networks

The interference-tolerant CR networks allow CR users to operate on frequency bands assigned to the primary system as long as the total interference power received at the primary receivers remains below a certain threshold [14–17]. As a new metric to assess the interference at primary receivers, the concept of interference temperature [2] was proposed by the Federal Communications Commission (FCC) in 2002. Similar to the concept of noise temperature, interference temperature measures the power and bandwidth occupied by interference. Moreover, the concept of interference temperature limit [2] was introduced to characterize the "worst-case" interfering scenario in a particular frequency band and at a particular geographic location. CR transmissions in a given band are considered to be "harmful" only if they would raise the interference temperature above this limit. Unlike traditional transmitter-centric approaches that seek to regulate interference indirectly by controlling the emissions of interfering transmitters, the interference temperature concept takes a receiver-centric approach and aims to directly manage interference at primary receivers. Recently, in 2007, the FCC has abandoned its use of "interference temperature" due to current difficulties in implementing this concept. However, the philosophy behind it is still valid, and this concept is still widely used to facilitate the research of interference-tolerant CR systems.

Thanks to the more fundamental receiver-centric approach for interference management, interference-tolerant CR is expected to achieve a much better spectrum utilization than noninterfering CR. However, it is more difficult to implement because identifying the interference temperature limit is much more difficult than sensing the frequency voids. Typically, the primary network needs to be aware of the interference-tolerant CR network and inform CR transmitters about the interference levels perceived by primary receivers. In this case, a real-time feedback mechanism is essential, and modifications of the primary network is usually inevitable.

To implement the concept of the interference temperature limit, it can be specified in terms of the following three interference constraints: the peak interference power constraint [15–17], average interference power constraint [15–17], and interference outage constraint [19, p. 135].

- The peak interference power constraint bounds the peak power of the interference perceived by primary receivers. It is suitable if the interference power at primary receivers is known in a real-time fashion to CR transmitters, which can then adjust their transmit power accordingly to fulfill the constraint. In practice, CR transmitters can obtain or estimate the interference levels at primary receivers by means of a common control channel [20], primary receiver detection [21], or a direct primary receiver feedback. When the primary receivers are strictly passive and therefore "hidden" from the CR network, it is more reasonable to use the average interference power constraint or interference outage constraint.
- The average interference power constraint puts limits on the average power of the interfering signal. It is appropriate when the Quality-of-Service (QoS) of the primary network is determined by the average signal-to-interference-and-noise ratio (SINR), for example, for delay-insensitive primary services.
- The interference outage constraint limits the probability that the interference power exceeds a certain threshold, that is, the probability that the primary service is interrupted due to strong instantaneous interfering signals. This constraint is more appropriate when the QoS of the primary system depends on the instantaneous SINR, for example, for delay-sensitive primary services.

2.1.3 Central Access CR Networks

In central access CR networks, fixed CR BSs are deployed as the infrastructure to provide direct (single-hop) connection to CR users [6]. To establish a wide-area CR network with an acceptable infrastructure investment, central access CR network usually requires long-range CR access capability to provide a satisfactory coverage with a reasonable BS density. Such a long-range requirement may limit the application of central access CR networks in many frequency bands due to the inevitable constraints on CR transmit powers.

A practical example of a central access CR network is the IEEE 802.22 system [9,10], where the CR network is formed by BSs and customer-premises equipments (CPEs). A centralized approach is pursued in the standard by allowing a BS to manage its own cell and control the medium access for all the CPEs attached to it. In addition to conventional BS functionalities, the CR BS has additional functions such as acting as the central coordinator for CR channel discovery.

2.1.4 Ad Hoc CR Networks

Unlike central access CR networks, ad hoc CR networks require no infrastructure and allow CR users to communicate with each other through ad hoc connections [6]. Ad hoc CR networks usually use less transmit power because only short-to-medium transmit ranges are required to facilitate communications among neighbor users. As CR networks are power-limited in nature, the ad hoc CR network is envisioned as a popular form for future CR networks.

There are many potential applications for the ad hoc CR networks. For example, it can be used as a local area wireless network (WLAN) to provide local broadband wireless connectivity. Alternatively, it can be used to establish a CR mesh network [6] to provide a wide service coverage by transmitting data in a multi-hop fashion. Such a CR mesh network can be deployed in dense urban areas, adding capacity to the existing infrastructure to cope with high traffic load [6].

2.1.5 Capacity Analysis: The State of the Art and Motivation

Our study in this chapter focuses on the interference-tolerant CR networks, which describe a long-term vision of CR-based spectrum-sharing systems and has attracted huge research interest. So far, researches on interference-tolerant CR systems have mainly focused on the theoretical aspects. The purpose is to understand the system limits and long-term potentials. To this end, some pioneering information theoretic work on the channel capacities of CR channels were presented in [22–24].

Under an average received interference power constraint at primary receivers, the CR channel capacities in different additive white Gaussian noise (AWGN) channels in the absence of fading were studied in [15,16]. In [17], the channel capacities were derived in different fading environments under both average and peak interference power constraints. Based on the assumption that the CR transmitter has perfect channel state information (CSI), the channel capacities were formulated as a maximization problem. However, the capacity analysis carried out in [17] was limited to the link level of a CR network, where only one CR transmitter communicates with one CR receiver in the presence of a single or multiple primary receivers. Because path loss was not taken into account, the capacities given in [17] serve as upper-bounds of the link-level capacities achievable by interference-tolerant CR networks.

Unlike the above works that consider only the link-level performance of CR networks, the focus of this chapter is on the system-level capacity of CR networks. Our previous work presented in [25,26] is extended in this chapter to show a more detailed investigation on the system-level capacities of two different CR networks. We adopt a system-level model consisting of multiple CR transmitters and multiple primary receivers. Furthermore, we consider the underlying channel model to include path loss, which preserves the topology information of the network. Such a system-level model allows us to study the impact of the network topology, characterized mainly by the densities of CR transmitters and primary receivers, on the network capacity. In this chapter, we only investigate the CR network capacities subject to average interference power constraints. Following a similar procedure, this work can be extended to evaluate the CR network capacities under other interference power constraints.

In this chapter, we consider two application scenarios of interference-tolerant CR networks. The first scenario, as illustrated in Figure 2.1, is to establish a central access CR network consisting of a CR BS located in the center of a circular cell and multiple CR users uniformly distributed in the cell. The architecture is similar to that proposed by the IEEE 802.22 standard working group [9], but the coexistence mechanism is completely different. Such a centralized CR network can be expected to provide wireless broadband access to mobile users. The second scenario, as illustrated in Figure 2.2, aims to utilize an ad hoc CR network to improve the cellular access performance and consequently the capacity of a cellular system,

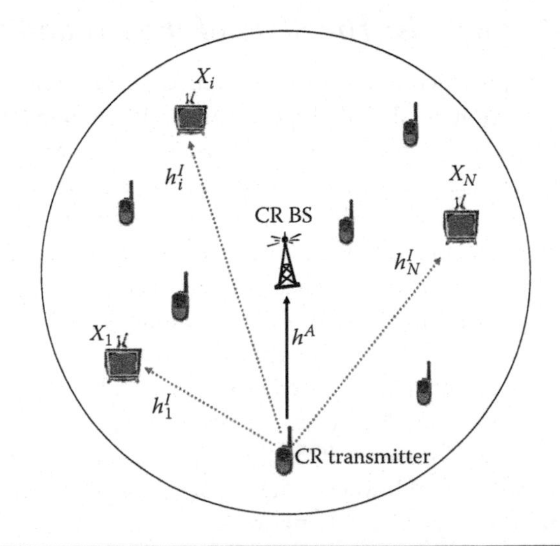

Figure 2.1 Central access cognitive radio network.

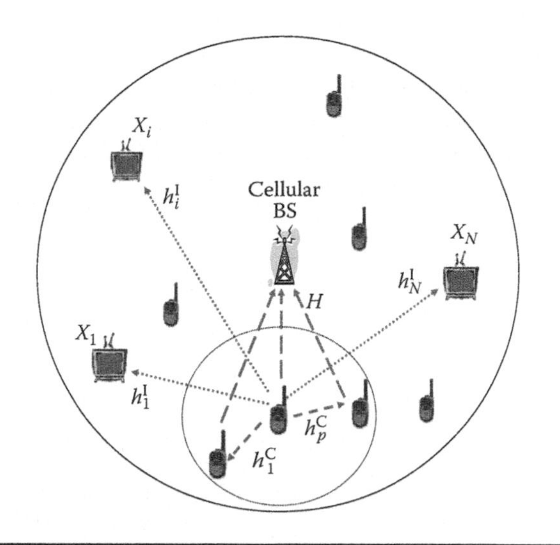

Figure 2.2 Cooperative cognitive radio network.

for example, universal mobile telecommunications system (UMTS). The mobile terminals are assumed to be dual-mode devices that can operate in both cellular bands and CR bands simultaneously. Both mobile CR users and primary users are assumed to be uniformly distributed in a circular cell centered at the cellular BS. The ad hoc CR network is used to help a target mobile terminal communicate and cooperate with neighboring mobile terminals in CR bands to form a virtual antenna array (VAA) [27]. The VAA will then communicate with the cellular BS antenna array in cellular bands to create a virtual multiple-input-multiple-output (MIMO) system. In what follows, the uplink capacities of both central access and cooperative CR networks are analyzed and compared.

2.2 Transmit Power Control

As illustrated in Figures 2.1 and 2.2, for both CR scenarios, multiple CR users are assumed to be uniformly distributed in a circular cell with radius R. Additionally, we assume that N primary receivers, denoted as X_i ($i = 1, \ldots, N$), are also uniformly distributed in the cell. Within the CR network, we assume that CR users transmit in orthogonal channels to avoid mutual interferences. In this chapter, we use a time division multiple access (TDMA) scheme, which implies that only one target CR user is scheduled to transmit in a given time slot. Although such a TDMA scheme does not necessarily achieve optimum spectrum efficiency, it leads to a simple and practical power control scheme, which is explained in detail subsequently.

Because only the target CR user is allowed to transmit at any given time, it is the only interference source to the primary network. We refer the underlying channels from the CR transmitter to primary receivers as interference channels. The instantaneous channel power gains from the scheduled CR transmitter to the ith primary receiver is denoted as h_i^I. In addition, let P be the transmit power of the target CR user. Under a certain constraint on the average received interference power at all N primary receivers, we have

$$PE\left\{h_i^I\right\} = P\bar{h}_i^I \leq I_0, \quad i = 1, \ldots, N \tag{2.1}$$

where

$E\{\cdot\}$ is the statistical average operator

I_0 is the maximum average interference power that the primary receivers can tolerate

For simplicity, we assume that the averaged interference channel gain $\bar{h}_i^I$ within a given time slot can be described by the path loss expressed by [19]

$$\bar{h}_i^I = E\left\{h_i^I\right\} = \frac{(h_c h_p)^2}{(d_i)^\alpha} \tag{2.2}$$

where

d_i is the distance between the target CR transmitter and the ith primary receiver

$\alpha = 4$ is the path loss factor

h_c and h_p are the antenna heights of the CR transmitter and the primary receivers, respectively

In this chapter, we assume that $h_c = h_p = 1.5$ m. From (2.1) and (2.2), we define

$$P_{\max} = \frac{I_0(d_{\min})^4}{(h_c h_p)^2} \tag{2.3}$$

as the maximum allowable transmit power, where $d_{\min} = \min\{d_i\}$ stands for the distance between the CR transmitter and the nearest primary receiver. It is noted that $d_{\min} \in [0, R + r]$ always holds, where r is the distance between the target CR transmitter and the cell center. We also assume that the target CR transmitter can accurately estimate $P_{\max}$, which can be obtained by either listening to a common control channel [28] or using certain feedback power control mechanisms [29]. In what follows, we first derive the probability density function (PDF) $f_{P_{\max}}(x)$ of $P_{\max}$.

As shown in (2.3), the random variable (RV) $P_{\max}$ is expressed as a function of another RV $d_{\min}$. To calculate $f_{P_{\max}}(x)$, we should first get the cumulative density function (CDF) $F_{d_{\min}}(d)$ of $d_{\min}$, which can be derived from the proposed geometric method detailed in the appendix. The resulting CDF $F_{d_{\min}}(d)$ is given by

$$F_{d_{\min}}(d) = 1 - \left[\frac{S(d)}{\pi R^2}\right]^N, \quad 0 \leq d \leq R + r \tag{2.4}$$

where

$$S(d) = \begin{cases} \pi R^2 - \pi d^2, & d \in [0, R-r] \\ \pi R^2 - \pi d^2 + S_1 - S_2, & d \in (R-r, \sqrt{R^2 - r^2}] \\ \pi R^2 - S_2 - S_1, & d \in (\sqrt{R^2 - r^2}, \sqrt{R^2 + r^2}] \\ S_2 - S_1, & d \in (\sqrt{R^2 + r^2}, R + r]. \end{cases} \tag{2.5}$$

In (2.5), S_1 and S_2 are given by

$$S_1 = d^2(\theta_1 - \sin\theta_1 \cos\theta_1) \tag{2.6}$$

$$S_2 = R^2(\theta_2 - \sin\theta_2 \cos\theta_2) \tag{2.7}$$

respectively, with

$$\theta_1 = \cos^{-1}\left|\frac{d^2 + r^2 - R^2}{2dr}\right| \tag{2.8}$$

$$\theta_2 = \cos^{-1}\left|\frac{R^2 + r^2 - d^2}{2Rr}\right| \tag{2.9}$$

The values of θ_1 and θ_2 are assumed to be in the interval $[0, \pi/2]$. From (2.3) and (2.4), it can easily be shown that the CDF $F_{P_{\max}}(x)$ of $P_{\max}$ is given by

$$F_{P_{\max}}(x) = F_{d_{\min}}\left[\left(\frac{h_c^2 h_p^2 x}{I_0}\right)^{1/4}\right], \quad 0 \le x \le X \tag{2.10}$$

where $X = I_0(R+r)^4/(h_c h_p)^2$ represents the upper limit of $P_{\max}$ when $d_{\min}$ in (2.3) takes the largest value $R + r$. From (2.10), the PDF of $P_{\max}$ is given by $f_{P_{\max}}(x) = F_{P_{\max}}(x)/dx$.

Let $P_{\lim} = I_0(R+R)^4/(h_c h_p)^2$ denote the maximum possible transmit power under any circumstance of primary user locations. Figure 2.3 shows the theoretical PDF $f_{P_{\max}}(x)$ on a $\log_{10}$ scale as a function of the normalized power $x/P_{\lim}$ with $r/R = 0.5$ and different values of N. The results agree with our intuition that $P_{\max}$ decreases with the increase of the number of primary users N. In Figure 2.4, the theoretical PDF $f_{P_{\max}}(x)$ on a $\log_{10}$ scale is illustrated with $N = 10$ and different values of r/R. The results demonstrate that with a larger r/R, that is, when a CR transmitter is closer to the cell edge, the maximum allowable transmit power $P_{\max}$ becomes higher. In Figures 2.3 and 2.4, simulation results are also shown to justify the theoretical derivation of $f_{P_{\max}}(x)$. The comparison of Figures 2.3 and 2.4 tells us that PDF $f_{P_{\max}}(x)$ reacts more dramatically to the change of N. This means that

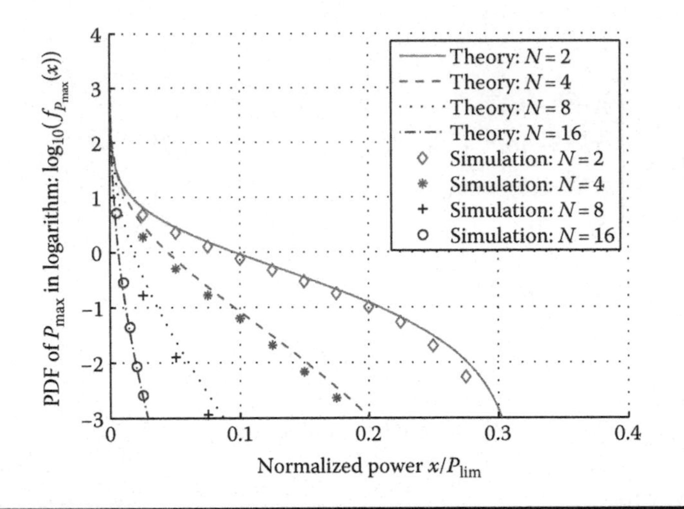

Figure 2.3 The PDF of P_{max} in logarithm $\log_{10}(f_{P_{max}}(x))$ with $r/R = 0.5$ and different values of N.

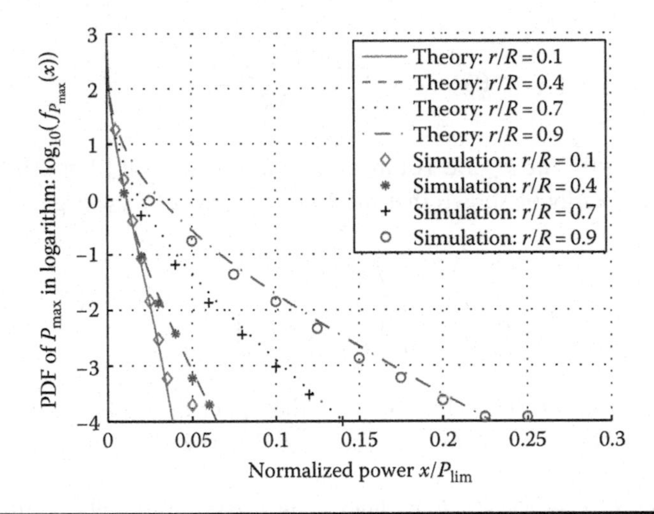

Figure 2.4 The PDF of P_{max} in logarithm $\log_{10}(f_{P_{max}}(x))$ with $N = 10$ and different values of r/R.

if compared with r/R, N is a dominant parameter that governs the CR transmit power constraints.

In Sections 2.3 and 2.4, we analyze and compare uplink channel capacities of two CR scenarios by assuming that the target CR terminal is always able to transmit with its maximum allowable power P_{max} determined by the primary network. It should be noted that, in practice, the transmit powers of CR terminals are also limited by their own device capabilities. There might be cases where the primary users are far away from CR users, such that P_{max} is larger than the realistic transmit power of the CR user constrained by its own device capability. Therefore, the derived results in Sections 2.3 and 2.4 should be considered as upper-bounds on the capacity of CR networks without considering practical device limitations.

2.3 Capacity Analysis of a Central Access Cognitive Radio Network

2.3.1 System Model

In this section, we consider a scenario where CR is used to establish a central access network with a BS and multiple CR users. The scenario is illustrated in Figure 2.1, where a CR BS is located at the center of the cell. To communicate with the CR BS, the target CR user transmits at its maximum allowable power P_{max}.

2.3.2 Capacity Analysis and Numerical Results

The channel from the CR transmitter to the CR BS is defined as the CR access channel. The underlying instantaneous channel power gain is denoted by h^{A}. It follows that the instantaneous uplink channel capacity is given by

$$C_{\text{CA}} = W \log_2 \left(1 + \frac{P_{\text{max}} h^{\text{A}}}{I_{\text{N}}}\right) \tag{2.11}$$

where

W is the signal bandwidth

I_{N} is noise plus interference power at the CR BS

The access channel gain h^{A} can be written as the product of three parts [19]

$$h^{\text{A}} = g_{\text{p}}^{\text{A}} g_{\text{s}}^{\text{A}} g_{\text{m}}^{\text{A}} = \frac{h_{\text{b}}^2 h_{\text{c}}^2}{r^4} g_{\text{s}}^{\text{A}} g_{\text{m}}^{\text{A}} \tag{2.12}$$

where

g_{p}^{A}, g_{s}^{A}, and g_{m}^{A} represent the power gains of path loss, shadowing, and multipath fading, respectively

r is the distance between the CR BS and target CR transmitter

h_{b} and h_{c} are the antenna heights of the CR BS and CR transmitter, respectively

In this chapter, we assume $h_{\text{b}} = 30$ m and $h_{\text{c}} = 1.5$ m. The shadowing factor g_{s}^{A} is a RV with a log-normal PDF given by [19]

$$f_{g_{\text{s}}^{\text{A}}}(x) = \frac{10}{\ln 10 \sqrt{2\pi}\delta_{\text{s}} x} \exp\left\{-\frac{(10 \log_{10} x)^2}{2\delta_{\text{s}}^2}\right\} \tag{2.13}$$

where δ_s is the shadowing standard deviation ranging from 5 to 12 dB, and $\delta_s = 8$ dB is taken as a typical value in macro-cell environments [19]. We further assume that multipath fading follows a Rayleigh distribution. Correspondingly, g_m^A follows an exponential distribution and its PDF is given by [19]

$$f_{g_m^A}(x) = \frac{1}{2\delta_m^2} \exp\left\{-\frac{x}{2\delta_m^2}\right\} \tag{2.14}$$

where δ_m is the standard deviation of the underlying real Gaussian process and is normalized to $\sqrt{2}/2$ here. The substitution of (2.3) and (2.12) into (2.11) results in

$$C_{CA} = W \log_2 \left(1 + \frac{I_0}{I_N} \frac{h_b^2}{h_p^2 r^4} d_{min}^4 g_s^A g_m^A\right) \tag{2.15}$$

where d_{min}, g_s^A, and g_m^A are independent RVs with PDFs given by $f_{d_{min}}(x) = dF_{d_{min}}(x)/dx$, $f_{g_s^A}(x)$ in (2.13), and $f_{g_m^A}(x)$ in (2.14), respectively. The ergodic capacity, that is, the mean value of C_{CA}, is calculated as

$$E\{C_{CA}\} = \int_{g_m^A} \int_{g_s^A} \int_{d_{min}} C_{CA} f_{d_{min}}(d_{min}) f_{g_s^A}\left(g_s^A\right) f_{g_m^A}\left(g_m^A\right) \, dd_{min} \, dg_s^A \, dg_m^A \tag{2.16}$$

Figure 2.5 shows the normalized ergodic capacity or the spectrum efficiency $E\{C_{CA}\}/W$ as a function of the primary user number N with $I_0/I_N = 1$, $R = 1000$ m,

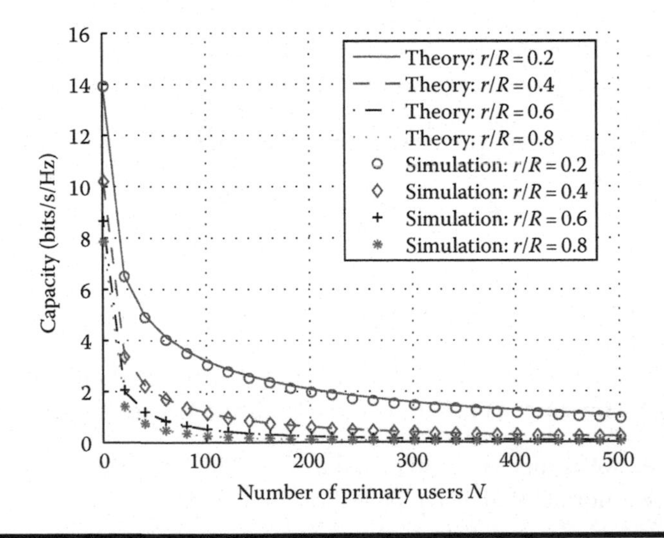

Figure 2.5 The uplink normalized ergodic capacity of the CR-based central access network as a function of N with different values of r/R ($I_0/N_0 = 1$, $R = 1000$ m).

and different values of r/R. Corresponding simulation results were also obtained by averaging over 10,000 realizations of the instantaneous capacity calculated from (2.11). The theoretical results obtained from the numerical integration agree very well with the simulation results. From Figure 2.5, we have the following observations. Given the number of primary users N, the ergodic capacity of the uplink CR channel decreases quickly with increasing r/R. Given r/R, the capacity decreases dramatically as N increases. Only with a small number of primary users N, a large capacity can be achieved. This demonstrates that the capacity provided by a CR-based central access network is significantly restricted by the number of primary users. As a result, such application is more suitable for less-populated rural areas, where the density or the number of primary receivers is relatively low.

2.4 Capacity Analysis of a Cooperative Cognitive Radio Network

2.4.1 System Model

The second CR scenario we consider in this section is the so-called CR-assisted virtual MIMO communication network. The purpose of utilizing CR here is to improve the cellular access ability of a cellular system, for example, UMTS. This is different from the first CR scenario, where only a central access CR network and the primary network coexist. The scenario is shown in Figure 2.2. The cellular BS is equipped with an antenna array located in the center of a cell with radius R. The mobile terminals are dual-mode devices capable of operating in both cellular bands and CR bands simultaneously. We assume that there are M mobile CR users and N primary users, both uniformly distributed in the cell. The basic idea behind this scenario is to first utilize an ad hoc CR network for helping a target mobile transmitter to cooperate with neighboring mobile terminals in CR bands to form a VAA. The VAA will then communicate with the cellular BS antenna array in cellular bands. The CR-assisted virtual MIMO system is expected to greatly improve the spectrum utilization efficiency and system capacity.

2.4.2 Cooperative Communications and Signaling

The current and future cellular networks are challenged by users' increasing demand of high quality and high data rate multimedia services. MIMO is envisioned as a key technology to meet this challenge [30,31]. By deploying multiple antennas at both transmitter and receiver ends, an MIMO system promises significant increase in capacity [37]. It has been shown that the channel capacity increases linearly with the number of antenna pairs in spatially dispersive channels [30,37]. However, it is still not feasible to implement a large number of antennas into small-size mobile terminals with sufficient decorrelation among antenna elements. Virtual MIMO communication [27] was proposed as an alternative that emulates an MIMO system by coordinating multiple single-antenna users to form a VAA.

Here, the VAA will be established in CR bands instead of cellular bands. This can greatly relieve the congestion problem of cellular bands. Once the target transmitter is allocated with the CR bands, it first determines the maximum allowable transmit power $P_{\max}$. Then, the CR transmitter broadcasts in the allocated CR band and cooperates with neighboring users that happen to be inside a circle with radius $\hat{R}$ centered on the CR transmitter. Consequently, the number of cooperating users is a random number. Let us denote the number of transmit antenna elements in the VAA as n_{T}. In this chapter, we assume that $1 \leq n_{\mathrm{T}} \leq 8$, corresponding to the number of antenna elements likely to be equipped at the cellular BS. This means that even though there are more than seven other CR users inside the circle, the target CR transmitter will only cooperate with seven CR users. We assume that the pth $(1 \leq p \leq n_{\mathrm{T}} - 1)$ antenna of the VAA is from the pth cooperating user, while the n_{T}th antenna is from the target transmitter.

Let us denote the PDF of n_{T} as $f_{n_{\mathrm{T}}}(n)$. When $1 \leq n \leq 7$, $f_{n_{\mathrm{T}}}(n)$ can be considered as the probability that $n-1$ out of M CR users are located within the cooperation area with radius $\hat{R}$. When $n = 8$, $f_{n_{\mathrm{T}}}(n)$ is the probability that more than seven CR users are located within the cooperation area. Using basic combinatorial mathematics, it can easily be shown that $f_{n_{\mathrm{T}}}(n)$ is related to R, $\hat{R}$, and the total number of CR users M as follows:

$$f_{n_{\mathrm{T}}}(n) = \begin{cases} \langle M-1, n-1 \rangle \left(\dfrac{\hat{R}^2}{R^2}\right)^{n-1} \left(1 - \dfrac{\hat{R}^2}{R^2}\right)^{M-n}, & 1 \leq n \leq 7 \\ 1 - \displaystyle\sum_{k=1}^{7} \langle M-1, k-1 \rangle \left(\dfrac{\hat{R}^2}{R^2}\right)^{k-1} \left(1 - \dfrac{\hat{R}^2}{R^2}\right)^{M-k}, & n = 8 \end{cases}$$

where the operator $\langle \cdot, \cdot \rangle$ calculates the binomial coefficient.

The signaling for establishing an ad hoc CR VAA is more complex than the signaling of a CR-based central access network. Suppose that the information vector intended to be transmitted at time t from the constructed VAA in cellular bands is $\mathbf{s}^t = \left[s_1^t, \ldots, s_{n_{\mathrm{T}}}^t\right]^{\mathrm{T}}$, where $[\cdot]^{\mathrm{T}}$ represents the transpose of a matrix. At time t, the target transmitter will only transmit the symbol $s_{n_{\mathrm{T}}}^t$. At the previous time slot $t-1$, the target user has to transmit the information symbols s_p^t $(p = 1, \ldots, n_{\mathrm{T}} - 1)$ to the corresponding $n_{\mathrm{T}} - 1$ cooperating users in CR bands. To communicate with $n_{\mathrm{T}} - 1$ cooperating CR users, the available CR bandwidth W is divided into $n_{\mathrm{T}} - 1$ channels based on orthogonal frequency division. Also, the maximum allowable power $P_{\max}$ of the target transmitter is allocated to each channel with equal power of $P_{\max}/(n_{\mathrm{T}} - 1)$. Let us define the channels from the target user to cooperating users as "cooperation channels." The pth cooperation channel gain is denoted as h_p^{C}.

The received symbol $\hat{r}_p^t$ at the pth cooperating user is therefore given by

$$\hat{r}_p^t = \sqrt{\frac{P_{\max}}{n_{\mathrm{T}} - 1}} \sqrt{h_p^{\mathrm{C}}} e^{j\theta} s_p^t + \sqrt{\frac{I_{\mathrm{N}}}{n_{\mathrm{T}} - 1}} \hat{n}_p^t \tag{2.17}$$

where

$\hat{n}_p^t$ is AWGN with unit power

$e^{j\theta}$ represents the random phase shift introduced by the cooperation channel

It is noted that in (2.17), I_{N} is the interference plus noise power at the CR receivers, and the interference is assumed to obey a Gaussian distribution. The cooperation channel h_p^{C} is modeled to account for the effects of path loss, shadowing, and multipath fading. The log-normal shadowing and Rayleigh fading are assumed to be the same as described by (2.13) and (2.14), respectively. We adopt the free-space path loss model by using the expression $\left(\lambda_{\mathrm{c}}/(4\pi l_p)\right)^2$ because it is more accurate for a mobile-to-mobile channel [33]. Here, λ_{c} is the carrier wavelength and l_p is the distance between the target CR transmitter and the pth cooperation user. In this chapter, we assume that the underlying CR networks operate in VHF/UHF TV broadcasting bands between 54 and 862 MHz (or 47–910 MHz modified by the IEEE 802.22 Project Authorization Request) [9,10]. In this chapter, we take an example value of $\lambda_{\mathrm{c}} = 60$ cm, corresponding to a carrier frequency of 500 MHz.

In the literature, there are several methods for the cooperating user to retransmit (relay) the information, including direct amplifying [34], decoding and remodulating [35], and waveform compression [36]. Here, we use a direct amplifying method. The retransmitted symbol $\hat{s}_p^t$ from the pth cooperating user is a linear scaling of $\hat{r}_p^t$ given by

$$\hat{s}_p^t = \frac{\hat{r}_p^t}{\sqrt{\frac{P_{\max}}{n_{\mathrm{T}}-1}} \sqrt{h_p^{\mathrm{C}}} e^{j\theta}} = s_p^t + \frac{\hat{n}_p^t e^{-j\theta}}{\sqrt{\hat{\rho}_p}} \tag{2.18}$$

where $\hat{\rho}_p = P_{\max} h_p^{\mathrm{C}} / I_{\mathrm{N}}$ is the received SINR. Such scaling in (2.18) means that the VAA transmits the "effective signal" s_p^t from all antennas with identical powers. The signal vector actually transmitted from the VAA at time t is then given by $\hat{\mathbf{s}}^t = \left[\hat{s}_1^t, \ldots, \hat{s}_{n_{\mathrm{T}}-1}^t, s_{n_{\mathrm{T}}}^t\right]^{\mathrm{T}}$. We can rewrite (2.18) in a vector form as follows:

$$\hat{\mathbf{s}}^t = \mathbf{s}^t + \hat{\mathbf{n}}^t \tag{2.19}$$

where $\hat{\mathbf{n}}^t = \left[\hat{n}_1^t/\sqrt{\hat{\rho}_1}, \ldots, \hat{n}_{n_{\mathrm{T}}-1}^t/\sqrt{\hat{\rho}_{n_{\mathrm{T}}-1}}, 0\right]^{\mathrm{T}}$. Once the signaling procedure is clear, we will drop the index t subsequently because it only denotes a specific time instant and does not affect the subsequent derivations.

The VAA with n_{T} transmit antennas and the cellular BS array with n_{R} receive antennas will then form an $n_{\mathrm{R}} \times n_{\mathrm{T}}$ virtual MIMO communication link in cellular bands. Let us denote the $n_{\mathrm{R}} \times n_{\mathrm{T}}$ virtual MIMO channel transfer matrix as **H**. With this $n_{\mathrm{T}} \times 1$ transmitted signal vector $\hat{\mathbf{s}} = [\hat{s}_1, \ldots, \hat{s}_{n_{\mathrm{T}}-1}, s_{n_{\mathrm{T}}}]^{\mathrm{T}}$ from the VAA, the $n_{\mathrm{R}} \times 1$ received signal vector $\mathbf{y}$ at the cellular BS is given by [30]

$$\mathbf{y} = \sqrt{\frac{E_{\mathrm{s}}}{n_{\mathrm{T}}}}\mathbf{H}\hat{\mathbf{s}} + \tilde{\mathbf{n}} = \sqrt{\frac{E_{\mathrm{s}}}{n_{\mathrm{T}}}}\mathbf{H}\mathbf{s} + \mathbf{n} \tag{2.20}$$

where

E_{s} denotes the total average transmit symbol energy of the "effective" signal

$E_{\mathrm{s}}/n_{\mathrm{T}}$ represents the average transmit energy per symbol per antenna

$$\mathbf{n} = \sqrt{\frac{E_{\mathrm{s}}}{n_{\mathrm{T}}}}\mathbf{H}\hat{\mathbf{n}} + \tilde{\mathbf{n}} \tag{2.21}$$

represents the effective noise, which includes the noise $\tilde{\mathbf{n}}$ in the cellular channel and regenerative noise $\sqrt{E_{\mathrm{s}}/n_{\mathrm{T}}}\mathbf{H}\hat{\mathbf{n}}$ from the CR channel. The noise power in the cellular channel is denoted as Ω_0. The elements of the cellular virtual MIMO channel matrix **H** are modeled as the composite of a log-normal shadowing process with a standard deviation of 8 dB and an independent Rayleigh fading process with a standard deviation of $\sqrt{2}/2$ for the underlying real Gaussian process.

2.4.3 Capacity Analysis

Uplink capacity of the virtual MIMO channel is defined as the maximum of the mutual information $I(\mathbf{s};\mathbf{y})$ between vectors **s** and **y** [37]. It is noted that $I(\mathbf{s};\mathbf{y})$ is given by [30]

$$\begin{aligned} I(\mathbf{s};\mathbf{y}) &= H(\mathbf{y}) - H(\mathbf{n}) \\ &= \log_2\left[\det(\pi e \mathbf{R}_{yy})\right] - \log_2\left[\det(\pi e \mathbf{R}_{nn})\right] \\ &= \log_2\left[\det(\mathbf{R}_{nn}{}^{-1}\mathbf{R}_{yy})\right] \end{aligned} \tag{2.22}$$

where

$H(\cdot)$ denotes the differential entropy of a vector

$\det(\cdot)$ calculates the determinate of a matrix

$\mathbf{R}_{yy}$ and $\mathbf{R}_{nn}$ are the covariance matrices of **y** and **n**, respectively

$(\cdot)^{-1}$ gives the inverse of a matrix

From (2.20), the covariance matrix of $\mathbf{y}$ can be expressed as

$$\mathbf{R}_{yy} = E\left\{\mathbf{y}\mathbf{y}^{\mathrm{H}}\right\} = \frac{E_{\mathrm{s}}}{n_{\mathrm{T}}}\mathbf{H}\mathbf{R}_{ss}\mathbf{H}^{\mathrm{H}} + \mathbf{R}_{nn} \tag{2.23}$$

where

$(\cdot)^{\mathrm{H}}$ denotes the complex transpose of a matrix

$\mathbf{R}_{ss} = E\left\{\mathbf{s}\mathbf{s}^{\mathrm{H}}\right\}$ is the covariance matrix of $\mathbf{s}$

Substituting (2.23) into (2.22) yields [32]

$$I(\mathbf{s};\mathbf{y}) = \log_2\left[\det\left(\mathbf{I}_{n_{\mathrm{R}}} + \frac{E_{\mathrm{s}}}{n_{\mathrm{T}}}\mathbf{R}_{nn}{}^{-1}\mathbf{H}\mathbf{R}_{ss}\mathbf{H}^{\mathrm{H}}\right)\right] \tag{2.24}$$

where $\mathbf{I}_{n_{\mathrm{R}}}$ denotes a $n_{\mathrm{R}} \times n_{\mathrm{R}}$ identity matrix. From (2.21), the covariance matrix of $\mathbf{n}$ can be calculated by

$$\mathbf{R}_{nn} = \Omega_0\mathbf{I}_{n_{\mathrm{R}}} + \frac{E_{\mathrm{s}}}{n_{\mathrm{T}}}\mathbf{H}\mathbf{R}_{\hat{n}\hat{n}}\mathbf{H}^{\mathrm{H}} = \Omega_0\mathbf{G} \tag{2.25}$$

with

$$\mathbf{G} = \mathbf{I}_{n_{\mathrm{R}}} + \frac{E_{\mathrm{s}}}{\Omega_0 n_{\mathrm{T}}}\mathbf{H}\mathbf{R}_{\hat{n}\hat{n}}\mathbf{H}^{\mathrm{H}}. \tag{2.26}$$

In (2.26), E_{s}/Ω_0 is the received SNR at the cellular BS and $\mathbf{R}_{\hat{n}\hat{n}} = E\left\{\hat{\mathbf{n}}\hat{\mathbf{n}}^{\mathrm{H}}\right\}$ is given by

$$\mathbf{R}_{\hat{n}\hat{n}} = \frac{N_0}{P_{\max}}\mathbf{diag}\left[\left(h_1^{\mathrm{C}}\right)^{-1},\ldots\left(h_{n_{\mathrm{T}}-1}^{\mathrm{C}}\right)^{-1},0\right] \tag{2.27}$$

where the operator $\mathbf{diag}\,[\mathbf{x}]$ denotes a matrix whose diagonal entries are taken from the vector $\mathbf{x}$ while other entries are zero.

Under the constraint of constant effective signal power, the normalized uplink virtual MIMO channel capacity is given by [30]

$$C_{\mathrm{VM}} = \max_{\mathrm{Tr}(\mathbf{R}_{ss})=n_{\mathrm{T}}} I\left(\mathbf{s};\mathbf{y}\right) \tag{2.28}$$

where $\mathrm{Tr}(\mathbf{R}_{ss}) = n_{\mathrm{T}}$ indicates that the trace of the matrix $\mathbf{R}_{ss}$, that is, the sum of the diagonal elements, equals n_{T}. When the channel $\mathbf{H}$ is completely unknown to the transmitter, the vector $\mathbf{s}$ may be chosen to be statistically non-preferential, that is, $\mathbf{R}_{ss} = \mathbf{I}_{n_{\mathrm{T}}}$ [30]. This implies that the signals are independent and equally powered at the transmit antennas. From (2.23) and (2.27), the normalized capacity of the virtual MIMO channel in the absence of channel knowledge at the transmitter can be obtained as

$$C_{\mathrm{VM}} = \log_2\left[\det\left(\mathbf{I}_{n\mathrm{R}} + \frac{E_{\mathrm{s}}}{n_{\mathrm{T}}}\mathbf{R}_{nn}{}^{-1}\mathbf{H}\mathbf{H}^{\mathrm{H}}\right)\right]$$

$$= \log_2\left[\det\left(\mathbf{I}_{n\mathrm{R}} + \frac{E_{\mathrm{s}}}{\Omega_0 n_{\mathrm{T}}}\mathbf{G}^{-1}\mathbf{H}\mathbf{H}^{\mathrm{H}}\right)\right] \tag{2.29}$$

It should be noted that the classical MIMO channel capacity is given by [30]

$$C_{\mathrm{MIMO}} = \log_2\left[\det\left(\mathbf{I}_{n\mathrm{R}} + \frac{E_{\mathrm{s}}}{\Omega_0 n_{\mathrm{T}}}\mathbf{H}\mathbf{H}^{\mathrm{H}}\right)\right] \tag{2.30}$$

The comparison of (2.29) and (2.30) demonstrates that the virtual MIMO channel capacity differs from the classical MIMO channel capacity by an additional matrix of $\mathbf{G}^{-1}$.

There are a number of parameters that could affect the instantaneous channel capacity C_{VM}. The system parameters include the cell radius R, the cooperation range $\hat{R}$, and the value of I_0/I_{N}. Here, we assume that $R = 1000$ m, $\hat{R} = 20$ m, and $I_0/I_{\mathrm{N}} = 1$. Other relevant parameters include the received SNR E_{s}/Ω_0 at the cellular BS, the maximum allowable CR transmit power P_{max}, and the VAA antenna numbers n_{T}. It is important to mention that the RVs P_{max}, n_{T}, and $\mathbf{H}$ are independent. Taking the mean value of C_{VM} and C_{MIMO} over fading channels $\mathbf{H}$ results in the normalized ergodic virtual MIMO channel capacity $E\{C_{\mathrm{VM}}\}$ and real MIMO channel capacity $E\{C_{\mathrm{MIMO}}\}$, respectively.

2.4.4 Results and Discussions

Figure 2.6 shows the numerical results of $E\{C_{\mathrm{VM}}\}$ as a function of the average received SNR with different values of the minimum distance d_{min} and antenna pairs. For comparison purposes, the corresponding results of $E\{C_{\mathrm{MIMO}}\}$ are also shown in the figure. Clearly, with the increase of antenna pairs, both the real and virtual MIMO channel capacities increase. For any given multiple antenna pairs ($n_{\mathrm{T}} = n_{\mathrm{R}} > 1$), a relatively large d_{min} ($d_{\mathrm{min}} = 400$ m as an example here) makes the resulting ergodic virtual MIMO channel capacity approach closely to the corresponding real MIMO channel capacity, that is, $E\{C_{\mathrm{VM}}\} \approx E\{C_{\mathrm{MIMO}}\}$. It is noted that their capacities tend to be very close when SINR in the CR-relaying channels goes to infinity. With a smaller d_{min} (e.g., $d_{\mathrm{min}} = 100$ m), the virtual MIMO channel capacity is much smaller than the real MIMO capacity. For a special case, when $n_{\mathrm{T}} = n_{\mathrm{R}} = 1$, that is, single-input single-output (SISO) case, $E\{C_{\mathrm{VM}}\} = E\{C_{\mathrm{MIMO}}\}$ holds because the CR relaying channel is not involved. The multiplexing gain of an MIMO system, also known as the gain in terms of the number of degrees of freedom, can be observed as the slope of capacity curves at

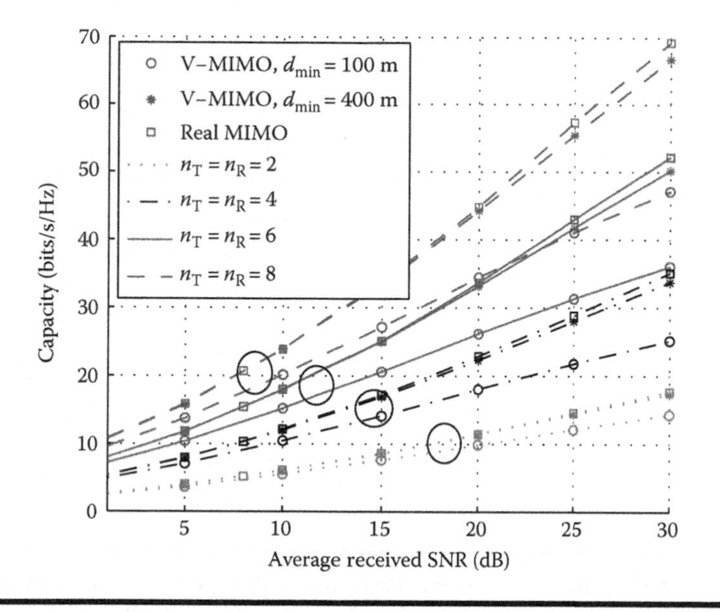

Figure 2.6 The uplink normalized ergodic capacity of the CR-assisted virtual MIMO network as a function of the average received E_s/N_0 ($I_0/N_0 = 1$).

the high-SNR regime [30]. From Figure 2.6, it is obvious that the multiplexing gain of the virtual MIMO system is reduced as d_{min} decreases.

To understand better the effect of d_{min} on the virtual MIMO channel capacity, in Figure 2.7 we show the numerical results of $E\{C_{VM}\}$ as a function of d_{min} with a fixed SNR $E_s/\Omega_0 = 8$ dB. For $n_T = n_R > 1$, the virtual MIMO channel capacity increases very fast with increasing d_{min} when d_{min} is relatively small, for example, $d_{min} < 150$ m. When d_{min} becomes relatively large, the virtual MIMO channel capacity increases slowly with the increase of d_{min} and gradually approaches the real MIMO channel capacity. Subsequently, we will fix $E_s/\Omega_0 = 8$ dB and further investigate the influence of the number of primary receivers N and the number of CR users M on the channel capacity.

To study the influence of N on the virtual MIMO channel capacity, let us consider d_{min} as a RV related to the number N of primary users and the relative position r/R of the CR transmitter in the cell. Suppose that we have $r/R = 0.5$. The average of $E\{C_{VM}\}$ over d_{min} results in the ergodic virtual MIMO channel capacity $\bar{E}\{C_{VM}\}$ as a function of N, that is,

$$\bar{E}\{C_{VM}\} = \int_{d_{min}} E\{C_{VM}\} f_{d_{min}}(x)\mathrm{d}x \tag{2.31}$$

Figure 2.8 shows $\bar{E}\{C_{VM}\}$ as a function of N with different antenna pairs (n_T, n_R). For $n_T = n_R > 1$, with the increase of N the virtual MIMO channel capacity reduces

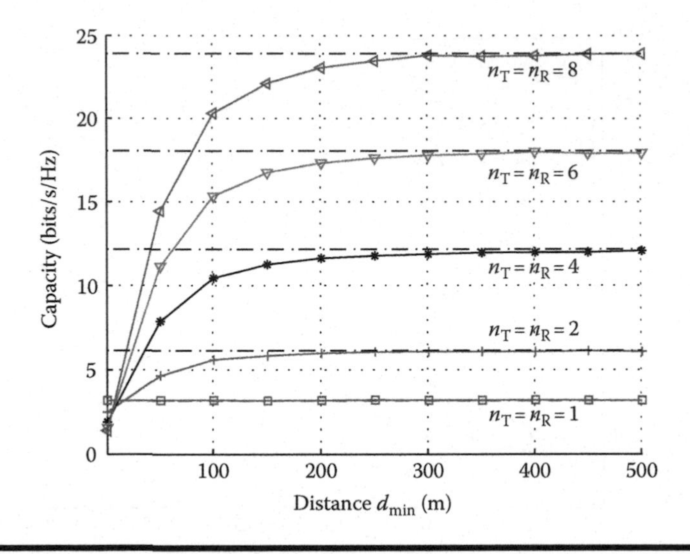

Figure 2.7 The uplink normalized ergodic capacity of the CR-assisted virtual MIMO network as a function of d_{min} ($I_0/N_0 = 1$, $E_s/N_0 = 10$ dB).

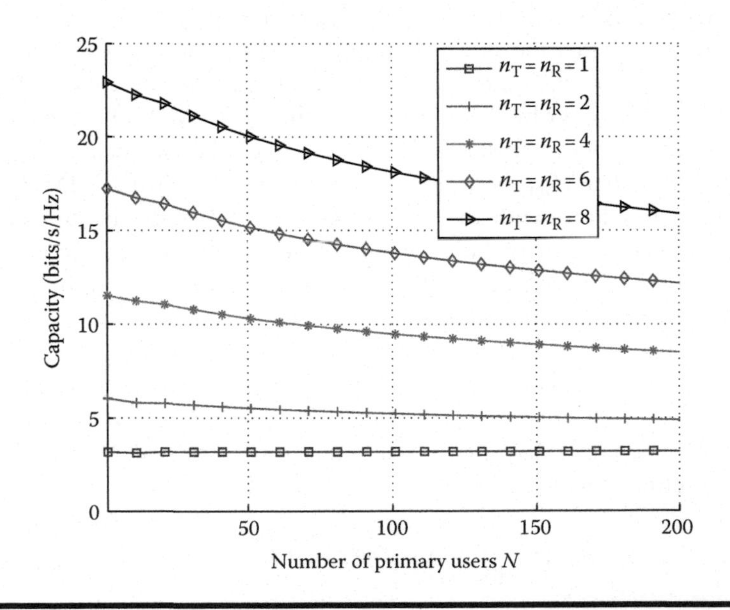

Figure 2.8 The uplink normalized ergodic capacity of the CR-assisted virtual MIMO network as a function of N ($I_0/N_0 = 1$, $E_s/N_0 = 10$ dB, $r/R = 0.5$).

gradually, but not dramatically. When $n_T = n_R = 1$, the capacity does not change with N.

If the whole cell is considered, the distance r between the target CR transmitter and cell center is an RV with a distribution $f_r(r) = 2\pi r/(\pi R^2) = 2r/R^2$. Moreover, n_T is an RV with its PDF given by (2.17). The average of $\bar{E}\{C_{VM}\}$ over r and n_T will result in the ergodic cell capacity $\vec{E}\{C_{VM}\}$ as a function of N and M, that is,

$$\vec{E}\{C_{VM}\} = \int_{r,n_T} \bar{E}\{C_{VM}\} f_r(r) f_{n_T}(n) \mathrm{d}r\mathrm{d}n \tag{2.32}$$

The numerical results of $\vec{E}\{C_{VM}\}$ are demonstrated in Figure 2.9 as a function of N with different values of M. For comparison purposes, in Figure 2.9 we also present the capacities of an SISO channel ($n_T = n_R = 1$) and a real MIMO channel with $n_T = n_R = 8$. The gap between the virtual MIMO channel capacity and SISO capacity can be viewed as the capacity gain achieved by the CR-assisted virtual MIMO network. With the increase of M, the virtual MIMO channel capacity increases and gradually approaches to the real MIMO channel capacity. This is due to the fact that when more CR users are located in the cell, there is a higher probability that more CR users will be within the cooperation range. Consequently, the number of antennas for establishing the VAA will also increase, which will

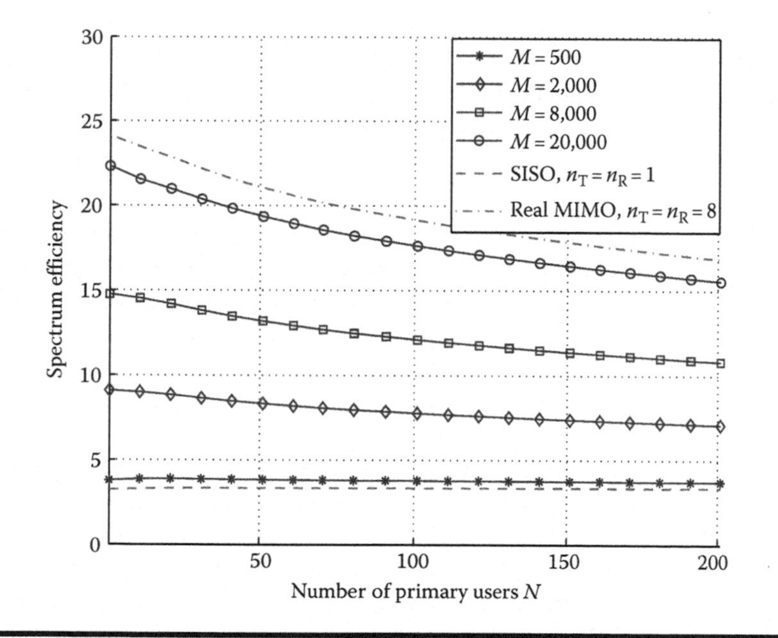

Figure 2.9 The uplink normalized ergodic capacity of the CR-assisted virtual MIMO network as a function of N with different values of M ($I_0/N_0 = 1$, $E_s/N_0 = 10$ dB, $\hat{R} = 20$ m, $R = 1000$ m).

further increase the virtual MIMO channel capacity. Similar to the conclusion obtained from Figure 2.8, we can see from Figure 2.9 that the virtual MIMO channel capacity decreases slowly as N increases.

Also, it is noted that the capacity reduction of the CR-assisted virtual MIMO channel is not so sensitive to the increase of the number of primary users N. This is different from the CR-based central access network, the capacity of which decreases dramatically with the increase of N, as shown in Figure 2.5. On the other hand, a larger number of CR users M results in a higher virtual MIMO channel capacity. This demonstrates that a CR-assisted virtual MIMO network is more suitable for urban areas, where a high density of CR users exists, despite the fact that the number of primary receivers N might also be large.

2.5 Conclusions and Open Issues

In this chapter, we have studied the system-level capacities of interference-tolerant CR networks under average received interference power constraints. Two CR scenarios, namely, central access CR networks and cooperative CR networks, have been studied as potential applications of CR networks. Our analysis has shown that the central access CR network is more suitable for less-populated rural areas, while the cooperative CR network performs better in urban environments.

System-level capacity analysis is indispensable for the strategic planning and economical study of CR networks. The objective for such analysis is to establish tractable and realistic system models that can be used to provide fundamental guidelines to CR network designs. Despite the initial efforts shown in this chapter, there are still many open research problems, some of which are as follows:

- How is the performance of CR networks affected by the characteristics of primary networks such as the density and spatial distribution pattern of primary users? In this chapter, we have used a simple model that assumes uniform primary user distribution in a circular cell. More complicated stochastic models, such as models based on Poisson point processes or clustering models, can be used to give more realistic descriptions of the spatial distribution of primary users.
- Given certain received interference power constraints, what is the performance limit of CR networks in terms of system capacity and communication range? To protect the primary systems, various received interference power constraints can be specified. In this chapter, we have only considered the average received interference power constraints, while others such as peak received interference power constraints and interference outage constraints should also be investigated.
- How can CR networks support multiple users and different QoS requirements? Ideally, a CR network should be able to deliver a multitude of different

services ranging from QoS-guaranteed services like high-quality voice-calls to best-effort services like web surfing. It is likely that a CR network should be designed differently and will have different capacities to support different types of services. Moreover, the systems we have considered in this chapter use TDMA as the multiple access scheme. CR networks based on other multiple access schemes can be further studied and compared.

- Is it better to establish a pure CR network or use CR networks in a framework of multiple radio access technologies (RATs)? As suggested by our study in this chapter, the performance of a pure CR network (e.g., the central access CR network) can be poor due to the received interference power constraints. In contrast, using CR in the context of multiple RATs (e.g., the CR-assisted virtual MIMO communication) seems to be a promising application because we can jointly optimize the physical layer design of a radio access network given different, possibly complementary radio resources (e.g., bandwidth-limited licensed cellular spectrum and power-limited CR spectrum). The CR-assisted virtual MIMO communication studied in this chapter is just one example of CR application in the multiple RAT context. A wide variety of other cooperative communication schemes can be further investigated. It is envisioned that further information-theoretic studies can provide more insight to guide the design of CR networks in a multiple RAT environment.

2.6 Appendix: Derivation of (2.4) through (2.9)

In this appendix, we use a geometric method to derive the CDF $F_{d_{\min}}(d)$ of $d_{\min}$ shown in (2.4) through (2.9). As illustrated in Figure 2.10, we use a circular area C_1 with radius R to represent a cell of the CR network. The center of a cell is denoted as O_1. Multiple CR users and N primary receivers are uniformly distributed within

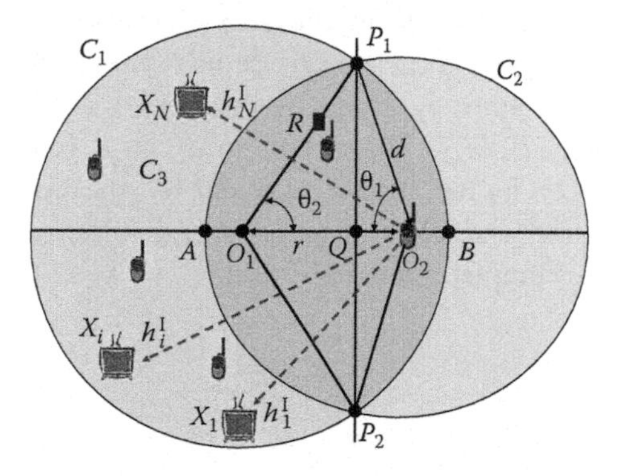

Figure 2.10 The proposed geometric method to calculate the CDF $F_{d_{\min}}(d)$ of $d_{\min}$.

the cell. The target CR transmitter is located in O_2 with the distance of r from the cell center O_1. The CDF $F_{d_{\min}}(d)$ of $d_{\min}$ is the probability that $d_{\min} \leq d$ holds. If we plot another circle C_2 with the radius d centered on the target CR transmitter O_2, the CDF $F_{d_{\min}}(d)$ can be considered as the probability that at least one primary receiver is located within the circle C_2 and certainly within C_1. Let us use $S(d)$ to represent the area of the proportion C_3 that is within C_1 but outside C_2. The probability of the random event that all N primary receivers are located within C_3 can be calculated by $\left[S(d)/(\pi R^2)\right]^N$. The probability of the complementary event, that is, at least one primary receiver is located outside C_3 but within C_1, is the CDF $F_{d_{\min}}(d)$ of $d_{\min}$. It follows that we have $F_{d_{\min}}(d) = 1 - \left[S(d)/(\pi R^2)\right]^N$, as given by (2.4). The remaining task is to calculate the area $S(d)$ of the proportion C_3.

It is noted that the radius d of the circle C_2 is a RV ranging from 0 to $R + r$. When d increases from 0 to $R + r$, $S(d)$ will correspondingly decrease from πR^2 to 0. Only when $d \in [0, R - r]$, C_2 is completely included in C_1. If $d > R - r$, the two circles C_1 and C_2 will intersect at two points P_1 and P_2. As clearly shown in Figure 2.10, the line P_1P_2 is perpendicular to the line AB with the intersection Q. When $d \in (R - r, \sqrt{R^2 - r^2}]$, Q is located between B and O_2. When $d \in (\sqrt{R^2 - r^2}, \sqrt{R^2 + r^2}]$, Q is between O_2 and O_1. When $d \in (\sqrt{R^2 + r^2}, R+r]$, Q is between O_1 and A. According to the interval in which d is located, the corresponding area $S(d)$ can be calculated using basic geometric methods. The final result is shown in (2.5). The derivation of $S(d)$ in (2.5) for each region of d is quite similar. In what follows, we will only show how to derive $S(d)$ when $d \in (\sqrt{R^2 - r^2}, \sqrt{R^2 + r^2}]$, that is, Q is located between O_2 and O_1.

As shown in Figure 2.10, the triangle $O_1O_2P_1$ has side lengths r, d, and R. Based on the law of cosines, the angles θ_1 and θ_2 can easily be calculated, which are given by (2.8) and (2.9), respectively. The area of the sector $O_2P_1AP_2$ is given by $2\pi d^2 \frac{\theta_1}{2\pi} = d^2\theta_1$. The area of the triangle $O_2P_1P_2$ is given by $d \sin\theta_1 d \cos\theta_1 = d^2 \sin\theta_1 \cos\theta_1$. The area S_1 of the segment P_1QP_2A is obtained by subtracting the area of the triangle $O_2P_1P_2$ from the area of the sector $O_2P_1AP_2$, that is, $S_1 = d^2(\theta_1 - \sin\theta_1 \cos\theta_1)$, as given by (2.6). Similarly, S_2 is the area of the segment P_1QP_2B, which can be obtained by subtracting the area of the triangle $O_1P_1P_2$ from the area of the sector O_1P1BP_2. This results in $S_2 = R^2(\theta_2 - \sin\theta_2 \cos\theta_2)$, as given by (2.7). When Q is located between O_2 and O_1, the area $S(d)$ is obtained by subtracting S_1 and S_2 from the area of the circle C_1, that is, $S(d) = \pi R^2 - S_1 - S_2$, as shown in (2.5). This completes the derivation.

References

1. Federal Communication Commission, Spectrum policy task force report, Washington DC, FCC 02-155, Nov. 2, 2002.
2. Federal Communications Commission, Unlicensed operation in the TV broadcast bands, Washington, DC, ET Docket No. 04-186, 2004.

3. Federal Communication Commission, Facilitating opportunities for flexible, efficient, and reliable spectrum use employing cognitive radio technologies, NPRM & Order, Washington, DC, ET Docket No. 03-108, FCC 03-322, Dec. 30, 2003.
4. J. Mitola and G. Maguire Jr., Cognitive radio: Making software radios more personal, *IEEE Pers. Commun. Mag.*, 6(6), 13–18, Aug. 1999.
5. H. H. Chen and M. Guizani, *Next Generation Wireless Systems and Networks*, Chichester, U.K.: John Wiley & Sons, 2006.
6. I. F. Akyildiz, W. Y. Lee, M. C. Vuran, and S. Mohanty, Next generation dynamic spectrum access/cognitive radio wireless networks: A survey, *Comput. Netw.*, 50(13), 2127–2159, Sept. 2006.
7. S. Haykin, Cognitive radio: Brain-empowered wireless communications, *IEEE J. Sel. Areas Commun.*, 3(2), 201–220, Feb. 2005.
8. C.-X. Wang, H. H. Chen, X. Hong, and M. Guizani, Cognitive radio network management: Tuning in to real-time conditions, *IEEE Veh. Technol. Mag.*, 3(1), 28–35, Mar. 2008.
9. IEEE P802.22, Functional requirements for the 802.22 WRAN standard, IEEE 802.22-05/0007r48, Nov. 29, 2006.
10. IEEE Standards Association, Project Authorization Request, Information technology—Telecommunications and information exchange between systems—Wireless Regional Area Networks (WRAN)—Specific requirements—Part 22: Cognitive Wireless RAN Medium Access Control (MAC) and Physical Layer (PHY) specifications: Policies and procedures for operation in the TV Bands. http://www.ieee802.org/22/802-22-PAR.pdf.
11. S. A. Jafar and S. Srinivasa, Capacity limits of cognitive radio with distributed and dynamic spectral activity, *Proceedings of the IEEE ICC'06*, Istanbul, Turkey, June 2006, pp. 5742–5747.
12. T. A. Weiss and F. K. Jondral, Spectrum pooling: An innovative strategy for the enhancement of spectrum efficiency, *IEEE Radio Commun.*, 42(3), 8–14, Mar. 2004.
13. S. Mangold, A. Jarosch, and C. Monney, Operator assisted cognitive radio and dynamic spectrum assignment with dual beacons—Detailed evaluation, *Proceedings of First International Conference on Communication Systems Software and Middleware*, New Delhi , India, Jan. 2006, pp. 1–6.
14. D. Ugarte and A. B. McDonald, On the capacity of dynamic spectrum access enable networks, *Proceedings of the IEEE International Symposium on New Frontiers in Dynamic Spectrum Access Networks*, Baltimore, MD, Nov. 2005, pp. 630–633.
15. M. Gastpar, On capacity under received-signal constraints, *42th Annual Allerton Conference on Communication, Control, and Computing*, Monticello, IL, Oct. 2004, pp. 1322–1331.
16. M. Gastpar, On capacity under receive and spatial spectrum-sharing constraints, *IEEE Trans. Inf. Theory*, 53(2), 471–487, Feb. 2007.
17. A. Ghasemi and E. S. Sousa, Capacity of fading channels under spectrum-sharing constraints, *Proceedings of the IEEE ICC'06*, Istanbul, Turkey, June 2006, pp. 4373–4378.
18. J. Neel, R. M. Buehrer, B. H. Reed, and R. P. Gilles, Game theoretic analysis of a network of cognitive radios, *Proceedings of Midwest Symposium on Circuits and Systems*, Tulsa, Oklahoma, Aug. 2002, pp. 409–412.

19. G. L. Stüber, *Principles of Mobile Communication*, 2nd edn., Boston, MA: Kluwer Academic Publishers, 2001.
20. S. Mangold, A. Jarosch, and C. Monney, Operator assisted cognitive radio and dynamic spectrum assignment with dual beacons—Detailed evaluation, *Proceedings of the First International Conference on Communication Systems Software and Middleware*, Delhi, India, Jan. 2006, pp. 1–6.
21. B. Wild and K. Ramchandran, Detecting primary receivers for cognitive radio applications, *Proceedings of the IEEE DySPAN'05*, Baltimore, MD, Nov. 2005, pp. 124–130.
22. N. Devroye, P. Mitran, and V. Tarokh, Achievable rates in cognitive radio channels, *IEEE Trans. Inf. Theory*, 52(5), 1813–1827, May 2006.
23. N. Devroye, P. Mitran, and V. Tarokh, Limits on communications in a cognitive radio channel, *IEEE Commum. Mag.*, 44(6), 44–49, June 2006.
24. S. Srinivasa and S. A. Jafar, The throughput potential of cognitive radio: A theoretical perspective, *IEEE Commun. Mag.*, 45(5), 73–79, May 2007.
25. X. Hong, C.-X. Wang, H. H. Chen, and J. Thompson, Performance analysis of cognitive radio networks with average interference power constraints, *Proceedings of IEEE ICC'08*, Beijing, China, May 2008, pp. 3578–3582.
26. C.-X. Wang, X. Hong, H. H. Chen, and J. Thompson, On capacity of cognitive radio networks with average interference power constraints, *IEEE Trans. Wireless Commun.*, 8(4), 1620–1625, April 2009.
27. M. Dohler, Virtual Antenna Array, PhD thesis, King's College London, London, U.K., 2003.
28. S. Kim, H. Jeon, H. Lee, and J. S. Ma, Robust transmission power and position estimation in cognitive radio, *Proceedings of the International Conference on Information Networking* (*ICOIN'07*), Estoril, Portugal, Jan. 2007.
29. H. Islam, Y.-C. Liang, and A. T. Hoang, Joint power control and beamforming for secondary spectrum sharing, *Proceedings of the VTC-Fall'07*, Baltimore, MD, Sept. 2007, pp. 1548–1552.
30. A. Paulraj, R. Nabar, and D. Gore, *Introduction to Space-Time Wireless Communications*, Cambridge, MA: Cambridge University Press, 2003.
31. C.-X. Wang, X. Hong, H. Wu, and W. Xu, Spatial temporal correlation properties of the 3GPP spatial channel model and the Kronecker MIMO channel model, *EURASIP J. Wireless Commun. Netw.*, 2007, 9, 2007, Article ID 39871, doi:10.1155/2007/39871.
32. F. R. Farrokhi, G. J. Foschini, A. Lozano, and R. A. Valenzuela, Link-optimal space-time processing with multiple transmit and receive antennas, *IEEE Commun. Lett.*, 5(3), 85–87, Mar. 2001.
33. T. J. Harrold, A. R. Nix, and M. A. Beach, Propagation studies for mobile-to-mobile communications, *Proceedings of the IEEE VTC'01-Fall*, Atlantic City, NJ, Oct. 2001, pp. 1251–1255.
34. J. N. Laneman, G. W. Wornell, and D. N. C. Tse, An efficient protocol for realizing cooperative diversity in wireless networks, *Proceedings of the IEEE International Symposium on Information Theory*, Washington, DC, June 2001, pp. 294.

35. T. E. Hunter and A. Nosratinia, Diversity through coded cooperation, *IEEE Trans. Wireless Commun.*, 5(2), 283–289, Feb. 2006.
36. T. M. Cover and A. A. El Gamal, Capacity theorems for the relay channel, *IEEE Trans. Inf. Theory,* 25(5), 572–584, Sept. 1979.
37. I. E. Telatar, Capacity of multi-antenna Gaussian channels, Technical Report, AT & T Bell Labs, June 1995.

Chapter 3

Power Control for Cognitive Radio Ad Hoc Networks

Lijun Qian, Xiangfang Li, John Attia, and
Deepak Kataria

Contents

Several recent measurement reports show that the assigned spectrum from 0 to 3 GHz are highly underutilized. To achieve much better spectrum utilization and viable frequency planning, cognitive radios (CRs) are under development to dynamically capture the unoccupied spectrum. Although Federal Communication Commission (FCC) proposes spectrum sharing between a legacy TV system and a CR network to increase spectrum utilization, one of the major concerns is that the interference from the CR network should not violate the quality-of-service (QoS) requirements of the primary users. Specifically, can secondary users (CR network) even operate without causing excessive interference to primary users (TV users)? Furthermore, can certain QoS for secondary users be provided under such constraints? So far, most of the previous works address these two issues by time sharing the spectrum between the TV system and the CR network.

In this chapter, we consider the scenario where the CR network is formed by secondary users with low-power wireless devices and when both systems are operating simultaneously. In order for the spatial spectrum sharing to sustain, a framework for power control in CR networks is proposed such that the energy efficiency of the secondary users is maximized and the QoS of both the primary users and the secondary users are guaranteed. The feasibility conditions and the centralized and distributed solutions of the power control problem are derived for CR networks employing Time Division Multiple Access (TDMA), Code Division Multiple Access (CDMA), and Orthogonal Frequency Division Multiple Access (OFDMA), respectively. The achievable data rate of the secondary users is obtained, and insights are provided for CR network design and deployment. Because the co-channel interference are from heterogeneous systems, a joint power control and admission control procedure is suggested such that the priority of the primary users is always ensured. The proposed schemes are evaluated through extensive simulations. The advantages and disadvantages of temporal and spatial spectrum sharing are discussed and open problems are identified.

3.1 Introduction

Although the U.S. government frequency allocation data [1] shows that there is fierce competition for the use of spectra, especially in the bands from 0 to 3 GHz, it is pointed out in several recent measurement reports that the assigned spectra are highly underutilized [2,3]. The discrepancy between spectrum allocation and spectrum use suggests that "spectrum access is a more significant problem than physical scarcity of spectrum, in large part due to legacy command-and-control regulation that limits the ability of potential spectrum users to obtain such access" [2]. To achieve much better spectrum utilization and viable frequency planning, CR are under development to dynamically capture the unoccupied spectrum [4,5].

It is envisioned that suitably designed CRs have the potential for creating a next-generation adaptive wireless network in which a single universal radio device is capable of operating in a variety of spectrum allocation and interference conditions by selecting appropriate physical and network layer parameters often in collaboration with other radios operating in the same region [6]. Besides an economical solution to 3G and beyond, CRs provide a practical solution for the ubiquitous deployment of uncoordinated wireless ad hoc networks with increased network capacity and user performance [7].

There are a lot of ongoing standardization efforts related to CRs, such as the IEEE 802.22 [8] and SCC41 (formerly known as P1900) [9]. IEEE 802.22 wireless regional area networks (WRANs) with a CR-based air interface for use by license-exempt devices on a noninterfering basis in VHF and UHF (54-862 MHz) bands. It will be the first CR-based international standard with tangible frequency bands for its operation [10]. IEEE 802.22 standard specifies the air interface, including Physical (PHY) and Media Access Control (MAC) layers, of fixed point-to-multipoint WRANs. Operating on a strict noninterference basis in spectrum assigned to, but unused by, the incumbent licensed services requires a new approach using purpose-designed CR techniques that will permeate the PHY and MAC layers [11]. On the other hand, the Standards Coordinating Committee (SCC) 41 is focused on dynamic spectrum access networks and has several standards currently in development [12]. They are well known for their CR activities and have a broader scope than the previous coexistence-oriented activities that are being conducted in IEEE 802.

The FCC has recognized the promising technique and is pushing to enable a full realization of the technique. As the first step, the FCC proposes to experiment unlicensed cognitive sharing in the TV bands (the VHF and UHF bands) [13–15]. The TV bands are chosen due to the better penetration of the frequency band, "strong" received signal of the primary TV users, and TV transmitters are left on more or less continuously, and infrequently change location or frequency [16].

Despite the advantages of using the TV bands for unlicensed cognitive spectrum sharing, there are some concerns to be solved first to convince the FCC to finally open the TV bands. First, can secondary users (CR network) even operate without causing excessive interference to primary users (TV users)? Second, can certain quality-of-service (QoS) for secondary users be provided under such constraints? So far, most of the previous works address these two issues by time sharing the spectrum between the TV system and the CR network. In this case, there will be no co-channel interference, and it is suitable for secondary users with high transmission power (e.g., higher power "fixed/access" unlicensed devices that may provide wireless Internet access [14,15]). One of the main difficulties is to detect the presence of the TV signals accurately. Much work has been done in this area, such as [17,18] and the references therein.

In this chapter, we consider a different case where the TV system and the CR network are ON simultaneously and they share the same spectrum through space separation. This case is mainly studied through MAC design, such as in [19]. Power control is only applied to address the nonintrusion to the services of the primary users [20], but not the QoS of the secondary users. We argue that the QoS of the secondary users is also very important [21]. If the capacity for the secondary users is not enough to realize their required QoS after meeting the QoS constraints of the primary users, that channel might not be a good opportunity for secondary users to access.

According to the recent suggestions from the FCC [14,15], two distinct types of unlicensed broadband devices may be used in the TV bands. One category will consist of lower-power "personal/portable" unlicensed devices. The second category will consist of higher-power "fixed/access" unlicensed devices that may provide wireless Internet access. This chapter will consider the spectrum-sharing problem for the first category, and we focus on the case where both the TV system and the CR network operate simultaneously. The power control problem becomes tougher than that in cellular systems or pure wireless ad hoc networks because the interference tends to be more difficult to model and control in two heterogeneous systems. In this chapter, we try to provide some preliminary analysis and design to address the two issues mentioned in the previous paragraph when two heterogeneous systems operate in the same frequency band at the same time. Given the QoS requirements of the primary (TV) users and secondary users, a power control framework is proposed to address the spectrum sharing between the two systems while considering the unique requirements of high spectral utilization and power efficiency of the low-power wireless devices that have very limited resources. Specifically, a power control problem of the secondary users is formulated to maximize the energy efficiency of the secondary users and reduce the harmful interference to the primary users who have absolute priority. QoS guarantee of the secondary users is also included in the problem formulation. Feasibility conditions for the power control problem are highlighted and the corresponding joint power control and admission control procedures are provided. The results obtained on power control policies will provide insights for network designers as well as policy makers.

It is worth pointing out that although using CR in a low-power wireless device may consume some energy due to spectrum sensing, it is necessary in the environment where the spectrum shortage is a serious concern. In addition, a secondary user only needs to sense the spectrum when it has new traffic demand. In the proposed design, these new secondary users will search for an appropriate band for data transfer without interfering with the legacy system as well as the existing communications of the existing secondary users. This in turn will improve the energy efficiency of the CR network.

Power control in CR networks has attracted a lot of attention recently [22–28]. In [22], a CR network in which a set of base stations make opportunistic spectrum access to support fixed-location wireless subscribers within their cells is considered. A downlink channel/power allocation scheme that maximizes the number of supported subscribers is obtained by solving a mixed-integer linear programming problem. However, the formulation in [22] is not applicable to the case when primary users employ OFDMA and spread transmissions over multiple carriers. An opportunistic power control strategy for the cognitive users is proposed in [23], which serves as an alternative way to protect the primary users transmission and to realize spectrum sharing between the primary user and the cognitive users. The key feature of the proposed strategy is that, via opportunistically adapting its transmit power, the cognitive user can maximize its achievable transmission rate without degrading the outage probability of the primary user. In [24], a transmit power control system using fuzzy logic is proposed. With the built-in fuzzy power controller, a CR is able to opportunistically adjust its transmit power in response to the changes of the interference level to the primary user, the distance to primary user, and its received power difference at the base station while satisfying the requirement of sufficiently low interference to primary users. In [25], the transmit power of the CR is controlled by using the side information of spectrum sensing to minimize the interference to the primary users. In [26], the optimal power control in a CR network is modeled as a concave minimization problem, and an improved branch and bound algorithm is proposed. A utility function–based approach is also applied to the power control problem in peer-to-peer CR networks [27]. Energy efficiency maximization is considered in [28] for OFDMA CR ad hoc networks through subcarrier and power allocation. Given the data rate requirement and maximal power limit, a constrained optimization problem is formulated for each individual CR user to minimize the energy consumption per bit over all selected subcarriers, while avoiding the introduction of harmful interference to the existing users. Because of the multidimensional and non-convex nature of the problem, a fully distributed subcarrier selection and power allocation algorithm is proposed by combining an unconstrained optimization method with a constrained partitioning procedure.

The chapter is organized as follows. Section 3.2 provides the model of spectrum sharing of a CR network with a TV broadcast system, and the associated power control problem is formulated. The solution of the power control problem for a single secondary transmitter–receiver pair is given in Section 3.3. Both centralized and distributed power control algorithms are provided for the case of multiple secondary user pairs in Section 3.4. The effectiveness of the proposed schemes is tested through simulations in Section 3.5. Section 3.6 provides discussions, open problems, and suggestions on future research topics. Section 3.7 contains the concluding remarks. Throughout the chapter, we provide case studies for secondary users employing OFDMA. It also made clear that the proposed power control framework applies to CR users using other physical layer technologies such as TDMA or CDMA.

3.2 Model and Problem Formulation

Given an existing TV station with transmission power p_{TV}, the effective receiving range is D. The effective receiving range is defined by the successful decoding of the TV signals, that is, the received signal-to-interference-plus-noise ratio (SINR) should be above a given threshold (10 dB or higher [16], which will depend on the type of TV station) such that the received TV signal is decodable. Note that the data of transmission power and effective receiving range of TV stations are publicly available, such as in [29,30]. It is assumed that the secondary users locate in an $l \times l$ square area. The center of the CR network is d meters away from the nearest primary receiver. The distance from the TV station to the ith secondary receiver is h_i. y_i is the distance from the ith secondary transmitter to the TV receiver at the border of the TV coverage area. An example of the model is given in Figure 3.1, where only one pair of secondary users are shown. Note that although the effective receiving range of the TV station may not overlap with the transmission range of the CR network, the transmissions in both systems still cause nonnegligible co-channel interferences to the receivers of the other system. For instance, if both systems are ON simultaneously, the transmission from the secondary users will cause interference at the primary receivers and may cause the received TV signals degraded and become unacceptable. Hence, the co-channel interference is the major barrier for the successful coexistence of the two systems.

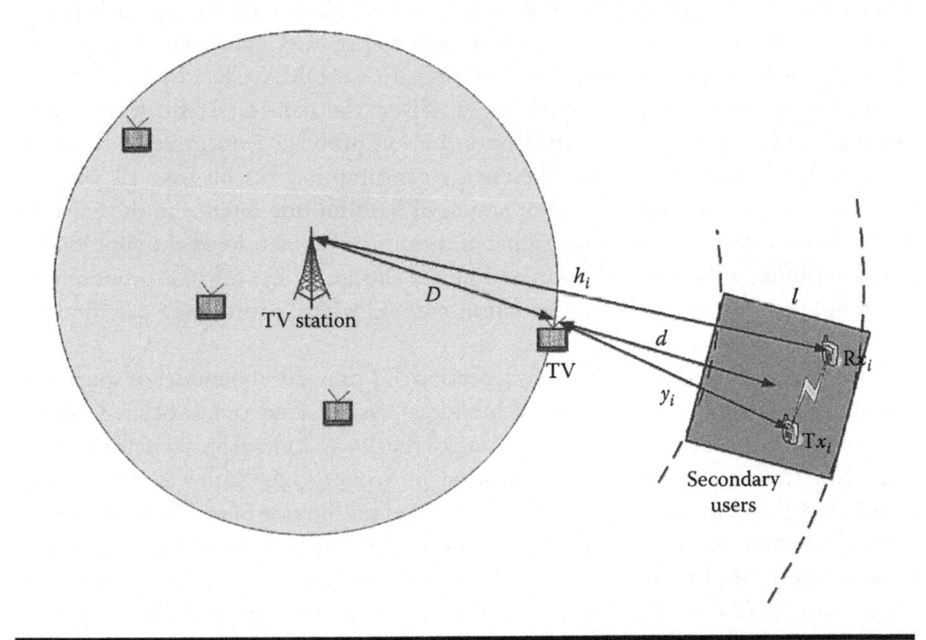

Figure 3.1 An example of spectrum sharing of a CR network with a TV broadcast system.

In this chapter, we address the interference problem by considering the QoS at both the primary receivers and the secondary receivers in terms of the received SINR. Suppose there are totally N pairs of secondary users, and $p_{i,\mathrm{sec}}$ is the transmission power of the ith transmitter. Define the SINR at the mth primary receiver as $\gamma_{m,\mathrm{TV}}$, and the SINR at the ith secondary receiver as $\gamma_{i,\mathrm{sec}}$, the power control problem for energy efficiency maximization and interference suppression is formulated as follows:
(P.1)

$$\min \sum_{i=1}^{N} p_{i,\mathrm{sec}} \tag{3.1}$$

subject to

$$\gamma_{m,\mathrm{TV}} \geq \gamma_{\mathrm{TV}}^{\mathrm{tar}}, \quad \forall m \tag{3.2}$$

$$\gamma_{i,\mathrm{sec}} \geq \gamma_{i,\mathrm{sec}}^{\mathrm{tar}}, \quad i = 1, \ldots, N. \tag{3.3}$$

$$p_{\mathrm{sec}}^{\min} \leq p_{i,\mathrm{sec}} \leq p_{\mathrm{sec}}^{\max}, \quad i = 1, \ldots, N. \tag{3.4}$$

where $\gamma_{\mathrm{TV}}^{\mathrm{tar}}$ and $\gamma_{i,\mathrm{sec}}^{\mathrm{tar}}$ are the target SINRs for the primary receivers and the secondary receivers, respectively. $p_{\mathrm{sec}}^{\min}$ and $p_{\mathrm{sec}}^{\max}$ are the minimum and maximum allowable transmission powers of the secondary users. These are "hard" limits including many considerations such as safety and hardware limitations that are set by the standard organization or government agencies [15]. The objectives of power control in a CR network are to maximize the energy efficiency of the secondary users and suppress harmful interference to both the primary users and the secondary users. This can be achieved by minimizing the total transmission power of the secondary users (Equation 3.1), while guaranteeing both the QoS of the primary users (Equation 3.2) and the QoS of the secondary users (Equation 3.3).

3.2.1 Case Study: CR Users Employing OFDMA

In this case study, we show how the general model applies to CR users employing OFDMA. The reasons of choosing OFDMA are threefold: First, it is observed that signal energy across the TV band is not uniform [31]. Hence, it is beneficial to divide the entire TV band into many subbands so that the CR users may choose to transmit only in those subbands that are lightly loaded. Second, each CR user may only need to detect a subset of bands rather than the entire TV band, thus simplifying the detection design. Third, when a large amount of data needs to be transmitted in a timely fashion, many CR users may need to transmit data simultaneously to the access point. In this case, it is necessary to avoid or reduce the interference from concurrent transmissions to achieve high power efficiency. OFDMA is well suited because it is agile in selecting and allocating subbands dynamically and facilitates decoding at the receiving end of each subband [32].

It is assumed that an access point is located at the center of the area, and it is d meters away from the nearest primary receiver. The access point connects to the Internet through wireline connections. The CR network has N energy-constrained nodes indexed by $\mathcal{N} := \{1, 2, \ldots, N\}$. The nodes are equipped with small and unreplenishable energy reserves. However, it is assumed that the access point does not have an energy constraint. The system is assumed to be a time-slotted OFDMA system with fixed time slot duration T_S. Slot synchronization is assumed to be achieved through a beaconing mechanism (as in IEEE 802.11). A node either transmits data to the access point, receives data from the access point or sleep in each time slot. Before each time slot, a guard interval is inserted to achieve synchronization, perform spectrum detection as well as resource allocation (based on the proposed scheme).

At the physical layer, the uplink channel is assumed to be a frequency-selective Rayleigh fading channel, and the entire spectrum is appropriately divided into M subcarriers with each subcarrier experiencing flat Rayleigh fading [33]. Inter-carrier interference (ICI) caused by frequency offset of the side lobes pertaining to transmitter i is not considered in this book (which can be mitigated by windowing the OFDM signal in the time domain or adaptively deactivating adjacent subcarriers [34]).

Given a time slot, primary (TV) users and some other CR nodes may already have occupied some subcarriers of the system. If there is a CR node that wants to start a new transmission in this time slot, it first needs to detect the available subcarriers and only employs the available subcarriers that will not interfere with the primary (TV) users and the existing CR nodes. We label the subcarrier set available to CR node i after spectrum detection by $\mathcal{L}_i \subset \{1, 2, \ldots, M\}$. Let $\mathbf{G} := \left\{G_i^k,\ i \in \mathcal{N},\ k \in \mathcal{L}_i\right\}$ denote the subcarrier fading coefficient matrix, where G_i^k stands for the sub-channel coefficient gain from CR node i to the access point over subcarrier $k \in \mathcal{L}_i$. $G_i^k = \left|H_i^k(f)\right|^2$, where $\left|H_i^k(f)\right|$ is the transfer function [35]. We further assume that $\mathbf{G}$ adheres to a block fading channel model that remains invariant over blocks (coherence time slots) of size T_S and uncorrelated across successive blocks. The noise is assumed to be additive, white and Gaussian (AWGN), with variance σ^2 over all subcarriers.

The corresponding power control problem in this case is

$$\min \left[\sum_{i=1}^{N} \sum_{k \in \mathcal{L}_i} p_i^k \right] \tag{3.5}$$

subject to

$$\gamma_{m,\mathrm{TV}} \geq \gamma_{\mathrm{TV}}^{\mathrm{tar}}, \quad \forall m \tag{3.6}$$

$$\sum_{k \in \mathcal{L}_i} \gamma_i^k \geq \gamma_i^{\mathrm{tar}}, \quad i = 1, \ldots, N. \tag{3.7}$$

$$p^{\min} \leq \sum_{k \in \mathcal{L}_i} p_i^k \leq p^{\max}, \quad i = 1, \ldots, N. \tag{3.8}$$

The subscript "sec" is dropped for simplicity of presentation. In the current network setting, only the uplink power control is considered. Because the access point has much more resources and centralized control, the downlink power control is expected to be simpler and thus it is omitted here.

3.3 Power Control for a Single Secondary Transmitter–Receiver Pair

In this section, a simple case where there is only one secondary transmitter will be considered. We will first check the feasibility of the power control problem (P.1). We assume that the received power is only a function of the transmitted power and path loss, that is, the fading effects (shadowing and small-scale fading) are omitted for now. We further assume that the path loss factor from the TV transmitter is α_1, and the path loss factor from the CR transmitter is α_2. Because the antenna height of the TV transmitter is usually several hundred meters higher [29] than that of the CR transmitters, it is expected that the path loss factor from the TV transmitter (α_1) will be better (smaller) than the path loss factor from the CR transmitter (α_2). The interference between the primary users and the secondary users depends on many factors such as modulation schemes and waveform design, and we assume the orthogonality factors to be f_1 and f_2, respectively. The orthogonality factor can be defined by the cross-correlation between the waveforms of the primary users and the secondary users.

Based on the above assumptions, the SINR of the TV receiver at the worst location of the TV coverage area is (please refer to Figure 3.1)

$$\gamma_{\mathrm{TV}} = \frac{p_{\mathrm{TV}}/D^{\alpha_1}}{f_2 p_{\mathrm{sec}}/y^{\alpha_2} + \sigma^2} \tag{3.9}$$

and the SINR of the secondary receiver is

$$\gamma_{\mathrm{sec}} = \frac{p_{\mathrm{sec}}/r^{\alpha_2}}{f_1 p_{\mathrm{TV}}/h^{\alpha_1} + \sigma^2} \tag{3.10}$$

where

r is the distance between the secondary transmitter and the secondary receiver
σ^2 is the background noise

To satisfy the two constraints on the primary and secondary SINR values, inequality (3.2) and (3.3), we need

$$p_{\text{sec}} \leq \left[\frac{p_{\text{TV}}}{D^{\alpha_1}\gamma_{\text{TV}}^{\text{tar}}} - \sigma^2\right] y^{\alpha_2}/f_2, \tag{3.11}$$

and

$$p_{\text{sec}} \geq (f_1 p_{\text{TV}}/h^{\alpha_1} + \sigma^2)\gamma_{\text{sec}}^{\text{tar}} r^{\alpha_2}. \tag{3.12}$$

If the power control problem is feasible, Equations 3.11, 3.12, and 3.4 have to be satisfied simultaneously.

THEOREM 3.1 Given the transmission power of the primary transmitter (p_{TV}) and the background noise (σ^2), the target SINR values of the primary receiver and the secondary receiver ($\gamma_{\text{TV}}^{\text{tar}}$ and $\gamma_{\text{sec}}^{\text{tar}}$), and the distances ($D$, y, h, r), the feasibility condition of the power control problem (P.1) for a single secondary transmitter is

$$\max\left\{p_{\text{sec}}^{\min}, \underline{p}_{\text{sec}}\right\} \leq p_{\text{sec}} \leq \min\left\{\bar{p}_{\text{sec}}, p_{\text{sec}}^{\max}\right\} \tag{3.13}$$

where $\bar{p}_{\text{sec}} = \left[\frac{p_{\text{TV}}}{D^{\alpha_1}\gamma_{\text{TV}}^{\text{tar}}} - \sigma^2\right] y^{\alpha_2}/f_2$ and $\underline{p}_{\text{sec}} = (f_1 p_{\text{TV}}/h^{\alpha_1} + \sigma^2)\gamma_{\text{sec}}^{\text{tar}} r^{\alpha_2}$.

The feasibility condition given in Theorem 3.1 may be interpreted as follows:

COROLLARY 3.1 Define two transmission power sets, $S_1 = \{p_{\text{sec}}^{\min} \leq p_{\text{sec}} \leq p_{\text{sec}}^{\max}\}$, and $S_2 = \{\underline{p}_{\text{sec}} \leq p_{\text{sec}} \leq \bar{p}_{\text{sec}}\}$, the power control problem (P.1) for a single secondary transmitter is feasible iff $S_1 \cap S_2 \neq \emptyset$.

One possible case of feasible transmission power of the secondary user is shown in Figure 3.2. If the feasibility condition (inequality (3.13)) is satisfied, the optimal transmission power of the secondary user is $\max\left\{p_{\text{sec}}^{\min}, \underline{p}_{\text{sec}}\right\}$. If the minimum allowable transmission power is 0, the optimal transmission power of the secondary user is $\underline{p}_{\text{sec}}$.

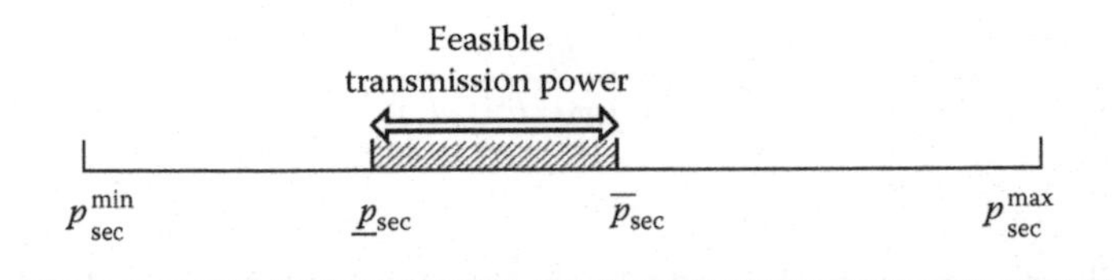

Figure 3.2 Feasible transmission power of the secondary user.

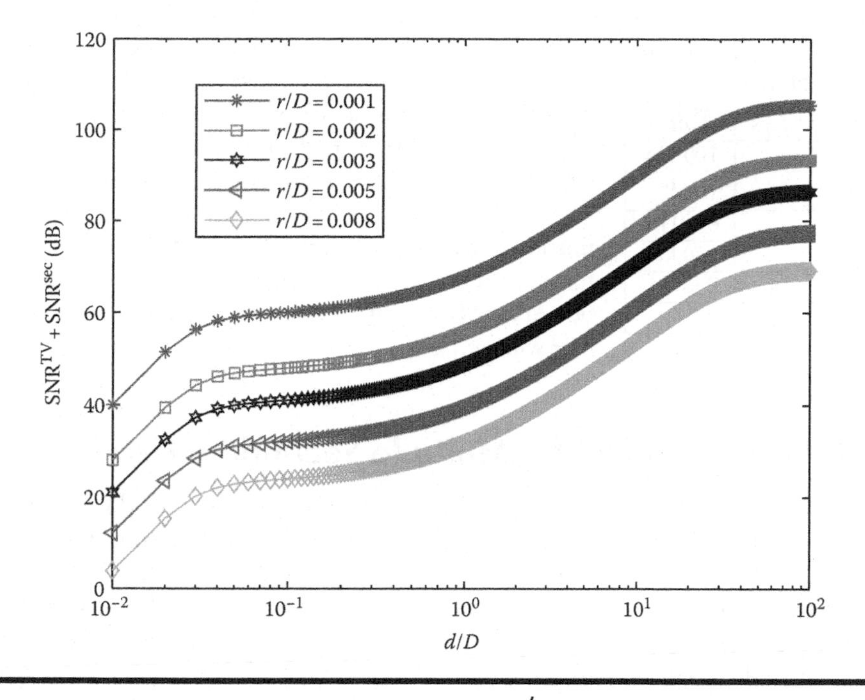

Figure 3.3 The sum SINR values (in dB) versus $\frac{d}{D}$ and $\frac{r}{D}$, $f_1 = 1, f_2 = 1$.

If the interference is dominant, that is, if $f_2 p_{\mathrm{sec}}/y^{\alpha_2} \gg \sigma^2$ and $f_1 p_{\mathrm{TV}}/h^{\alpha_1} \gg \sigma^2$, which is usually the case, the sum of the SINR (in dB) of the TV receiver at the border of the TV coverage area and the SINR of the secondary receiver can be expressed as:

$$\gamma_{\mathrm{TV}}^{\mathrm{dB}} + \gamma_{\mathrm{sec}}^{\mathrm{dB}} \approx \alpha_1 \frac{h}{D}(\mathrm{dB}) + \alpha_2 \frac{y}{r}(\mathrm{dB}) - [f_1 + f_2](\mathrm{dB}). \tag{3.14}$$

The achievable SINR of the secondary users can be estimated by substracting $\gamma_{\mathrm{TV}}^{\mathrm{tar}}$ from the sum of the SINR.

It is observed that the sum of these two SINR values (in dB) is only a function of the relative distances. One sample simulation result is plotted in Figure 3.3. The parameters used in the simulation are given in Table 3.1, and it is assumed that $h \approx D + d$ and $y \approx d$ as $d \gg l$. It is observed that the distance between the secondary transmitter and the secondary receiver, r, has the dominant effect on the sum of the SINR values. For example, if r decreases from 300 m $\left(\frac{r}{D} = 0.005\right)$ to 60 m $\left(\frac{r}{D} = 0.001\right)$, the gain of the sum of the SINR values is about 30 dB. In addition, if r is large, say r is 480 m $\left(\frac{r}{D} = 0.008\right)$, even if the secondary user is far away from the TV coverage area $\left(\text{say, } \frac{d}{D} = 1\right)$, the sum of the SINR values is still very low, about 30 dB. In other words, if the required primary SINR is 34 dB, the maximum achievable SINR for the secondary user is about −4 dB. The results

Table 3.1 Simulation Parameters

Parameters	*Value*
p_{TV}	100 kW
σ^2	10^{-14}
W	6 MHz
D	120 km
l	2 km
α_1	3
α_2	4

suggest that only low-power secondary users with short-range transmissions (low-power personal/portable devices [15]) are allowed when the primary users are ON. This also calls for multi-hop communications rather than single-hop long-range transmissions in the CR network.

We would like to point out that although the transmission powers are not explicitly included in the formula for the sum SINR, they indeed will determine the proportion of the SINR that the primary user and the secondary user will get.

3.4 Power Control for Multiple Secondary Users

In this section, we are going to provide both centralized and distributed solutions to the power control problem (P.1). To evaluate the interference and solve the power control problem, we assume that the distances such as d and y_i can be estimated accurately. Indeed, geolocation devices (e.g., GPS), control signals, or spectrum sensing may be applied to detect the primary transmissions and get an accurate estimate of the distances [15].

3.4.1 Centralized Solution

The SINR of the TV receiver at the worst location of the TV coverage area is

$$\gamma_{\text{TV}} = \frac{p_{\text{TV}}/D^{\alpha_1}}{f_2 \sum p_{i,\text{sec}}/y_i^{\alpha_2} + \sigma^2} \tag{3.15}$$

The SINR of the ith secondary receiver is

$$\gamma_{i,\text{sec}} = \frac{g_{ii}p_{i,\text{sec}}}{\sum_{j\neq i} g_{ij}p_{j,\text{sec}} + f_1 p_{\text{TV}}/h_i^{\alpha_1} + \sigma^2} \tag{3.16}$$

where g_{ij} is the link gain from the jth secondary transmitter to the ith secondary receiver.

The following theorem gives the feasibility condition of the power control problem (P.1).

THEOREM 3.2 The power control problem (P.1) is feasible for all N simultaneous transmitting–receiving pairs of secondary users within the same channel as long as

1. The matrix $\left[I - \Gamma_{\text{sec}}^{\text{tar}} Z\right]$ is non-singular (thus invertible);
2. The transmission power vector p_{sec}^{*} satisfies inequality (3.4) element-wise, where

$$p_{\text{sec}}^{*} = \left[I - \Gamma_{\text{sec}}^{\text{tar}} Z\right]^{-1} u, \tag{3.17}$$

matrix Γ^{tar} is a diagonal matrix

$$\Gamma_{\text{sec}\,ij}^{\text{tar}} = \begin{cases} \gamma_{i,\text{sec}}^{\text{tar}} & i = j \\ 0 & \text{otherwise} \end{cases}, \tag{3.18}$$

matrix Z is the following nonnegative matrix

$$Z_{ij} = \begin{cases} \frac{g_{ij}}{g_{ii}} & i \neq j \\ 0 & i = j \end{cases}, \tag{3.19}$$

u is the vector with elements

$$u_i = \gamma_{i,\text{sec}}^{\text{tar}} \eta_i^2 / g_{ii}, \quad i = 1, 2, \ldots, N \tag{3.20}$$

and

$$\eta_i^2 = f_1 p_{\text{TV}} / h_i^{\alpha_1} + \sigma^2. \tag{3.21}$$

3. The transmission power vector p_{sec}^{*} also satisfies the following inequality:

$$\frac{p_{\text{TV}}/D^{\alpha_1}}{f_2 \sum p_{i,\text{sec}}^{*}/y_i^{\alpha_2} + \sigma^2} \geq \gamma_{\text{TV}}^{\text{tar}}. \tag{3.22}$$

Proof A target SINR vector γ^{tar} is achievable for all simultaneous transmitting–receiving pairs of secondary users within the same channel if the following conditions are met [36,37]

$$\gamma_{i,\text{sec}} \geq \gamma_{i,\text{sec}}^{\text{tar}} \tag{3.23}$$

$$p \geq 0 \tag{3.24}$$

where p is the vector of transmitting powers. Define η_i^2 as in Equation 3.21. Replacing $\gamma_{i,\text{sec}}$ with Equation 3.16 and rewriting the above conditions in matrix form gives

$$[I - \Gamma^{\text{tar}} Z] p \geq u \tag{3.25}$$

$$p \geq 0 \tag{3.26}$$

where matrix Γ^{tar}, matrix Z, and vector u are defined in Equations 3.18 through 3.20, respectively.

It is shown in [37] that if the system is feasible, the matrix $[I - \Gamma^{\text{tar}} Z]$ must be invertible and the inverse should be element-wise positive, thus proving part (1) of the theorem.

It is also shown in [37] (Proposition 2.1) that if the system is feasible, there exists a unique (Pareto optimal) solution that minimizes the transmitted power. This solution is obtained by solving a system of linear algebraic equations:

$$[I - \Gamma^{\text{tar}} Z] p^* = u \tag{3.27}$$

To satisfy the constraints (3.2) and (3.4) in the power control problem (P.1), the transmission power vector p^*_{sec} must satisfy the inequality (3.4) element-wise and the inequality (3.22), thus proving the theorem. ■

The above proof highlighted the centralized solution to the problem (P.1). Although it seems that the power control problem (P.1) is similar to that in cellular systems [38] and in wireless ad hoc networks [39], the power control problem considered here addressed interference from heterogeneous systems and an additional constraint (3.2) has to be satisfied, and the interference between primary and secondary users has to be taken into account in the problem formulation. It also calls for a joint design of power control and admission control for the CR network such that the QoS of the primary users is ensured all the time. The procedures of joint power control and admission control is summarized below.

3.4.1.1 Joint Power Control and Admission Control

1. Solve the transmission power vector p^*_{sec} using Equation 3.17.
2. Check whether the transmission powers are within limit, that is, $p^{\min}_{\text{sec}} \leq p^*_{i,\text{sec}} \leq p^{\max}_{\text{sec}}, \forall i$? If Yes, goes to the next step; otherwise, the power control problem (P.1) is not feasible. Remove the jth secondary user that has the largest $\sum_{i=1}^{N} [Z_{ij} + Z_{ji}]$ and return to Step 1 with reduced number of transmitters.
3. Check whether the transmission powers satisfy inequality (3.22). If Yes, set the transmission power vector as p^*_{sec}; otherwise, the power control problem (P.1) is not feasible. Remove the secondary user that requires the largest transmission power ($p = \max \left\{ p^*_{i,\text{sec}} \right\} \; \forall i$) and return to Step 1 with reduced number of transmitters.

The block diagram of the proposed joint power control and admission control is given in Figure 3.4. It is worth pointing out that Steps 2 and 3 implement admission control for the secondary users. When the power control problem (P.1) is not feasible, the secondary user that caused the worst interference should be silenced. The central

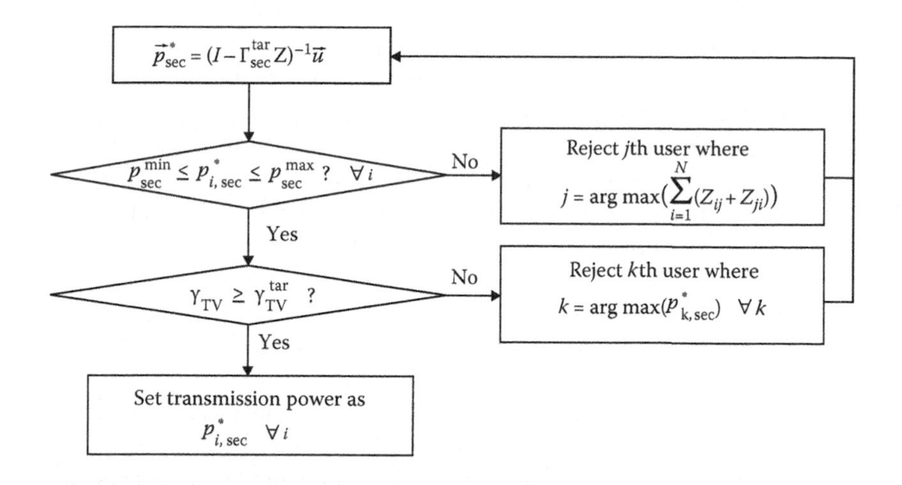

Figure 3.4 The joint power control and admission control.

controller can verify the transmission power limits in a straightforward way in Step 2 after solving p^{*}_{sec} using Equation 3.17. The worst interferer to other secondary users inside the CR network is the one that has the largest row and column sum of matrix Z. In Step 3, given that p_{TV}, γ^{tar}_{TV}, and D are publicly available data, and y_i can be estimated accurately, the central controller can verify the inequality (3.22). This time the worst interferer to the primary receivers is the one that has the largest transmission power because all the secondary transmitters have more or less the same distance to the primary receivers. In a CR network with centralized management, such as in a cluster-based architecture, the above procedures may be implemented.

3.4.2 Distributed Solution

The centralized solution (Equation 3.17) needs a central controller and global information of all the link gains, and centralized power control requires extensive control signaling in the network, and it is difficult to be implemented in practice, especially for an infrastructure-less wireless ad hoc network. Therefore, a distributed implementation that use only local information to make a control decision is proposed for realistic scenarios.

Distributed power control schemes may be derived by applying iterative algorithms to solve Equation 3.27. For example, using the first-order Jacobian iteration [40], the following distributed power control scheme is obtained:

$$p_{i,\text{sec}}(k+1) = \min\left\{\frac{\gamma^{\text{tar}}_{i,\text{sec}}}{\gamma_{i,\text{sec}}(k)} p_i(k), p^{\max}_{\text{sec}}\right\}, \quad i = 1, 2, \ldots, N. \tag{3.28}$$

Note that each node only needs to know its own received SINR at its designated receiver to update its transmission power. This is available by feedback from the receiving node through a control channel. As a result, the algorithm is fully distributed. Convergence properties of this type of algorithms were studied by Yates [38,41]. An interference function $I(p)$ is standard if it satisfies three conditions: positivity, monotonicity, and scalability. It is proved by Yates [41] that the standard iterative algorithm $p(k+1) = I(p(k))$ will converge to a unique equilibrium that corresponds to the minimum use of power. The distributed power control scheme (Equation 3.28) is a special case of the standard iterative algorithm.

Because the Jacobi iteration is a fixed-point iterative method, it usually has slow convergence speed to the sought solution. However, we select Equation 3.28 as the power control algorithm in CR networks due to its simplicity. Other advanced algorithms with faster convergence speed can be found in [36,42].

The distributed power control algorithm given in Equation 3.28 does not enforce the QoS requirement of the primary users represented by the inequality (3.22). Thus, the secondary users applying Equation 3.28 alone may violate the QoS requirement of the primary users. To address this issue, we propose two possible solutions. The first solution is a direct solution, where a "genie" is placed near the primary receiver at the border of the TV coverage area. The genie will monitor the interference level and inform the secondary users (such as using a beacon signal) if the interference level is too high, and the QoS requirement of the primary users will be violated. One possible implementation of the genie is a secondary user that happens to locate inside the TV coverage area. The second solution is an indirect solution. Assume that $y_i \approx y_j = d$, $\forall i \neq j$,* then the inequality (3.22) may be written as

$$\sum_i p_{i,\mathrm{sec}} \leq \left[p_{\mathrm{TV}}/\left(D^{\alpha_1}\gamma_{\mathrm{TV}}^{\mathrm{tar}}\right) - \sigma^2\right]\frac{d^{\alpha_2}}{f_2}. \tag{3.29}$$

Suppose that all secondary users that are planning to transmit will report to a manager their respective transmission power, $p_{i,\mathrm{sec}}$ for user i, the manager will be able to verify the QoS requirement of the primary users by checking the inequality (3.29).

3.4.3 Case Study: Power Control with Best Subband Selection for CR Users Employing OFDMA

Before we discuss how the obtained results may apply to power control for CR users employing OFDMA, it is worth pointing out that the results obtained so far are applicable to CR users employing TDMA and CDMA in a straightforward manner. For instance, in the case of TDMA, only one secondary user is allowed to transmit during one time slot; the results of the single secondary transmitter receiver pair in

* This assumption is expected to be true most of the time, because typically the secondary users must reside far away enough from the TV coverage area.

Section 3.3 give the optimal power control for one TDMA CR network. The results in Section 3.4 correspond to the power control of co-channel secondary users in multiple TDMA CR networks.

3.4.3.1 *Best Subcarrier Selection*

The SINR on subcarrier k of the TV receiver at the worst location of the TV coverage area is

$$\gamma_{\mathrm{TV}}^{k} = \frac{p_{\mathrm{TV}}^{k}/D^{\alpha_1}}{f_2 \sum_{i=1}^{N} \left(p_i^k / y_i^{\alpha_2}\right) + \sigma^2}, \tag{3.30}$$

where p_{TV}^{k} is the portion of transmission power of the TV station on subcarrier k. The received SINR on subcarrier k at the access point for the ith CR node is

$$\gamma_i^k = \frac{G_i^k p_i^k}{\sum_{j \neq i} G_j^k p_j^k + f_1 p_{\mathrm{TV}}^{k}/h_i^{\alpha_1} + \sigma^2} \tag{3.31}$$

A necessary condition for the optimal solution of the power control problem is given by

$$\sum_{k \in \mathcal{L}_i} \gamma_i^k = \gamma_i^{\mathrm{tar}}, \quad i = 1, \ldots, N. \tag{3.32}$$

In general, there are $N \times M$ variables to be determined. If $p_i^l = 0$, subcarrier l is not selected by CR node i. However, the problem is under-determined because there are only N equations. Obtaining the optimal solution would require an exhaustive search, which has high computational complexity. Motivated by the energy-efficient resource allocation scheme proposed for multi-carrier cellular networks [43], a best subcarrier selection approach is adopted here. Given the available subcarrier set of each CR node, the best subcarrier (in terms of the largest channel gain) may be selected to transmit data to the access point [43]. An example of spectrum sharing between a legacy TV system and a CR OFDMA network when employing best subcarrier selection is shown in Figure 3.5.

Denote the best subcarrier for CR node i as $i*$. With the best subcarrier selection, that is, each CR node only uses the best subcarrier to transmit data to the access point, γ_{TV} and γ_i^{i*} are given by the following equations:

$$\gamma_{\mathrm{TV}} = \frac{p_{\mathrm{TV}}/D^{\alpha_1}}{f_2 \sum_{i=1}^{N} \left(p_i^{i*} / y_i^{\alpha_2}\right) + \sigma^2} \tag{3.33}$$

$$\gamma_i^{i*} = \frac{G_i^{i*} p_i^{i*}}{\sum_{j*=i*, j \neq i} G_j^{j*} p_j^{j*} + f_1 p_{\mathrm{TV}}^{i*}/h_i^{\alpha_1} + \sigma^2} \tag{3.34}$$

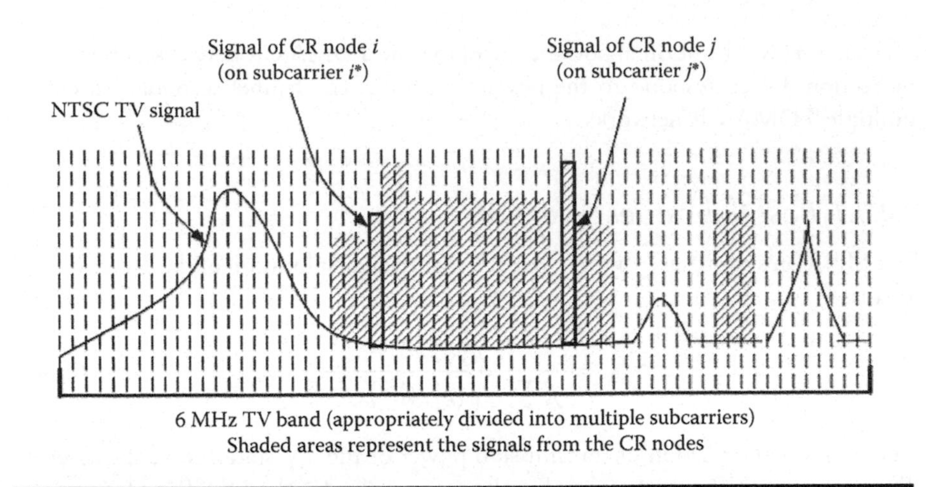

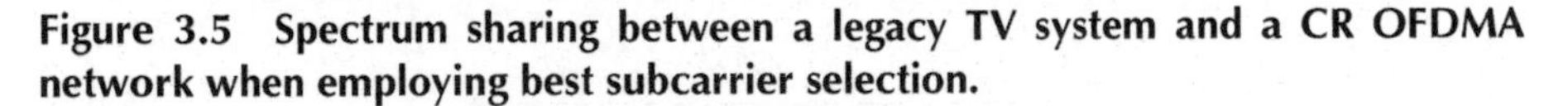

Figure 3.5 Spectrum sharing between a legacy TV system and a CR OFDMA network when employing best subcarrier selection.

3.4.3.2 Nonoverlapping Subcarrier Selection

We will first study the case where each CR node only uses the best subcarrier to transmit data to the access point, and their respective selected subcarriers are nonoverlapping. In other words, for any transmitting CR node i

$$p_i^k = \begin{cases} p_i^{i*} & k = i* \\ 0 & \text{otherwise} \end{cases}, \tag{3.35}$$

and

$$j* \neq i*, \quad \forall j \neq i, \text{ and } i, j \in \mathcal{N}. \tag{3.36}$$

This may be achieved by centralized coordination through the access point or using a rendezvous channel for coordination among all the transmitting CR nodes.

Under this assumption, Equations 3.33 and 3.34 are reduced to

$$\gamma_{TV}^k = \begin{cases} \frac{p_{TV}^k/D^{\alpha_1}}{f_2 p_i^{i*}/y_i^{\alpha_2}+\sigma^2} & k = i* \\ \frac{p_{TV}^k/D^{\alpha_1}}{\sigma^2} & \text{otherwise} \end{cases}, \tag{3.37}$$

$$\gamma_i^{i*} = \frac{G_i^{i*} p_i^{i*}}{f_1 p_{TV}^{i*}/h_i^{\alpha_1} + \sigma^2} \tag{3.38}$$

To satisfy the two constraints on the primary and secondary SINR values, we need

$$p_i^{i*} \leq \left[\frac{p_{TV}^{i*}}{D^{\alpha_1}\gamma_{TV}^{tar}} - \sigma^2 \right] y_i^{\alpha_2}/f_2, \tag{3.39}$$

and

$$p_i^{i*} \geq \left(f_1 p_{\mathrm{TV}}^{i*}/h_i^{\alpha_1} + \sigma^2\right)\gamma_i^{\mathrm{tar}}/G_i^{i*}. \tag{3.40}$$

If the power control problem is feasible, Equations 3.39, 3.40, and 3.8 have to be satisfied simultaneously. It is now clear that Theorem 3.1 applies to CR OFDMA users employing nonoverlapping best subcarrier selections. A typical case of feasible transmission power of a CR node i satisfies $p^{\min} \leq \underline{p}_i \leq \bar{p}_i \leq p^{\max}$, and the optimal transmission power is $\underline{p}_i$. It is also observed that the distance from the CR user to the access point, r_i, has the dominant effect on the sum of the SINR values. Again, this suggests that only low-power CR users with short-range transmissions are allowed when the primary users are ON.

3.4.3.3 Overlapping Subcarrier Selection

In the following section, we relax our assumptions by allowing CR nodes to have overlapped subcarriers. This models the situation where the CR nodes are not coordinated and some of them may choose the same subcarrier. However, we still maintain that each CR node only selects the best subcarrier for its uplink data transmissions.

In this case, Theorem 3.2 gives the feasibility condition of the power control problem (P.1) when best subcarrier selection is applied and overlapped subcarriers may exist, with the adaptation of the following parameters: matrix Z is the following $N \times N$ nonnegative matrix,*

$$Z_{ij} = \begin{cases} \frac{G_j^{j*}}{G_i^{i*}} & i \neq j,\ j* = i* \\ 0 & \text{otherwise} \end{cases}, \tag{3.41}$$

u is the vector with elements

$$u_i = \gamma_i^{\mathrm{tar}} \eta_i^2 / G_i^{i*}, \quad i = 1, 2, \ldots, N \tag{3.42}$$

and

$$\eta_i^2 = f_1 p_{\mathrm{TV}}^{i*}/h_i^{\alpha_1} + \sigma^2. \tag{3.43}$$

3.5 Simulation Results

In this section, the performance of the proposed power control algorithm is examined. It is assumed that a group of $N = 50$ transmitting–receiving pairs

* When the respective selected subcarriers of all the CR nodes are nonoverlapping, the matrix Z becomes a zero matrix and Theorem 3.2 gives the same result as described in Theorem 3.1.

(100 secondary users) using low-power devices are communicating with each other in a 2000 × 2000 m area. They share the same spectrum with a TV system, and the TV station is located $D + d$ meters away. The initial transmission power of the CR nodes are randomly chosen between $p^{\min} = 0$ and $p^{\max} = 100$ mW. The rest of the parameters are given in Table 3.1.

3.5.1 Baseline Evaluation

In this part of the simulations, the locations of the transmitting–receiving pairs are chosen such that $r_{ij} > 3r_{ii}$ to ensure the feasibility of the power control problem, where r_{ij} is the distance from the jth transmitter to the ith receiver and $g_{ij} = 1/r_{ij}^{\alpha_2}$.

The average achievable SINR value of the secondary users $\left(\gamma_{\text{sec}}^{\text{avg}}\right)$ versus d/D is shown in Figure 3.6. It is observed that $\gamma_{\text{sec}}^{\text{avg}}$ increases monotonically with d as expected. It is also shown that the gain in $\gamma_{\text{sec}}^{\text{avg}}$ decreases when d increases, because the interference between the two systems plays too less a role in the achievable SINR value when they are further away. When $d/D > 2$, $\gamma_{\text{sec}}^{\text{avg}}$ is pretty much limited by the interference of its own system.

In the following part of the simulation, $d = 0.5D$ and the distributed power control algorithm, Equation 3.28, is applied. The convergence of the mean square error of the secondary user's SINR $\left(e_{\text{sec}}^2 = E\left[\left(\gamma_{\text{sec}} - \gamma_{\text{sec}}^{\text{tar}}\right)^2\right]\right)$ is given in Figure 3.7. It is observed that the power control algorithm converges very fast (in about 10 steps). Similarly, the convergence of the transmission power of some randomly chosen

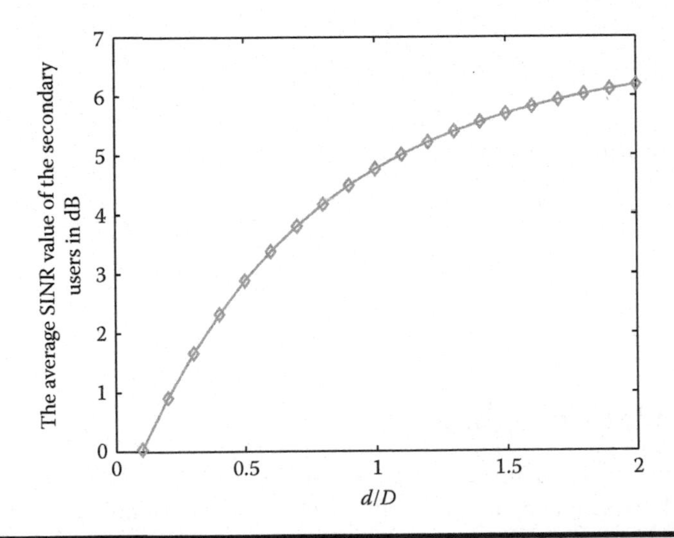

Figure 3.6 The average achievable SINR value of the secondary users $\left(\gamma_{\text{sec}}^{\text{avg}}\right)$ versus d/D.

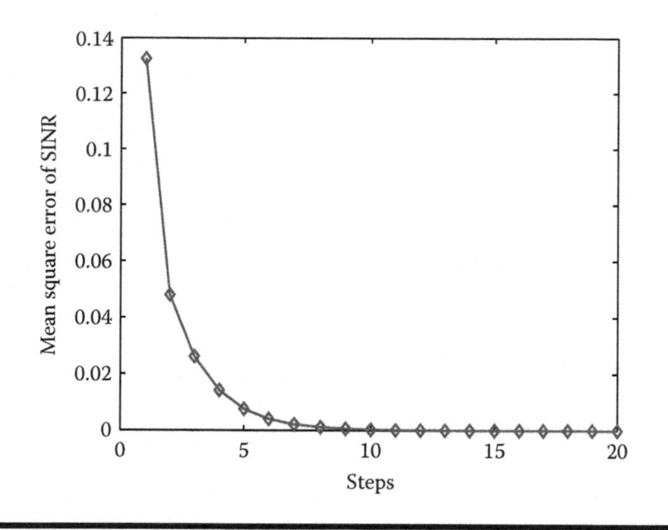

Figure 3.7 The convergence of the mean square error of the secondary user's SINR.

secondary users is shown in Figure 3.8. The minimum SINR value of the primary users during the power control process of the secondary users is shown in Figure 3.9. It is confirmed that the QoS of the primary users is not violated during the power control process.

3.5.2 CR Network with OFDMA Users

In this part of the simulations, the performance of the proposed subcarrier selection and power control scheme is examined. It is assumed that the CR OFDMA network has 100 CR nodes and the entire spectrum was partitioned into 2000 subcarriers. The available subcarriers is a random variable uniformly distributed between 200 and 1000. Each CR node selects its best subcarrier for data transmission to the access point and in each time slot there are between 10 and 100 simultaneously transmitting CR nodes (depending on whether they have data to send).

In the following part of the simulation, $d = 0.5D$ and the distributed power control algorithm is applied. The convergence of the mean square error of the received SINR at the access point $\left(e^2 = E\left[\left(\gamma_i - \gamma_i^{\text{tar}}\right)^2\right]\right)$ is given in Figure 3.10. It is observed that the power control algorithm converges in one step when the number of simultaneously transmitting CRs ($n = 10$) is much smaller than the number of available subcarriers ($k = 200$). As a result, the subcarriers chosen by different CRs are not overlapped and the distributed power control algorithm gives the result as described in Theorem 3.1 in one step (no iteration required). When the number of simultaneously transmitting CRs increases, the power control algorithm

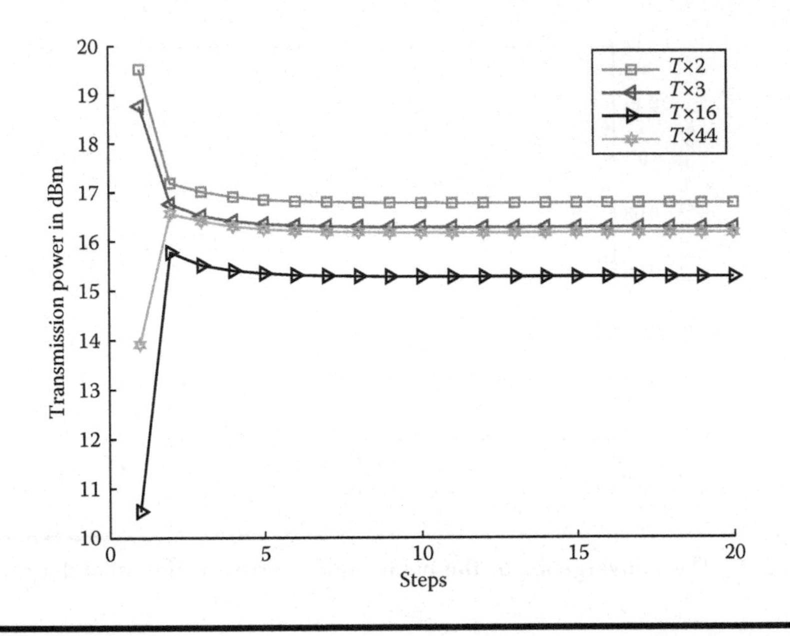

Figure 3.8 The convergence of the transmission power of the secondary users.

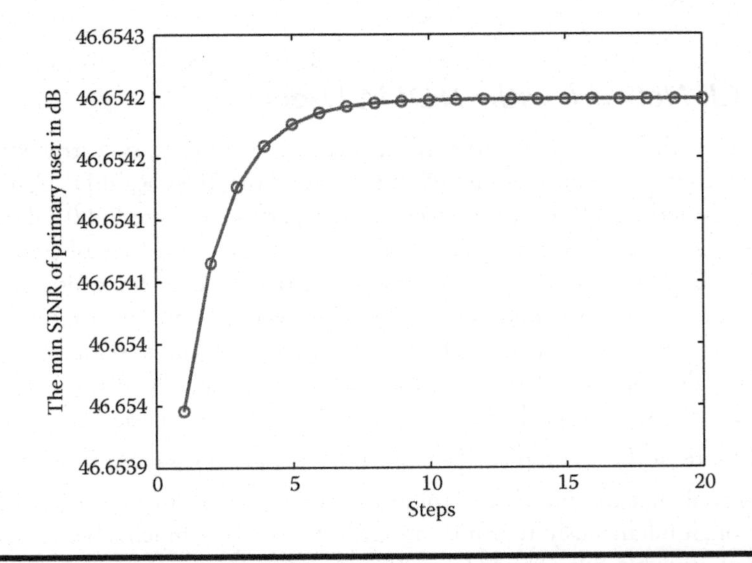

Figure 3.9 The minimum SINR value of the primary users during the power control process of the secondary users.

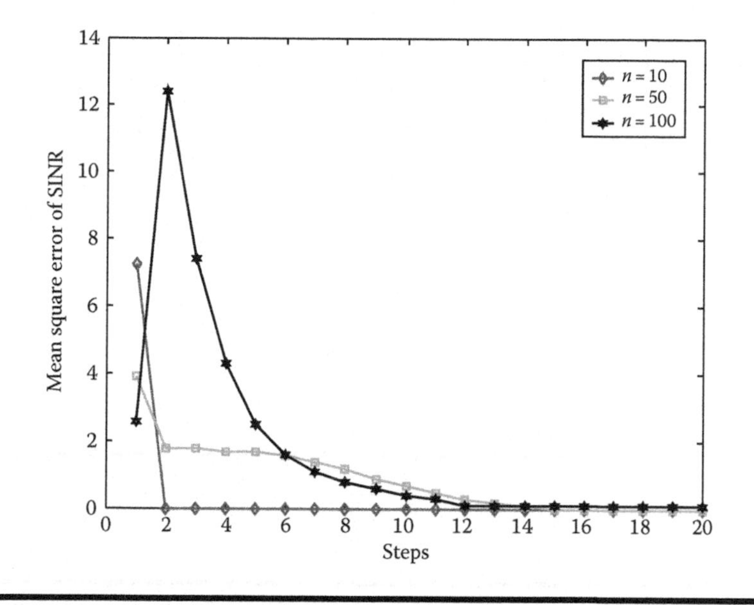

Figure 3.10 The convergence of the mean square error of the received SINR at the access point of all the CR users (*n*: number of simultaneously transmitting CR users; $k = 200$: number of available subcarriers).

still converges fast (in about 14 steps). However, when all the CRs are transmitting simultaneously ($n = 100$), a closer look shows that the mean square error e^2 does not go to zero but remains at 0.1. This means that not all CRs have satisfactory SINRs. It can be better understood by looking at the convergence of the transmission power of some randomly chosen CRs in Figure 3.11. CR user 2 needs to share the same subcarrier with three other CR users, and so does CR user 3; thus they both transmit at maximum power due to high co-channel interference (but still do not meet the SINR target). CR user 16 occupies the subcarrier by itself; thus the power converges in a single step as discussed before. CR user 44 needs to share the same subcarrier with only one other CR user, and the power converges in about 14 steps.

The minimum SINR value of the primary users during the power control process of the CRs is shown in Figure 3.12. It is confirmed that the QoS of the primary TV users is not violated during the power control process.

It is expected that better energy efficiency will be achieved when the simultaneously transmitting CR nodes have the least overlapped subcarriers. Hence, it is informative to estimate the probability of subcarrier overlapping. Assuming that there are n simultaneously transmitting CR nodes and k available subcarriers, the probability that none of the CR nodes select the same subcarrier is given by

$$Q(k, n) = k!/((k - n)!k^n) \leq e^{-n(n-1)/2k}. \tag{3.44}$$

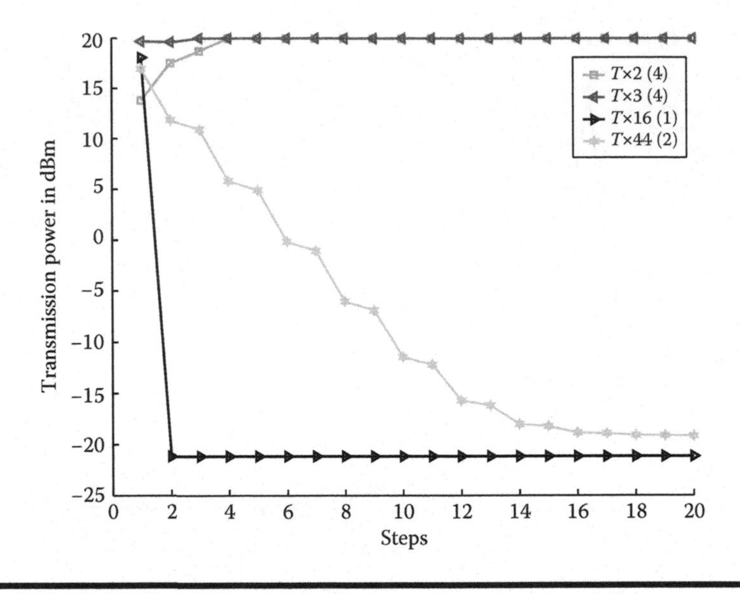

Figure 3.11 The convergence of the transmission power when 100 CR users transmit simultaneously (Txi (.): transmission power of the *i*th CR user (the number of CR users sharing the same subcarrier)).

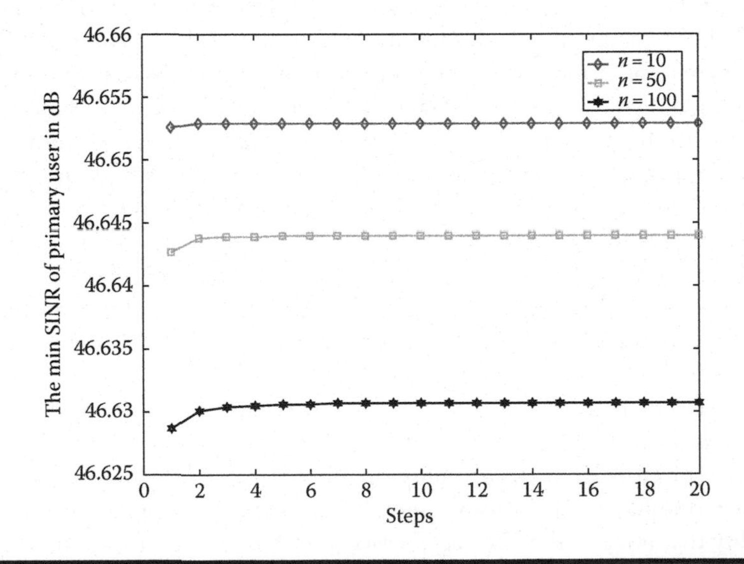

Figure 3.12 The minimum SINR value of the primary users during the power control process of the CR users (*n*: the number of simultaneously transmitting CR users).

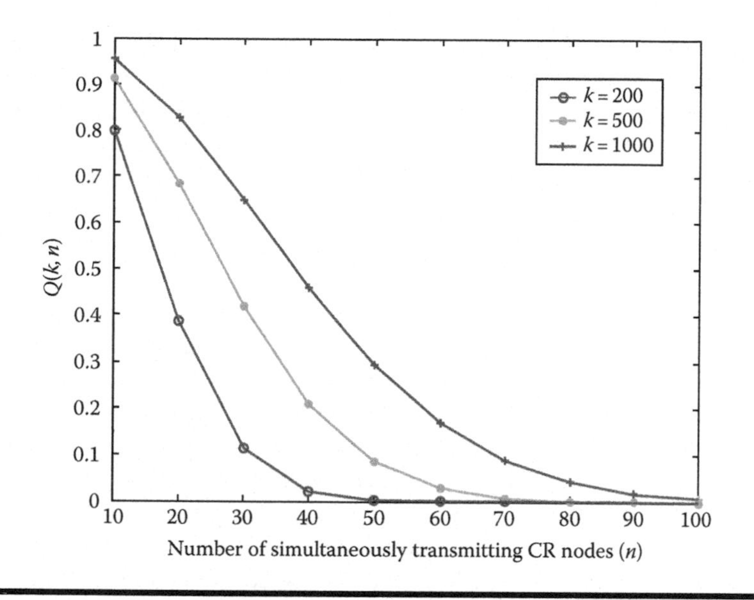

Figure 3.13 The probability that none of the CR users select the same subcarrier ($Q(k,n)$) versus the number of simultaneously transmitting CR user (n) and available subcarriers (k), assuming best subcarrier selection.

It is observed from Figure 3.13 that $Q(k, n)$ decreases as n increases or k decreases, as expected. It may serve as a baseline for further derivation and analysis of the magnitude of co-channel interference.

3.6 Discussions and Open Problems

To fulfill the requirements of vast deployment in terms of thousands of wireless ad hoc networks with millions of low-power wireless devices [44], and in many cases they are colocated with many other wireless systems, spectrum shortage is a serious concern. Contrary to many previous approaches of temporal spectrum sharing between legacy systems and CR networks, spatial spectrum sharing provides another degree of freedom to increase spectrum utilization. While temporal spectrum sharing is suitable for CR users with high transmission power, spatial spectrum sharing is appropriate for secondary users with low-power portable/personal devices.

In spatial spectrum sharing, CR users may get relatively moderate data rate compared to that in temporal spectrum sharing; however, it is usually enough for most applications on low-power portable/personal devices. Furthermore, an advantage of using spatial spectrum sharing is that the CR users can operate almost continuously in time provided that the interference level is not too high; this would be the case most of the time when the network is not heavily loaded, while the CR users using temporal spectrum sharing must remain silent as long as the primary users are operating.

There are many issues that remain to be investigated in the future. First of all, the distributed joint power control and admission control should be designed carefully such that the signaling and control overhead will not be overwhelming. Second, the performance of the joint power control and admission control for the infeasible cases should be evaluated. In addition, the effects of inaccurate estimates of the distances have to be justified.

In general, although rudimentary cognitive capabilities such as spectrum sensing and dynamic spectrum allocations as well as power control techniques already exist, it appears that the existing standards have not yet risen to the point of being cognitive [11]. But the promise and potential value of such techniques is clearly recognized, and almost all existing and future wireless standards are trying to incorporate CR, dynamic spectrum access, and coexistence techniques.

There also exist many challenges in the implementation of practical portable CRs, such as power efficiency, size, and cost. There are many research projects for building CR prototypes, such as the KU Agile Radio [45], Virginia Tech Chameleonic Radio [46], and Microsoft Research KNOWS [47], just to name a few. One of the recent efforts of building a portable software radio prototype using primarily commercial off-the-shelf components and open-source software is given in [48]. It demonstrated that a general purpose processor–based software radio could be built in a portable form factor offering reasonable runtime when powered from an internal battery. However, these prototypes are mainly for research and development purposes, and they still have a long way to go before we could have a cognitive handset that is as small as today's cell phone and yet has all the cognitive capabilities envisioned by researchers.

3.7 Conclusions

In this chapter, a power control framework is proposed for a CR network that operates simultaneously in the same frequency band with a TV system. Both centralized and distributed solutions are given to maximize the energy efficiency of the CR network and provide QoS support for both primary and secondary users. In addition, the feasibility condition is derived and a joint power control and admission control procedure is suggested such that the priority of the primary users is ensured all the time. Furthermore, the proposed power control and admission control procedure may be combined with MAC design to enhance the promise of nonintrusion to the primary system during spectrum sharing.

It is demonstrated that the proposed power control framework may be applied to CR users employing different physical layer technologies such as TDMA, CDMA, and OFDMA. Case studies for CR users employing OFDMA are provided. It is noticeable that the uneven energy distribution across certain spectra is not limited to

TV bands; the so-called discontiguous spectrum operation [49] is observed in many other systems as well. Hence, the proposed architectural design and power control scheme may be well suited for many other situations and be easily generalized to fit various scenarios in practice.

Acknowledgment

This research work is supported in part by the National Science Foundation under award 0531507.

References

1. United States Frequency Allocations. Available at: http://www.ntia.doc.gov/osmhome/allochrt.pdf
2. Federal Communication Commission, Spectrum Policy Task Force Report. ET Docket No. 02-135, Washington, DC, Nov. 2002.
3. M. A. McHenry, NSF spectrum occupancy measurements project summary, Shared Spectrum Company, Technical Report, Aug. 2005. Available at: http://www.sharedspectrum.com/inc/content/measurements/nsf/NSF_Project_Summary.pdf
4. I. J. Mitola et al., Cognitive radio: Making software radios more personal, *IEEE Personal Communications*, 6(4), 13–18, Aug. 1999.
5. J. Mitola, Cognitive radio: An integrated agent architecture for software defined radio, Doctoral dissertation, Royal Institute of Technology (KTH), Stockholm, Sweden, 2000.
6. D. Raychaudhuri, Adaptive wireless networks using cognitive radios as a building block, *MobiCom 2004 Keynote Speech*, Philadelphia, PA, Sept. 2004.
7. I. Akyildiz, W. Lee, M. C. Vuran, and S. Mohanty, NeXt generation/dynamic spectrum access/cognitive radio wireless network: A survey, *Computer Networks*, 50(13), 2127–2159, 2006.
8. IEEE 802.22 WG. Available at: http://www.ieee802.org/22/
9. IEEE SCC41. Available at: http://www.scc41.org
10. C. Cordeiro et al., IEEE 802.22: The first worldwide wireless standard based on cognitive radios, *Proceedings of the 1st IEEE Symposium on New Frontiers in Dynamic Spectrum Access Networks*, Baltimore, MD, 2005, pp. 328–337.
11. M. Sherman et al., IEEE standards supporting cognitive radio and networks, dynamic spectrum access, and coexistence, *IEEE Communications Magazine*, 46(7), 72–79, July 2008.
12. R. V. Prasad et al., Cognitive functionality in next generation wireless networks: Standardization efforts, *IEEE Communications Magazine*, 46(4), 72–78, Apr. 2008.
13. FCC, Facilitating opportunities for flexible, efficient, and reliable spectrum use employing cognitive radio technologies, notice of proposed rule making and order, FCC 03-322. Available at: http://hraunfoss.fcc.gov/edocs_public/attachmatch/FCC-03-322A1.pdf

14. FCC, Unlicensed operation in the TV broadcast bands, ET Docket No. 04-186; additional spectrum for unlicensed devices below 900 MHz and in the 3 GHz band, ET Docket No. 02-380, FCC 04-113. Available at: http://hraunfoss.fcc.gov/edocs_public/attachmatch/FCC-04-113A1.pdf
15. FCC, Unlicensed operation in the TV broadcast bands; additional spectrum for unlicensed devices below 900 MHz and in the 3 GHz band, FCC 06-156. Available at: http://hraunfoss.fcc.gov/edocs_public/attachmatch/FCC-06-156A1.pdf
16. M. Marcus, Unlicensed cognitive sharing of TV spectrum, *IEEE Communications Magazine*, 43(5), 24–25, May 2005.
17. A. Sahai, N. Hoven, S. M. Mishra, and R. Tandra, Fundamental tradeoffs in robust spectrum sensing for opportunistic frequency reuse, Technical Report, 2006. Available at: http://www.eecs.berkeley.edu/sahai/Papers/CognitiveTechReport06.pdf
18. X. Liu and S. Shankar, Sensing-based opportunistic channel access, *ACM Journal on Mobile Networks and Applications* (*MONET*), 11(1), 577–591, May 2006.
19. Q. Zhao, L. Tong, A. Swami, and Y. Chen, Decentralized cognitive MAC for opportunistic spectrum access in ad hoc networks: A POMDP framework, *IEEE JSAC*, 25(3), 589–600, Apr. 2007.
20. N. Hoven and A. Sahai, Power scaling for cognitive radio, *WirelessCom 05 Symposium on Emerging Networks, Technologies and Standards*, Maui, HI, June 2005.
21. M. Andrews, L. Qian, and A. Stolyar, Optimal utility based multi-user throughput allocation subject to throughput constraints, *Proceedings of IEEE INFOCOM*, Miami, FL, Mar. 2005.
22. A. Hoang and Y. Liang, Downlink channel assignment and power control for cognitive radio networks, *IEEE Transactions on Wireless Communications*, 7(8), 3106–3117, Aug. 2008.
23. Y. Chen et al., On cognitive radio networks with opportunistic power control strategies in fading channels, *IEEE Transactions on Wireless Communications*, 7(7), 2752–2761, July 2008.
24. H. Le and Q. Liang, An efficient power control scheme for cognitive radios, *IEEE Wireless Communications and Networking Conference* (*WCNC*), Kowloon, Hong Kong, 2007.
25. K. Hamdi, W. Zhang, and K. B. Letaief, Power control in cognitive radio systems based on spectrum sensing side information, *IEEE International Conference on Communications* (*ICC*), Glasgow, U.K., 2007.
26. X. Wang and Q. Zhu, Power control for cognitive radio base on game theory, *International Conference on Wireless Communications, Networking and Mobile Computing*, Shanghai, China, 2007.
27. N. Gatsis, A. G. Marques, and G. B. Giannakis, Utility-based power control for peer-to-peer cognitive radio networks with heterogeneous QoS constraints, *IEEE International Conference on Acoustics, Speech and Signal Processing* (*ICASSP*), Las Vegas, NV, 2008.
28. S. Gao, L. Qian, and D. R. Vaman, Distributed energy efficient spectrum access in wireless cognitive radio sensor networks, *IEEE Wireless Communications and Networking Conference* (*WCNC*), Las Vegas, NV, 2008.

29. FCC, KRON-TV at San Francisco, CA. Available at: http://www.fcc.gov/fcc-bin/tvq?list=0&facid=65526
30. FCC, KRON-TV service contour map. Available at: http://www.fcc.gov/fcc-bin/FMTV-service-area?x=TV282182.html
31. M. A. McHenry, The probe spectrum access method, *Proceedings of IEEE Dyspan*, Baltimore, MD, 2005, pp. 346–351.
32. J. Bazerque and G. Giannakis, Distributed scheduling and resource allocation for cognitive OFDMA radios, *IEEE CROWNCOM*, Orlando, FL, 2007.
33. S. Kondo and B. Milstein, Performance of multicarrier DS CDMA systems, *IEEE Transactions on Communications*, 44(2), 238–246, Feb. 2001.
34. T. Weiss, J. Hillenbrand, A. Krohn, and F. K. Jondral, Mutual interference in OFDM-based spectrum pooling systems, *IEEE 59th Vehicular Technology Conference*, Vol. 4, Milan, Italy, May 2004, pp. 1873–1877.
35. J. G. Proakis, *Digital Communications*, 4th edn., McGraw-Hill, New York, 2000.
36. G. Foschini and Z. Miljanic, A simple distributed autonomous power control algorithm and its convergence, *IEEE Transactions on Vehicular Technology*, 42(4), 641–646, Nov. 1993.
37. D. Mitra, An asynchronous distributed algorithm for power control in cellular radio systems, *Proceeding of 4th WINLAB Workshop of 3rd Generation Wireless Information Networks*, New Brunswick, NJ, Oct. 1993, pp. 177–186.
38. S. Grandhi, J. Zander, and R. Yates, Constrained power control, *International Journal of Wireless Personal Communications*, 1(4), 257–270, Apr. 1995.
39. T. ElBatt and A. Ephremides, Joint scheduling and power control for wireless ad-hoc networks, *Proceedings of IEEE INFOCOM*, New York, 2002, pp. 976–984.
40. D. Bertsekas and J. Tsitsiklis, *Parallel and Distributed Computation: Numerical Methods*, Prentice Hall, Englewood Cliffs, NJ, 1989.
41. R. Yates, A framework for uplink power control in cellular radio systems, *IEEE JSAC*, 13(7), 1341–1348, Sept. 1995.
42. X. Li, Uplink power control and scheduling in CDMA systems, MS thesis, WINLAB, Rutgers University, Rutgers, NJ, 2003.
43. F. Meshkati, M. Chiang, V. Poor, and S. Schwartz, A game-theoretic approach to energy-efficient power control in multicarrier CDMA systems, *IEEE JSAC*, 24(6), 1115–1129, June 2006.
44. Technical Document on Overview—Wireless, Mobile and Sensor Networks. GDD-06-14. Available at: http://www.geni.net/GDD/GDD-06-14.pdf
45. KU Agile Radio. Available at: http://agileradio.ittc.ku.edu
46. Virginia Tech Chameleonic Radio. Available at: http://www.ece.vt.edu/swe/chamrad
47. Microsoft Research KNOWS project. Available at: http://research.microsoft.com/netres/projects/KNOWS
48. M. Dickens, B. Dunn, and J. N. Laneman, Design and implementation of a portable software radio, *IEEE Communications Magazine*, 46(8), 58–66, Aug. 2008.
49. S. Seidel, T. Krout, and L. Stotts, An adaptive broadband mobile ad-hoc radio backbone system, *DARPA WAND Project*. Available at: www.darpa.mil/sto/solicitations/WAND/pdf/DARPA_NC_Radio_System.pdf

PROTOCOLS AND ECONOMIC APPROACHES

Chapter 4

Medium Access Control in Cognitive Radio Networks

Jie Xiang and Yan Zhang

Contents

Medium Access Control (MAC) protocols play an important role in cognitive radio (CR) networks. In this chapter, we first introduce the challenges in designing CR MAC protocols. Then, we classify the state-of-the-art CR MAC protocols according to the spectrum-sharing modes into two major types, i.e., overlay MAC and underlay MAC. In overlay MAC, secondary users (SUs) opportunistically access the licensed spectrum not occupied by primary users (PUs) and should vacate the spectrum when PUs return. On the contrary, in underlay MAC, SUs can continue using the spectrum when PUs return, but the interferences from SUs to PUs should be carefully controlled under the predefined interference thresholds. After reviewing the CR MAC protocols, we conclude this chapter by identifying some open research issues in the realization of CR MAC protocols.

4.1 Introduction

In CR networks, the spectrum can be divided into several channels, either nonoverlapping or partially overlapping. When we say a channel is available for a SU to use, it means either no PU works on that channel or the interference from this SU to the active PUs is tolerable. MAC protocols are used to utilize these available channels. In general, the number of available channels can be more than one. Thus, the CR networks are similar to multi-channel (MC) wireless networks. However, there are many differences between CR networks and MC wireless networks. For example, the available channels at each node are equal and fixed in MC wireless networks whereas they are variable in CR networks. Therefore, the MAC protocols in MC wireless networks cannot be applied directly to CR networks.

The CR MAC acts as a bridge between the CR physical layer and the CR network layer. On the one hand, it can utilize the spectrum-sensing results from the CR physical layer, characterize the channels, and decide which channel to use and when to access. On the other hand, it can help the CR network layer to decide the routing path by reporting the characteristic information and the list of available channels. Also, the CR network layer can tell the CR MAC to choose a suitable channel for a dedicated quality-of-service (QoS) requirement. In general, the CR MAC should support the following two functions.

- *Interference control and avoidance for PUs*: This is the premise that SUs can share the spectrum with PUs. There are two modes for spectrum sharing between SUs and PUs. One is called overlay, wherein SUs should vacate the channel as soon as the PUs return. The other one is called underlay, wherein SUs can work in the same channel with PUs as long as the interference from SUs to PUs is no more than the predefined threshold.
- *Collision avoidance amongst SUs*: Because different SUs may coexist, collisions may happen if they simultaneously move to and use the same spectrum band according to their spectrum-sensing results. Thus, the CR MAC should control the spectrum access of different SUs to avoid the collisions.

The organization of this chapter is as follows. In Section 4.2, we describe the research and design challenges for CR MAC protocols. In Section 4.3, we classify the existing CR MAC protocols into two types, i.e., overlay MAC and underlay MAC. Then we discuss the overlay and underlay MAC protocols in Sections 4.4 and 4.5, respectively. Finally, we draw our conclusions in Section 4.6.

4.2 Research and Design Challenges

There are many challenges in designing a CR MAC protocol, such as channel definition, channel availability, channel heterogeneity, channel quality, common control channel problem, and multi-channel hidden terminal problem.

4.2.1 Channel Definition

A channel in CR networks is always assumed as a spectrum unit in the literature. But there has been no definition about the bandwidth of a channel yet. This issue was first addressed by Akyildiz et al. in [1]. Later Xu et al. studied the optimal channel bandwidth problem in [2] to maximize the SUs' throughput. Generally, the wider bandwidth the channel has, the more capacity the channel gets. However, in this situation, the channel-switching probability may increase because the probability for PUs' return to a wider range of spectrum could be higher than that to a smaller one. The increased channel-switching operations will then cause additional overheads, like switching delay, which would reduce the SUs' throughput. Therefore, the definition of channel bandwidth influences the performance of a CR MAC protocol.

Another uncertainty in defining a channel is whether it is overlapping. When the available spectrum is divided into several channels, these channels could be nonoverlapping or partially overlapping. Two channels are said to be nonoverlapping when they are separated by at least 25 MHz [3]. Using nonoverlapping channels can eliminate the interference between different channels, but may result in a waste of spectrum. On the contrary, using partially overlapped channels can improve spectrum utilization, and this is not always harmful according to the study in [4]. An example of using partially overlapped channels is the IEEE 802.11 standard [5], where the 2.4 GHz industrial, scientific, and medical bands (ISM band) are divided into several partially overlapped channels with a bandwidth of 22 MHz and only a 5 MHz space between two neighboring central frequencies. Although channel overlapping can increase the number of channels and improve spectrum utilization, the adjacent SUs that are using the partially overlapped channels may cause interference to each other. Moreover, in this case, the interference to any PU on a certain channel should include all the transmission of SUs on the partially overlapped channels. Therefore, channel overlapping will influence the two major functions of CR MAC, i.e., the interference control and avoidance for PUs and the collision avoidance (CA) amongst SUs.

The above-mentioned issues mainly focus on channels divided by a continuous spectrum. However, a channel constructed with discrete subcarriers is possible by the orthogonal frequency-division multiplexing (OFDM) modulation scheme in the physical layer, which has been widely used in IEEE 802.11a/g and IEEE 802.16 standards [6]. It can also be applied in CR networks and can affect the design of CR MAC protocols.

4.2.2 Channel Availability and Heterogeneity

A channel is said to be available to SUs when it is not occupied by any PUs (in the overlay spectrum-sharing mode) or the interference from SUs to PUs is under a tolerable threshold (in the underlay spectrum-sharing mode). The arbitrary activities of PUs result in a dynamic nature of channel availability. In the literature, most of the work such as [7–11] assumes that the channel usage pattern of PUs follows an independent and identically distributed ON/OFF random process, where the ON period represents that the channel is occupied by PUs while the OFF period represents that the channel is available to SUs.

The channel availability of SUs on different locations may be distinct from each other because the activity of PUs could be different. Even in the same geometrical area, SUs may have different available channels because of hardware limitations such as sensing constraints (different SUs may be capable of sensing different ranges of the spectrum) and transmission constraints (the radios of different SUs may be capable of transmitting on different ranges of spectrum) [12]. This phenomenon would result in the problem of channel heterogeneity where SUs have different available channels at a certain time [13]. In this heterogeneous situation, neighboring SUs should negotiate a common channel to communicate with each other before data transmission. The design of the CR MAC should take into account this issue.

4.2.3 Channel Quality

The quality of wireless channels varies over time, space, and frequency. The following parameters were addressed in [1].

- *Interference*: Because channels are shared by different SUs, some channels may be more crowded compared to others. Therefore, an SU using the same transmission power on different channels may result in different signal-to-interference-and-noise ratios (SINRs) on its destination SU receiver. A higher SINR would bring a higher throughput to the SU. Moreover, considering the protection of PUs in the underlay spectrum-sharing mode, the allowed interference on different channels may be different. Therefore, the allowed transmission power of an SU should be controlled and may be different on different channels.
- *Path loss*: The path loss is related to the distance between the SU transmitter and the receiver, as well as the channel central frequency. The path loss increases while the distance and frequency increase. Therefore, an SU transmitter may increase its transmission power to compensate for the increased path loss to its destination SU receiver. However, this may cause higher interference to other SUs and PUs.
- *Wireless link errors*: The link errors using different channels depend on the modulation scheme as well as the interference at the SU receiver.
- *Holding time*: The holding time of a channel refers to the expected time duration for which SUs can work on this channel. Because the activities of PUs may be different on each channel, the holding time would be changed accordingly. The longer the holding time, the better the channel quality would be.

The channel quality can be characterized by the above parameters jointly. Then, the CR MAC can use it as a metric for channel selection and access strategy.

4.2.4 Common Control Channel Problem

Neighboring SUs in a CR network can communicate with each other directly only if they work on a common channel. But before the communication, they do not know which channel can be used on each other. So, they need to exchange messages to know the available channels on each other. Thus, a common channel can be chosen based on their agreement. But the exchanged messages require a common control channel. This is called the common control channel problem, as addressed in [14]: "a channel is required to choose a channel."

In [15], the authors analyzed the design requirement of CR networks and suggested to distinguish the control channel and data traffic channels. A simple solution is to have a dedicated common control channel. This channel is a dedicated licensed spectrum band to SUs for the exchanging of control messages; thus, it will not be interrupted by PUs. In the literature, several works hold this assumption, such as [7,12,16–19]. But this assumption has several drawbacks as follows:

- *License fee*: A license fee is required to get the licensed band. Therefore, it would be expensive to build such a CR network.
- *Saturation*: This dedicated channel can be saturated easily if many SUs contend the control channel for their own traffic. Therefore, it would be the bottleneck of the network throughput.
- *Security*: It is feasible for adversaries to attack SUs by forging control messages to the control channel. This may cause saturation of the control channel that results in denial-of-service (DoS). These forged control messages can also cause communication disruptions and gain unfair advantages in resource allocation [20].

Another solution is to choose a control channel among the available channels such as in [14,21,22]. There are several challenges in this case. First, the channels used by SUs have to be vacated when PUs are detected. Therefore, the control channel should be the most reliable channel that cannot be interrupted frequently. Second, it is sometimes not feasible to select a common control channel for the whole network due to the channel heterogeneity problem we have mentioned.

4.2.5 Multi-Channel Hidden Terminal Problem

The multi-channel hidden terminal problem in traditional MC wireless networks was well discussed in [23]. Because the CR network is a kind of MC wireless networks, this problem could also happen in this network. Consider a general CR network in Figure 4.1 consisting of five SUs, A, B, C, D, and E. Suppose that each SU is equipped with only one CR transceiver. Each SU cannot transmit and receive at the same time. There are several available channels. One of the channels is a common control channel for exchanging control messages such as request to send (RTS) and clear to send (CTS). The contention on the common control channel is similar to the IEEE 802.11 distributed coordination function (DCF) [5]. All the other channels are for data. When SU A wants to transmit a packet to SU B, they exchange RTS/CTS messages on the control channel to reserve a data channel (DC). When sending an RTS, SU A puts its list of available channels. Upon receiving the RTS, SU B selects a channel and puts the selected channel in the CTS. Then, SU A and SU B switch to the agreed DC, exchange data frames, and end with an acknowledgment (ACK) message. After that, SU A and SU B switch to the common control channel for the next contention of data transmission.

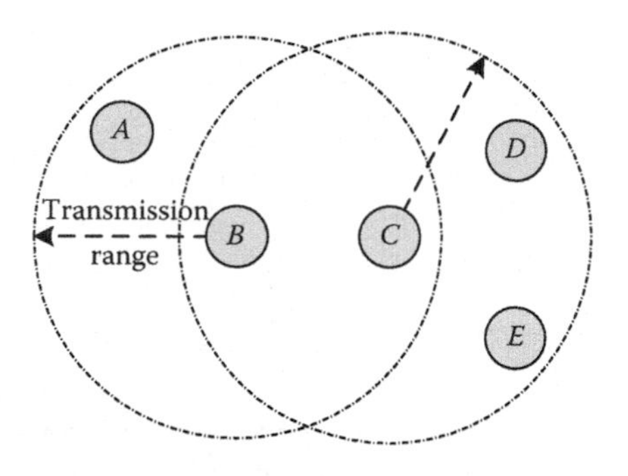

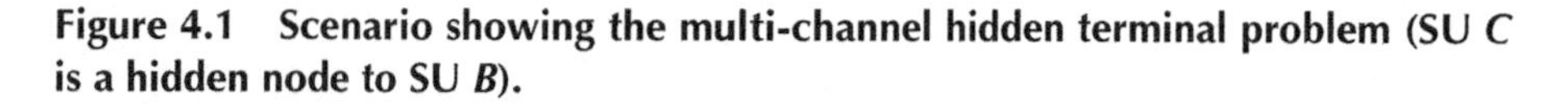

Figure 4.1 Scenario showing the multi-channel hidden terminal problem (SU *C* is a hidden node to SU *B*).

Suppose that SU *A* and SU *B* select a DC, say, 1, while SU *C* is transmitting data to SU *D* on another DC, say, 2. Because there is only one transceiver, SU *C* cannot receive the CTS information from *B* on the control channel. Therefore, after SU *C* finishes the transmission to SU *D*, SU *C* may choose to transmit the data to SU *E* on DC 1. Therefore, collision would happen at SU *B*. We say that SU *C* is a hidden node to SU *B*.

The above problem occurs due to the fact that SUs may listen to different channels, which makes it difficult to use virtual carrier sensing to avoid the hidden terminal problem.

4.3 Classification

Existing works on MAC protocols for CR networks can be classified by the following four metrics: architecture, spectrum-sharing behavior, spectrum-sharing mode, and access mode.

According to different architectures, MAC protocols can be classified into two categories, i.e., centralized and distributed. In centralized MAC, there is a central controller (e.g., base station [BS] or access point [AP]) to coordinate the channel access of SUs, for example, in the IEEE 802.22 draft standard [24] and the Dynamic Spectrum Access Protocol (DSAP) [16].

Regarding to different spectrum-sharing behaviors, MAC protocols can be classified into another two categories, i.e., cooperative MAC and noncooperative MAC. In a cooperative MAC, the SUs work cooperatively to maximize a predefined network utility, such as throughput. In a noncooperative MAC, each SU works independently and tries to maximize its own utility.

As for different spectrum-sharing modes, there are two kinds of modes in CR networks between SUs and PUs. One is called the overlay mode, wherein SUs can only use the spectrum that is not occupied by PUs [1,25]. The other one is

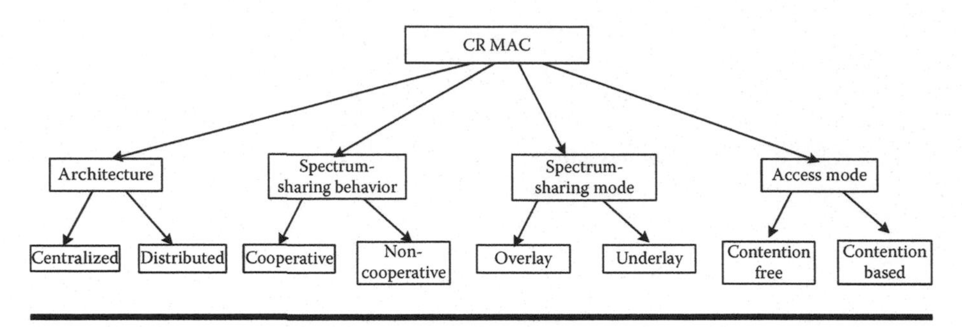

Figure 4.2 Classification of CR MAC protocols.

called the underlay mode, wherein SUs and PUs can coexist and share the same spectrum with each other, provided the interferences from SUs to PUs are under the predefined thresholds. The CR MAC layer protocols can be classified according to the spectrum-sharing modes into the following two categories: overlay MAC and underlay MAC.

According to the spectrum access mode, the CR MAC can be classified into two categories, i.e., contention-free and contention-based MAC protocols. In contention-free CR MAC, SUs access the spectrum according to the time slots in a frame structure. In contention-based CR MAC, SUs contend a channel for transmission opportunities, for example, Carrier Sense Multiple Access with Collision Avoidance (CSMA/CA).

Moreover, there are many other factors in different CR MAC protocols, for example, the number of transceivers and the requirement of a common control channel. Some MAC protocols require only one transceiver, while some others require two or more transceivers. Some MAC protocols require a common control channel, while some others do not.

Figure 4.2 shows the classification of CR MAC protocols. Note that a given MAC protocol may belong to more than one category, because the above categories are not independent of each other. For instance, IEEE 802.22 MAC is not only an overlay MAC but also a centralized, contention-free MAC. For the sake of convenience in discussion, we have broadly arranged the CR MAC protocols in overlay and underlay classes.

4.4 Overlay Mode MAC Protocols

With overlay mode MAC protocols, SUs dynamically access the licensed spectrum when it is not used by PUs. In literature, lots of research works have been done to study the MAC protocols in the overlay mode. We classify the overlay mode MAC protocols into two subcategories: centralized MAC and distributed MAC.

4.4.1 Centralized MAC

The centralized MAC protocols require a central controller, for example, a BS, to coordinate the channel access for SUs. In this category, we review two CR MAC protocols, i.e., IEEE 802.22 MAC [24] and DSAP [16].

4.4.1.1 IEEE 802.22

Currently, the IEEE 802.22 standard is still in the draft stage [24]. It is designed for wireless regional area networks (WRANs), which are operated in the VHF/UHF TV broadcast bands from 54 to 864 MHz, depending on the regulations around the world, and can cover a rural area up to 100 km in radius [26]. In the IEEE 802.22 standard, an SU is called a customer premises equipment (CPE), while a PU is called an incumbent equipment, such as an analog TV, a digital TV, and a wireless microphone. In a WRAN, a central BS coordinates the medium access of a number of associated CPEs. The BS has a spectrum manager function that can use the inputs from the spectrum-sensing function, geolocation, and the incumbent database to decide on the available channels. The IEEE 802.22 will define a single air interface based on a 2048 carrier orthogonal frequency-division multiple access (OFDMA) scheme [26]. It supports three different channel bandwidths, i.e., 6, 7, and 8 MHz, according to the regulation of TV channels all over the world. There are four different lengths of the cyclic prefix (as the symbol duration), i.e., 1/4, 1/8, 1/16, and 1/32, to allow different channel delay spreads while efficiently utilizing the spectrum.

In IEEE 802.22, the downstream (DS) data from the BS to CPEs are scheduled over consecutive MAC slots, while the upstream (US) channel capacity from CPEs to the BS is shared by CPEs based on a demand-assigned multiple access (DAMA) scheduling scheme [26].

The IEEE 802.22 MAC employs a hierarchical frame and superframe structure, as shown in Figure 4.3. Every superframe contains 16 frames with a 10 ms size each. The first frame contains a superframe preamble, a frame preamble, and a superframe control header (SCH), while the rest 15 frames start with only a frame preamble. Every frame is divided into a DS subframe and an US subframe with an adaptive boundary in between. The DS subframe contains a frame control header (FCH), a DS/US MAP, an US channel descriptor (UCD), a DS channel descriptor (DCD), and CPE bursts, where FCH contains the size of the DS/US MAP fields together with channel descriptors. The DS/US MAP gives the scheduling information for CPE bursts. The US subframe contains the information as follows: The ranging that informs the distance from the BS, the bandwidth request, the urgent coexistence situation (UCS) notification that informs CPEs about the incumbents that are just detected, and several CPE bursts. Following the US subframe there is a self-coexistence window (SCW), and time buffers that are used to absorb the delay because of propagation and the initial ranging process. Every superframe contains

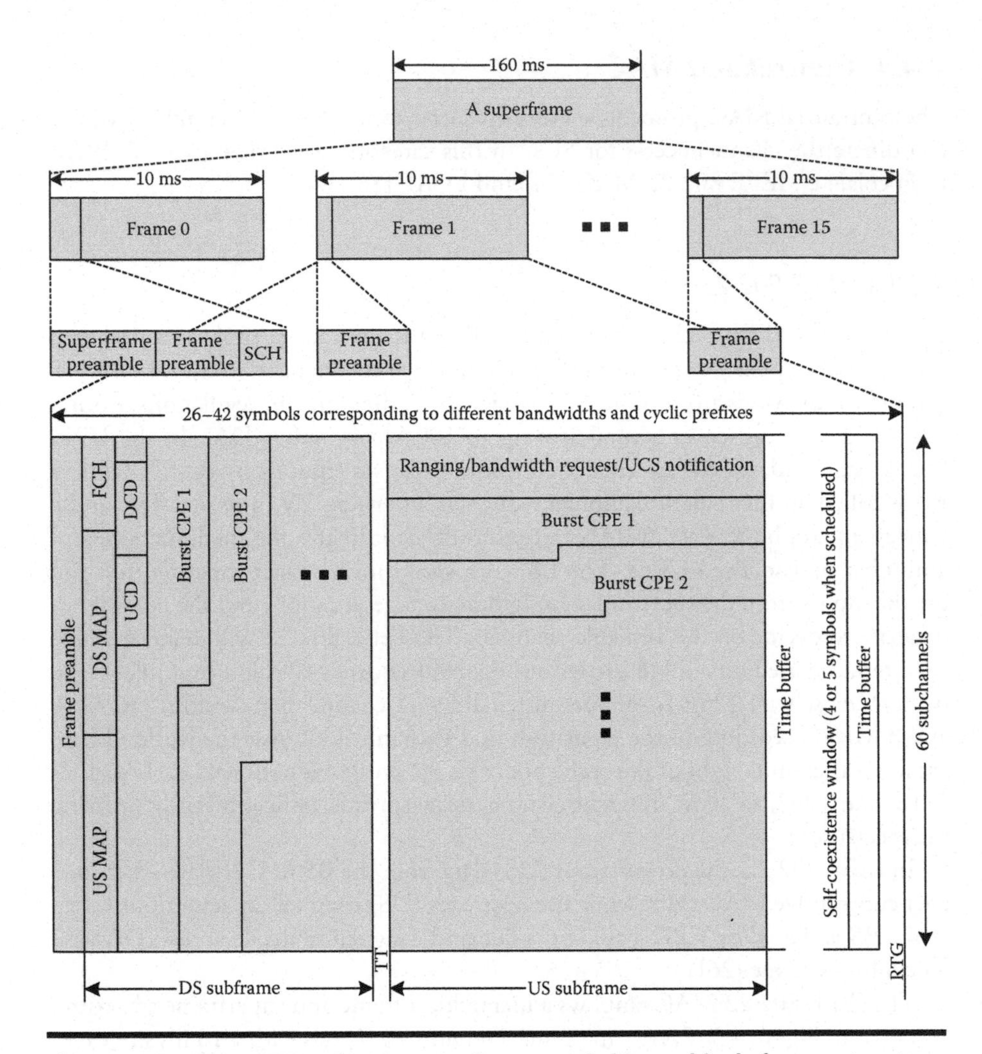

Figure 4.3 The IEEE 802.22 superframe and hierarchical frame structure. (Adapted from Stevenson, C. et al., *IEEE Commun. Mag.,* 47, 130, 2009.)

26–42 OFDM symbols corresponding to a bandwidth from 6 to 8 MHz and cyclic prefixes from 1/4 to 1/32. Every OFDM symbol consists of maximum 60 subchannels, where each subchannel has 28 subcarriers.

WRANs can be operated without a dedicated common control channel. Whenever a CPE is switched on, it scans all the channels in the licensed TV band to find out vacant channels. On the other hand, the BS broadcasts the above OFDMA superframes in the vacant channels. The CPE would choose one of the vacant channels and scan for the SCH information in the superframe. The duration for which a CPE stays in a channel is no less than the superframe duration of 160 ms.

Once the CPE receives the SCH, it acquires the channel and network information. After the above initialization, the CPE will receive and transmit data according to the DS and the US MAP, respectively, without contending channels with other CPEs. Therefore, this is a contention-free MAC.

The IEEE 802.22 MAC specifies a self-coexistence mechanism based on the Coexistence Beacon Protocol (CBP) to address the problem that multiple WRANs may operate in the same vicinity [26]. A typical CBP packet comprises a preamble, an SCH, and a CBP MAC packet data unit (PDU). It is delivered on the operating channel through beacon transmission in a dedicated SCW (see Figure 4.3) at the end of some frames. During a synchronized SCW, a BS or a CPE can either transmit CBP packets on its operating channel or receive CBP packets on any channel. Each WRAN system is required to maintain a minimum repeating pattern of SCWs in the active mode, and can reserve its own SCWs on the operating channel for exclusive CBP transmission or share the active SCWs with other co-channel neighbors through a contention-based access. A WRAN can capture CBP packets from the neighboring WRANs by knowing their SCW patterns.

To protect the incumbents, the BS and CPEs will temporarily stop transmission based on a quiet period (QP) scheme. During the QP, spectrum sensing is done by both the BS and CPEs. CPEs will report the spectrum-sensing result to the BS, while the BS will make the final decision about whether a channel is available or not.

There are several requirements to implement and deploy the IEEE 802.22 WRANs [26,27] as follows:

- *Antennas*: Each CPE requires two antennas, one directional and the other omnidirectional. The directional antenna is used for communication between the CPE and the BS, while the omnidirectional antenna is used for spectrum sensing.
- *Locations*: Each CPE is assumed to know its own location before communicating with the BS. This can be done by GPS devices.
- *Incumbent database*: The BS is assumed to access an incumbent database service, which provides accurate and up-to-date information.
- *Other WRANs database*: There is another database that contains other IEEE 802.22 WRANs in the area. This is used for the coexistence of multiple IEEE 802.22 systems.

4.4.1.2 Dynamic Spectrum Access Protocol

A DSAP was proposed in [16]. In the DSAP, a server collects the spectrum-sensing information and maintains a *RadioMap*, which holds the information about all SUs and channel conditions in the CR network. This server can allocate the channels to SUs in a similar manner as the Dynamic Host Configuration Protocols (DHCP) server, which provides IP address leases to hosts in a network.

Whenever an SU wants to communicate with other SUs, it will ask the server for a channel by broadcasting a *ChannelDiscover* message consisting of a MAC identifier, a location, radio capabilities, a destination SU identifier, and the desired lease options on a dedicated common control channel. Then the server will respond to the SU with a *ChannelOffer* message containing the server's choice of lease, which may be different from what the SU requested. Then, this SU can use this channel to its intent.

When a channel is going to be expired on an SU, this SU will send a *Channel-Request* message on the common control channel to the server; then the server will reply with a *ChannelACK* message to accept or decline the request. If accept, the server will send a *ChannelOffer* message indicating if the SU can continue using the channel or change to another new channel. If a channel becomes unavailable to SUs because of the return of PUs by spectrum sensing on the server, the server will send a *ChannelReclaim* message to the SUs on this channel to reassign or terminate the channel lease.

The communication between SUs and the server could be contention based. SUs can contend the common control channel with other SUs to exchange a *ChannelDiscover*/*ChannelOffer* message with the server. Moreover, the communication in each SU pair could be contention based, so that the server can win a transmission opportunity to send a *ChannelReclaim* message to the SU for channel reassignment on its operating channel.

The DSAP requires the following premises. (1) *Common control channel*: A dedicated common control channel is required by the DSAP to exchange control messages between the server and SUs. (2) *Antennas*: Each SU may be equipped with only one antenna, but the server has at least two antennas. One antenna works on the dedicated common control channel, while the other antennas can switch to any other channels to reach any SUs. (3) *Interference to PUs*: The returning PUs should be able to tolerate some interference during the time that the server sends a *ChannelReclaim* message to the SUs.

4.4.2 Distributed MAC

In contrast to the centralized MAC, distributed MAC protocols can work without a central coordinator. In this category, we review the following CR MAC protocols: Hardware Constrained MAC (HC-MAC) [12], Decentralized Cognitive MAC (DC-MAC) [17], Statistical Channel Allocation MAC (SCA-MAC) [18], Opportunistic Spectrum MAC (OS-MAC) [19], Cognitive MAC (C-MAC) [21], Synchronized MAC (SYN-MAC) [28], Opportunistic MAC (O-MAC) [7], and Efficient Cognitive Radio-EnAbled Multi-Channel MAC (CREAM-MAC) [22].

4.4.2.1 Hardware Constrained MAC

In [12], Jia et al. proposed an HC-MAC protocol to address the problem of hardware constraints during sensing and transmission. Specially, in HC-MAC, the authors considered a single transceiver that can sense a certain number of adjacent channels in a limited time and can aggregate a maximum number of spectrum fragments for transmission.

The time frame in HC-MAC consists of three consecutive phases of operations, i.e., contention, sensing, and transmission. Every operation has its own control messages on the common control channel. C-RTS/C-CTS are used to contend and reserve the common control channel, where the contention process is similar to the IEEE 802.11 DCF. S-RTS/S-CTS are used for exchanging the list of available channels in the sensing phase. T-RTS/T-CTS are used to notify the neighboring SUs about the completion of the transmission.

Whenever an SU is switched on or enters a CR network, it listens to the control channel for at least a duration of the maximal sensing and transmission time. During this time, if a C-RTS or a C-CTS is received, it will defer and wait for the T-RTS or the T-CTS. Then, it jumps to the contention phase if it has data to transmit after receiving T-RTS/T-CTS or a time threshold. The SU reserves time for the following sensing and transmission operations within its neighboring SUs by exchanging C-RTS/C-CTS messages with the intent SU on the control channel.

In the sensing phase, only the SU pairs that win the contention can start to sense the spectrum. The sensing time can be divided into several time slots. Each time slot consists of an actual spectrum sensing and negotiation between the sender and the receiver through the exchange of S-RTS/S-CTS. At the end of the negotiation, a decision to stop or continue sensing is made according to an optimal stopping rule proposed in [12]. This optimal stopping sensing rule is based on the solution of the problem of maximizing the effective data rate while considering the sensing and transmission constraints. To reduce the complexity, the authors proposed a k-stage look-ahead rule to decide at each stage whether to stop or to continue sensing.

In the transmission phase, the SU pair starts transmitting and receiving on the negotiated channels. After the transmission is finished, the SU sender sends a T-RTS message to the SU receiver on the common control channel. Then, the intent SU receiver, upon receiving the T-RTS, will reply with a T-CTS message. Other SUs overhearing these messages know the completion of the transmission and will contend the control channel with random backoff if they have data to transmit. Because SUs contend the common control channel before sensing and transmission, the multi-channel hidden terminal problem is eliminated.

There are several requirements for deploying HC-MAC. (1) *Common control channel*: HC-MAC requires a dedicated common control channel to exchange the control messages. (2) *Interference to PUs*: Because there is only one antenna on SUs, SUs cannot stop transmission during the transmission phase if any PUs return. Although it works in an overlay spectrum-sharing mode, it requires that PUs tolerate the interference from SUs during the whole transmission phase. (3) *Channel availability*: To make an optimal stopping in HC-MAC, the probability of channel availability is assumed to be equal for every channel and should be known before making the decision.

4.4.2.2 Decentralized Cognitive MAC

In [17], Zhao et al. proposed a DC-MAC protocol based on a Partially Observable Markov Decision Process (POMDP) framework. DC-MAC addressed the following issues: hardware constraint, sensing error, channel availability, and common control channel.

In DC-MAC, time is slotted and synchronized to each SU communication pair. It is assumed that SUs can obtain the slot information from PUs. Each SU pair accesses the available channels following a manner similar to that of CSMA. Specially, every time slot consists of consecutive operations, i.e., spectrum sensing, RTS/CTS exchange, data transmission, and ACK. In the spectrum-sensing phase, the SU pair (both the sender and the receiver) chooses a set of channels to sense, based on a randomized optimal strategy. Specially, the authors addressed the particular case that every SU is able to sense only one channel at the sensing phase because of hardware constraints. In this situation, the SU sender and receiver can select the same channel to sense for the sensing period. After the sensing phase, if the channel is available during the sensing period, the SU sender sends an RTS to the receiver after a random backoff time if no other SUs have already accessed the channel. The receiver, upon receiving the RTS, replies with a CTS message if this channel is also available at the receiver during the sensing period. Thereafter, the SU sender transmits data on this channel in the data transmission phase, and the receiver will send an ACK message to the sender to confirm the successful reception of the data.

In DC-MAC, any SU pair can synchronize without a common control channel by an initial handshake and synchronous spectrum hopping. The initial handshake is based on the assumption that every SU regularly monitors all the available channels whenever an SU sender transmits a handshake signal over one of its available channel. After this initialization, the SU pair follows the same spectrum-sensing and access strategy, and spectrum hopping is synchronized.

To develop the optimal spectrum-sensing and access strategy, it is assumed that PUs occupy n licensed channels following a discrete-time Markov process with 2^n states. In POMDP, every SU receives a reward at the end of a slot based on the channel it sensed and accessed on a state. The reward function is defined as the number of bits delivered, which is the throughput of the SU pair. The objective is to choose the optimal channel to sense and access to maximize the reward function. To reduce the complexity of the POMDP problem, the authors showed that a vector of conditioned probability based on the sensing and decision history for all channels that the channel is available at the beginning of a slot is a sufficient statistic for the optimal spectrum access strategy. If a channel is not sensed, the probability is updated according to the Markov chain. Considering the sensing errors, such as false alarm and miss detection, it is assumed that PUs can tolerate a given maximum collision rate from SUs. Then, the authors proposed a suboptimal greedy approach to maximize the per-slot throughput of SUs. In this approach, both the sender and the receiver in any SU pair can select the same optimal channel to sense, access, and get a reward, which is used for the input of the next slot.

From the above description of DC-MAC, any sender cooperates with its receiver to develop the optimal sensing and access strategy for every slot. Therefore, it is a cooperative MAC. Furthermore, the channel access in any slot is contention based, because every SU pair should contend for a transmission opportunity on the operating channel by a backoff time after a sensing phase.

There are several limitations of DC-MAC. (1) It is assumed that SUs can obtain the slot information from PUs. (2) The spectrum usage statistics of PUs remain unchanged for a certain number of slots. (3) The transition probabilities of the channel state are assumed to be known. (4) PUs are assumed to be able to tolerate a given maximum probability of collisions from SUs.

4.4.2.3 Statistical Channel Allocation MAC

A CR MAC protocol using statistical channel allocation, called SCA-MAC, was proposed in [18]. SCA-MAC requires a common control channel and has three major phases: (1) environment sensing and learning, (2) RTS/CTS exchange over the common control channel, and (3) DATA/ACK transmission over DCs.

In the first phase, SUs perform environment sensing and learning, where a channel allocation scheme based on successful rate prediction is processed. It is assumed that every SU has an optimal operating range, which specifies the proper spectrum range that it would search for transmission opportunities, and a maximum number of continuous channels, which can be used simultaneously. The main difference of SCA-MAC to other MAC protocols is the channel allocation strategy using the statistic information collected by continuous and periodic spectrum sensing. The optimal channel allocation strategy is to choose a set of DCs that can achieve a maximum successful rate, which is defined as the production of channel availability and spectrum hole sufficiency. Channel availability in SCA-MAC is the probability of successful channel allocation within the operating range of the SU receiver, while spectrum hole sufficiency is the joint probability that a specific length of the packet can fit in the constraint of maximum continuous channels that can be used by SUs. However, the calculation of channel availability and spectrum hole sufficiency requires not only the operating range and the maximum aggregated number of continuous DCs, but also the information of the utilization of PUs and the neighboring SUs, the average number of channels used by the neighboring SUs, and the packet length. At the end of this phase, SUs will select the channels with a successful rate higher than a predefined threshold, α, where $1 - \alpha$ is the maximum allowed probability of collisions from SUs to PUs. The higher the successful rate, the smaller the collision rate.

In the second phase, SUs follow a standard CSMA/CA scheme to contend for the common control channel and exchange RTS/CTS messages. At the beginning, every SU should listen to the common control channel before sending any data packet and wait until it becomes idle. Then the SU sender transmits an RTS message if the channel is idle after waiting for a distributed coordination function interframe space (DIFS) duration and a contention window (CW) period. The SU receiver, upon receiving the RTS message, checks the potential transmission opportunities based on the channel allocation strategy carried out in the first phase, and replies with a CTS message, which contains the information of the selected DCs and a collision avoidance (CA) window. If there is a collision on the RTS or the CTS message, the SU sender would repeat the negotiation process, but would double the CW size until it reaches the maximum.

In the third phase, the SU sender transmits data on the agreed DCs after a random time in the range of the CA window if the channel is still available. This CA window is set by the CTS message from the receiver in the second phase, and it is related to the number of neighboring SUs at the receiver. The SU receiver will reply with an ACK message to the SU sender after a Short Interframe Space (SIFS) if data are successfully received. If no ACK message is received at the SU sender, it means the transmission has failed, and the SU sender will go back to the second phase to contend the control channel.

The requirements of SCA-MAC are as follows. (1) *Common control channel*: SCA-MAC requires a common control channel for SUs to exchange control messages. (2) *Interference to PUs*: Although SCA-MAC works in an overlay mode, it requires that PUs tolerate the maximum probability of collision with SUs.

4.4.2.4 Opportunistic Spectrum MAC

An efficient MAC protocol called OS-MAC was proposed by Hamdaoui and Shin in [19]. In OS-MAC, each SU is assumed to be equipped with a single half-duplex transceiver; there is a dedicated common control channel and several nonoverlapping DCs with equal bandwidths.

OS-MAC is designed for the scenario that SUs form into different groups called SUGs. At the beginning, an SU can choose to create a new SUG or join an existing SUG by listening to the common control channel. All SUs in one SUG work in the same DC. At any time, only one SU can transmit while others are listening. The channel access scheme is the IEEE 802.11 DCF without RTS/CTS exchange. In any SUG, there is a delegated SU called the DSU, which is in charge of the information exchange with other DSUs from other SUGs.

In OS-MAC, time is divided into periods, which consist of three consecutive phases: select, delegate, and update. In the select phase, each SUG selects the best DC and uses it for communication until the end of the current period. In the delegate phase, a DSU is selected by a simple method as follows. Let all SUs contend for the DC in the manner of the IEEE 802.11 DCF; the first SU that successfully delivers a packet is automatically appointed as the DSU in this period. In the update phase, every DSU has its own time slot to transmit a control frame containing the traffic information of its DC. To guarantee that each DSU has its own time slot, the update phase window is divided according to the number of DCs. The number of SUGs is assumed to be no higher than the number of DCs. So, every active DSU can work on different DCs. After the update phase, the network jumps into the select phase, where these DSUs switch to their DCs and broadcast a control message containing all the channel traffic information. The sender in this SUG will decide the best channel based on a probability. Then, the sender broadcasts a control message to all SUs in the SUG containing the channel decision result. After receiving this message, all the SUs in the SUG will switch to the chosen DC.

As there is no specification about the protection of PUs in the above description of OS-MAC, the authors suggested some extensions to OS-MAC in [19]. In these extensions, noncooperative and cooperative approaches can be applied according to different situations. If PUs do not cooperate with SUs, a noncooperative approach will be applied, where all SUs will suspend their sessions, switch to the control channel, and wait until the next update phase to select a new DC upon detection of any return of PUs. On the other hand, in the cooperative approach, it is assumed that PUs allow SUs to continue using the spectrum after the detection of PUs for a short while. In this short duration, SUs can negotiate with each other and switch to a new DC.

In spite of the limitation of PUs' protections in OS-MAC, there are some other premises. (1) It requires a common control channel. (2) It is only efficient in the special scenario where SUs are divided into several groups. Each group follows a manner of one member talking and the others listening, that is, only one SU in a group can transmit while the other SUs receive. (3) Although it requires only a single half-duplex transceiver, spectrum sensing may need one or more antennas.

4.4.2.5 Cognitive MAC

Cordeiro and Challapali proposed a C-MAC protocol in [21]. In C-MAC, only a single half-duplex transceiver is required for every SU.

In C-MAC, each channel is logically divided into a structured superframe, as shown in Figure 4.4. Each superframe is comprised of a slotted beacon period (BP) and a data transfer period (DTP). Every DTP has a QP for in-band measurements. During a QP in a channel, SUs on other channels can perform out-of-band measurements. BPs and QPs across different channels are nonoverlapping; this is done by an inter-channel coordination mechanism. The first two slots of the BP are signaling slots for new SUs joining this channel, while the other slots are for SUs to transmit their own beacons. Thus, collisions between SUs are prevented. Because BPs on different channels are nonoverlapping, SUs can get the traffic information of all channels by listening to the BP of that channel. As a result of this beaconing approach, the SUs are synchronized in time, space, and frequency. Thus, it solves the multi-channel hidden terminal problem.

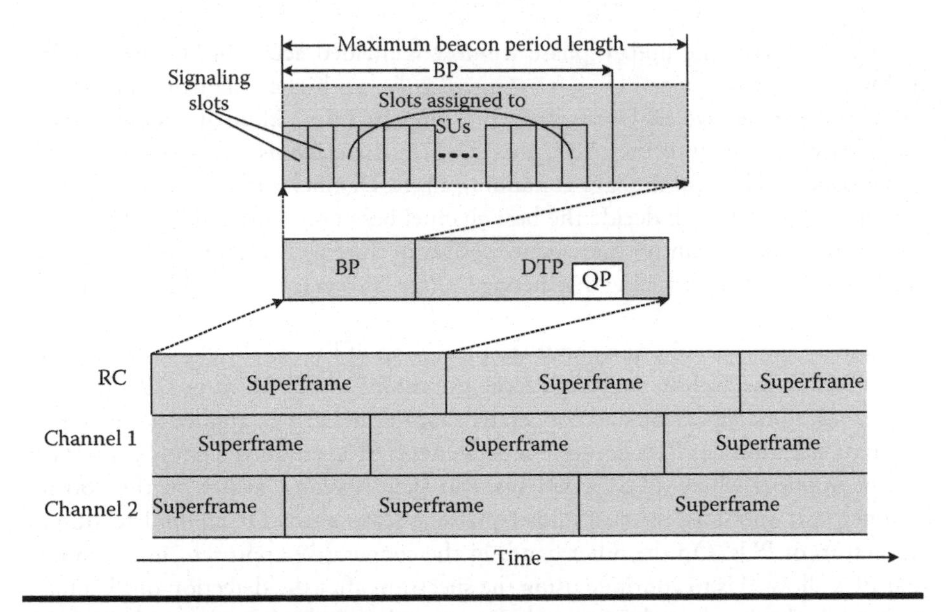

Figure 4.4 Superframe structure in C-MAC. (Adapted from Cordeiro, C. and Challapali, K., C-MAC: A cognitive MAC protocol for multi-channel wireless networks, in *The 2nd IEEE International Symposium on New Frontiers in Dynamic Spectrum Access Networks 2007* (*DySPAN 2007*), Dublin, Ireland, April 2007, pp. 147–157.)

Because there is no dedicated common control channel in C-MAC, the rendezvous channel (RC) (as shown in Figure 4.4) becomes the backbone of C-MAC. The RC is assigned as the most reliable channel out of all the available channels. The RC is used to coordinate SUs in different channels, reserve a multi-channel resource for SUs, and coordinate QPs for PU detection. With the RC, C-MAC can support networkwide group communication, self-coexistence, inter-channel synchronization, neighborhood discovery, and load balancing.

The selection of an RC is done as follows. Whenever an SU is switched on, it does spectrum sensing on all the channels and looks for beacon frames transmitted by other SUs. The SU stays on each available channel for at least one superframe length, so that it can receive at least one beacon frame. Suppose an SU receives a beacon frame on a channel, it reads the beacon frame header. If the bit of the RC field is set to one (meaning this is an RC), the SU may decide to join this BP by sending its own beacon during the signaling slots and then move to an assigned beacon slot. If the SU cannot find any RC after scanning all the channels, it will select a channel as an RC by itself and transmit beacons with the RC field set to one.

After the RC is set up, SUs can then exploit other available channels and tell other SUs about it by appending a channel switch information element during its beacon transmission on the RC. Once this SU hops to its new channel, it sets up the channel by transmitting its own beacon and starting a new sequence of recurring superframes. Note that this SU is required to switch back to the RC periodically to resynchronize.

There are also some limitations of C-MAC. (1) C-MAC assumes that all the SUs use the same RC to communicate. In practice, for a multi-hop network, it is sometimes impossible to find such a channel through all the links, because of channel heterogeneity. (2) Channel is not switched as soon as the detection of PUs return. The information of channel switching should first be sent on the BP until all SUs get this information, then the SUs decide to switch to a new channel simultaneously. (3) When the number of available channels increases, it may be impossible to make sure that the BPs of different channels are nonoverlapping for a limited superframe size.

4.4.2.6 Synchronized MAC

In [28], a SYN-MAC protocol was proposed by Kondareddy and Agrawal. The main idea behind SYN-MAC is the use of different time slots to represent different channels. All nodes should be synchronized by listening to the same channel at the beginning of each time slot. SYN-MAC assumes that every SU is equipped with two transceivers. One is used for exchanging control messages (control radio) as well as for spectrum sensing on every channel in the corresponding time slot, while the other one is used for both receiving and transmitting data (data radio).

At the beginning of network initialization, suppose that there are maximum N available channels; the first SU divides time equally into N time slots. Therefore, each time slot is corresponding to a channel. Then, the SU beacons on all its available channels at the beginning of the corresponding time slots. The other SUs tune their control radios to one of the available channels and listen for the beacon messages for a period of N slots. After receiving a beacon message, SUs exchange the information with the first SU about the available channels during this time slot. If it does not receive any beacon, it is considered to be the first node. After network initialization, all SUs are synchronized and have the information about their neighbors and respective channel sets.

There are four kinds of control messages in SYN-MAC. The first one is the notification for new SUs entering the already initialized network. The second one is used for SUs to notify their neighbors about the changing of their available channels. The third one is used for SUs to inform their neighbors about their actions, such as start, stop, or change their operational channel. The last one is used for data transmission between two neighboring SUs.

As in the example shown in Figure 4.5, time is divided into N slots for the maximum N channels. Any SU sender would wait for the time slot of a particular channel when it wants to send a data packet to an SU receiver on that channel. In that time slot, the channel access follows a manner similar to the IEEE 802.11

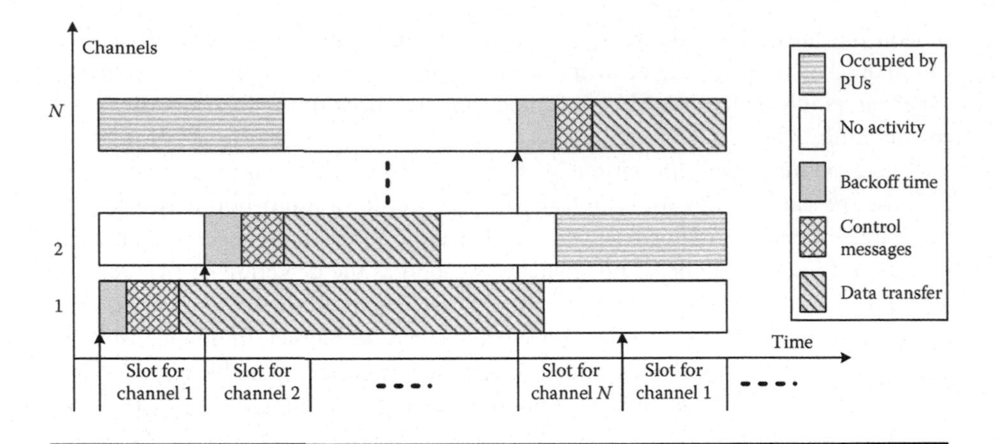

Figure 4.5 Illustration of the SYS-MAC protocol. (Adapted from Kondareddy, Y. and Agrawal, P., Synchronized MAC protocol for multihop cognitive radio networks, in *IEEE International Conference on Communications 2008* (*ICC '08*), Beijing, China, May 2008, pp. 3198–3202.)

DCF. After a random backoff time at the SU sender, they exchange control messages to establish the connection. Once the receiver confirms the transmission, the SU sender starts to transmit the data packet on this channel. Note that the transmission time could be larger than the time slot.

However, in the initialization period, SUs may not receive any beacon even if the first SU exists. Because the channel is randomly chosen, this channel may be not available on the first SU. Therefore, in [14], the authors point out that every SU should listen to every channel for a time of $N \times T$. This procedure can ensure that other SUs can receive beacons if there is any common channel between this SU and the first SU. Otherwise, this SU will believe that it is the first SU and will start beaconing on every available channel.

The requirements of SYN-MAC are as follows. (1) Every SU should be equipped with two transceivers. (2) All SUs should know the maximum number of available channels in advance.

4.4.2.7 Opportunistic MAC

In [7], Su and Zhang proposed an O-MAC protocol. In this chapter, we call it O-MAC for the sake of convenience. In O-MAC, a dedicated common control channel is required, and each SU is supposed to be equipped with two transceivers. One is fixed and works only on the common control channel (control transceiver), while the other one can switch to any channel and performs spectrum sensing and data transmission (CR transceiver).

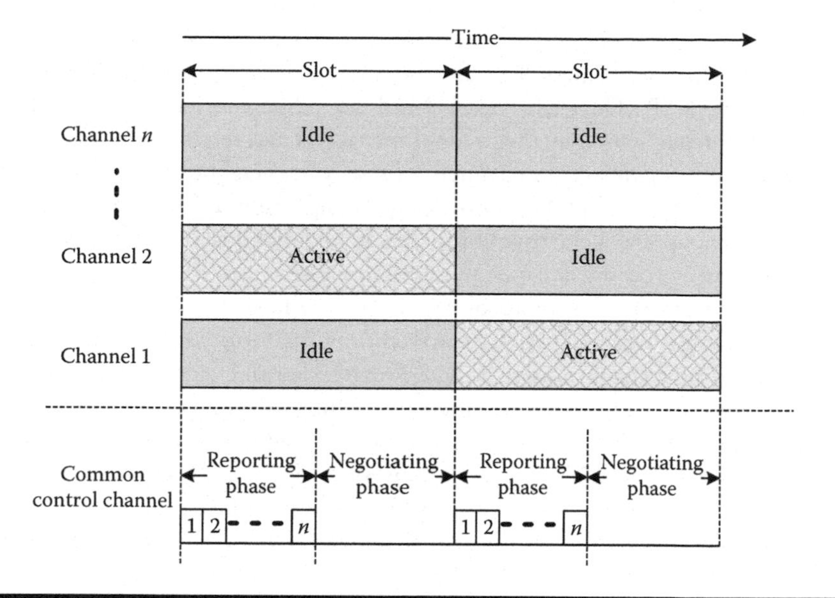

Figure 4.6 The principle of O-MAC. (Adapted from Su, H. and Zhang, X., *IEEE J. Sel. Areas Commun.*, 26, 118, 2008.)

The principle of the O-MAC protocol is shown in Figure 4.6. In the control channel, the time axis is divided into a number of time slots. Each time slot is divided into two phases: a reporting phase and a negotiating phase.

The reporting phase is used to inform which channel is idle, and it is further divided into n mini-slots, each of which corresponds to one of the n licensed channels. In this phase, every SU listens to the control channel from its own control transceiver, while its CR transceiver keeps on sensing on the primary channels. Considering hardware constraints, at each time slot, a CR transceiver can only sense on one channel. Once it detects a channel, e.g., channel j ($j = 1, \ldots, n$), as idle, it will send a beacon on the jth mini-slot on the control channel by the control transceiver. The neighboring SUs will receive the information and recognize that channel j is idle. Thereafter, the available channel list will be updated on these SUs.

In the negotiating phase, it employs a p-persistent CSMA protocol to contend the transmission opportunity. The SU sender listens to the control channel and waits until it becomes idle. Then, it transmits an RTS with probability p. If the SU receiver receives the RTS from its control transceiver, it checks the channel list and replies with a CTS. Then the SU sender, upon receiving the CTS, will tune its CR transceiver to the agreed DC to transmit data packets.

Two sensing policies called random sensing policy (RSP) and negotiation-based sensing policy (NSP) are used in the reporting phase. By using the RSP, each SU independently selects one of the n licensed channels with a probability of $1/n$ for sensing. When the number of SUs is smaller than the number of primary channels, not all the channels can be sensed in one reporting phase. Hence, the authors introduced the NSP. The main idea behind the NSP is to let SUs know which channels are already sensed by their neighboring SUs and select a different channel to sense. The information of the channels sensed by the neighboring SUs is obtained by exchanging RTS/CTS packets in the negotiating phase. Each RTS/CTS packet has a byte of a special field containing the sensed-channel information. In this fashion, all primary channels can be sensed in a limited number of slots.

The limitations of this protocol lay in the following. (1) It requires two transceivers for each SU. (2) A dedicated common control channel is needed to associate with the control transceiver. (3) The number of channels should be configured in advance, because the parameter *n* cannot be changed after the SUs are running. Therefore, the scalability of this MAC protocol is low. (4) Every SU should be synchronized, because every SU uses the same reporting phase and negotiating phase. (5) It assumes that all the SUs have the same channel availability, which may not be suitable for a large CR network, where channel heterogeneity happens.

4.4.2.8 Efficient Cognitive Radio-Enabled Multi-Channel MAC

In [22], Su and Zhang made a modification on O-MAC and proposed a CREAM-MAC protocol. In CREAM-MAC, the common control channel is not only a pre-allocated static channel but also can be a dynamically selected channel from the available channels.

To overcome the problem of spectrum sensing of the *n* channels, CREAM-MAC assumes that each SU is equipped with *n* sensors. Therefore, *n* channels can be sensed by every SU simultaneously. Therefore, the reporting phase in Figure 4.6 can be removed. The control channel only works on channel negotiation. It also improves the channel-negotiating scheme by introducing an additional pair of control messages, namely, channel state transmitter/channel state receiver (CST/CSR), besides RTS/CTS. CST and CSR contain the available channel list at the SU sender and receiver, respectively. By these modifications, SUs can work independently without knowing the slots of primary networks. Thus, a new working principle is illustrated in Figure 4.7.

Because the instantaneous sensing result may be not accurate, the authors assume that PUs can tolerate a maximum tolerable interference period. Therefore, in CREAM-MAC, SUs cannot continuously transmit on the channel longer than the time threshold.

The requirements for CREAM-MAC are as follows. (1) Every SU is equipped with a CR transceiver and multiple sensors. (2) Although, it does not require a static common control channel, the authors assumed that the dynamic common control channel should always be reliable and available.

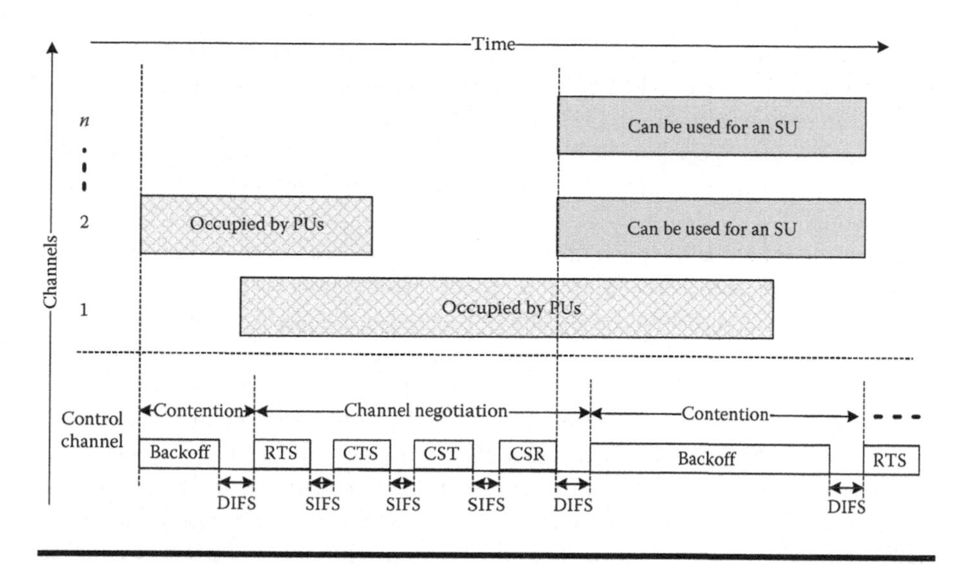

Figure 4.7 Illustration of the CREAM-MAC protocol. (Adapted from Su, H. and Zhang, X., CREAM-MAC: An efficient cognitive radio-enabled multi-channel MAC protocol for wireless networks, in *2008 International Symposium on a World of Wireless, Mobile and Multimedia Networks* (*WoWMoM 2008*), Newport Beach, CA, June 2008, pp. 1–8.)

4.5 Underlay Mode MAC Protocols

Using the underlay mode MAC protocols, SUs share the spectrum with PUs provided the interferences to the PUs are less than the predefined tolerable thresholds. Comparing with the MAC protocols in the overlay mode, there are fewer MAC protocols in the underlay mode. Moreover, most of the work in the underlay mode focuses on code division multiple access (CDMA) networks, such as [29–31].

4.5.1 Centralized MAC

In this category, medium access is controlled by a central controller, for instance, a BS. The BS will collect the information of PUs' geolocations, channel state, and interference threshold. Normally, there is a common control channel for the control messages between the BS and SUs. The problem is to maximize the network utility, e.g., throughput, while guaranteeing the interference to the PUs is under a tolerable threshold. Sometimes, the QoS of SUs is considered in the formulation. The key issues in this category are channel allocation, power control, and admission control problems.

For the CR power control problem, in [32–35], the authors investigated a particular scenario of one SU link and one PU link. In [36,37], the authors investigated the power control problem in a multiple PUs scenario with the objective to maximize the weighted sum rate of all SUs. The authors proposed a suboptimal solution in [36] and an optimal solution in [37]. In [37–40], the authors explored the multiple-channel scenarios. Specially, they assumed that each SU can use more than one channels simultaneously and tried to find out the transmission power of each SU on each channel.

When the interference constraints cannot be guaranteed by allowing all the SU links, admission control is required in addition to power control. In [41], the authors proposed a distributed constrained power control algorithm and found the optimal link subset to achieve the maximum revenue with the help of a potential game. In [42], the authors modeled a smooth optimization problem and proposed a minimal SINR removal algorithm to search the optimal set of SUs. However, all of these studies assumed only a single PU in the system. In [30,31,43–45], the authors explored the scenario with multiple PUs. For example, in [30], J. Xiang et al. studied the problem of how to coordinate SUs to access the licensed channel to the BS to maximize the total revenue to the network operator, and proposed a joint admission and power control scheme using a minimal revenue efficiency removal algorithm.

4.5.2 Distributed MAC

In this category, Shao-Yu Lien et al. proposed a Carrier Sensing–Based Multiple Access protocol for CR networks in [46]. In this chapter, we call it CSMA-MAC for the sake of convenience. It was designed for the scenario that secondary and primary networks have their own BS. SUs and PUs contend for the channel to transmit data to their own BS. It assumes that PUs use a classical CSMA protocol to access channels. To guarantee that PUs have a higher priority than SUs to access the channel, in CSMA-MAC, SUs use a longer carrier-sensing period than PUs. It also assumes that the physical layer of secondary networks can support adaptive modulation and coding (AMC), so that a low level of modulation scheme, which can tolerate high interference, will be applied when the signal is weak. On the contrary, when the signal is strong, a high level of modulation scheme with a high data rate will be applied.

There are four major operations in this MAC protocol. (1) When an SU wants to transmit a packet, if the channel is busy, the SU will sense the channel for a duration. If the channel is either idle or busy at the end of the sensing duration, the SU sends an RTS message to the secondary BS to contend the channel. (2) If an SU receives an RTS message, it computes the feasible transmission power and rate. If a feasible transmission power and rate can be obtained, the SU will send this information by a CTS message to the SU transmitter. Otherwise,

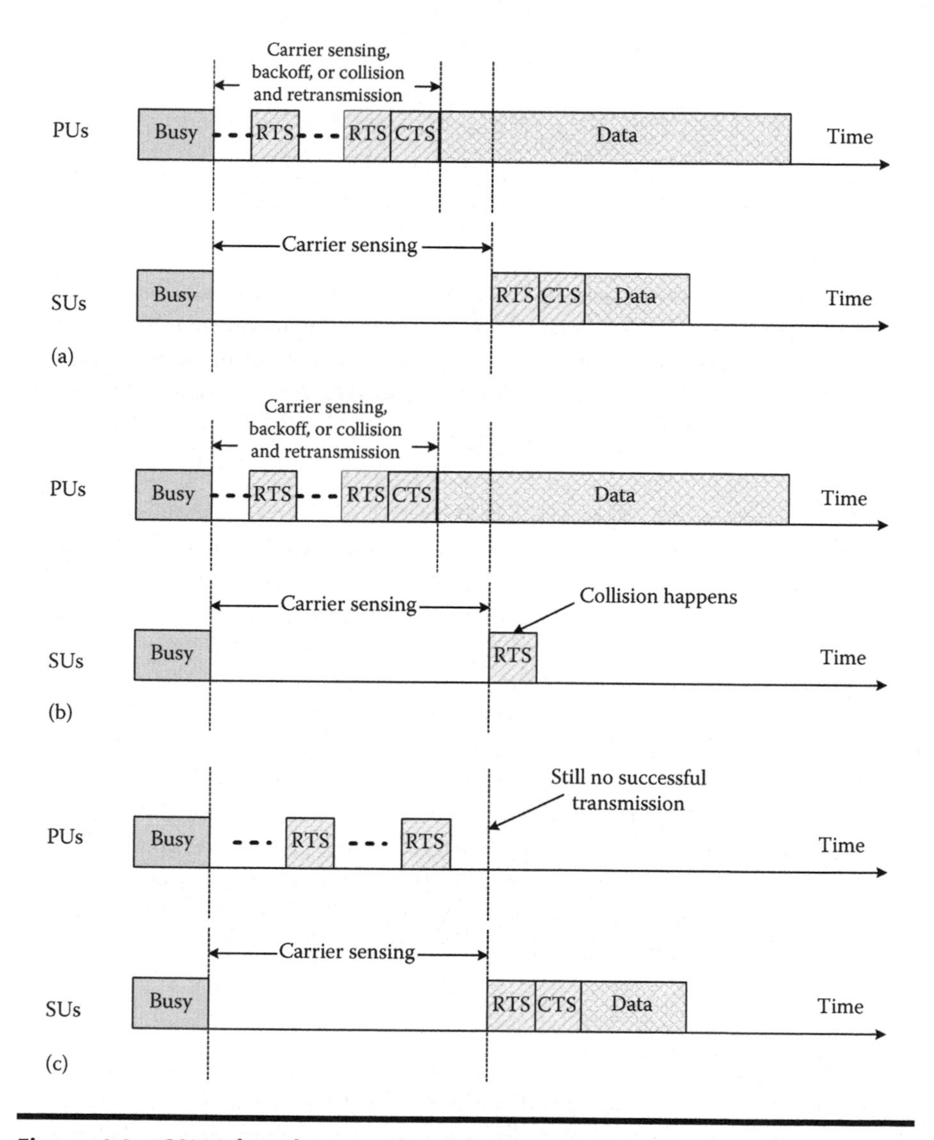

Figure 4.8 CSMA-based protocols with four-way handshaking procedure. (a) SU can transmit a packet with feasible power and data rate. (b) RTS messages are collided, SU should wait for the next transmission opportunity. (c) SU can transmit without power control when PU fails to receive CTS from the BS.

(continued)

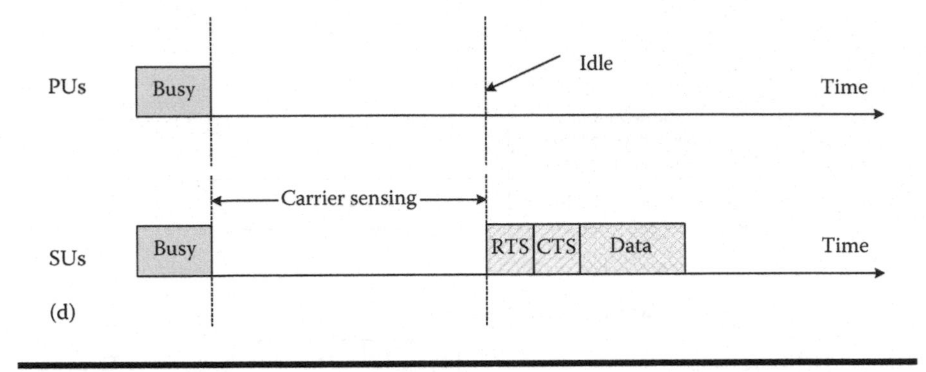

Figure 4.8 (continued) (d) SU can transmit without power control while channel is idle from PUs. (Adapted from Lien, S.-Y. et al., Carrier sensing based multiple access protocols for cognitive radio networks, in *IEEE International Conference on Communications 2008* (*ICC '08*), Beijing, China, May 2008, pp. 3208–3214.)

the CTS message is not responded. (3) If an SU receives a corresponding CTS message, it transmits the data packet with the power and rate carried in the CTS message. Otherwise, the SU transmitter shall wait for the end of the next carrier-sensing period to send another RTS message. (4) When an SU receives a data packet, it replies the SU transmitter with an ACK message. The detailed behavior of this MAC protocol is shown in Figure 4.8.

There are several limitations for this protocol. (1) This protocol is designed for the scenario where SUs and PUs use CSMA protocols to contend the channels. It is invalid for the scenario where PUs do not use CSMA protocols. (2) The SUs should know the carrier-sensing duration of the PUs in advance, because the secondary networks have a longer carrier-sensing duration than the primary networks. (3) The secondary BS should know the interference information from the primary networks to calculate the feasible transmission power for the SUs. Otherwise, the interference from SUs to PUs may exceed the tolerable threshold.

4.6 Conclusions

In this chapter, we have investigated the problem of MAC in CR networks. The state-of-the-art MAC protocols have been reviewed and summarized in Table 4.1. We have addressed the following MAC design issues: channel definition, dynamic channel availability and heterogeneity, channel quality, common control channel problem, and multi-channel hidden terminal problem. Besides these issues, several MAC protocols addressed the problem of hardware constraints, for example, HC-MAC, DC-MAC, and SCA-MAC.

Although we distinguish the overlay MAC from the underlay MAC, there are some overlay MAC protocols that cannot eliminate the interference to PUs. For

Table 4.1 Summary of State-of-the-Art CR MAC Protocols

MAC	*Architecture*	*Spectrum-Sharing Mode*	*Common Control Channel*	*Number of Transceivers*	*Access Mode*
IEEE 802.22	Centralized	Overlay	No	2	Contention free
DSAP	Centralized	Overlay	Yes	1 (server requires 2)	Contention based
HC-MAC	Distributed	Overlay	Yes	1	Contention based
DC-MAC	Distributed	Overlay	No	≥ 1	Contention based
SCA-MAC	Distributed	Overlay	Yes	≥ 1	Contention based
OS-MAC	Distributed	Overlay	Yes	1	Contention based
C-MAC	Distributed	Overlay	No	1	Contention free
SYN-MAC	Distributed	Overlay	No	2	Contention based
O-MAC	Distributed	Overlay	Yes	2	Contention based
CREAM-MAC	Distributed	Overlay	Yes	1 (plus additional sensors)	Contention based
CSMA-MAC	Distributed	Underlay	No	1	Contention based

example, in the DSAP, the returning PUs may receive interference from SUs in a duration of the *ChannelReclaim* message; in HC-MAC, SUs cannot stop transmission during the transmission phase if any PUs return; in DC-MAC and SCA-MAC, PUs are assumed to be able to tolerate a maximum allowed collision rate from SUs.

In future, there are several directions for CR MAC protocols to provide the following requirements, such as security and heterogeneous coexistence.

As for security issues, it is feasible for adversaries to attack SUs by forging control messages to the control channel. This may cause saturation of the control channel that results in DoS. These forged control messages can also cause communication disruptions and gain unfair advantages in resource allocation [20].

The self-coexistence issue has been addressed in the IEEE 802.22 MAC and C-MAC protocols; however, the coexistence for heterogeneous CR networks has not been studied in most of the CR MAC protocols. In [47], He et al. considered the scenario where several different types of CR networks coexisted and proposed a Transmission Opportunity (TXOP)-Based Spectrum Access Control Protocol. In this situation, different CR networks may have different channel definitions; thus, spectrum coordination between different CR networks would be more difficult than in one type of CR network.

Abbreviation List

MAC	Medium Access Control
CR	cognitive radio
SUs	Secondary Users
PUs	Primary Users
QoS	Quality-of-Service
MC	Multi-channel
OFDM	Orthogonal Frequency Division Multiplexing
ISM	Industrial Scientific and Medical
SINR	Signal-to-Interference-and-Noise-Ratio
DoS	Denial-of-Service
DCF	distributed coordination function
CTS	clear-to-send
RTS	request-to-send
ACK	acknowledgment
BS	base station
DSAP	Dynamic Spectrum Access Protocol
AP	access point
DSAP	Dynamic Spectrum Access Protocol
WRANs	Wireless Regional Area Networks
CPE	customer premises equipment
OFDMA	orthogonal frequency-division multiplex access
DS	downstream
US	upstream
DAMA	demand-assigned multiple access
DCD	DS Channel Descriptor
FCH	frame control header
UCD	US Channel Descriptor
UCS	urgent coexistence situation
CBP	coexistence beacon protocol
PDU	Packet Data Unit
HC-MAC	Hardware Constrained MAC
DC-MAC	Decentralized Cognitive MAC
SCA-MAC	Statistical Channel Allocation MAC
OS-MAC	Opportunistic Spectrum MAC

C-MAC	Cognitive MAC
SYN-MAC	Synchronized MAC
O-MAC	Opportunistic MAC
CREAM-MAC	Efficient Cognitive Radio-EnAbled Multi-Channel MAC
POMDP	Partially Observable Markov Decision Process
DIFS	Distributed Coordination Function Interframe Space
CW	Contention window
CA	collision avoidance
SIFS	Short Interframe Space
DTP	data transfer period
QP	quiet period
BP	beacon period
RSP	random sensing policy
NSP	negotiation-based sensing policy
AMC	adaptive modulation and coding
SCW	self-coexistence window

References

1. I. F. Akyildiz, W.-Y. Lee, M. C. Vuran, and S. Mohanty, Next generation/dynamic spectrum access/cognitive radio wireless networks: A survey, *Computer Networks*, 50(13), 2127–2159, 2006.
2. D. Xu, E. Jung, and X. Liu, Optimal bandwidth selection in multi-channel cognitive radio networks: How much is too much? in *3rd IEEE Symposium on New Frontiers in Dynamic Spectrum Access Networks (DySPAN 2008)*, Chicago, IL, Oct. 2008, pp. 1–11.
3. P. Fuxjager, D. Valerio, and F. Ricciato, The myth of non-overlapping channels: Interference measurements in ieee 802.11, in *Fourth Annual Conference on Wireless on Demand Network Systems and Services (WONS '07)*, Obergurgl, Austria, Jan. 2007, pp. 1–8.
4. A. Mishra, V. Shrivastava, S. Banerjee, and W. Arbaugh, Partially overlapped channels not considered harmful, *ACM SIGMETRICS Performance Evaluation Review*, 34(1), 63–74, 2006.
5. IEEE standard for information technology-telecommunications and information exchange between systems-local and metropolitan area networks-specific requirements – Part 11: Wireless LAN Medium Access Control (MAC) and Physical Layer (PHY) specifications, *IEEE Std 802.11-2007 (Revision of IEEE Std 802.11-1999)*, pp. C1–1184, 12 2007.
6. IEEE standard for local and metropolitan area networks Part 16: Air interface for broadband wireless access systems, *IEEE Std 802.16-2009 (Revision of IEEE Std 802.16-2004)*, pp. C1–2004, 29 2009.
7. H. Su and X. Zhang, Cross-layer based opportunistic MAC protocols for QoS provisionings over cognitive radio wireless networks, *IEEE Journal on Selected Areas in Communications*, 26(1), 118–129, Jan. 2008.

8. D. Djonin, Q. Zhao, and V. Krishnamurthy, Optimality and complexity of opportunistic spectrum access: A truncated markov decision process formulation, in *IEEE International Conference on Communications 2007 (ICC '07)*, SECC, Glasgow, Scotland, June 2007, pp. 5787–5792.
9. Q. Zhao, L. Tong, and A. Swami, A cross-layer approach to cognitive MAC for spectrum agility, in *Conference Record of the Thirty-Ninth Asilomar Conference on Signals, Systems and Computers (ACSSC '05)*, Pacific Grove, CA, Oct. 28, 2005–Nov. 1, 2005, pp. 200–204.
10. H. Su and X. Zhang, Cognitive radio based multi-channel MAC protocols for wireless Ad Hoc networks, in *IEEE Global Telecommunications Conference 2007 (GLOBECOM '07)*, Washington, DC, Nov. 2007, pp. 4857–4861.
11. H. Kim and K. Shin, Efficient discovery of spectrum opportunities with MAC-layer sensing in cognitive radio networks, *IEEE Transactions on Mobile Computing*, 7(5), 533–545, May 2008.
12. J. Jia, Q. Zhang, and X. Shen, HC-MAC: A hardware-constrained cognitive MAC for efficient spectrum management, *IEEE Journal on Selected Areas in Communications*, 26(1), 106–117, Jan. 2008.
13. M. Ma and D. Tsang, Impact of channel heterogeneity on spectrum sharing in cognitive radio networks, in *IEEE International Conference on Communications 2008 (ICC '08)*, Beijing, China, May 2008, pp. 2377–2382.
14. Y. Kondareddy, P. Agrawal, and K. Sivalingam, Cognitive radio network setup without a common control channel, in *IEEE Military Communications Conference, 2008 (MILCOM 2008)*, San Diego, CA, Nov. 2008, pp. 1–6.
15. P. Pawelczak, R. Venkatesha Prasad, L. Xia, and I. Niemegeers, Cognitive radio emergency networks—Requirements and design, in *First IEEE International Symposium on New Frontiers in Dynamic Spectrum Access Networks, 2005 (DySPAN 2005)*, Baltimore, MD, Nov. 2005, pp. 601–606.
16. V. Brik, E. Rozner, S. Banerjee, and P. Bahl, Dsap: A protocol for coordinated spectrum access, in *2005 First IEEE International Symposium on New Frontiers in Dynamic Spectrum Access Networks (DySPAN 2005)*, Baltimore, MD, Nov. 2005, pp. 611–614.
17. Q. Zhao, L. Tong, A. Swami, and Y. Chen, Decentralized cognitive MAC for opportunistic spectrum access in ad hoc networks: A POMDP framework, *IEEE Journal on Selected Areas in Communications*, 25(3), 589–600, Apr. 2007.
18. A. Chia-Chun Hsu, D. Weit, and C.-C. Kuo, A cognitive MAC protocol using statistical channel allocation for wireless ad-hoc networks, in *IEEE Wireless Communications and Networking Conference (WCNC 2007)*, Hong Kong, China, Mar. 2007, pp. 105–110.
19. B. Hamdaoui and K. Shin, OS-MAC: An efficient MAC protocol for spectrum-agile wireless networks, *IEEE Transactions on Mobile Computing*, 7(8), 915–930, Aug. 2008.
20. K. Bian and J.-M. Park, MAC-layer misbehaviors in multi-hop cognitive radio networks, in *2006 US–Korea Conference on Science, Technology, and Entrepreneurship (UKC2006)*, Teaneck, NJ, Aug. 2006.

21. C. Cordeiro and K. Challapali, C-MAC: A cognitive MAC protocol for multi-channel wireless networks, in *The 2nd IEEE International Symposium on New Frontiers in Dynamic Spectrum Access Networks 2007 (DySPAN 2007)*, Dublin, Ireland, Apr. 2007, pp. 147–157.
22. H. Su and X. Zhang, CREAM-MAC: An efficient cognitive radio-enabled multi-channel MAC protocol for wireless networks, in *2008 International Symposium on a World of Wireless, Mobile and Multimedia Networks (WoWMoM 2008)*, Newport Beach, CA, June 2008, pp. 1–8.
23. J. So and N. H. Vaidya, Multi-channel mac for ad hoc networks: Handling multi-channel hidden terminals using a single transceiver, in *Proceedings of the 5th ACM International Symposium on Mobile Ad Hoc Networking and Computing (MobiHoc '04)*, Roppongi Hills, Tokyo, Japan, ACM Press, New York, 2004, pp. 222–233.
24. IEEE 802.22/D1.0 draft standard for wireless regional area networks Part 22: Cognitive wireless RAN medium access control (MAC) and physical layer (PHY) specifications: Policies and procedures for operation in the TV bands, *IEEE Draft Standard 802.22*, Apr. 2008.
25. Q. Zhao and B. Sadler, A survey of dynamic spectrum access, *IEEE Signal Processing Magazine*, 24(3), 79–89, May 2007.
26. C. Stevenson, G. Chouinard, Z. Lei, W. Hu, S. Shellhammer, and W. Caldwell, IEEE 802.22: The first cognitive radio wireless regional area network standard, *IEEE Communications Magazine*, 47(1), 130–138, Jan. 2009.
27. C. Cordeiro, K. Challapali, and D. Birru, IEEE 802.22: An introduction to the first wireless standard based on cognitive radios, *Journal of Communications*, 1(1), 38–47, April 2006.
28. Y. Kondareddy and P. Agrawal, Synchronized MAC protocol for multi-hop cognitive radio networks, in *IEEE International Conference on Communications 2008 (ICC '08)*, Beijing, China, May 2008, pp. 3198–3202.
29. M. H. Islam, Y.-C. Liang, and A. T. Hoang, Distributed power and admission control for cognitive radio networks using antenna arrays, in *2nd IEEE International Symposium on New Frontiers in Dynamic Spectrum Access Networks (DySPAN 2007)*, Dublin, Ireland, Apr. 17–20, 2007, pp. 250–253.
30. J. Xiang, Y. Zhang, and T. Skeie, Joint admission and power control for cognitive radio cellular networks, in *11th IEEE Singapore International Conference on Communication Systems (ICCS 2008)*, Guangzho, China, Nov. 2008, pp. 1519–1523.
31. L. B. Le and E. Hossain, Resource allocation for spectrum underlay in cognitive radio networks, *IEEE Transactions on Wireless Communications*, 7(12), 5306–5315, Dec. 2008.
32. Y. Chen, G. Yu, Z. Zhang, H. Hwa Chen, and P. Qiu, On cognitive radio networks with opportunistic power control strategies in fading channels, *IEEE Transactions on Wireless Communications*, 7(7), 2752–2761, July 2008.
33. X. Kang, Y.-C. Liang, A. Nallanathan, H. Garg, and R. Zhang, Optimal power allocation for fading channels in cognitive radio networks: Ergodic capacity and outage capacity, *IEEE Transactions on Wireless Communications*, 8(2), 940–950, Feb. 2009.

34. Y. Ma, H. Zhang, D. Yuan, and H.-H. Chen, Adaptive power allocation with quality-of-service guarantee in cognitive radio networks, *Computer Communications*, 2009.
35. R. Zhang, Optimal power control over fading cognitive radio channels by exploiting primary user CSI, 2009. [Online]. Available: http://arxiv.org/abs/0804.1617
36. L. Zhang, Y.-C. Liang, and Y. Xin, Joint beamforming and power allocation for multiple access channels in cognitive radio networks, *IEEE Journal on Selected Areas in Communications*, 26(1), 38–51, Jan. 2008.
37. L. Zhang, Y. Xin, Y.-C. Liang, and H. V. Poor, Cognitive multiple access channels: Optimal power allocation for weighted sum rate maximization, *accepted to IEEE Transactions on Communications*, Sept. 2008.
38. P. Setoodeh and S. Haykin, Robust transmit power control for cognitive radio, *Proceedings of the IEEE*, 97(5), 915–939, May 2009.
39. Y. Che, J. Wang, J. Chen, W. Tang, and S. Li, Hybrid power control scheme in hierarchical spectrum sharing network for cognitive radio, *Physical Communication Special Issues on Cognitive Radio Networks: Algorithms and System Design*, 2(1–2), 73–86, 2009.
40. Y. Wu and D. Tsang, Distributed power allocation algorithm for spectrum sharing cognitive radio networks with qos guarantee, in *The 28th IEEE Conference on Computer Communications (INFOCOM 2009)*, Rio de Janeiro, Brazil, Apr. 2009, pp. 981–989.
41. Y. Xing, C. N. Mathur, M. Haleem, R. Chandramouli, and K. Subbalakshmi, Dynamic spectrum access with qos and interference temperature constraints, *IEEE Transactions on Mobile Computing*, 6(4), 423–433, Apr. 2007.
42. L. Zhang, Y.-C. Liang, and Y. Xin, Joint admission control and power allocation for cognitive radio networks, in *IEEE International Conference on Acoustics, Speech and Signal Processing (ICASSP 2007)*, Honolulu, HI, Vol. 3, Apr. 15–20, 2007, pp. 673–676.
43. H. Islam, Y. chang Liang, and A. Hoang, Joint power control and beamforming for cognitive radio networks, *IEEE Transactions on Wireless Communications*, 7(7), 2415–2419, July 2008.
44. D. Kim, L. Le, and E. Hossain, Joint rate and power allocation for cognitive radios in dynamic spectrum access environment, *IEEE Transactions on Wireless Communications*, 7(12), 5517–5527, Dec. 2008.
45. J. Xiang, Y. Zhang, T. Skeie, and J. He, Qos aware admission and power control for cognitive radio cellular networks, *Wireless Communications and Mobile Computing*, 2009.
46. S.-Y. Lien, C.-C. Tseng, and K.-C. Chen, Carrier sensing based multiple access protocols for cognitive radio networks, *IEEE International Conference on Communications 2008 (ICC '08)*, Beijing, China, May 2008, pp. 3208–3214.
47. J. He, Y. Zhang, D. Kaleshi, A. Munro, and J. McGeehan, Dynamic spectrum access in heterogeneous unlicensed wireless networks, *International Journal of Autonomous Adaptive and Communication Systems*, 1(1), 148–163, 2008.

Chapter 5

Cross-Layer Optimization in Cognitive Radio Networks

Christian Doerr, Dirk Grunwald, and
Douglas C. Sicker

Contents

When placed in its operating environment, a cognitive radio commands over a large number of sensors and system parameters to find the configuration maximizing its operation. This flexibility, however, makes the search for the best configuration complex and highly expensive, as the system parameters may influence each other, thus leading to an exponential if not factorial search complexity.

This chapter introduces the use of fractional factorial designs to reduce this search complexity and enable adaptation even in a very dynamic environment, and discusses how this strategy—encapsulated in a framework called the "rapid adaptation architecture"—can be used to enhance a variety of state-of-the-art cognitive radio control algorithms.

The work described in this chapter is an extension of previous findings presented in [1] and [2], augmented with further performance characterizations and an in-depth analysis of performance improvements gained by the rapid adaptation architecture.

5.1 Introduction

When positioned in its application context, a cognitive radio has a large variety of sensors from which it can obtain information about its environment and a large variety of parameters that it can modify and tune to adapt to its overall behavior. Depending on the underlying physical layer adaptation capabilities and the number and the type of parameters and configurations from which the cognitive radio system can choose, there will typically be a large quantity of parameters influencing the cognitive radio's operation and resulting performance. The difficult task that the cognitive radio faces is to identify those parameters that will likely have the largest impact on its performance, quantify their influence on the cognitive radio's operation, and select a technically feasible combination to meet or exceed given performance requirements.

To select a parameter combination that will maximize its utility in the current operation context, the cognitive radio needs to evaluate all variables it can select. This evaluation process could be done either experimentally, where the cognitive radio reconfigures to the particular configuration at hand and runs an experimental test determining its performance value, or theoretically, where a utility function would directly tell the cognitive radio the usefulness of a given configuration without a reconfiguration and experimental evaluation.

Thus, selecting a cognitive radio configuration suffers from one or more of these three problems:

1. *Every reconfiguration and evaluation of a parameter configuration uses resources.* To determine the goodness of any given configuration, a cognitive radio needs to invoke a mechanism for evaluation. If the radio reconfigures itself to evaluate the new configuration using a benchmark suite, measuring the goodness of the configuration will consume energy, as well as incur computational cost and lost airtime for changing its parameters and running a test estimating the usefulness of the parameters. If a well-defined fitness function does exist for all configurations possible, this fitness has to be called for every configuration to be evaluated, creating computational cost, thus resulting in slower reconfiguration time and energy consumption.
2. *A configuration setting that turns out to be unsuitable might disconnect the link.* Very often it is hard to estimate the boundary between a beneficial configuration that minimizes energy consumption and interference to other stations, and a configuration that does not maintain connectivity to the network. If a cognitive radio modifies its transmit power parameter, for example, its objective might be to use the least energy possible. When generating possible parameter configurations, several might turn out to be energy efficient but not adequate to maintain a link of sufficient quality to other stations in the network. In this situation, evaluating or using such a configuration will create problems for users of a cognitive radio.
3. *Many configurations evaluated do not yield any useful information.* Many configurations that will be evaluated contain no information, that is, the effect of the reconfiguration of a particular parameter might (1) not be distinguishable from noise, (2) yield very little outcome and thus not justify the expense generated by the search process, or (3) even degrade performance.

It is therefore critical that a cognitive radio system conducts the search for a parameter configuration satisfying the users/applications service requirements in the most efficient way possible, as any parameter search and evaluation step uses resources (energy or lost bandwidth).

This needs to be as efficient as possible in the search, that is, commit the least resources as possible while maintaining a configuration that will maximize the radio's operation at any given time, and becomes even more critical due to two additional complications:

- First, due to environmental influences or other radio devices operating on the same frequency, channel conditions are highly time-variant and may require a fast adaptation rate. It is therefore important that a cognitive radio's search algorithm finds or converges to a solution fast enough to benefit from it before a change in the radio's environment makes another reconfiguration necessary.
- Second, as the cognitive radio system might have a large number of parameters it can adapt to, there exist many different variables that may influence the radio's performance, which need to be investigated during a reconfiguration search. Worse yet, as these variables may interact and influence one another, there actually exists an exponential, if not factorial, number of possible cross-interactions between factors that need to be studied and analyzed as well. This will make the search for an adaptation very expensive, even if no actual experimental runs are being conducted (Figure 5.1).

These complications will be discussed in more detail in the following sections.

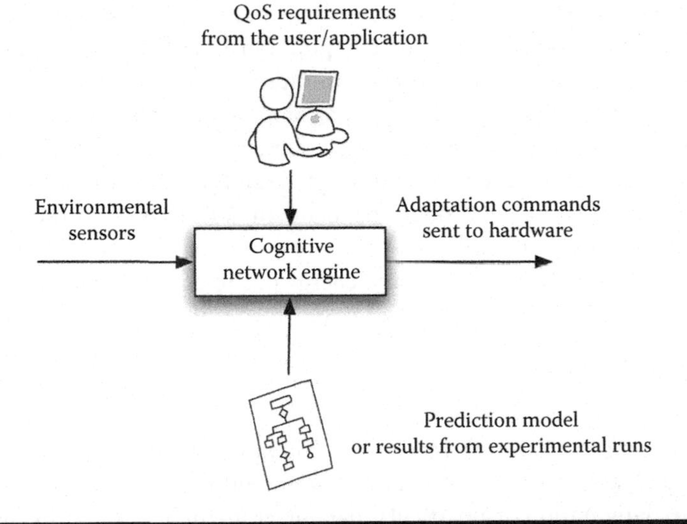

Figure 5.1 The control algorithm inside the cognitive radio selects an appropriate configuration based on the sensed environmental conditions, the QoS requirements from the user or application and, based on information from a theoretical prediction model or experimental runs, sends adaptation commands to the underlying hardware.

5.1.1 Varying Channel Conditions Require Frequent Retraining

This section elaborates on the previous argument that channel conditions are time-variant and discusses the implications for a cognitive radio's adaptation search. Figure 5.2 depicts long-term noise floor measurements sampled by a high-rate wide-band spectrum analyzer for the 54–88 MHz TV band and the 2390–2500 MHz

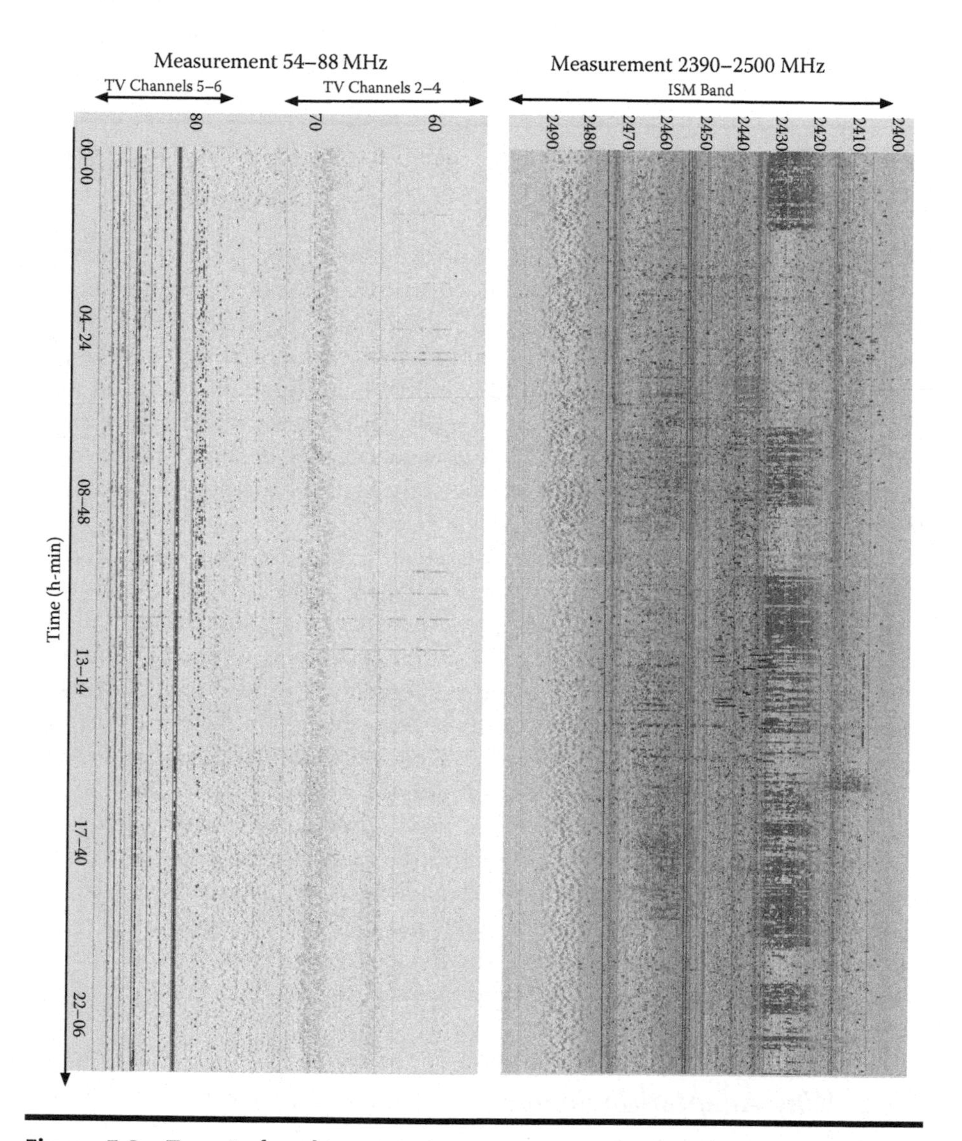

Figure 5.2 Twenty-four-hour spectrum measurements on the 54–88 MHz TV band and 2400–2500 MHz ISM band.

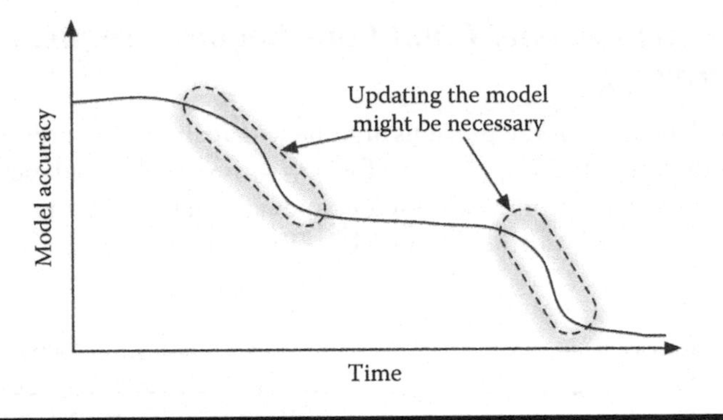

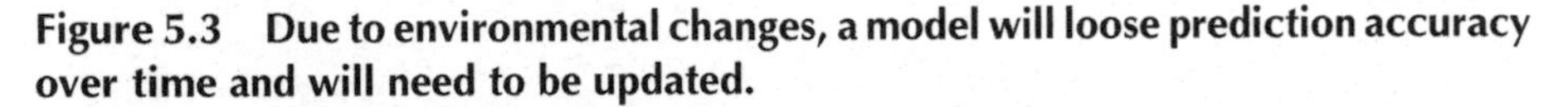

Figure 5.3 Due to environmental changes, a model will loose prediction accuracy over time and will need to be updated.

ISM band. These bands were monitored over a 24 hour period, and each colored dot represents the average noise floor on a particular frequency in three minute intervals. Each band was measured with 501 data points, so that the measurements that were taken with 0.0678 and 0.2195 MHz windows, respectively, confirm the frequently cited observation that channel conditions are highly fluctuant and may rapidly change between low and high noise levels. This means that a cognitive radio system that has just finished experimental tests assessing the quality of the channel may soon need to rerun those experiments as the channel conditions could have dramatically changed within short periods of time.

This time variance will not only be an issue for a cognitive radio system basing on experimental assessments of its environment, but will likewise affect systems that are using theoretical prediction models to select configurations. Independent of the actual approach to collect and store data about the environmental conditions of the radio and the impact of its parameters, the information encoded in the model is actually only predicting the behavior of the system correctly in the exact situation in which it was collected and derived earlier. Any environmental derivation or state change within the cognitive radio itself will degrade the prediction accuracy of the model and may lead to suboptimal, if not false, adaptation decisions. Figure 5.3 visualizes this issue schematically. As over time the system's setup and environment develops, the prediction capabilities of the model decrease. This performance loss will be more dramatic during times of large and dynamic change, during which actions should be taken to update the prediction model of the system to incorporate the most recent environmental change and maintain the cognitive radio's ability to make correct configuration decisions.

5.1.2 Why Adaptation Searches Can Be Expensive

This section elaborates on the argument that adaptation searches can be expensive and describes a selection of current, state-of-the-art cognitive radio control

Table 5.1 Overview of Different Cognitive Radio Control Algorithms and the Configurations Evaluated

Methodology	*Factors Explored*	*Type of Evaluation*	*# of Factors → # of Evaluations*
Genetic algorithm [3]	Power, frequency, pulse shape, symbol rate, modulation	Fitness func./ reconfiguration	5 → 1500
Game theory [4]	Frequency	Utility function	1 → 120 (*)
DOE [5]	Power, frequency, MTU, FEC, SelQ, data rate	Reconfiguration	6 → 288
RSM [6]	QoS architecture, routing and MAC protocol, load, speed of moving nodes	Reconfiguration	5 → 240

algorithms. For each of these algorithms, the complexity of search is investigated to show the computational expenses generated by each system during the adaptation process.

A summary of all algorithms discussed in the section is presented in Table 5.1. This table lists the general search methodology used for each contribution along with all factors that were the degrees of freedom in the search. As every parameter configuration can be evaluated in many ways, for example, by reconfiguring the radio and running a short test suite or determining the configuration using a fitness function, the table also specifies the type of evaluation used. To allow for a comparison of computational complexity, Table 5.1 also lists the number of evaluations necessary in the process of the search.

5.1.2.1 Genetic Algorithms

The first approach of cognitive radio control algorithms in prior works uses genetic algorithms to generate, evaluate, and select configurations. For its search and optimization process, the genetic algorithm creates a set of chromosomes where each chromosome encodes a certain parameter configuration. In each iteration, the genetic algorithm evaluates each chromosome according to its "fitness," thus creating a mapping between the configuration and a value determining its usefulness. After the evaluation, chromosomes with a higher utility are chosen overproportionally as parents of the next generation; additional features such as mutation and crossover introduce diversity and randomization to open up new directions for the search process. This process is repeated for a certain number of generations. Thus, the overall set of parameter configurations to be evaluated consists of the number of chromosomes multiplied by the number of generations the algorithms will execute.

In [3], Rondeau et al. use a genetic algorithm to select the optimal configuration among the five parameters: power, frequency, pulse shape, symbol rate, and signal modulation. Their search employs 30 chromosomes in parallel and terminates after 50 generations; thus a total of 1500 configuration evaluations need to be determined. These parameter settings however are evaluated off-line, that is, are evaluated using a fitness function and the radio is not being reconfigured to all 1500 configurations during the search. Still, all 1500 configurations are generated and need to be evaluated using a fitness function.

A similar approach using genetic algorithms is also done by Rieser et al. [7]; but in this report the authors do not comment about the number of chromosomes and generations utilized to find the most optimal configuration.

5.1.2.2 *Statistical Methods: Design of Experiments and Response Surface Methodology*

The second approach used to determine the configuration of a cognitive radio is applying the design of experiments (DOE) methodology to explore the parameter space. Here, the experimenter enumerates all possible parameter combinations and exhaustively tests the available parameter space. While the number of evaluations needed in this approach will increase exponentially, a search through DOE will eventually find the best configuration of a cognitive radio. More interesting in this approach, however, is that along with the best configuration, the DOE process will also explain why it is the best configuration, that is, which parameter settings contributed most to the success of this configuration that gives the experimenter further insights into the dynamics of the underlying system and can be used to create an abstract rule base to be used later by other cognitive radios to select the best possible configuration without continuously repeating the same set of experiments.

Vadde and Syiortiuk [6] estimate the effect of the QoS architecture, routing and MAC protocol, offered load and node mobility on average delay, real-time throughput, and total throughput. Their parameters can take between two and five levels, and they evaluate a total of 240 reconfigurations in a simulation environment by running benchmarks on each parameter configuration.

In a similar experiment, Weingart et al. [5] search through all settings of power, frequency, MTU size, forward-error correction (FEC), selective queueing (SelQ), and data rate to find the best possible configuration in each of their test scenarios. Because the number of levels assigned to those parameters can vary between two, three, or four settings, this leads to a total of 288 reconfigurations performed by the cognitive radio.

5.1.2.3 Game Theory

A third approach performs configuration management using game theory. In this class of previous work, each cognitive radio analyzes the choices for configuring its parameters using a utility function; therefore the utility function has to be called once for every permutation of parameters. The configuration period, however, is typically asynchronous, that is, each radio makes one choice at a time in a round-robin fashion. As the choices other players make typically influence one's own configuration thus creating the need for updating a prior configuration due to the changed environment, an equilibrium state is usually reached after a number of iterations. In these cases where the algorithm converges before reaching its maximum number of iterations, we note the number of evaluations with an asterisk at which the authors have determined a steady state experimentally.

In an experiment by Nie and Comaniciu [4], the authors formulate a utility function to assign frequencies in a network of cognitive radios. Each function evaluates its choice, that is, which frequency to transmit on, based on the maximal possible utility it can get according to the predefined utility function that is modeled so that it will have globally maximum SINR within the network. In their experiment, each radio can choose one of four possible frequencies and the algorithm will stop after a maximum of 120 iterations; however, the algorithm converges to a steady state after about 30 iterations. Thus, a total of 120 configurations need to be evaluated using a utility function.

5.1.2.4 Synopsis

In summary, it can be concluded that all available systems for the adaptation of a cognitive radio network conduct parameter searches that require extensive reconfiguration or large-scale evaluations using a utility function (which needs to be updated and maintained over time as discussed before). If each reconfiguration and subsequent test is assumed to only consume one second of airtime, this would mean that a cognitive radio assessing the environmental conditions experimentally using the DOE or response surface methodology would spend four minutes in a training period. These training times, which would need to be regularly repeated whenever the application context or environmental conditions have significantly changed, render the radio inoperable for actual payload traffic and block the channel currently being assessed for other stations in the network to be utilized as well, thus consuming a significant amount of resources for both the individual radio and the surrounding network.

This issue becomes even worse when viewed in context with the dynamic behavior of the channel conditions. As shown in Figure 5.2, the noise floor and the availability of a channel may vary drastically over a period of only a few minutes. In the worst case this can mean that a cognitive radio just having finished an adaptation search may need to redo the search over again, as the environment could have changed enough to render the previously obtained measurements useless, thus, in highly fluctuant environments, never converging to a stable solution that provides good performance over an extended period of time.

5.2 Reducing the Adaptation Search through Fractional Factorial Designs

As a means to address these challenges of adaptation search, this section describes how to efficiently configure cognitive radios using fractional factorial designs. To set the stage for the method, this section will first discuss the general reasoning behind this technique and then show, based on a literature survey of prior research in cognitive radios, that these prerequisites also apply for the area of wireless networks. Finally, an example of constructing a sparse parameter set using fractional factorial designs is presented.*

Consider a system with three input parameters A, B, and C. To analyze the behavior of such a system, an experimenter (or an automated system) can modify each of the input parameters and observe the output of the system. For simplification, in this discussion, it will be assumed that each input parameter can only be varied between a "high" and a "low" setting and that the system generates a single quantity as output that needs to be maximized. The experimenter is interested in a linear model that can explain and quantify the behavior of the system.

In the simplest case, assuming that every parameter is independent and there exist no interactions between the main variables, the system can be defined in a linear model as $r = w_A * f_a + w_B * f_b + w_C * f_c$, where the overall response of the system is only determined by the input values of A, B, and C (either on/off or low/high) and each input value contributes with a different weight w_i to the overall output of the system. However, if one imagines parameter A to model the power output of the cognitive radio and parameter B to be its data rate, it becomes evident that parameters A and B cannot be assumed to operate in isolation. Thus, the model needs to be populated with additional terms that account for any interactions

* See [8] for a detailed discussion of fractional factorial designs.

between any set of variables, thus

$$r = w_A * f_a + w_B * f_b + w_C * f_c \\ + w_{A:B} * f_{A:B} + w_{A:C} * f_{A:C} + w_{B:C} * f_{B:C} \\ + w_{A:B:C} * f_{A:B:C}$$

or in the general case:

$$r = w_0 + \sum_i w_i f_i \quad \forall i \in \text{set of all variable permutations}$$

5.2.1 Factor Interactions

To fully explore the behavior of the system, it will be necessary to choose all combinations of possible input variables for all three factors. For three parameters, this will require the generation and the evaluation of 2^3 inputs. After the evaluation, the system's reaction to all input parameters and all possible interactions between the input factors can be determined. Thus, the effects of A, B, C, A:B, A:C, B:C, and A:B:C on the output of the system can be derived. The number of parameter combinations to generate and test grows exponentially in this design (2^n), while the number of factor interactions that one is able to determine follows the binomial coefficient. For n number of input parameters, there will be

$$\binom{n}{k} = \frac{n!}{k!\,(n-k)!}$$

k factor interactions.

Thus, to determine the effect of power, frequency, data rate, MTU, FEC, SelQ, and modulation on the throughput of a cognitive radio, $2^7 = 128$ combinations would need to be evaluated. As a result, the linear model explaining the system's behavior could predict one average throughput together with seven main effects, 21 two-factor, 35 three-factor, 35 four-factor, 21 five-factor, 7 six-factor, and 1 seven-factor interactions.

Box et al. [8] note that higher-order factor interactions in most domains are typically not statistically significant, and even if they are significant they only contribute to a very small fraction of the system's overall response. It therefore makes economical sense to design these experiments in such a way that it is possible to obtain first estimates about the influence and significance of the factors by running only a fraction of the experiments necessary for a full analysis. If such a prescreening reveals further need for investigation, a fractional factorial design can be scaled up to gain precision by adding experiments to the previously ran set. Because in the configuration of cognitive radio networks one is typically interested in a summary of the rough magnitude of impact of the factors instead of an exact regression model, determining cognitive radio configurations can benefit from using fractional factorial designs.

5.2.2 Factor Interactions in Wireless Networks

A review of prior studies in wireless networks also confirms the general observation that most multifactor interactions are not distinguishable from noise and that typically only main effects and two- or three-factor interactions are statistically significant.

In the analysis of the experiments run by the DOE approach, Weingart et al. [9] report that only four main factors and five two-factor interactions were statistically significant in an ANOVA analysis to explain the average throughput of a cognitive radio. In an additional regression model to explain throughput, a selection of main factors and two-factor interaction achieved near perfect explanatory precision ($r^2 = 0.97$).

Vadde et al. [10] analyzed factor interactions for throughput and latency based on the response surface method. All response functions explaining average throughput and average delay contained only main factors and two-factor interactions that were statistically significant, besides one model containing a three-factor interaction as well. The three models also achieved high to near perfect precision ($r^2 = \{0.90, 0.77, 0.97\}$).

Similar results were obtained for an earlier study of mobile ad hoc networks by Vadde and Syrotiuk [6]. When determining average latency, real-time throughput, and total throughput for a network of 30 mobile nodes, all factor interactions of more than three factors proved not to be statistically significant. When it comes to lower-order factor interactions, in a model for average latency, main factors and two-factor interactions contributed most to the explanatory power with the single three-factor interaction only accounting for 3 percent of the average delay. Real-time throughput was also dominated by main and two-factor interactions. Here, the single statistically significant three-factor interaction accounted for about 1 percent of the observed real-time throughput and for the model explaining total throughput no statistically significant three-factor interaction was present at all.

Thus, neglecting higher-order factor interactions for an initial screening of factors through fractional factorial designs can also be assumed as safe in the area of wireless networks. This issue will be further addressed and discussed in Section 5.3 where the amount of precision lost when not accounting for interactions of a higher order is quantified in a model study.

5.2.3 Creating Sparse Designs

To visualize how such sparse designs can be created, this section will extend the example from Section 5.2.1, but use five parameters instead of three. A full factorial design, which is shown in Figure 5.4a, would consist of $2^5 = 32$ experiments for parameters with two settings each. If each parameter configuration is evaluated by either reconfiguring and initiating a test transmission or by evaluating the configuration through a utility or fitness function, the cognitive radio can get a value of utility y_i, such as throughput, latency, and bit-error rate, associated with each

#	A	B	C	D	E
1	–	–	–	–	–
2	+	–	–	–	–
3	–	+	–	–	–
4	+	+	–	–	–
5	–	–	+	–	–
6	+	–	+	–	–
7	–	+	+	–	–
8	+	+	+	–	–
9	–	–	–	+	–
10	+	–	–	+	–
11	–	+	–	+	–
12	+	+	–	+	–
13	–	–	+	+	–
14	+	–	+	+	–
15	–	+	+	+	–
16	+	+	+	+	–
17	–	–	–	–	+
18	+	–	–	–	+
19	–	+	–	–	+
20	+	+	–	–	+
21	–	–	+	–	+
22	+	–	+	–	+
23	–	+	+	–	+
24	+	+	+	–	+
25	–	–	–	+	+
26	+	–	–	+	+
27	–	+	–	+	+
28	+	+	–	+	+
29	–	–	+	+	+
30	+	–	+	+	+
31	–	+	+	+	+
32	+	+	+	+	+

(a)

#	A	B	C	D	E
1	–	–	–	–	+
2	+	–	–	–	–
3	–	+	–	–	–
4	+	+	–	–	+
5	–	–	+	–	–
6	+	–	+	–	+
7	–	+	+	–	+
8	+	+	+	–	–
9	–	–	–	+	–
10	+	–	–	+	+
11	–	+	–	+	+
12	+	+	–	+	–
13	–	–	+	+	+
14	+	–	+	+	–
15	–	+	+	+	–
16	+	+	+	+	+

(b)

#	A	B	C	D	E
1	–	–	–	+	+
2	+	–	–	–	–
3	–	+	–	–	+
4	+	+	–	+	–
5	–	–	+	+	–
6	+	–	+	–	+
7	–	+	+	–	–
8	+	+	+	+	+

(c)

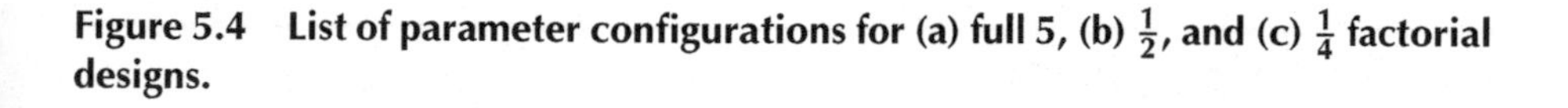

Figure 5.4 List of parameter configurations for (a) full 5, (b) $\frac{1}{2}$, and (c) $\frac{1}{4}$ factorial designs.

parameter configuration i. These results can then be used to estimate the effects of each factor and every possible factor combination, thus it is possible to analyze the overall average and the effects of A, B, C, D, E, A:B, A:C, A:D, A:E, B:C, B:D, B:E, C:D, C:E, D:E, A:B:C, A:B:D, A:B:E, A:C:D, A:C:E, A:D:E, B:C:D, B:C:E, C:D:E, A:B:C:D, A:B:D:E, B:C:D:E, A:B:C:E, A:C:D:E, and A:B: C:D:E on the system's output.

To determine the net effect of A, for example, one would multiply the utility value of each of the 32 runs by A's sign in that particular column and divide it by 16, the total number of positive signs. Thus, the net effect of A $= (-y_1 + y_2 - y_3 + y_4 - \cdots + y_{32})/16$. This approach works analogously for all factor interactions, for example, A:B $= ((-1)(-1)y_1 + (+1)(-1)y_2 + (-1)(+1)y_3 + \cdots + (+1)(+1)y_{32})/16$. This method of generating columns that encode inter-factor interactions is further visualized in the example in Figure 5.5.

As noted earlier, it is very likely that most higher-order factor interactions, especially those involving more than two factors, are either statistically not significant or contribute only a small fraction to the overall effect. For the task of reconfiguring, a cognitive radio is trying to pick all "low hanging fruit" first (i.e., find and optimize those factors that make the largest impact as early as possible), it makes sense to modify the design to investigate main and lower-order factors first.

To achieve a first reduction by half, a $\frac{1}{2}$ factorial design, as shown in Figure 5.4b, will be created. As in the example in Figure 5.4a, the factors A, B, C, and D still follow a full factorial design, but factor E was replaced by the procedure of multiplying the signs of columns that was used to calculate the inter-factor interactions above. As there are now four independent variables, the factor E can be set to the multiple of ABCD [8, ch. 12], that is, E $= (-1)(-1)(-1)(-1) = +1$ in row 1. By evaluating 16 instead of 32 parameter configurations, a factor 2 speedup is gained, but this introduced some confounding of variables.

Figure 5.5 shows which precision was lost in return. As you will notice, the signs for the inter-factor correlation ABC and the inter-factor correlation DE are identical. This is due to the fact that column E was generated through a linear

A	B	C	D	E	A:B	...	D:E	A:B:C
−	−	−	−	+	+	...	−	−
+	−	−	−	−	−	...	+	+
−	+	−	−	−	−	...	+	+
+	+	−	−	+	+	...	−	−
−	−	+	−	−	+	...	+	+
+	−	+	−	+	−	...	−	−
−	+	+	−	+	−	...	−	−
+	+	+	−	−	+	...	+	+

Figure 5.5 Higher-order factor interactions.

combination of other variables, E = ABCD. Thus, the main effect E is now confounded with the effect of the four-factor interaction A:B:C:D, which is expected to be not distinguishable from noise. This generator E = ABCD further declares which other variables are now confounded due to the speedup. By multiplying each side by D,* the generator can be transformed to DE = ABC. Thus, the two-factor interaction D:E was also confounded with the three-factor interaction A:B:C. Similar confounding interactions of this generator are AE = BCD, BE = ACD, and CE = ABD. These interactions are referred to as "aliases," as the interaction B:C:D will be exactly the same as A:E, for example, thus being its alias.

By using the $\frac{1}{2}$ fractional design in Figure 5.4b instead of the full design in Figure 5.4a, the number of configurations that need to be evaluated is reduced by a factor of 2, while it is still possible to estimate the average and the magnitude and direction of five main factors and 10 two-factor interactions. At this point, however, the effect of a set of three-factor interactions becomes indistinguishable from their corresponding two-factor interactions, but this is taken into account knowingly based on the assumption that the magnitude of higher-order factor interactions is likely to be either low or statistically not significant. Knowing how much explanatory power each of the main factors and confounded factor interactions account for, the control algorithm can configure the cognitive radio based on these results (the magnitude will guide the system to tune the most important parameters first and the sign will show which direction to turn them to) and create additional configurations to differentiate between certain confounded interactions of interest if they show statistical significance.

Because each fractional factorial design is a true subset of a full factorial design,† running a fractional design first to estimate the most important factors will not loose any information or create additional overhead even if the control algorithm decides later on to scale the experiments up to a full factorial design.

This methodology can also be repeated to create smaller designs, for example, the $\frac{1}{4}$ fractional design shown in Figure 5.4c. Here, the generators D = AB and E = AC were used to estimate the five main factors and their inter-factor interactions with a total of eight runs. These generators then also determine which inter-factor correlations will be confounded.

Based on the number of factors involved, one can create fractional factorial designs that will confound only factors of a certain order with each other, for example, designs where only factor interactions of order three or higher are subject to confounding (which according to previous work are rarely statistically significant). The type of fraction to use in practice will depend upon the number of factors subject to evaluation and the level of confounding one is willing to take into account to achieve a $\frac{1}{2^i}$ reduction of elements to evaluate.

---

* $D^2 = (\pm 1)^2 = 1$, so D can be omitted on the right side of the equation.

† Note that row 1 in (b) equals row 17 in (a), 2(b) = 2(a), 3(b) = 3(a), 4(b) = 20(a), etc.

In a deployment, the fractional factorial design could be used either as a standalone technique or as a pre-filtering device for other control algorithms as for example the DOE, response-surface method, or game-theoretic approaches. In the former case, it could be either used after a full factorial design to do repeated, expedited online training or if proper generators have been chosen up front to speed up a complete training. In the latter case, it would select those experimental configurations that could support a prescreening according to the fractional factorial design methodology for immediate evaluation, while deferring others for later when additional precision is needed. Thus, this methodology can be used to enhance and speed up other control algorithms as well.

5.3 Applicability of Fractional Factorial Designs for Cognitive Radio Networks: A Proof of Concept

This section demonstrates the utility of fractional factorial designs for optimizing the parameter search of a cognitive radio and provides a proof of concept implementation and evaluation by comparing the speedup and the loss of accuracy of $\frac{1}{2}$ and $\frac{1}{4}$ factorial designs in contrast to a full factorial design. This evaluation is based on the setup described in [9], which is shown in Figure 5.6: the cognitive radio placed in the middle is sending a large packet data stream (1400 byte packets at 384 kbps) and a VoIP stream (120 byte packets at 96 kbps) to the server on the left. The jammer on the right is transmitting continuously at the same frequency with constant power. To determine the best configuration in this scenario, the cognitive radio will alternate through all configurations and complete a six minute test on each configuration to measure overall throughput and latency. To measure the effect on throughput and latency of each parameter, each configuration is fixed during each test run. The experiment contained seven binary factors that could be changed between the low/high levels as shown in Figure 5.7.

Note that the use of a simulation instead of a hardware implementation is intentional and beneficial to this evaluation: As this evaluation is determining the usefulness of a statistical analysis method, using a simulation will create data that is reproducible and clear from any environmental influence introduced by the RF environment or the hardware equipment. That way, any loss in accuracy between the

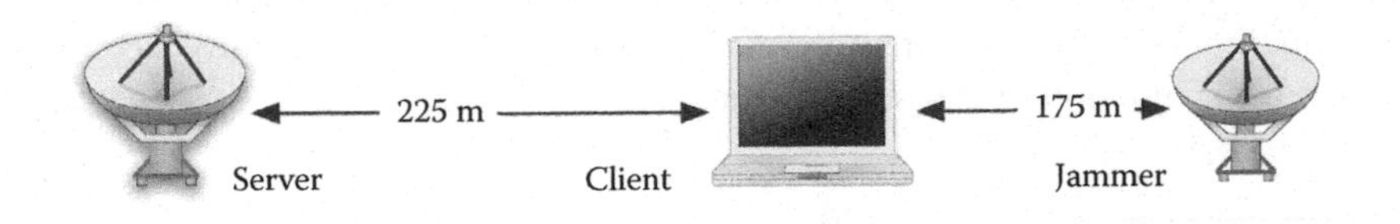

Figure 5.6 Setup for the evaluation.

Factor 1: selective queueing = {off, on}
Factor 2: ARQ = {off, on}
Factor 3: frame size = {2048 bits, 18432 bits}
Factor 4: data rate = {1 Mbps, 11 Mbps}
Factor 5: power = {5 mW, 100 mW}
Factor 6: FEC = {off, on}
Factor 7: jammer power = {1.2 W, 36 mW}

Figure 5.7 Parameters along with their corresponding low/high configuration values.

different factorial designs can entirely be accounted for to that particular fractional design in question and not to any fluctuations in the environment.

Full factorial design: For the full factorial analysis, all 2^7 combinations are evaluated. Table 5.2 shows the estimates of effects on throughput for the full factorial as well as the half and quarter factorial design. The analysis identified four statistically significant variables for throughput ($p < 0.01$), which are set in bold typeface and are those factors in Figure 5.8b with a standard deviation larger than 0.5. For Table 5.2 and this analysis, all interactions of three or more factors are omitted as these 102 higher-order factor interactions together accounted for less than 2 percent of the total variance.

Half factorial design: The half factorial was created according to the procedure in Section 5.2.3. A full factorial design was generated for the factors 1 through 6, and where factor 7 is chosen to be the dependent variable, as a multiple of the signs of the six independent factors: $7 = 123456$. For this generator, 64 runs were generated and evaluated.

Quarter factorial design: For the quarter factorial design, factors 4 and 6 were picked to be the dependents. A full factorial design on the variables 1, 2, 3, 5, and 7 was generated, thus creating 32 runs. The generators were chosen to be $6 = 135$ and $4 = 237$.

Comparison: As a performance evaluation, the cognitive radio system estimated the accuracy of half and quarter factorial designs in comparison to the results from a full factorial design for the two target variables overall throughput and latency. These results are discussed in detail in the following sections.

5.3.1 Estimating Throughput

Table 5.2 lists the estimated effects on throughput of main and two-factor interactions for the full design together with a $\frac{1}{2}$ and $\frac{1}{4}$ factorial design. These effects are also plotted in Figure 5.8a. As shown in the figure, throughput is dominated by the influences of four main effects and only two two-factor interactions. The

Table 5.2 Estimated Factor Effects on Throughput in Terms of Explained Variance for Full, Half, and Quarter Factorial Designs

Factor	*Full*	*Half*	*Quarter*	*Alias to*
1	0.2	0.2	0.1	
2	0.0	0.0	0.0	
3	0.1	0.1	0.1	
4	0.5	0.5	0.1	
5	**89.8**	**85.1**	**79.4**	
6	0.7	0.7	0.1	
7	**1.6**	**1.5**	**0.6**	
1:2	0.0	0.0	0.1	
1:3	0.0	0.0	0.9	
1:4	0.0	0.0	0.4	
1:5	0.0	0.0	0.2	
1:6	0.0	0.0	0.0	
1:7	0.0	0.0	0.1	
2:3	0.0	0.0	0.2	
2:4	0.0	0.0	0.0	
2:5	0.0	0.0	0.0	
2:6	0.0	0.0	0.0	
2:7	0.0	0.0	0.0	
3:4	0.0	0.0	—	2:7
3:5	0.0	0.0	—	1:6
3:6	0.0	0.0	—	1:5
3:7	0.0	0.0	—	2:4
4:5	0.8	0.8	1.8	
4:6	0.0	0.0	0.3	2:3

Table 5.2 (continued) Estimated Factor Effects on Throughput in Terms of Explained Variance for Full, Half, and Quarter Factorial Designs

Factor	*Full*	*Half*	*Quarter*	*Alias to*
4:7	0.0	0.0	—	
5:6	**1.9**	**1.8**	—	1:3
5:7	0.0	0.0	0.0	1:5
6:7	**2.1**	**2.0**	**3.4**	
$\sum$	98.5	93.3	90.1	
r^2	0.99	0.99	0.99	

throughput of the cognitive radio is almost exclusively determined by its own power level (factor 5), secondary whether in the presence of a highly jamming node the cognitive radio uses FEC (factor 6:7), tertiary whether it uses FEC in conjunction with a high transmission power (factor 5:6), and only quaternary dependent on the power level of the jamming node (factor 7). Note that both the half and quarter factorial designs return the same results in terms of the importance of factors and their relative impact of the change. Thus, after running any of the factorial designs the control algorithm knows which of the cognitive radio's knobs to turn in which direction and what effect it can assume to get.

Evaluating a half factorial instead of a full factorial design would give the cognitive radio a factor 2 speedup, as instead of 128 only 64 evaluations needed to be computed. The direction and relative magnitude of all main effects and two-factor interactions was estimated correctly by the half factorial design, that is, providing equal results as the full factorial design as can be seen in Figure 5.8a. The only difference to this rule is effects that are not distinguishable due to confounding from noise, for example, factors 2, 1:4, or 2:5. Here, the analysis on the reduced data set did introduce errors that however are still part of the noise. The total estimate came within 0.005 percent of the estimate from the full factorial design.

Evaluating a quarter factorial instead of a full factorial design would provide the radio with a factor 4 speedup, as only 32 instead of 128 parameter configurations are part of the test. As for the half-factorial design, the relative magnitude and direction of all main effects and all two-factor effects that were statistically significant in the full and half design matched the results from the quarter factorial design. However, those effects that were not distinguishable from noise were also overestimated by the half factorial design, especially factor interaction 1:3. Due to these errors

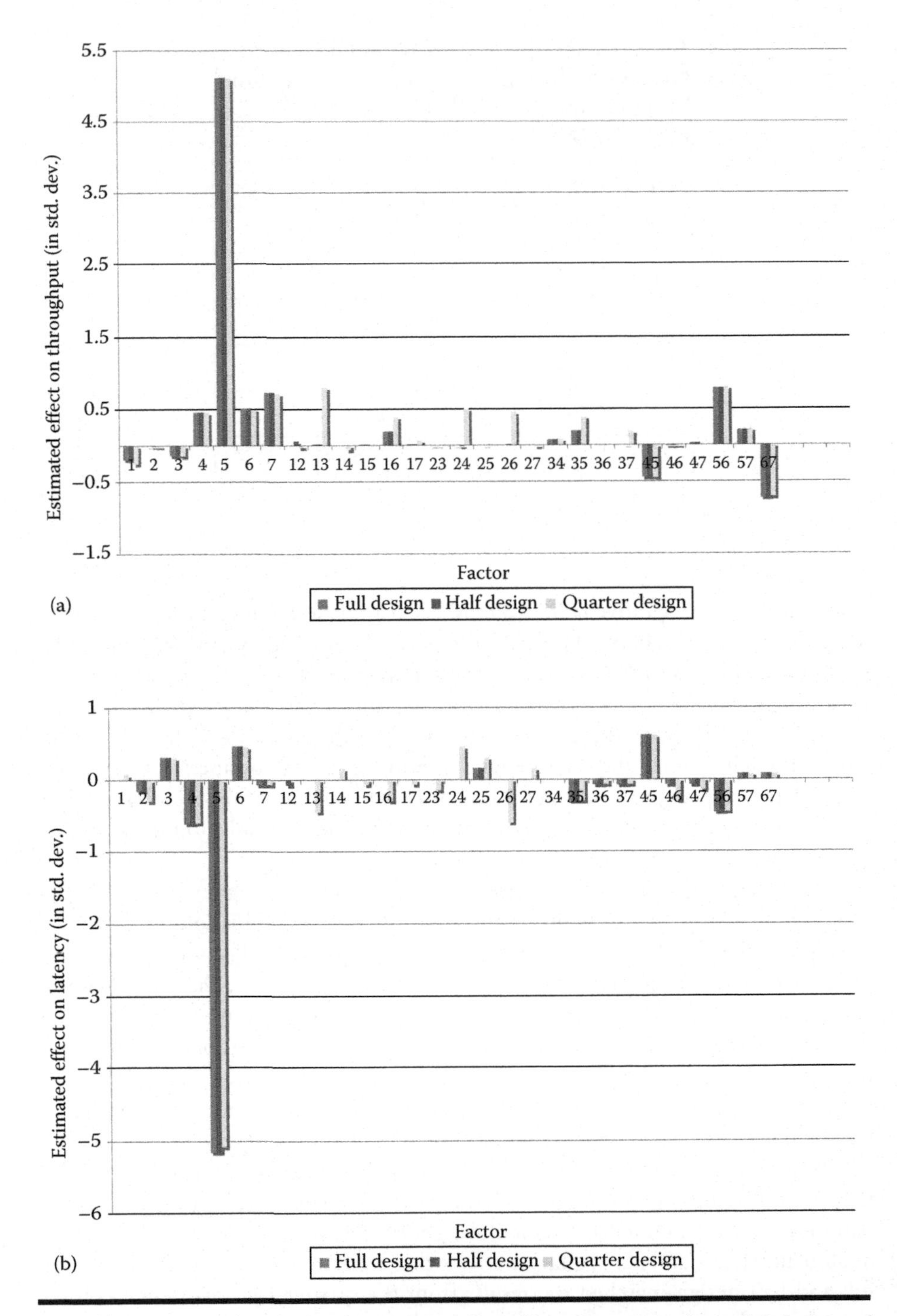

Figure 5.8 Estimated effects between full, half, and quarter factorial design of factors (in standard deviations). (a) On throughput and (b) on latency.

in noisy factor interactions, the total estimate was within 28 percent of the estimates of the full factor design while still predicting all statistically significant factors correctly.

If the general magnitude of the values are taken into account, the accuracy of the estimates grows. If considering effects bigger than 0.1 standard deviations, a half factorial design estimates those factors with only 0.0009 percent deviation and a quarter factorial design within 29 percent deviation. For factors larger than 0.2 standard deviations, half factor and quarter factor designs score within 0.0008 percent and 0.0003 percent, respectively, of the full factor design. Factors bigger than 0.5 standard deviations are estimated correctly within 0.0004 percent and 0.0005 percent for the two designs.

5.3.2 Estimating Latency

In a second analysis the study is replicated to determine the accuracy of a half and quarter design to predict latency in comparison to a full factorial design. These effects are plotted in Figure 5.8b.

As for the previous evaluation of throughput, the same observations hold for the accuracy of fractional factorial designs on the estimates of latency. More specifically, latency is also primarily dominated by main effects and few two-factor interactions, more specifically 3, 4, 5, 6, 4:5, and 5:6. For all factors, the half factorial design estimated both the general direction of the factors' influences and their overall magnitude very precisely. Using a half factorial design would estimate the factors within 0.01 percent of the target values obtained by a full factorial design. This accuracy would also grow when only considering factors of a certain strength, thus eliminating statistically nonsignificant factors as part of the noise. When only considering factors larger than 0.1 standard deviations, its estimates are within 0.005 percent and for factors larger than 0.2 and 0.5 standard deviations, a half factorial design would gain an additional order of magnitude in accuracy and score within 0.0005 percent of the full factorial design while only evaluating 50 percent of its samples.

Similar performance results were obtained for the quarter factorial design. As for the estimates on throughput, a quarter factorial design estimated correctly the general direction and the magnitude of all statistically significant factors. As it is only evaluating 25 percent of the samples of the full factorial design, it did overestimate the effect of those factors that were part of the noise and thus scored within 21 percent of the full factorial design. When only factors bigger than a certain threshold were taken into account, this also increased its overall accuracy for the quarter design. For factors bigger than 0.1 standard deviations, accuracy was within 21 percent, when considering only factors bigger than 0.2 standard deviations, a quarter factorial design estimated the factors within 0.014 percent of the estimates of a full factorial design.

5.4 Automatic Design and Implementation of Fractional Factorial Designs: The Rapid Adaptation Architecture

After the general feasibility of using fractional factorial designs to improve cognitive radio adaptation was demonstrated, this section will now focus on an automatic application of this concept within a cognitive radio system. The system presented in this section, the rapid adaptation architecture, can be used in conjunction with existing control algorithms and will operate as a statistical pre- and post-processor to enhance and speed up the search of the control algorithm for a feasible adaptation. This section first presents an architectural overview of the rapid adaptation architecture and discusses in what ways it can be used to enhance existing cognitive radio control algorithms, showing its minimal impact design and describes three ways by which the system can speed up the adaptation process of a cognitive radio algorithm.

5.4.1 General System Architecture

Figure 5.9 shows the overall system architecture from a high-level overview: part (a) of the figure depicts the current situation where the cognitive radio is directly adapting the hardware without the use of the rapid adaptation architecture and part (b) displays the situation where a cognitive radio uses the system to increase its adaptation performance.

As can been seen in the figure, in the traditional cognitive radio design, the control algorithm is embedded inside the network stack, where it can monitor the characteristics of incoming and outgoing traffic flows and correspondingly adapt the system's parameters to any occurring changes. After the deployment and repeatedly afterward during its normal operation, the control algorithm will

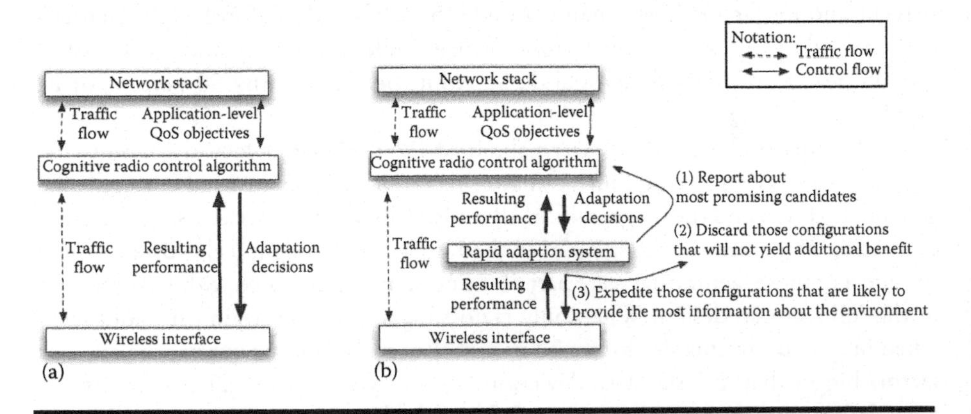

Figure 5.9 Cognitive radio adaptation (a) with and (b) without the rapid adaptation system.

need to perform link tests to learn which parameter configurations will currently provide the most utility. To get this information, the control algorithm will issue a set of reconfiguration instructions to the wireless interface, run link tests for each configuration, and collect the resulting performance statistics afterward.

In the system using the rapid adaptation architecture (Figure 5.9b), the general components and their tasks stay exactly the same, except that the rapid adaptation system is inserted in between the control algorithm and the wireless interface. To minimize or even completely eliminate the need for modifications, the rapid adaptation system works as a transparent layer to both the control algorithm and the wireless interface: to the control algorithm it will look and behave like a wireless interface, receiving adaptation instructions and reporting back performance statistics, and to the wireless interface the rapid adaptation system will look like a control algorithm, issuing adaptation decisions and receiving performance characteristics using the same interfaces and protocols as the original system. As all interfaces and protocols remain the same, the layer can directly be inserted into a cognitive radio without making any changes necessary. Once deployed inside the stack it will intercept both adaptation instructions and performance statistics, automatically perform a statistical analysis in the background, and design and execute adaptation improvements.

5.4.2 Enhancing the Adaptation Search

The rapid adaptation system can provide three different services to the cognitive radio control algorithm as can be seen in Figure 5.9b:

- First, the system can create recommendations for the control algorithm about the most promising candidate configurations that the control algorithm should explore first. This recommendation list can be used by the control algorithm to prioritize or weight more promising configurations higher and thus increase the rate of performance improvements in each iteration.
- Second, when the control algorithm has generated a set of parameter configurations that need to be evaluated (e.g., the DOE approach has enumerated all parameter combinations), the rapid adaptation system can discard those configurations that are based on previous adaptation results and will not provide additional benefit or information. This will slim down the total amount of configurations that are to be tested and reduces the total training time.
- Third, when a (reduced) set of candidate configurations finally needs to be evaluated through the wireless interface, the rapid adaptation system will dynamically reorder the candidate configurations before proctoring each link test. More specifically, it can expedite those configurations that are likely to provide most of the information about the environmental conditions and hold back those that will most likely provide little more information at this point. This dynamic reordering process is driven by a mathematical

and statistical analysis of the candidate configurations and previously learned results; formally speaking the rapid adaptation system will prioritize those configurations that will create a vector basis for the adaptation search space and hold back those configurations that differ only in few respects to previously tested candidates.

The exact functioning of these three services and how they might be used to enhance current cognitive radio control algorithm will be explained in the following based on the example of the genetic algorithm and DOE methodology.

5.4.2.1 Feature 1: Directing the Adaptation Search

A variety of current cognitive radio control algorithms make limited, if any, use of previously learned information to direct their adaptation search. One algorithm that does use feedback on previous configuration performances is the genetic algorithm, but the rapid adaptation architecture can be used to improve the generation of new candidate configurations.

At a high level, the genetic algorithm works as follows. First, a set of candidate configurations ("chromosomes") is created using a random generator. Each of these candidate configurations is then evaluated in terms of its utility using a so-called fitness function. For the next iteration of the algorithm, a new list is derived based on the list evaluated in the last round, where the relative fitness influences the probability that a configuration will be reused for the new list. To avoid local maxima and to find alternative configurations not present in the previous generation, the algorithm lightly randomizes the elements in each iteration ("mutation" and "crossover"). The development of parameter configurations is therefore influenced by the previously obtained performance values, as highly scoring parameter configurations are more likely to be kept and further tested in the future.

With the use of the information automatically collected by the rapid adaptation architecture, this process, however, can be informed and guided even further. As the rapid adaptation system intends to identify those parameters and parameter interactions that are contributing most to the system's overall performance, the tool can directly identify which parts of the chromosome are contributing the most. This allows that not only the chromosome as a whole, but also the individual parts that are contributing most to the chromosome's success can be weighted.

Figure 5.10 visualizes this concept. Instead of viewing each chromosome as a black box and reusing the parameter configuration for the next generation proportional to its relative fitness, the statistical analysis of the rapid adaptation architecture can directly identify which parts of the chromosome yielded the most influence on the result, thus allowing the genetic algorithm to allocate resources for further development of the most significant parts of the configuration chromosome and arriving at higher performance levels earlier.

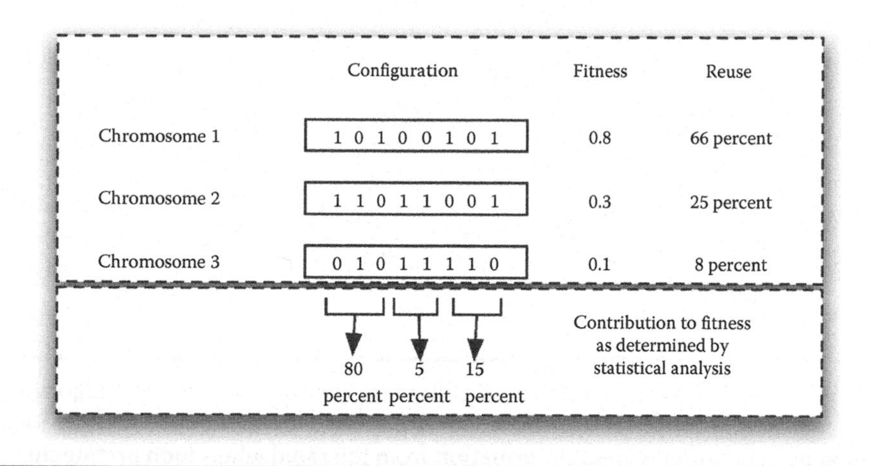

Figure 5.10 The statistical analysis performed by the rapid adaptation architecture can directly identify and account which parts of the chromosome are providing most to its overall success.

This recommendation capability of the system can also be used directly to (partially) generate candidate configurations that could lead to the highest performance improvements. As the tool knows which parameters are most important to the radio's success, this information can be directly taken into account by the genetic algorithm during the chromosome generation, as for example preset and fixed parts. This ability becomes even more important if there exist certain highly relevant factor interactions. Consider, for example, that the adaptation tool has identified that some high gain can be obtained if parameters A and C are both turned on simultaneously while neither A or C independently seem to have a large effect. Aware of these factor interactions, the genetic algorithm can be directed to overproportionally include such simultaneous activations and thus more efficiently optimize them by using relative fitness alone.

Figure 5.11 shows this functionality schematically. Instead of only weighting parameter configurations based on their relative fitness, the rapid adaptation system can propose a direction to search into, thus guiding the adaptation search more efficiently.

The performance improvement gained by this functionality will be further analyzed and evaluated in Section 5.5.

5.4.2.2 Feature 2: Removing Configurations without Information Gain

Certain currently used control algorithms may also contain some amount of redundancy, either within an iteration where a genetic algorithm, for example, might test two exactly same candidate configurations, or between iterations where the control

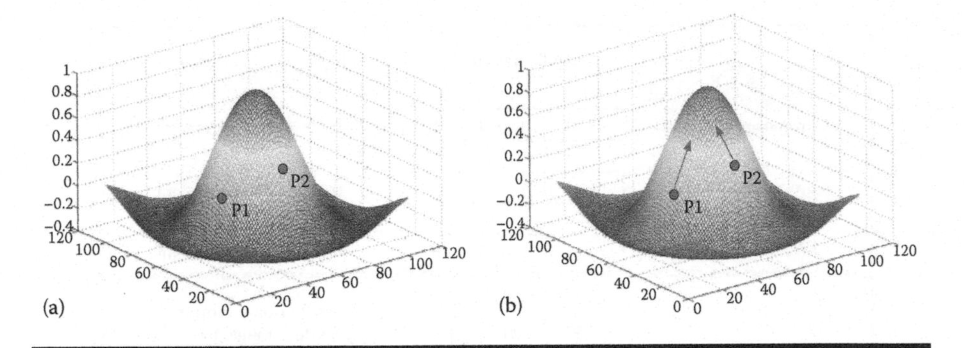

Figure 5.11 The rapid adaptation architecture provides the control algorithm with information on how to proceed in a parameter search. (a) Configuration candidate generation without information from the rapid adaptation architecture. (b) Configuration candidate generation with information from the rapid adaptation architecture.

algorithm might retest a configuration without the environment having changed. The second functionality of the rapid adaptation architecture is therefore to search for those configurations that will not likely provide any additional insight into the current environmental conditions, and remove them to save resources.

This removal can be done in two distinct ways: First, the rapid adaptation architecture may simply eliminate the configuration test before being executed by the hardware and leave it to the control algorithm to construct a model without this information. Second, the system may withhold the configuration and report back either cashed performance results that were obtained during a recent evaluation of that particular configuration or a theoretical estimate using the tool's previously collected information and its internal statistical model.

The performance improvement gained by this functionality will also be further quantified and evaluated in Section 5.5.

5.4.2.3 Feature 3: Expediting Promising Configurations

Of those configurations generated by the cognitive radio control algorithm that have passed the selection process of the rapid adaptation tool, not all will result in the same amount of information gain when being executed by the hardware. As the control algorithm typically does not depend on a specific order in which these disjoint configurations are being evaluated, but just needs to know the corresponding performance results for each configuration, the rapid adaptation tool has some additional opportunity to optimize this process.

Instead of testing the elements in a random order or according to the pattern used by the control algorithm to generate them, the tool internally creates a set of fractional factorial designs based on the number of parameters in the system and the maximum level of confounding the user is willing to accept. Based on these automatically calculated fractional factorial designs, the system reorders all configurations to be tested so that those corresponding to the fractional designs are evaluated first, thus ensuring that enough information to create a robust model is collected as soon as possible.

Figure 5.12 shows this process schematically for a system with four parameters. The control algorithm has created a series of configuration tests of which after removing redundancy 16 tests need to be performed by the wireless interface. Instead of executing these tests in the order they were received, the rapid adaptation system creates a half factorial design with the generator $D = ABC$ and concurrently a quarter factorial design with the generators $C = -B$ and $D = ABC$. Because the quarter factorial design is a true subset of the half factorial design, which itself is again a subset of all the configurations handed to the system by the control algorithm, the adaptation architecture creates no additional testing overhead if focusing first on the configurations contained in the fractional designs. It therefore reorders the complete list of configurations in such a way that first only those contained in the smallest factorial design (in this example the quarter factorial design, indicated in light gray) are evaluated, thus allowing a first assessment of the linear environmental model. After these initial tests, all configurations that are part of the next larger factorial design are evaluated (in this case, the half factorial design, as indicated in dark gray), which only requires a small incremental effort but still saves a significant amount of air time in comparison to the full list and allowing even more robust

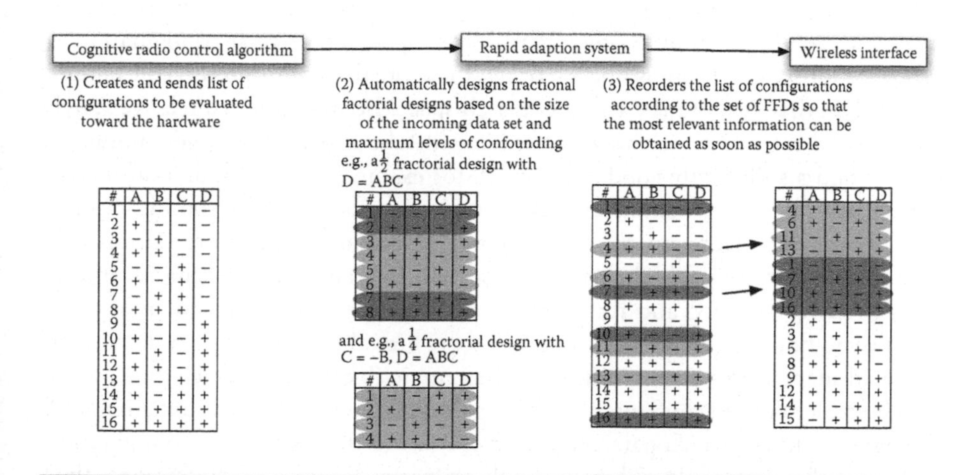

Figure 5.12 By reordering and expediting the subsets of configuration lists that correspond to a fractional factorial design, the rapid adaptation system can identify important factors earlier while not introducing additional configuration overhead.

system predictions. This scaling is continued until the entire list has been processed or the rapid adaptation architecture decides or is instructed to stop the evaluation.

The performance improvement gained by this functionality will also be further measured and evaluated in Section 5.5.

5.5 Enhancing Current Control Algorithms through the Rapid Adaptation Architecture

In this section, the performance gained through the three functionalities of the rapid adaptation architecture will be analyzed and evaluated. As genetic algorithms and the DOE approach have been frequently used for cognitive radio adaptation, these two standard algorithms will be equipped with the automatic analysis and improvement capabilities of the rapid adaptation system, and it will be quantified to what extent current state-of-the-art systems may be improved in terms of adaptation speed, accuracy, and resource consumption.

The evaluation setup for all three performance analyses is the same as the setup described in Section 5.3. The cognitive radio is internally structured according to the interaction diagram in Figure 5.9b, where for the first and second functionalities a genetic algorithm as described in [3] and for the third functionality the DOE algorithm as described in [5] is used.

5.5.1 Guiding the Adaptation Search

As discussed earlier, the genetic algorithm directs its search process based on the relative success of its chromosomes, that is, the higher a particular configuration as expressed through a chromosome is scoring, the more the algorithm will concentrate its search in that area of the parameter search space. Because it uses a relative search approach, there can be instances where it will take a relatively large amount of time to find a sufficiently good solution, if, for example, the configuration success depends on only a few parameters or parameter interactions and their specific settings. To enhance the adaptation search, the rapid adaptation architecture can provide additional insight to guide the parameter search. Instead of only basing its process on the relative ordering and thus relying on chance as encoded in mutations and crossover to find the configuration that will score best, the statistical preprocessing of the rapid adaptation system can instruct the genetic algorithm in which direction to search from its current configurations. Using the previously evaluated configurations, the system can find which parameters matter most to the success of a configuration and guide the control algorithm to search primarily along these dimensions, thus converging earlier and to higher results.

Figure 5.13 shows the best scoring parameter combination as well as the average configuration scores for the genetic algorithm using 50 chromosomes in parallel. The graph shows the development of the prediction of the genetic algorithm over

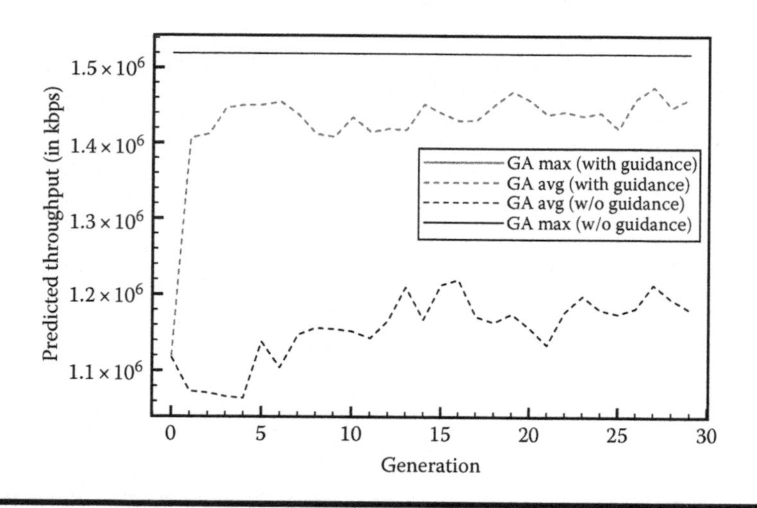

Figure 5.13 Prediction level (maximum/average) of a genetic algorithm with 50 chromosomes with and without guidance by the rapid adaptation system.

the course of its 30 generations; the red lines represent a genetic algorithm enhanced by the rapid adaptation tool and the black lines without its assistance. The evaluation was conducted in an off-line experiment where the hardware layer and fitness function that would have provided the performance results were replaced by a lookup mechanism that contained throughput measurements for all possible parameter combinations, thus allowing that the performance results were deterministic and all observed differences can be entirely accounted for to the search algorithms. While both systems identify the configuration yielding the maximal possible throughput, the average score of the unassisted control algorithm is generally lower. One can clearly observe the guidance of the statistical framework as during the first few generations the search is directed to the most important parameters, from which only small adjustments, finetuning the configurations, are being made, thus staying at a generally higher performance level. It is important to note, however, that in general (and ignoring its probabilistic nature) the genetic algorithm will also eventually converge to the maximum level as shown in Figure 5.14. This may, however, require an extensive amount of time, which is reduced through the direction of the rapid adaptation system.

This very large influence of guidance to the overall search success of the genetic algorithm therefore naturally raises the question of the system's robustness to noisy and wrong information. In other words, given that the guidance by the rapid adaptation architecture improves the average solution fidelity significantly, how will the system behave when the statistical preprocessing analyses noisy, erroneous data and therefore issues wrong guidance information to the control algorithm. Figure 5.15 displays the algorithm's robustness to noisy information over time and for varying noise levels.

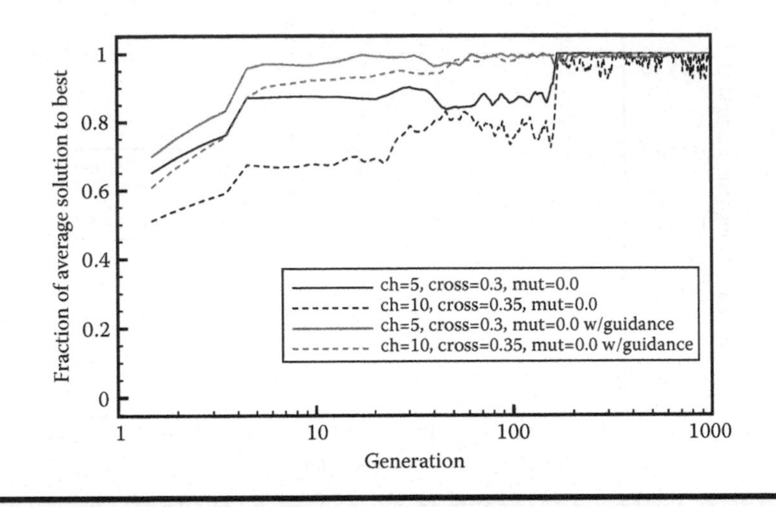

Figure 5.14 Long-term performance of a selection of guided and unguided genetic algorithms.

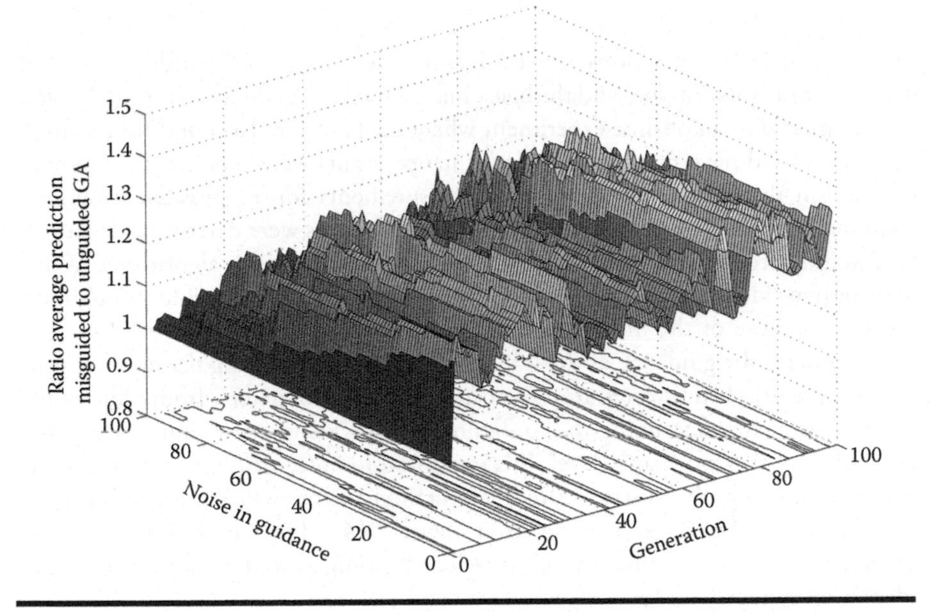

Figure 5.15 Impact of noise in guidance information on the average performance of a misguided genetic algorithm compared to an unguided genetic algorithm.

Confirming the findings seen earlier in Figure 5.13, Figure 5.15 also shows that the impact of guidance manifests early on within the first few iterations of the search and lead to a prolonged advantage yielding between 10 percent and 30 percent higher average findings. Once, however, the outcome of the statistical preprocessor is obfuscated with noise, the general average solution fidelity of the now misguided, compared to the previously correctly guided, genetic algorithm drops as much as 25 percent. Yet, the genetic algorithm guided with high levels of noise thus leading to (partially) wrong predictions still scores higher in most cases than a genetic control algorithm having no guidance at all. This impact is further visualized in Figure 5.16, which indicates that even moderate levels of random errors introduced in the data from the rapid adaptation architecture do not impact the overall performance of a guided genetic algorithm enough to fall behind an unguided system. The figure shows the number of configurations over 100 generations that a GA guided with wrong information will configure the cognitive radio parameter worse than a GA without any intuition (thus equalling a ratio <1 in Figure 5.15) replicated and averaged over 100 starting configurations. The data, however, also indicates that there exists a critical threshold after which the performance drastically degrades.

The reason for these high levels of robustness and the ability of the genetic algorithm to find the best possible configuration even without any guidance stems from two causes: first, the massive parallel exploration of the parameter search space and second, the internal relative weighting of the chromosomes between iterations.

The influence of the large-scale parallel exploration becomes visible in Figure 5.17. The figure displays the percentage of throughput achieved by the currently best-performing chromosome over the course of 30 generations, where the number of chromosomes working in parallel is varied between 1 and 10. As can be seen in the graph, the unguided control algorithm always scores significantly worse than the genetic algorithm possessing additional information. As soon as enough chromosomes are being used, however, ten in this example, the amount of parallel exploration is high enough to identify the best configuration, that is, with

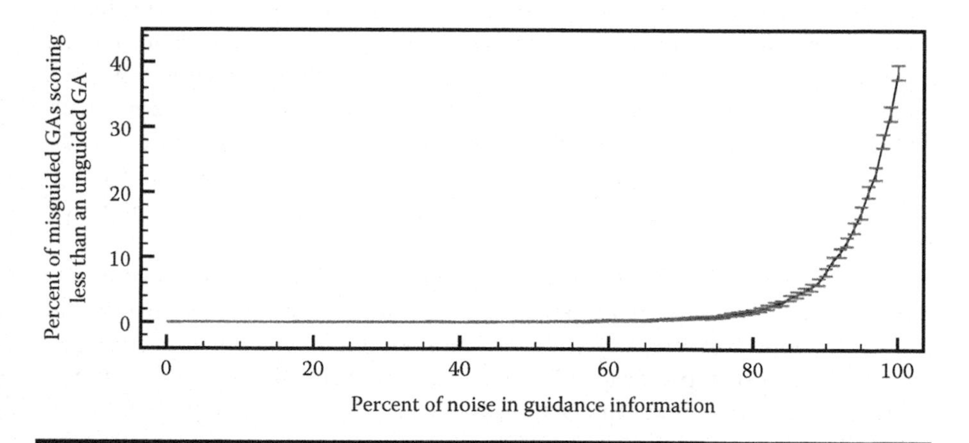

Figure 5.16 Percentage of less scoring iterations due to incorrect adaptation search information. Average number and error bars for 30 chromosomes with 100 replications.

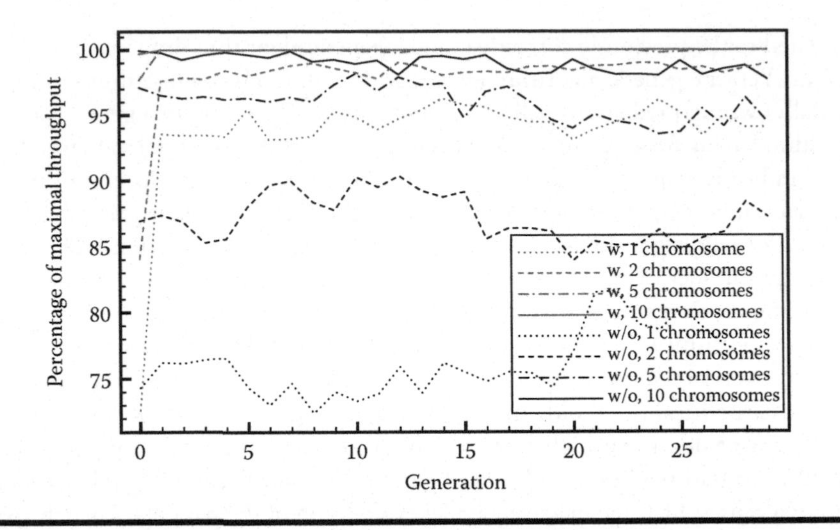

Figure 5.17 Maximal prediction of a genetic algorithm with varying chromosome sizes with and without guidance by the adaptation tool.

enough chromosomes and small enough search space there will likely exist a permutation that randomly selects the best configuration. In these cases, the statistical preprocessing can be used to reduce the number of chromosomes that need to be evaluated, thus reducing resource consumption, while still finding the significant variables in the set.

This is further aggravated by the fact that actually meaningful factor effects and factor interactions in the solution space are sparse, which is also the reason for the system's robustness to noise. As was shown in the literature survey in Section 5.2 and validated through the proof-of-concept implementation in Section 5.3, there exist only a few factors and factor interactions in the theoretically exponentially large search space that are statistically significant. Thus, in a practical system with a very limited number of statistically significant factor interactions, either a very large number will be needed to exhaustively cover the search space or some hints will required where to look first.

Figure 5.18 explains why in such a situation even highly noisy information can speed up the parameter search process. Without any background information, the genetic algorithm will spread its k probes (chromosomes) randomly over the entire d-dimensional search space, which is theoretically 2^d elements large. As only a small number of factors and factor interactions d_r are statistically relevant with $d_r \ll 2^d$, the likelihood to discover any one of these randomly (without guidance) is $\frac{d_r k}{2^d}$. As discussed earlier, this process is optimized through the rapid adaptation tool by steering the chromosome into the direction where the most important factor (interactions) is predicted based on previous data (see Figure 5.18a).

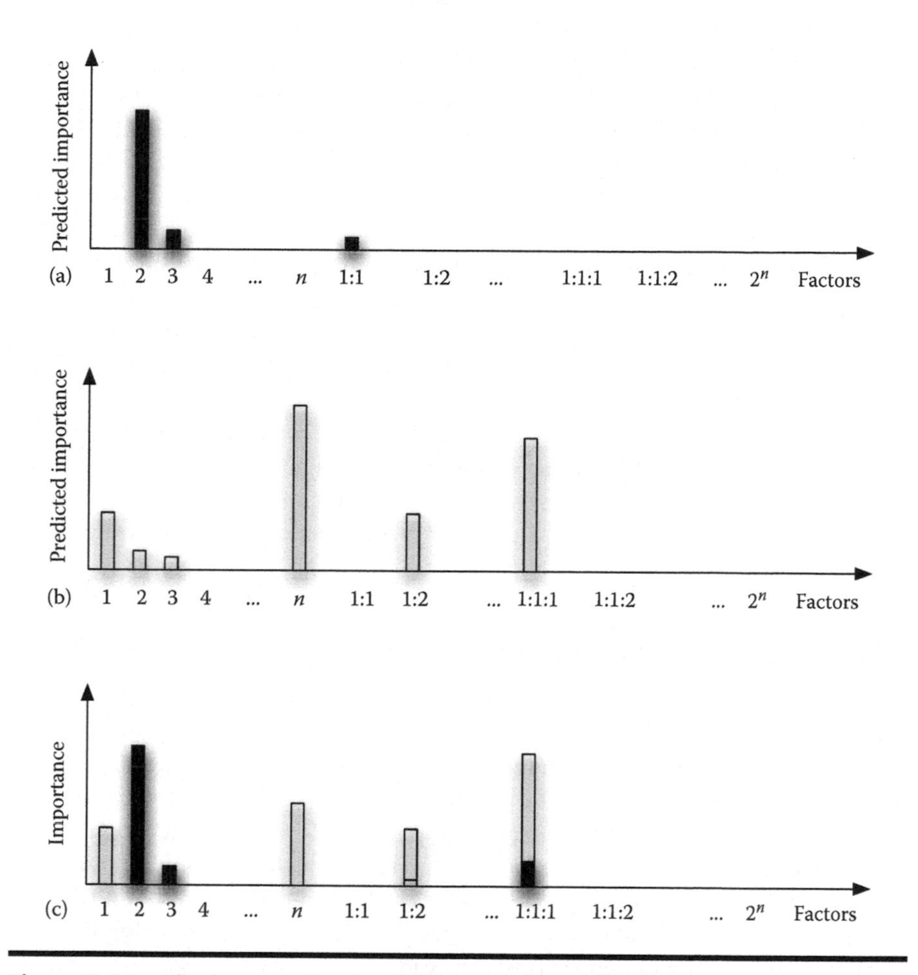

Figure 5.18 The impact of noisy guidance information is low through the combination of performance prediction by the rapid adaptation architecture and actual measurements by the genetic control algorithm. (a) Predicted performance in a noise-free situation. (b) Misguided performance predictions due to noise. (c) Overlay between predicted and actually achieved performance.

Once, however, the performance prediction is subject to large amounts of noise (Figure 5.18b), many insignificant factors will be estimated to have a large effect (similar to the issue of the amplification of noise in highly confounded factorial designs), while the effect of the truly important factors will also be misjudged and likely be underestimated. Yet, unless the noise will completely eliminate these true effects from the fractional factorial design calculation, these factors will still be flagged as significant, however to a much lesser degree. Due to the proportional weighting the enhanced GA will be biased into the wrong direction of the overestimated factors (factors n, 1:1:1, 1:2, and 1 in the Figure 5.18b) and some probes will explore the other, less-important regions of the parameter search space. Once in the next generation the chromosomes are reproduced based on their actually achieved performance (Figure 5.18c), the combination of performance prediction and actually achieved performance will direct the system toward the correct parameters for further exploration.

The sparseness of the statistically significant parameter space is therefore also one of the underlying reasons why even a misguided control algorithm does score better than an unguided one: in a situation where the genetic control algorithm possesses not enough probes to exhaustively cover the entire search space, statistical data obfuscated with 80 percent noise still direct a small share of chromosomes into the right direction. As the enhanced genetic algorithm then determines the further direction of the search based on both suggested direction and the relative fitness (see Figure 5.10), the small number of probes that did go into the right direction will dominate the fitness in a sparse search space high enough to let the algorithm converge to the right solution and still faster if the genetic algorithm did not have any information at all.

5.5.2 Removing Redundancy

The previous discussion also points out in which additional way the performance of cognitive radio control algorithms might be improved. Given that the genetic algorithm identifies the best configuration only if enough probes are being used in parallel, one can assume that the algorithm will repeatedly evaluate very similar or even the exactly same parameter configurations, even if limited change has taken place in the environment. This redundancy, which may only require dispensable evaluation of a fitness function or even dispensable configuration tests on the medium, can be eliminated by analyzing previous requests and using internal prediction and caching to remove this overhead.

Figure 5.19 quantifies the amount of redundancy created by the genetic algorithm for varying chromosome sizes and the number of generations during every single iteration of the algorithm. Even if a simple mechanism as caching is used, considerable amounts of evaluation may be saved and the system performance can therefore be accelerated. Additional efforts might even be saved if only a smaller amount of all configurations is being tested at any given time and for the remaining, untested, configurations predictions are made based on the internally used fractional factorial designs.

5.5.3 Faster Learning and Higher Model Robustness

The DOE approach was used to evaluate the gain of the third functionality of the rapid adaptation architecture, that is, the automatic reordering of configurations to be tested according to fractional factorial designs. As discussed before, for this functionality, the adaptation tool will create a set of fractional designs that are built

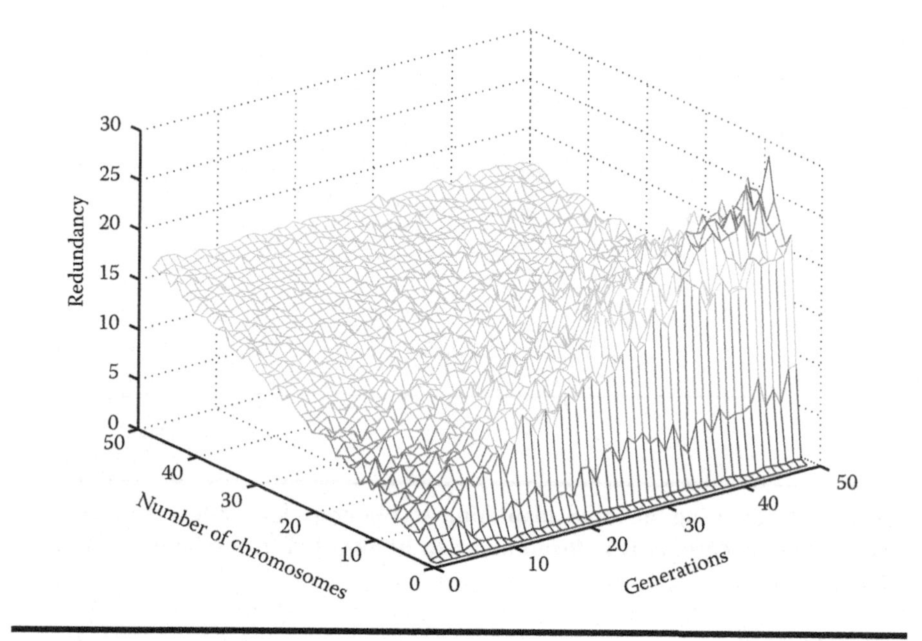

Figure 5.19 Redundancy of parameter evaluations of a genetic algorithm with varying chromosome sizes and generations.

on each other and use these cascading designs to select those configurations first that will likely give the highest information gain.

The benefit that the use of the rapid adaptation architecture will provide when combined with other algorithms, such as the DOE approach, is schematically visualized in Figure 5.20. At certain times, indicated as t_1 and t_2 in the figure, there will a rise the need to update the cognitive radio's linear model explaining the environment, due to sudden decreases in the model's accuracy. These drops in prediction accuracy result from environmental changes such as node mobility or usage changes and require that the cognitive radio's control algorithm initiates actions to adjust its computational model to incorporate these recent changes in the environmental conditions. These retraining events can be started when the control algorithm notices a certain deviation from the predicted system performance, say the model predicts a certain throughput/latency given the current configuration but the radio actually achieves only 80 percent of that predicted performance, as well as in regular intervals.

For the example of the DOE methodology, these retraining events require the evaluation of the cognitive radio's parameter search space using link tests, that is, for k parameters 2^k different configurations need to be evaluated through an actual experiment; during this time t_a the link is busy and cannot be used for the exchange of payload. After the retraining is complete and the model is updated to capture the

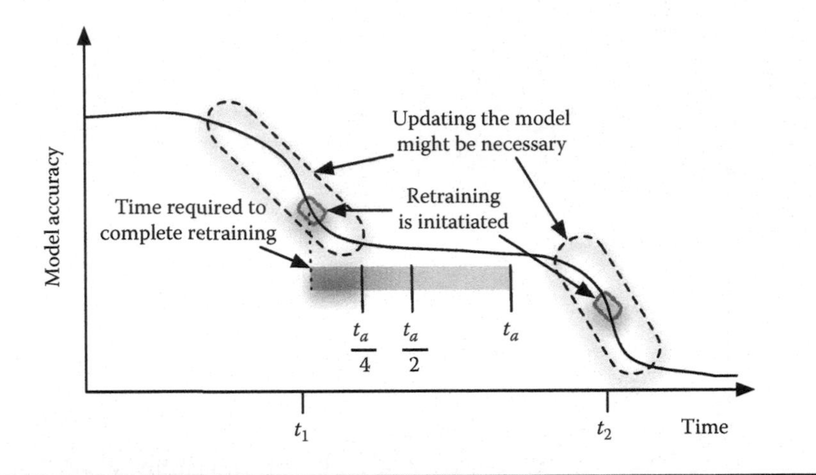

Figure 5.20 By using the rapid adaptation architecture for model retraining, the time for model learning is reduced and can be devoted to servicing the radio's actual demands.

current environmental conditions at time $t_1 + t_a$, the radio can now select the best possible parameter configuration until other environmental changes require another retraining period at time t_2. It is therefore in the radio's best interest to minimize the time t_a spent in the retraining backoff, so that more time is left to actually use the information during the time until the next training period $[t_1 + t_a, t_2]$. The use of the rapid adaptation architecture achieves this goal by only evaluating a specifically chosen subset of the parameter configurations, thus only consuming $\frac{t_a}{2}$, $\frac{t_a}{4}$, or less time in the training backoff.

In this evaluation the rapid adaptation system created three fractional designs: (1) a half factorial design with the generator G = ABCDEF, (2) a quarter factorial design with the generators G = ABCDEF and F = −AD, and (3) an eighth factorial design with the generators G = ABCDEF, F = −AD, and E = −BC. As these factorial designs are true subsets of each other and of the complete parameter list as specified by the DOE control algorithm, this experimental setup allows to investigate how well the linear model may be predicted using reordered, smaller parameter configuration lists. This prediction quality was analyzed in terms of the model's prediction level, that is, the highest scoring configuration that may be identified given the previously evaluated tests, and the model's prediction error, that is, the average offset between the actual performance of configurations compared to a predicted performance level based on the statistical model of the rapid adaptation tool.

Figures 5.21 and 5.22 show the system's prediction level and prediction error, respectively, for the control algorithm running without any intervention and when equipped with the rapid adaptation architecture running one of the three fractional factorial designs. As can be seen in Figures 5.21 and 5.22, both the prediction

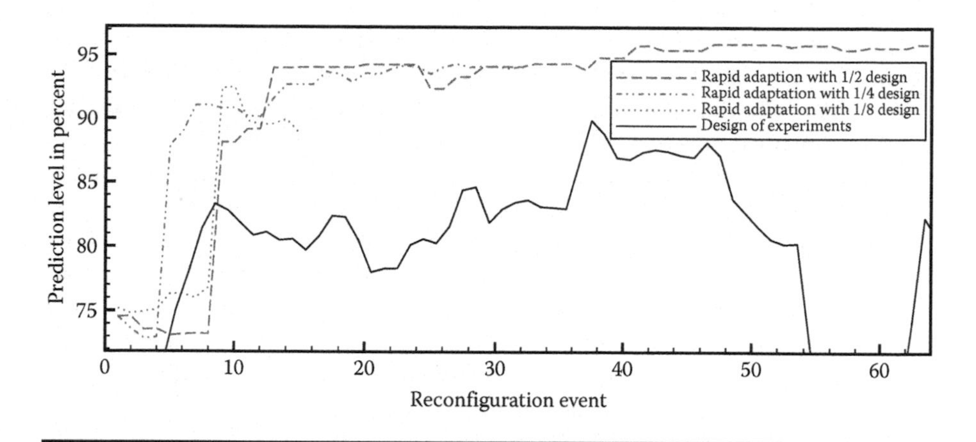

Figure 5.21 Feasible level of prediction of the DOE methodology with and without the intermediate rapid adaptation architecture.

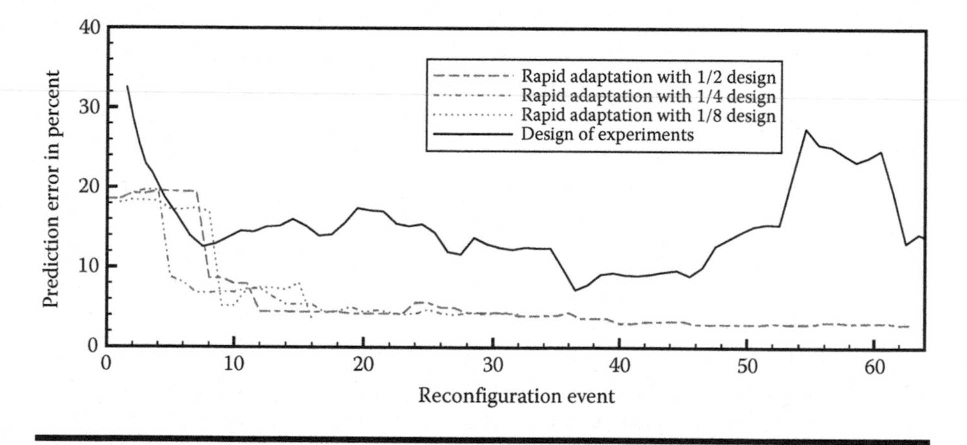

Figure 5.22 Prediction error of the DOE methodology with and without the intermediate rapid adaptation architecture.

level and the error are generally better for the cognitive radio system running the adaptation architecture together with the control algorithm. When it comes to the system's capability to predict and identify highly scoring configurations, the statistical preprocessor has the advantage of selecting those runs first that will provide the largest amount of information gain, thus reaching high prediction levels early on. An additional advantage of this selection and reordering process is that the obtained model is also highly robust, as most of the environmental information is contained in the earliest trials and later configurations only provide minor incremental refinements to an already established model. In contrast to this, the cognitive radio system without any statistical preprocessing executes and evaluates the parameter tests in no particular or random order, thus not controlling whether tests that may provide a large amount of information or incremental information are conducted first. This results in the following: the prediction level is also generally lower and the overall error of the prediction is larger than that in a controlled approach.

Besides a lower prediction level and higher error terms, the variability of the prediction level for the uncontrolled system is also significantly higher. The reason for this is that the control algorithm directly conducting the hardware tests may choose an unfortunate structure for its runs, for example, executing highly intercorrelated batches of configuration runs at certain times. If the parameters varied in such batches, however, are only lightly important to the overall configuration success, such unordered runs may build robust but irrelevant correlation models that are highly degraded or even contradicted by newly incoming data.

5.6 Summary

Figure 5.12 summarizes the improvements to link-level adaptation that can be gained through the application of fractional factorial design as provided through the rapid adaptation architecture. Instead of incorporating the information learned through the individual link-level tests in a final integration step as, for example, done in the DOE [5] (see (1) in Figure 5.12), the rapid adaptation architecture moves to an "online" optimization where information collected over the course of the link-layer tests is immediately used to (a) enhance the cognitive radio's configuration and (b) further direct the adaptation search. Thus, the cognitive radio's linear prediction model can be updated earlier on, but is subject to varying levels of prediction errors (see (2) in Figure 5.12).

As discussed in Section 5.5.3, this issue is then further remediated through the rapid adaptation tool by selecting the adaptations according to a certain scheme that maximizes the expressiveness of the results obtained from the configuration tests and minimizes the correlations and overlap between configurations, thus leading to a faster and more stable model update (see (3) in Figure 5.12).

References

1. C. Doerr, D. Grunwald, and D. Sicker. Optimizing for sparse training of cognitive radio networks. *Proceedings of the First International Workshop on Cognitive Wireless Networks* (CWNets), Vancouver, BC, Canada, 2007.
2. C. Doerr, D. Sicker, and D. Grunwald. Enhancing cognitive radio algorithms through efficient, automatic adaptation management. *IEEE Vehicular Technology Conference* (*VTC*), Singapore, 2008.
3. T. W. Rondeau, B. Le, C. J. Rieser, and C. W. Bostian. Cognitive radios with genetic algorithms: Intelligent control of software defined radios. *SDR Forum Technical Conference*, Phoenix, AZ, 2004.

4. N. Nie and C. Comaniciu. Adaptive channel allocation spectrum etiquette for cognitive radio networks. *ACM MONET (Mobile Networks and Applications)*, 11(6):779–797, 2006.
5. T. Weingart, D. C. Sicker, and D. Grunwald. Evaluation of cross–layer interactions for reconfigurable radio platforms, *IEEE Technology and Policy for Accessing Spectrum* (*TAPAS*), August 2006.
6. K. K. Vadde and V. R. Syrotiuk. Factor interaction on service delivery in mobile ad hoc networks. *IEEE Journal on Selected Areas in Communications*, 22:1335–1346, September 2004.
7. C. J. Rieser, T. W. Rondeau, C. W. Bostian, and T. M. Gallagher. Cognitive radio testbed: Further details and testing of a distributed genetic algorithm based cognitive engine for programmable radios. *Proceedings of the IEEE Military Communications Conference*, Monterey, CA, 2004.
8. G. E. P. Box, W. G. Hunter, and J. S. Hunter. *Statistics for Experimenters*. John Wiley & Sons, New York, 1978.
9. T. Weingart. A method for dynamic reconfiguration of a cognitive radio system. PhD thesis, University of Colorado at Boulder, Boulder, CO, 2006.
10. K. K. Vadde, V. R. Syrotiuk, and D. C. Montgomery. Optimizing protocol interaction using response surface methodology. *IEEE Transactions on Mobile Computing*, 5(6):627–639, 2006.

Chapter 6

Security in Cognitive Radio Networks

Jack L. Burbank

Contents

The concepts of cognitive radio and cognitive radio networking are exciting new fields of research and development that have the potential to change the very way in which we think about, design, develop, and evaluate communications networks. One area that is particularly disrupted by cognitive radio networking is that of wireless security. The cognitive radio paradigm introduces entirely new classes of security threats and challenges, and providing strong security may prove to be the most difficult aspect in establishing the long-term viability of cognitive radios. This chapter examines the issue of security in cognitive radio networks, discussing the key challenges of, and new potential threats and nefarious tactics introduced by, cognitive radio networks. The current security posture of the emerging IEEE 802.22 cognitive radio standard is examined. This chapter concludes by discussing promising research across multiple technical disciplines that can be brought to bear on this issue, and outlines areas of future research.

6.1 Introduction and Background

The one undeniable driving factor in the communications industry is the consumer's insatiable appetite for bandwidth and wireless access to the Internet. It is this insatiable appetite that, along with the corresponding desire of the communications industry to feed this appetite, is the real driving factor behind cognitive radio and other related concepts. As consumer demands continue to increase, both in terms of bandwidth and performance, traditional network design methods have found it difficult to keep pace. The core problem is that high-bandwidth access and wireless access are contradictory goals as the wireless environment is inherently resource constrained and unreliable. In fact, the one undeniable constraint in providing the type of wireless capability that is demanded by users, commercial and military alike, is spectrum. Spectrum is a precious resource, and there is simply not enough of it to meet the needs of today's and tomorrow's user base. This problem is exacerbated by the outdated way in which we manage spectrum. Regulatory agencies, such as the Federal Communications Commission (FCC), allocate spectrum for particular types of services that is then licensed to bidders for a fee. These allocations and licenses are static in nature, which means that this spectrum is unavailable for use, even if those who own the rights to this spectrum do not use it. This has led to considerable inefficiency in spectrum utilization and has created an unnecessary shortage of spectrum. This issue has been temporarily alleviated by providing for the availability of spectrum for unlicensed usage and has fueled the global deployment of 802.11-based technology. However, these unlicensed frequency bands are becoming overpopulated, and interference has grown to be a significant deployment constraint. All of these factors have led to the need to make dramatic changes in the spectrum regulatory process, as existing practices and policies are not capable of scaling with demand. This, in part, has led to the concept of **cognitive radio**.

6.2 Overview of Cognitive Radio

Despite the growing body of work on the topics of cognitive radio and cognitive radio networking, there is yet to emerge a common understanding for what exactly is a cognitive radio. One can typically ask the same question to multiple knowledgeable subject matter experts and get very different answers, because many of the relevant terms often have definitions that are not universal. However, there are some high-level attributes of cognitive radios that are generally agreed upon by the larger community [1]:

1. The cognitive radio **adapts** to its environment to meet requirements and goals.
2. It maintains **awareness** of surrounding and internal state.
3. It uses **reasoning** on ontology or observations to adjust adaptation goals.
4. It exhibits **learning** from past performance to recognize conditions and enable faster reaction times.
5. It **plans** to determine future decisions based on anticipated events.
6. It **collaborates** with other devices to build greater collective observations and knowledge.

The FCC defines cognitive radio as "A radio system whose parameters are based on information in the environment external to the radio system." The National Telecommunications and Information Association (NTIA) has proposed to define cognitive radio as "A radio or system that senses its operational electromagnetic environment and can dynamically and autonomously adjust its radio operating parameters to modify system operations, such as maximize throughput, mitigate interference, facilitate interoperability, and access secondary markets." IEEE 1900.1 defines cognitive radio as

1. "A type of *Radio* in which communication systems are aware of their environment and internal state and can make decisions about their radio operating behavior based on that information and predefined objectives. The

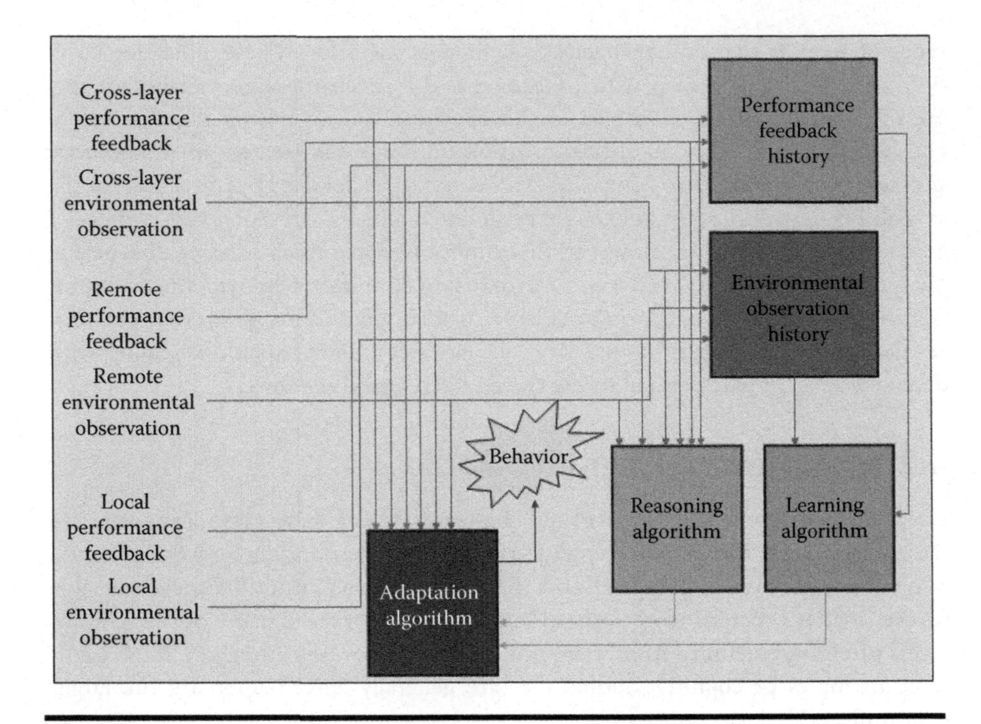

Figure 6.1 Generalized cognitive radio decision-making process.

environmental information may or may not include location information related to communication systems.

2. Cognitive Radio, as defined in (1), that utilizes software defined radio (SDR), adaptive radio, and other technologies to automatically adjust its behavior or operations to achieve desired objectives."

Note that the features called out in [6] are not specified by any of these "official" definitions, but are often discussed in the context of cognitive radio. However, the community remains divided on how many of these features a radio must possess before it is considered cognitive.

Equal confusion surrounds the basic characteristics of a cognitive radio. Therefore, an attempt is made to provide definitions that enable consistent discussion. For further discussion, consider Figure 6.1 that illustrates the general cognitive-networking process.

6.2.1 Adaptation

Adaptation suggests that some aspect of the radio's behavior is temporal (i.e., non-static) and is a function of some adaptation criteria. Examples of adaptation include adaptive modulation and coding (AMC), adaptive rate control, and adaptive power control.

6.2.2 Awareness

Awareness suggests that a radio builds knowledge of external factors that may influence performance along with relevant internal factors. Examples of external factors are environmental measurements such as spectral congestion and received signal strength. Examples of internal factors are performance measurements made by the cognitive radio platform such as error rate or latency.

There will be varying degrees of awareness based on the complexity of internal and external factors utilized. In nontrivial cases, it is reasonable to assume that the adaptation criteria will incorporate these non-data input functions to allow awareness to influence adaptation.

A radio can maintain awareness without exhibiting any type of adaptivity (e.g., statistics gathering or audit logging for human use). However, these cases are of little interest in the context of this discussion and are considered to be always in coexistence. It is important to note that **adaptive radios are not necessarily cognitive**. There already exist many examples of adaptive radios and techniques that are not considered cognitive [2]. These devices simply adapt based on some predefined algorithm or rule-set that does not change over time. Consider the example of AMC that is based on statically defined thresholds of link quality. Here, the radio will make some measurement of link quality (e.g., received signal strength indication [RSSI]) and adjust its modulation and coding in a predefined manner based on measured RSSI and predefined RSSI thresholds. A more reasonable definition is that a **radio cannot be considered "cognitive" unless it employs some degree of reasoning and/or learning**; this position is consistent with that of the larger community. However, what constitutes reasoning and learning is still debatable.

6.2.3 Reasoning

Reasoning and learning seem to be the two cognitive functions that cause the greatest amount of confusion. Reasoning is the function of modifying a radio's adaptation algorithms and rules based on current awareness (awareness of external and internal factors) to best meet the goals of the radio. Algorithms and rules of the cognitive radio are the primary link back to the goals and requirements of the cognitive radio. Given a set of stimuli, the radio determines its behavior according to some function that is determined such that the resulting behavior will conform to expectation. So, reasoning is the process of using the current observations of environment and performance to modify the algorithmic rules that govern adaptation.

6.2.4 Learning

Learning introduces significant complexity as adaptation algorithms are now potentially a function of both current and previous instances of time, up to the maximum history of the learning process (i.e., the amount of time for which external and internal factors influence radio algorithms). This complexity can dramatically increase the difficulty of understanding and predicting the radio's behavior, creating a significant challenge in forming a stable control system.

6.2.5 Planning

Planning is similar to reasoning and learning, except that it makes an influence on the adaptation algorithm at a future time. Planning might be based on some history of the measured environment or performance. Planning might be the result of a priori knowledge of future events that will affect performance (e.g., 802.22 planning spectrum evacuations based on a known news event such as a political convention where TV transmitters will be in use). Planning might be the result of a new policy placed into the cognitive radio network, where the cognitive radio wants to provide a smooth transition between old and new policy paradigms.

6.2.6 Collaboration

Collaboration is perhaps one of the easier cognitive functions to understand. Here, the radio is combining its own awareness with the awareness of other radios within the cognitive radio network to determine behavior. If reasoning and learning processes are employed, then this composite awareness will affect the basic algorithms of the radio, both current and future.

6.2.7 Cross-Layer Design and Cognitive Radio

The general cognitive radio is even more complex, as this type of decision-making process can be taking place at multiple places in the protocol stack and that these layers might in fact be cooperating with each other in a cross-layer approach (depicted by the cross-layer inputs of Figure 6.1). This cross-layer approach also complicates security approaches, as now a threat can potentially execute indirect attacks, potentially influencing the behavior at one layer of the protocol stack by influencing observations made by another layer of the protocol stack.

6.2.8 Cognitive Radio Standardization and Development Efforts

There are currently multiple development activities toward a full cognitive radio. One such effort is the DARPA neXt Generation (XG), which aims to develop technology to utilize unused spectra, primarily for the U.S. military [3,4]. In the commercial domain, IEEE 802.22 is the primary commercial cognitive radio development activity, which aims to develop technologies to utilize unused television spectra for broadband wireless services [5]. Furthermore, the IEEE Standards Coordination Committee 41 (SCC41), formerly the IEEE P1900 Standards Group, was established in 2005 to develop supporting standards associated with next-generation radio and advanced spectrum management. The SCC41 aims to develop common terminology and development practices to aid in the development of cognitive-radio-networking standards and technologies. There are six working groups (WG) within SCC41:

- IEEE 1900.1—Working Group on Terminology and Concepts for Next Generation Radio Systems and Spectrum Management
- IEEE 1900.2—Working Group on Recommended Practice for Interference and Coexistence Analysis
- IEEE 1900.3—Working Group on Recommended Practice for Conformance Evaluation of SDR Software Modules
- IEEE 1900.4—Working Group on Architectural Building Blocks Enabling Network-Device Distributed Decision Making for Optimized Radio Resource Usage in Heterogeneous Wireless Access Networks
- IEEE 1900.5—Working Group on Policy Language and Policy Architectures for Managing Cognitive Radio for Dynamic Spectrum Access Applications
- IEEE 1900.6—Working Group on Spectrum Sensing Interfaces and Data Structures for Dynamic Spectrum Access and other Advanced Radio Communication Systems

6.2.9 Cognitive Radio: Not Just Dynamic Spectrum Access

It is important to understand that cognitive radio networking can encompass much more than just frequency agility. The wireless-networking community as a whole is slowly gravitating to cognitive approaches, perhaps unknowingly, to mitigate the harsh wireless environment. This is particularly true given the increasing interest in performance-sensitive applications, such as voice and video, combined with the increasing reliance on wireless connectivity. This is evident in the multitude of cross-layer designs that continue to appear in open literature and the number of technology proposals that utilize performance feedback in algorithms designed to optimize performance. So, while spectrum scarcity has been an initial motivating factor for cognitive radio and cognitive techniques, there are numerous factors that are motivating cognitive techniques across the protocol stack.

It is important to note that given these more general definitions, we can easily postulate "cognitive radios" that are unrelated to dynamic spectrum access (DSA). Let us once again consider the simple case of a radio that employs AMC. Now let us consider the case where the radio no longer employs static RSSI thresholds but, instead, dynamically sets these threshold values based on channel measurements or performance feedback. This would now represent a radio that employs reasoning. Consider the case of an adaptive technique that performs threshold detection to make adaptation decisions. Now, suppose that the threshold is dynamically calculated based on environment. By definition, that radio is performing reasoning. If that dynamic threshold determination is based on both current and past observations, then that radio is reasoning and learning. If that RSSI information is shared with neighboring nodes to improve their decision-making process, then the radio is collaborative. This example is provided to illustrate that **a DSA radio is not the only example of cognitive radio**. This is an important point to make, because many of the security concerns and mitigation approaches discussed later in this chapter are applicable across all cognitive mechanisms. So, this realization necessitates the development of generic solutions that can potentially solve entire classes of problems.

6.2.10 Cognitive Techniques Are Not Limited to Radio Layers

Another common misconception associated with cognitive radio is that cognitive techniques are limited to the radio layers of the protocol stack, such as the examples of DSA or AMC. Although these are certainly useful examples in illustrating cognitive techniques, one can postulate cognitive techniques at each and every layer of the protocol stack. Rather, cognitive techniques can indeed be applied across the entire protocol stack. Furthermore, these techniques could be combined to be interdependent on one another (e.g., adaptive rate control at the application layer based on the feedback from the network layer), quickly increasing complexity of the overall technique. It is fully expected that an increasing number of cognitive techniques will emerge across the entire protocol stack, particularly given the rising usage of performance-sensitive applications, such as voice and video, where consistent performance is required even when operating over a highly variable and an inherently unreliable wireless channel.

6.3 Overview of the IEEE 802.22 Security Model

Despite the tremendous amount of research that has already been conducted in the area of cognitive radio, security is an area that is only recently beginning to receive attention. This is true of open literature and is also true of ongoing standardization efforts. Consider the ongoing 802.22 standardization effort, where security as a whole has been somewhat considered an afterthought. And only recently have there been any signs of movement toward protection of the cognitive plane.

It is expected that 802.22 will leverage many features of the existing 802.16 security model. A recent version of the working draft 802.22 specification states [3] "The security sublayer is in many respects inspired by the IEEE 802.16/D12 draft."

This makes sense in many respects, because the 802.16 security model has evolved significantly since its original inception, and is generally considered to provide commercial-grade security. With that said, 802.16 access technologies do not consider the unique aspects of a cognitive radio network. The 802.22 WG has just recently started to consider the unique aspects of cognitive radio. A recent contribution to the IEEE 802.22 WG [6] contains proposed text being developed by an IEEE 802.22 security study group for the text to eventually be placed in the security section of the IEEE 802.22 specification. This proposed text contains placeholders for cognitive-unique security aspects.

6.4 Security Challenges in a Cognitive Radio Network

Cognitive concepts require security mechanisms whose scope extends beyond what is provided in legacy technologies, such as 802.16. In the case of 802.22, this traditional approach might make sense, as 802.22 may or may not eventually embrace or reflect the entirety of what it means to be cognitive. However, traditional security approaches will likely prove insufficient for the generalized cognitive radio network. As one considers the required security functions of a cognitive network, it is prudent to identify the key differences between cognitive and noncognitive networks. There are likely many areas of overlap for which existing security mechanisms may provide equivalent security for both cognitive and noncognitive networks. These are important to identify as they represent security mechanisms that may be borrowed from traditional noncognitive networks, allowing focus to be placed on security problems unique to cognitive networks.

There are two fundamental differences between a traditional wireless network and a cognitive radio network:

1. The potential far-reaching and long-lasting nature of an attack
2. The ability to have a profound effect on network performance and behavior through simple environmental manipulation (i.e., generation of signals)

In the cognitive radio network, locally collected and exchanged information is used to construct a perceived environment that will impact both current and future behaviors, as well as the behavior of those around them. The induction of an incorrectly perceived environment will cause the cognitive radio to adapt incorrectly, which affects short-term behavior. Unfortunately, the cognitive radio uses these experiences to reason fundamentally new behaviors, learning from these experiences to anticipate future actions. If the malicious attack perpetrator is clever enough to disguise their actions from being correctly detected and deduced as attack actions, they have the opportunity for a long-term impact on behavior. Furthermore, the cognitive radio collaborates with its fellow radios to determine behavior. Consequently, this provides an opportunity to propagate a behavior through the network in much the same way that a malicious worm propagates through a network.

It is important to note that there exist examples of long-lasting and far-reaching effects in traditional wireless networks. For example, consider the events immediately following the September 11 attacks in the United States. A sudden increase in offered traffic and other network changes caused massive outages in the U.S. cellular telephony infrastructure. There are numerous other examples that can be pointed to where effects in one portion of a network have negative and long-lasting consequences in other portions of the network. However, it is key to understand that these are examples of unintended consequences caused by the complexity of the network, making it difficult to fully understand the relationship between all aspects of the network. While the additional complexity of the cognitive network will certainly make it more difficult to achieve this type of understanding and likely lead to an increased risk of unintended consequences, the cognitive network allows for an adversary to launch systematic attacks against it to achieve far-reaching and long-lasting effects as opposed to "getting lucky" by inducing an unintended consequence in the network.

At this point in the still immature development of cognitive radio network security, it is important to step back and first understand the key fundamental issues:

1. What are the potential threats to a cognitive radio network?
2. What are the potential attacks against a cognitive radio network?
3. What is the likelihood of these threats and attacks?
4. What is the potential consequence of these attacks?

6.4.1 High-Level Security Goals

There is the obvious desire to provide basic network security services in any deployed network, such as identity confidentiality (protection of identity determination), user data confidentiality (protection of data compromise), reliability (protection of network availability to support data services), and accountability (ability of network to police itself). Furthermore, these basic services typically require that the network provide services such as authentication (ability to positively identify network entity) and authorization enforcement (privilege limitation based on identification). This is true in the conventional noncognitive network. This remains true in a cognitive network. In both cases, sufficient security mechanisms must be put into place to defend against the set of envisioned threats to achieve these goals.

6.4.2 Threats

Many threats are common to both cognitive and noncognitive networks. And it is clear that, in general, threats to noncognitive wireless networks are still of interest in the cognitive network. In general, the two primary types of threats that must be considered are (1) the outside threat (i.e., unauthorized user) attempting to inject energy into the victim network to achieve a desired goal and (2) the Byzantine threat (i.e., insider threat) attempting to use its privilege to achieve a desired goal.

6.4.2.1 Outside Threat

The outside threat could consist of an attacker attempting to inject energy into the cognitive radio network to induce some type of behavior. A jammer is a common type of this outside threat (typically focused on DoS). However, the outside threat could be attempting a more sophisticated approach, such as the injection of otherwise valid messages into the network for a desired effect, for example, unauthorized network entry, data integrity attacks, and DoS. Table 6.1 provides a non-exhaustive summary of the general types of attacks that could be launched by an outside threat against which the network must protect itself. Note that all the attacks identified in Table 6.1 are not elaborated in subsequent text, as the goal is not necessarily to detail how these attacks could be conducted. Rather, the goal is to indicate whether these attacks are equivalent across cognitive and noncognitive networks or whether there are potential differences in this area of attack. Entries in light gray indicate a

Table 6.1 Outsider Threat Attack Examples

Target of Attack	*Attack Type*	*Typical Desired Goal*	*Cognitive Target Network Applicability*
Network	Jamming	DoS	
Network	Jamming	Herding	
Network	Jamming	Learning	
Network	Message injection	Unauthorized network entry	
Network	Message injection	Integrity attack	
Network	Message injection	DoS	
Network	Eavesdropping	Data compromise	
Network	Eavesdropping	Enumeration	
End host	Payload delivery	DoS	
End host	Payload delivery	Unauthorized entry	
End host	Payload delivery	Malware installation	
End host	Payload delivery	Own the box	

lower level of potential vulnerability. Entries in dark gray indicate a higher level of potential vulnerability.

The cognitive network should be no more vulnerable to message injection attacks than noncognitive networks assuming equivalent cryptographic-based access control. The cognitive network should be no more vulnerable to integrity attacks than noncognitive networks assuming equivalent cryptographic-based integrity protection. Furthermore, the cognitive network should be no more vulnerable to eavesdropping than noncognitive networks assuming equivalent cryptographic-based confidentiality protection. However, the cognitive network is expected to be more vulnerable to jamming attacks than compared to noncognitive networks. Furthermore, members of a cognitive network will likely be more vulnerable to end-host attack vectors (computer network attack [CNA] vectors) aimed at cognitive platform software and network services than members of noncognitive networks.

6.4.2.1.1 Message Injection and Eavesdropping

Let us first consider areas in which cognitive and noncognitive networks are comparable: message injection network attacks and eavesdropping. These are both areas in which the protection schemes are common across the two types of networks, that is, protection mechanisms against these types of actions in the noncognitive network will provide equivalent protection in the cognitive network. Protection mechanisms against eavesdropping typically include transmission security (TRANSEC) and communications security (COMSEC) protection mechanisms to prevent an unauthorized third party from gaining access to information as it transits the network. This same approach can be applied to the cognitive network with comparable effectiveness. This is also true for message injection attacks. If the network provides proper authentication and encryption, the cognitive and noncognitive networks provide equivalent performance.

6.4.2.1.2 Jamming

Jamming is an area in which the cognitive network is likely to have an inherent disadvantage, as illustrated in the "chaser–jammer" scenario [7]. Here, the jammer can be attempting to achieve numerous goals, but for the sake of this chapter let us temporarily limit the discussion to DoS. In this case, the jammer is attacking the stability and security of the cognitive radio's adaptation algorithms, with the opportunity to inflict much greater damage than in the noncognitive case. Burbank et al. present a taxonomy of the types of attacks a jammer could launch against the cognitive radio network, including immediate DoS, network degradation (i.e., nuisance attack), jammer learning to enable future attack, and herding (forcing the cognitive network into a known state) (see [8] for a more detailed discussion on the jamming of cognitive radio networks).

6.4.2.1.3 Cognitive Network Host Attacks

Attacks targeting the actual member hosts of the cognitive network also have the opportunity for improved effectiveness as compared to the noncognitive network. This is due primarily to the complexity of the cognitive radio software on the end host itself. As the complexity of cognitive networks increases, algorithms implemented within the cognitive radio platform also increase in complexity. Additionally, it is envisioned that policy-based management approaches will be common in cognitive networks, further increasing the complexity of the software residing onboard the cognitive radio platform. This makes it increasingly difficult to precisely predict the behavior of a cognitive radio platform given a set of stimuli, in turn making it difficult to verify proper operation of the cognitive radio software. Like any other computer system, code complexity leads to errors and bugs, which form the basis of vulnerabilities and exploits. The increased code complexity inherent in the cognitive radio platform introduces security risk by creating opportunities for CNA operations beyond these vulnerabilities in a noncognitive radio platform.

6.4.2.2 Inside Threat

The Byzantine threat is another serious threat, particular in wireless networks, due to the distributed and often unseen peers of the network. Here, a valid user of the cognitive network is, unknowing to the rest of the cognitive network population, compromised. In much of existing research dealing with the Byzantine threat to wireless networks, the concern is the corruption of the routing system of the network by the compromised node. This particular threat is considered to be comparable between cognitive and noncognitive networks. However, the Byzantine threat could be substantially more problematic in a cognitive network that is employing collaboration. Here, the compromised node can spoof data to its neighbors in an attempt to destabilize or, otherwise, control or influence learning and reasoning algorithms. This is an area that warrants additional research.

6.5 Required Security Features of a Cognitive Radio Network

To mitigate malicious manipulation of cognitive radios and cognitive radio networks into forced behaviors, a cognitive radio must possess at least five key characteristics:

1. The ability to authenticate the local observations that are used to form a perception of its environments
2. The ability to strongly secure collaboration exchanges between cognitive radio elements
3. The ability to authenticate the validity of observations exchanged between cognitive radio elements

4. The ability to make secure, stable decisions based on local and remote observations
5. The ability to perform self-analysis of behavior

As important as individual cognitive radios possessing these characteristics is the ability of cognitive networks to provide a framework that ties all these individual protection mechanisms into a single coherent network security approach. However, this is an area that is receiving little attention to date in existing literature.

As the first line of defense of the network, the cognitive radio needs to be capable of judging whether what it is locally sensing is real or falsified. This goes far beyond protecting the network from injection of false messages, as is the focus of traditional network authentication mechanisms. Rather, this means that not only are network messages authenticated, but also that observations of physical phenomena are also authenticated. These physical phenomena include physical attributes of the environment, such as signal presence or channel quality, that do not lend themselves to traditional authentication mechanisms.

Because a cognitive radio utilizes not only its own observations as a basis for decision making but also the observations of others, there is the obvious need to authenticate the shared observations. This is particularly true given the distributed and unseen nature of its peer cognitive radios. The cognitive radio needs assurances that messages are indeed from who they claim they are from. This is similar in nature to the authentication of traffic in any wireless network, and as such is not necessarily unique to the cognitive radio paradigm. It is here that lessons can be drawn from the field of secure exchanges of routing information in ad hoc wireless networks, an area that continues to receive significant research activity [9].

Once the authenticity of the source of collaborative cognitive radio network messages has been established, the cognitive radio needs to be able to judge whether the observations that other cognitive radio elements within the cognitive radio network are reporting are real or falsified. This combined with the ability to establish the authenticity of the source is critical to preventing the propagation of attacker effects within the cognitive radio network. This is critical for two reasons: (1) to prevent the degradation of the network because of a spoofed cognitive radio element within the network and (2) to protect against the Byzantine attack. In this paradigm, the security of each node in the network is dependent upon the security of every other node in the network. A potentially useful reference to use in thinking about this problem is some of the security work that has taken place in the software-defined radio (SDR) community [10,11].

Even if mechanisms are put into place to perform the authentication of local observations, to perform the authentication of collaborative messages, and to determine the validity of remote observations conveyed via collaborative messages, the cognitive radio platform must still be prepared to properly operate in the presence of malicious information attempting to influence its decision-making process. This requires that all algorithms implemented in the cognitive radio platform be "hardened" to maximize stability and security (i.e., inability to manipulate or drive the platform into instability because of algorithmic flaws). This includes adaptation algorithms, learning and reasoning algorithms, and planning algorithms.

The cognitive radio must be able to determine whether it is acting erratically or logically. This self-check is critical to the long-term health of the cognitive radio network. If the long-term behavior of the cognitive radio has been affected by an attacker, the cognitive radio must be capable of identifying itself as an affected node and take self-corrective measures. This aspect is also important because of the envisioned long-term complexity of the cognitive radio platform itself. With increasing software complexity, it will be increasingly difficult to test all possible code execution paths to prevent software bugs. Thus, it is important that the cognitive radio platform itself is able to perform self-diagnosis to determine if internal problems are present, either because of observation corruption or errors in the algorithmic design or implementation.

6.6 Related Work and Open Issues

6.6.1 Authentication of Local Observations

The first primary problem area is that of the authentication of local observations that are used to form perceived environments. But how do we provide a mechanism to determine the authenticity of phenomena that have no inherent authenticable feature? The issue of determining the authenticity of the locally observed environment can perhaps leverage the work in the area of physical emitter classification and identification [12], particularly for the case of DSA techniques. But we now consider more sophisticated cognitive mechanisms that perhaps utilize the measured bit error rate (BER). How does the cognitive radio determine if measured statistics have been manipulated by a malicious outsider? This is a difficult problem that is currently unsolved.

A cognitive radio can be made aware of the relevant context of its usage (i.e., where am I located and what would I expect to sense given this location?). In the case of DSA techniques, where incumbents and secondary cognitive radio users are all fixed in location, such as the IEEE 802.22 deployment model, a commonly proposed approach to introducing security into the spectrum-sensing process is to take advantage of a priori knowledge of incumbent locations and received incumbent signal strength to determine if that signal is authentic. However, the jammer also has a priori knowledge of these fixed locations and can adjust its transmission accordingly, so that at the receiver it still mimics the incumbent. In the generalized case of a mobile cognitive radio network with potentially mobile incumbents, this approach is not practical. This can be generalized to any cognitive technique: a context provision mechanism for a single cognitive radio node that relies only on publicly known information provides little-to-no additional security because that information is also available to the adversary and can be spoofed.

Another class of approaches applicable to DSA techniques is to perform distributed spectrum sensing, so that any spectrum-sensing decision by an individual node is based on information from multiple cognitive radio network nodes [13]. This information is used to estimate the position of the incumbent based on the received signal and to determine if this estimated location is consistent with the known position of the incumbent. Because multiple cognitive radio nodes are used in sensing the incumbent, the confidence of the measurements taken by the respective cognitive radios may increase when compared to a single cognitive radio. This approach is significantly more promising than the case of non-distributed spectrum sensing with a very promising performance shown in [13]. However, the effectiveness of this approach is predicated on assumptions regarding the jammer. It was suggested in [14] that the efficacy of signal strength–based approaches may be sensitive to the accuracy of propagation models, which might make these methods impractical. High-fidelity RF propagation modeling is a historically daunting task, and it may not be realistic to expect high-fidelity, or even moderate-fidelity, RF-propagation-modeling capabilities to be available to a cognitive radio node. Other approaches [13] are based on the phase of the received signals. Phase-based location approaches have been shown to be effective, but are also sensitive to node geometry and can perform poorly in urban environments due to the complex propagation environment (e.g., multipath). This and other geolocation techniques add complexity to the cognitive radio in terms of additional algorithms, tight synchronization requirements, or multiple antenna apertures. Additional research is warranted. Furthermore, promising distributed sensing approaches that are presented in [15] should be considered.

One potential problem with distributed sensing approaches is the assumption that sensing data can be reliably shared. For example, if a cognitive radio is already under attack from a jammer, particularly in the case of an immediate DoS jammer scenario, it is not necessarily a reasonable assumption that the neighbor cognitive radio can deliver information to the cognitive radio under attack. However, this class of distributed sensing would appear to provide moderate protection against learning and herding, as collective information can be used to retrain algorithms and could make it significantly more difficult for the jammer. But, in general, the cognitive radio should be capable of overcoming some degree of the chaser–jammer scenario without the help of others. If the cognitive radio is capable of properly reacting to an attack, the radio could potentially require the jammer to increase power, eliminating the jammer's advantage. Without this ability, a cognitive radio may become isolated and cannot recover, because it requires unavailable information. The cognitive radio cannot obtain the required information to defeat the jammer because its communications are already being denied by a jammer who gains a power advantage by exploiting the cognitive radio feature set. The threat may have the objective of separating a cognitive radio node from the remainder of the cognitive radio network in terms of timing synchronization or frequency synchronization (e.g., current channel selection). So, while distributed sensing approaches are considered a potential key component of next-generation cognitive networks, individual protection mechanisms are also required, leading to a hybrid sensing architecture.

6.6.2 Verification of Trustworthiness of Remote Observations

Recalling the individual cognitive radio will eventually make adaptation and, perhaps, learning decisions based on both locally collected and remote observations. Thus, the cognitive radio platform needs to be capable of judging the validity of observations reported by other cognitive radio elements within the cognitive radio network. This capability is needed to protect against both the Byzantine threat and the threat of misinformation dissemination among the cognitive radio network users. This is an open area of research.

6.6.3 Secure Decision Making

Given a set of situational awareness data and desired goals, the algorithms present in the cognitive radio platform will attempt to make the best decision that will come closest to meeting the set of desired goals. Adaptation algorithms attempt to make the best decisions regarding radio behavior based on both local and remote awareness and performance feedback inputs. Reasoning algorithms attempt to make the best decision regarding current adaptation algorithms based on current awareness and performance feedback inputs, both local and remote. Learning algorithms attempt to make the best decision regarding current adaptation algorithms based on both current and past awareness and performance feedback inputs. Planning algorithms attempt to make the best decisions regarding future adaptation algorithms based on both current and past awareness and performance feedback inputs.

For DSA algorithms, this is an area that continues to receive significant attention. However, very little attention has been paid to the general problem by the cognitive-radio-networking community. There are examples of existing work in other technical disciplines that can be applied to this problem space. Several fundamental approaches to this optimization problem stem from artificial intelligence research, including machine learning, biologically inspired (genetic) algorithms, and game theoretical approaches. Barreno et al. [16] provide a very good treatment of the current state of security issues surrounding machine-learning algorithms. From these research communities, there is significant work that can be leveraged to begin developing security solutions for cognitive radio networks. For example, researchers have considered the issue of optimally combining advice from a set of experts (e.g., [17]), analogous to cognitive radios sharing their expert advice regarding their environment, and several solutions have been proposed that attempt to optimally combine these expert opinions in a way that is most beneficial. However, this begs the question, which cognitive radios in a community are considered "experts" as well as "trustworthy?" Perhaps this definition of "expert" is governed by a policy loaded into the cognitive radio. However, this approach is questionable for mobile ad hoc networks. Another approach is to build the ability into a cognitive radio to determine who it considers an "expert" as well as to build the mechanisms into the cognitive network to enable such determinations. Additionally, perhaps this process of determining "experts" is not done individually but rather nodes build "reputations" in the network, and the network as a whole gravitates toward decisions regarding who are "experts" and who are not.

Work in the data-mining research community aims to make an optimal decision when an adversary is attempting to corrupt the decision process with false information (e.g., [18]). Here, the adversary is attempting to influence the learning–adaptation cycle, and [15] shows that if the adversary and the learner have complete information about each other, then the learner can find a strategy to defeat the adversary's attempted adaptations. An adversary will likely require information regarding the cognitive radio's goals, methods, and techniques for adaptation, learning, and reasoning to be successful. Furthermore, this suggests that it is highly beneficial to precisely understand the threat and the types of tactics that would be employed by the threat. This is generally true, as threat analysis is a necessary early step in developing a strong system security posture. Additionally, this suggests that it is also beneficial to prevent the threat from gaining a full understanding of the cognitive network.

If learning and reasoning functions are improperly designed or configured, the jammer may be able to train the cognitive radio into preprogrammed behaviors (as mandated by the malicious threat) and can introduce biases into the cognitive radio's decision-making process that benefit the adversary for the future zero-day operation. In the case of learning and reasoning, cognitive radio complexity has both disadvantages and advantages. Like in any other computer system, the cognitive radio code complexity can be expected to lead to errors and bugs, forming the basis of vulnerabilities and exploits that introduce security risk. The cognitive radio jammer scenario resembles a cat-and-mouse scenario. Each attempts to gain insight into one another's behaviors and tactics. As is generally the case with security, the more the adversary knows regarding your tactics, the more effective they will be. So, from this perspective, it is desirable to (1) make algorithms complex enough so that they are not easily derivable through eavesdropping and stimulation, and (2) keep the details of these algorithms secret. It is undesirable for the threat to be able to predict the cognitive radio network behavior based on a given stimulus, as this may be useful information for an attack vector (e.g., a DSA-enabled network could perhaps be perfectly jammed if a jammer could perfectly predict its next operating frequency). Here, security by obfuscation is not recommended, but rather cryptographic protection of algorithm details. Concealment and protection of algorithm details could certainly strengthen the cognitive radio security posture, but understandably may not be desirable from a logistical and cost perspective.

6.6.4 Self-Diagnosis

The cognitive radio needs to be able to judge whether it is acting erratically or logically. This self-check is critical to the long-term health of the cognitive radio network. A reliable, isolated self-diagnosis may prove difficult to accomplish (if a cognitive radio's algorithms have been compromised, how much assurance can be placed that its self-diagnosis functionality is still trustworthy?) and that a distributed approach should instead be considered. Furthermore, a stand-alone approach may not be desirable because any self-diagnosis function would have to have ultimate access to the rest of the cognitive radio, which could in itself be dangerous.

6.6.5 Byzantine Protection

The Byzantine attack represents the case where a trusted node of the cognitive network, unbeknownst to the radio, becomes an adversary and represents the most difficult subset of this problem space. The Byzantine threat could be substantially more problematic in a cognitive network that is employing collaboration. Here, the compromised node can spoof data to its neighbors in an attempt to destabilize or, otherwise, control or influence learning and reasoning algorithms. In addition, the adversary now has potential access to cognitive algorithm software implementations that could perhaps be leveraged into advanced exploits against the cognitive network. Furthermore, the adversary now has potential access to a rich set of network state information. It is possible that increased platform protection is required to limit access to the algorithmic software and network state information. This is an area that warrants additional research.

There are lessons that can be drawn from existing work in the area of Byzantine routing (e.g., [19]). However, we must be careful not to create an overly paranoid network, where nodes are too quickly distrusted if inconsistent with expectations. This paranoia itself could be used against the cognitive radio by an attacker to cause a forced effect.

6.6.6 Attack Recognition

In the majority of cognitive-networking discussions, the radio is attempting to perceive one or more aspects of its environment, such as spectral population in the case of DSA-enabled radios, or performance-related data, such as signal quality in the case of AMC or error rate performance in the case of transport-layer adaptation. However, there are generally few existing discussions related to building a view of the safety of the environment. Is a particular node currently attacked by a threat? Is the cognitive network operating in an environment known to be hostile? What is the history of hostile acts taken against the network as a whole? Building this type of view of the environment and sharing this information throughout the cognitive network could help build a context that could potentially be integrated into the trust and respect of the collaborative information, into the belief in authenticity of the locally observed environment, and in the hardening of decision-making algorithms by introducing the concept of "caution" or "degree of alertness" based on the perceived safety of environment. As an example, perhaps the cognitive radio may use the Bayesian inference over a set of sensed data to estimate whether it is influenced by a threat and then respond to it accordingly. This may depend upon the historical collection of data in various conditions in an attempt to differentiate the true environment from the threat-induced spoofed environment.

6.7 Summary

Cognitive radio and cognitive radio networking represent one of the most exciting research and development opportunities in recent times. However, developing effective security solutions for these networks is expected to prove a daunting task. This chapter has examined the issue of security in cognitive radio networks, discussing new potential threats and nefarious tactics. Current development and standardization efforts have largely ignored security until very recently, failing to recognize the distinct security challenges introduced by the cognitive nature of the network. However, there exists promising research across multiple technical disciplines that can be brought to bear on this problem. However, much work remains before we have effective cognitive radio network security.

References

1. S. S. Raytheon, IEEE 802 tutorial: Cognitive radio, Presented at the IEEE 802 Plenary, July 18, 2005.
2. A. J. Goldsmith and S. B. Wicker, Design challenges for energy-constrained ad hoc wireless networks, *IEEE Wireless Communications Magazine*, 9(4): 8–27, August 2002.
3. The XG vision, Request for comments, Version 2.0, XG Working Group, http://www.darpa.mil/ato/programs/XG/rfcs.htm
4. The XG architectural framework, Request for comments, Version 2.0, XG Working Group, http://www.darpa.mil/ato/programs/XG/rfcs.htm
5. C. Cordeiro et al., IEEE 802.22: The First worldwide wireless standard based on cognitive radios, *2005 First IEEE International Symposium on New Frontiers in Dynamic Spectrum Access Networks (DySPAN)*, Baltimore, MD, November 8–11, 2005, pp. 328–337.
6. A. Mody, R. Reddy, T. Kiernan, and M. Sherman, Recommended text for section 7 on security in 802.22, IEEE 802.22–08/0174r07, September 2008.
7. J. L. Burbank, Security in cognitive radio networks: The required evolution in approaches to wireless network security, *Proceedings of the Third International Conference on Cognitive Radio Oriented Wireless Networks and Communications (CrownCom)*, Singapore, May 15–17, 2008.

8. J. L. Burbank, A. R. Hammons, and S. D. Jones, A common lexicon and design issues surrounding cognitive radio networks operating in the presence of jamming, *Accepted for Presentation at the 2008 IEEE Military Communications (MILCOM) Conference*, San Diego, CA, November 2008.
9. M. Zapata and N. Asokan, Securing ad hoc routing protocols, *ACM Workshop on Wireless Security (WiSe)*, Atlanta, GA, September 28, 2002.
10. High-level SDR security requirements SDRF-06-S-0002-V1.0.0, January 12, 2006.
11. Security considerations for operational software for software defined radio devices in a commercial wireless domain SDRF-04-P-0010-V1.0.0, October 27, 2004.
12. A. A. Tomko, C. J. Rieser, and L. H. Buell, Physical-layer intrusion detection in wireless networks, *Proceedings of the 2006 IEEE MILCOM Conference*, Washington, DC, October 2006, pp. 1–7.
13. R. Chen et al., Toward secure distributed spectrum sensing in cognitive radio networks, *IEEE Communications Magazine*, 46: 50–55, April 2008.
14. R. Chen, J.-M. Park, and J. H. Reed, Defense against primary user emulation attacks in cognitive radio networks, *IEEE Journal on Selected Areas in Communications*, 26(1): 25–37, January 2008.
15. T. C. Clancy and N. Georgen, Security in cognitive radio networks: Threats and mitigation, *Third International CrownCom*, Singapore, May 15–17, 2008.
16. M. Barreno et al., Can machine learning be secure? *Proceedings of the ACM Symposium on Information, Computer, and Communication Security*, Taipei, Taiwan, March 2006.
17. N. Cesa-Bianchi et al., How to use expert advice, *Journal of the ACM*, 44(3): 427–485, May 1997.
18. V. Vovk, Aggregating strategies, *Proceedings of the 7th Annual Workshop on Computational Learning Theory*, San Mateo, CA, 1990, pp. 371–383.
19. B. Awerbuch, D. Holmer, C. Nita-Rotaru, and H. Rubens, An on-demand secure routing protocol resilient to byzantine failures, *ACM Workshop on Wireless Security (WiSe)*, Atlanta, GA, September 28, 2002.

Chapter 7

Distributed Coordination in Cognitive Radio Networks

Christian Doerr, Douglas C. Sicker, and Dirk Grunwald

Contents

This chapter explores the problem of configuration and coordination of cognitive radio networks through the use of local control algorithms. To address this problem, this research examines the issue of channel assignment for cognitive radio networks, wherein different nodes must agree on which channels they will use to communicate. Through a series of theoretical analysis and hardware implementation, this work demonstrates that biologically inspired local control algorithms are a feasible and worthwhile avenue for cognitive radio coordination and shows promising prospects for other areas in wireless systems such as sensor and ad hoc networks. This research also demonstrates that local control based on biologically inspired algorithms is well suited for the coordination of cognitive radio nodes in heterogeneous environments.

7.1 Introduction

This chapter examines the application of efficient local control algorithms to the problem of managing and coordinating cognitive radio networks. These algorithms must be able to configure the radio to allow satisfactory communication with external nodes because the cognitive radio's objective is to exchange data. Correspondingly, this chapter explores whether local control approaches (specifically based on biologically inspired algorithms) can be used to efficiently coordinate a cognitive radio network without the use of global information, but by relying on information that can be collected independently by each station.

We begin by identifying a series of problems that affect cognitive radio networks, or collections of cognitive radios. All these problems have the characteristic of needing to reach a mutual consensus about a particular network characteristic or function. One example is channel assignment, wherein different nodes agree on which of the many channels they will use to communicate. Other problems involve channel-aware routing, interference reduction, and dynamic bandwidth allocation. In previous works, we showed that these problems can be formalized in the context of a standard centralized solution technique involving multi-commodity flows [5]. Although this technique produces the optimal solution, it is not suitable in practice. In this chapter, we describe a solution technique based on biologically inspired distributed control algorithms to coordinate cognitive radios in ad hoc non-infrastructure deployments. We first describe the biological basis for these methods and the algorithms that derive from them. We then provide specific details on mapping the biologically inspired algorithms to the link-configuration problem and to cognitive radio network algorithms mentioned earlier. This description is followed by a detailed performance evaluation using a hardware-based test network. In particular, this work will show that local control using biologically inspired algorithms exhibits a series of general properties that are very appealing for use in cognitive radio networks:

- Swarm behavior as a general strategy of local coordination leads to efficient outcomes compared to a reference solution using global information.
- The algorithm can be applied to a variety of problems in cognitive radio networks.
- A control algorithm based on local observation of its surroundings is able to coordinate with other incompatible control strategies, thus making it a viable solution in heterogeneous cognitive radio deployments.

The material presented in this chapter is a compilation of previous works presented in [4,5] and has been extended with more background information about the general methodology and foundation of the approach.

7.2 Problems Encountered by Cognitive Radio Networks

In this section, we discuss the fundamental problem of channel selection in a cognitive radio network, especially in the presence of heterogeneous environment

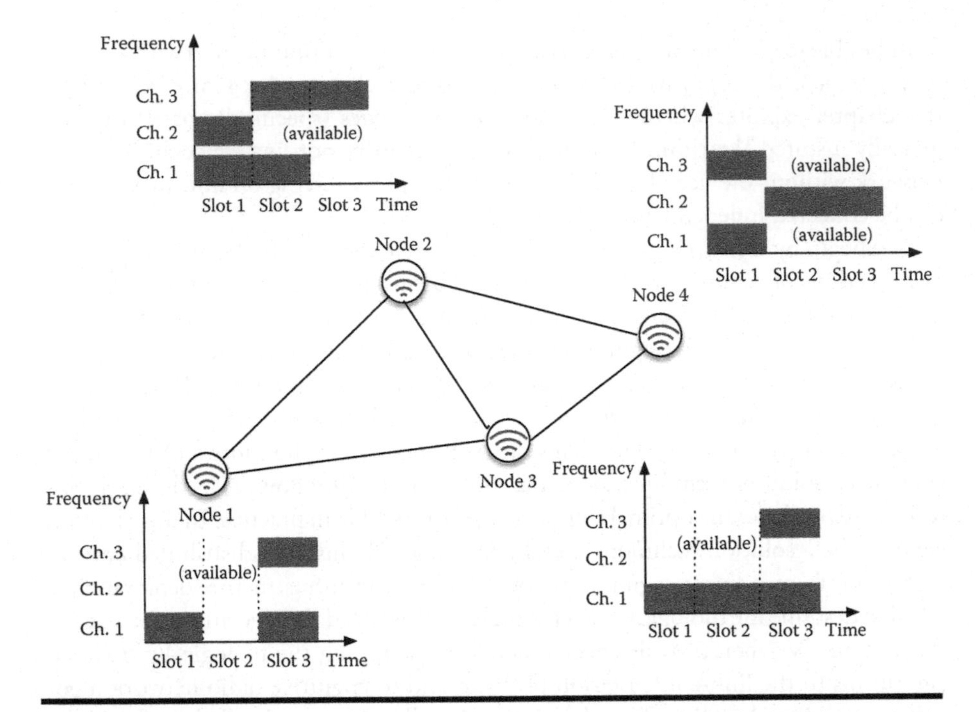

Figure 7.1 Distributed channel coordination of a cognitive radio network requires knowledge about the spectrum availability of each network node at any point in time.

conditions, and explore mechanisms that will allow efficient, decentralized coordination among network nodes.

Figure 7.1 visualizes the requirements and challenges of dynamic spectrum allocation in a small example: consider a small network of four cognitive radio nodes, of which nodes 1, 2, and 3 are all in range of each other and node 4 has a link only to nodes 2 and 3.

Following the paradigm of dynamic spectrum allocation, the wireless nodes are not assigned to a specific frequency, but rather sense the locally available spectrum at run-time to determine the set of unoccupied frequencies they can use. For simplification of the discussion, assume that each radio can only access three orthogonal channels and that spectrum availability changes in discrete time intervals over a three-iteration time horizon. This will most likely not be the case in practice, as each cognitive radio will have a large number of theoretically accessible frequencies at hand (which may even differ between each individual radio), spectrum availability will continuously change, and each cognitive radio will make decisions using its own independent clock. As can be seen in the figure, spectrum availability in the network is highly heterogeneous. Each node has a different set of available frequencies from any other node in the network, and this availability is also highly dynamic over time.

Despite the practical limitations of this example, it helps to formulate three requirements that a control algorithm for dynamic spectrum coordination in a cognitive radio network will need to fulfill:

Reaction to dynamic spectrum changes. First, the radio's control algorithm must be able to locally sense, identify, and make use of available frequencies for communication. As spectrum availability may change over time, the decision algorithm will need to repeatedly go through a learning and adaptation process, so that the radio always has at least one frequency at hand with which it can maintain connectivity with the network.

Propagation of relevant spectrum changes. Second, as each radio does not exist as an isolated entity but rather as a part of a larger network structure, the transmission frequencies must be chosen in such a way that a feasible link exists to a next-hop neighbor. In the example of Figure 7.1, node 1 has the option of selecting frequency 2 or 3 during the first time slot, but only channel 3 would give it connectivity to the remote station node 2.

As the available spectrum changes over time, new frequencies for communication will have to be selected by each node, even if the spectrum situation has not changed locally. In time slot 2, channel 3 becomes unusable to node 2, but not to both nodes 1 and 3. Nevertheless, it is beneficial for both these nodes to continue operation on a different frequency as well, as a different frequency (channel 2) offers better connectivity between these three nodes. Thus, a control algorithm for cognitive radio coordination must maintain links to neighboring nodes in the presence of spectrum heterogeneity and dynamic spectrum availability.

Dynamic, spectrum-aware routing of messages. Third, the control algorithm must consider these issues of dynamic spectrum availability at local and remote stations to make effective routing decisions. In traditional wireless networks (with statically allocated spectrum and all nodes operating on a fixed frequency), all communication messages can be delivered either through a direct link or through a series of hops from source to destination. In cognitive radio networks using dynamic spectrum allocation, however, the theoretical existence of a link or route (as two nodes are close enough to have a feasible communication between them) is not sufficient to actually deliver packets from source to destination, as there may not be a common frequency between the two nodes of a link to actually exchange data even though these nodes would theoretically have a wireless connection.

This issue is shown in Figure 7.1, where all packets to node 4 have to be routed through either node 2 or node 3. Although these two nodes are theoretically in range, their links are not usable all the time due to spectrum fluctuations. Specifically, during time slot 2 there exists a common frequency (channel 3) and therefore a feasible link between nodes 3 and 4, while during time slot 3 this link does not exist anymore and node 4 is now connected to node 2 using channel 1. Thus, for proper data delivery, the nodes in the network have to incorporate information about spectrum into their routing process to achieve dynamic, spectrum-aware routing of messages.

7.2.1 Spectrum-Aware Channel Selection: A Global Coordination Problem

To make efficient and effective decisions about local channel selection, link scheduling, and networkwide spectrum-aware routing, each node would ideally have global knowledge about the network topology and channel availability of every node.

Such global decision making could be achieved in two ways: (a) there exists a centralized entity to which all network nodes report their local spectrum availability and which then sends back instructions on how to configure each node and (b) each node obtains information about unused channels from all other nodes in the network to locally select its own channels.

This process would be difficult to implement in practice, as the outcome of the algorithm, a channel selection that would enable all nodes to communicate with each other, is actually the required precondition for the data dissemination phase serving as the input for the algorithm. In other words, to disseminate and collect data about remote spectrum availability there must be a frequency allocation that would enable all network nodes to communicate; however, this frequency allocation is the final outcome of the algorithm, but is required to supply its inputs.

To address this issue, one could, for example, set aside some dedicated coordination frequency on which such control information is exchanged. There exists a number of ways by which such a frequency could be chosen, for example, by setting a fixed frequency aside or by dynamically finding such a control channel. On the one hand, a fixed frequency (e.g., an out-of-band coordination channel similar to the proposed 50 kHz E^2R CPC [6]) would need to be known to each network node beforehand. In addition, such a channel must also be available at all times (which may only be guaranteed through licensed spectrum), as free access to this medium would be required for any initial coordination and therefore subsequent payload communication. However, settling on a fixed calling frequency would expose the system to vulnerabilities, as deliberate or accidental outage of this coordination frequency (as through interference, jamming, or congestion) might render the network inoperable. On the other hand, dynamically finding a coordination frequency would require each network node to independently search for such a channel, which may not be available in the network at all due to spectrum heterogeneity. Possible hybrid approaches also exist, where a unique, dedicated calling frequency (such as the CPC) is used to determine and find a dynamically selected coordination frequency.

7.3 Distributed Cognitive Radio Coordination through the Principles of Swarm Intelligence

The problem of distributed coordination facing a cognitive radio network is not too different from many coordination and cooperation tasks found in nature. In fact, there exist many instances in animal behavior where each individual of a larger group possesses only a piece of knowledge about the state of the system or the general environment, but where all pieces would be needed to obtain a good solution.

Such coordination and cooperation tasks with distributed information can be found in many instances:

- In a group of schooling fish, each individual knows about obstacles and dangers in its local environment, the area that can be directly assessed by its visual senses. Yet, as threats only seen by a certain individual can also be a danger for other members in the group, and vice versa, coordination and cooperation through efficient sharing of this information lets all members of the entire group avoid issues that may only be detected by remote individuals.
- When resources (such as food or nutrient concentrations) can be dispersed over an area that is too large for an individual alone to search (e.g., through chemotaxis), many species such as phytoplankton or Antarctic krill are known to form large structures and are therefore able to detect and follow the resource gradients [7,11] through the enlarged search area.
- After information about remote food resources have been obtained by scouts and brought back to the nest, foraging ants optimize the colony's transportation paths and worker allocation to achieve the most efficient resource consumption [1]. These optimization processes are formed through local interaction between single individuals that react to the presence and movement patterns of other workers around them and are communicated through pheromone trails so that previously uninformed individuals later passing by can be updated about evolved patterns [3].

As the task of distributed coordination is therefore not a technical problem per se, but can be found in many other instances in biological processes, it makes sense to study which approaches to coordination without global information that have evolved in these biological systems may be applicable to the technical problem of cognitive radio network coordination. This process of bio-mimicry, the analysis and adaptation of good solution concepts from nature to technical problem domains, has in recent years gained significant momentum as biologically inspired solutions have been shown to possess high levels of robustness and fault tolerance [9], while efficiently using available resources [13]. This study of a biological system and the process of adaptation to cognitive radio networks will be the focus of the following sections.

7.3.1 Requirements for Concept Transfer

After an analysis of the properties, requirements, and operating conditions of a variety of self-organizing systems in nature, it became apparent that the concept of swarm behavior as found in schooling fish or flocking birds is well suited to be applied to the control of cognitive radio networks. This is due to three properties of the environment, interaction, and sensing capabilities that both schooling fish or flocking birds and cognitive radio networks have in common:

7.3.1.1 The Surrounding Environment Is Memoryless

Radio spectrum, while being a limited resource, is inherently renewable. Unlike other resources that deplete when being utilized, radio spectrum is constantly available independent of whether or not it has previously been used. However, this poses a significant problem from a coordination perspective when information about the system state or environmental conditions is to be passed on. Unlike ant trails, for example, which can persist over a significant length of time even long after the originator of the trail or the environmental information that the trail was about have vanished, information communicated in the radio spectrum will not be available as soon as it has been transmitted.

While being a rather "philosophical" issue, this property has a significant implication as to whether certain biological coordination approaches, such as ant-inspired coordination through trail-markers, would work in this technological context. Other biological approaches, such as schooling and flocking, do share these characteristics traits, thus making them transferrable to spectrum systems.

7.3.1.2 State Information Can Be Directly Observed on a Limited Horizon

Even though an individual cognitive radio has not coordinated itself with its surrounding neighbors to actually form a cognitive radio network, it is still able to collect information about its local environmental conditions and, to a limited extent, infer about the environmental conditions or system state of its direct neighbors. If, for example, node 1's neighbor, node 3 (Figure 7.1), is only available on channels 2 and 3 and never transmits on channel 1, node 1 may conclude that at that point in time its local neighbor either does not have the capability to transmit on channel 1 or that this frequency might be occupied or not have sufficient fidelity to be used for communication. However, limited amounts of information can be gathered locally without the use of explicit communication.

7.3.1.3 Individuals Can Communicate through Actions

There exist two ways in which entities in the system can relay information—through direct communication or by executing actions that are observable to others. This implies that individuals do not have to rely on direct communication to propagate information, remedying the circular problem that a link is required to exchange information necessary to link maintenance. Instead of directly communicating information, individuals can use observable actions to inform others about their environment.

Through this inspection, it becomes apparent that many biological approaches, while also performing a coordination task without global information, are actually unsuited for transfer and application to the problem of distributed cognitive radio network configuration.

The following discussion will focus on the application of swarm behavior as found in schooling fish or flocking birds, as this biological behavior fulfills the aforementioned requirements and is a promising candidate for transferral.

7.3.2 Mechanisms of Swarm Behavior in Nature

The collective and seemingly complex behavior of "swarming" as expressed by schools of fish and flocks of birds is actually the result of a set of simple rules that are followed by each individual in the group. Yet, when each member observes these rules, everyone's local behavior and the resulting interactions between individuals give rise to a complex global behavior exhibiting properties beyond any behavioral response encoded in the original rule set. When working as a group, the group displays a globally well-coordinated behavior, allowing groups of up to one million fish to move in unison, aligning their direction and avoiding obstacles and predators as a cohesive structure.

This feature of swarming, the development of a unique behavior at a global level that is not encoded into the individual-level behavior, is commonly referred to as emergent behavior, because it only becomes visible through the interaction of many individuals at any given time and is not the immediate result of the actions of an individual member of the group.

This property of emergent behavior exists in two different dimensions: first, even though each individual observes a set of simple local rules, the interplay of many individuals creates a global structure that has complex dynamic behavior that was not present at the individual level. Second, while every individual is only aware of its immediate surroundings and interacts with its immediate neighbors, the group as a whole is able to react to influences that may occur anywhere in or near the group, even though most of the individuals are unaware of it. In particular, the latter property of this emergent behavior and the very simplicity of the rules make swarming a compelling solution for the coordination of cognitive radio networks. The algorithm's ability to globally react, even though only one node at a time possesses local information, provides an interesting solution to global coordination without communication. The algorithm's foundation on simple rules minimizes the algorithm's computational overhead and makes it deployable even on rudimentary hardware, offering wide applicability.

These rules can be titled and summarized as follows [12]:

- **Cohesion:** Move toward and stay with other members of your group
- **Obstacle avoidance:** Keep at a sufficient minimal distance from your neighbors and obstacles
- **Alignment:** Match your direction and speed to that of your neighbors

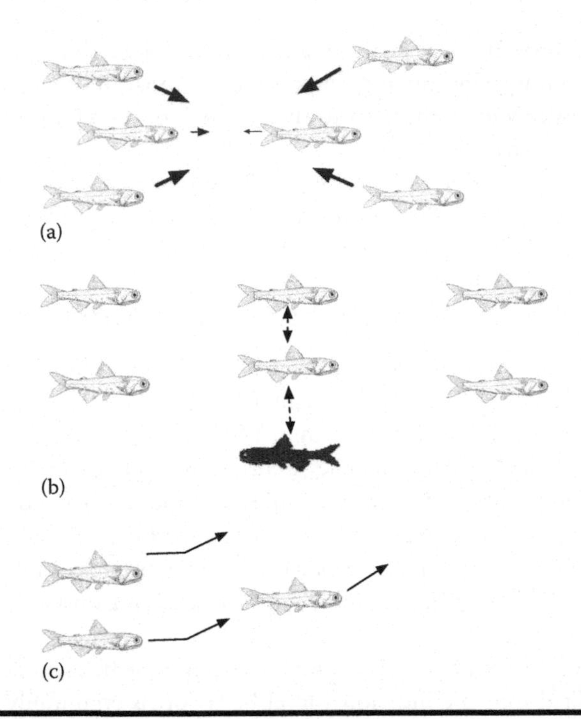

Figure 7.2 The global behavior of swarming is the result of three rules at work in every individual. (a) Cohesion: Moving toward other fish makes the group form a cohesive structure. (b) Obstacle avoidance: Individuals are maintaining a minimal distance to surrounding objects, thus obstacle information can propagate over many hops in the group. (c) Alignment: Matching the direction and speed to that of the neighbors will work predicatively to avoid collisions.

These three rules, which are visualized in Figures 7.2a through c, create a certain sub-behavior of swarming that works to form the emergent behavior described earlier. Each of these sub-behaviors is triggered only within certain ranges, as depicted in Figure 7.3. At relatively far distances r_a, individuals are attracted to remote peers and move toward them, which results in clusters and creates cohesion to the group. This tendency to move together is counteracted by the repulsion rule triggered at relatively short distances r_r, thus avoiding collisions between members of the group. At mid-distances r_o, each member will align its direction and speed to those around it, thus creating common movement and avoiding group members getting too close or distant, which would then affect the triggering of the cohesion and repulsion rules.

7.3.2.1 A Formalization of Swarm Behavior

These three sub-behaviors can easily be formalized mathematically for further analysis. For the following discussion, consider the existence of many groups (as there

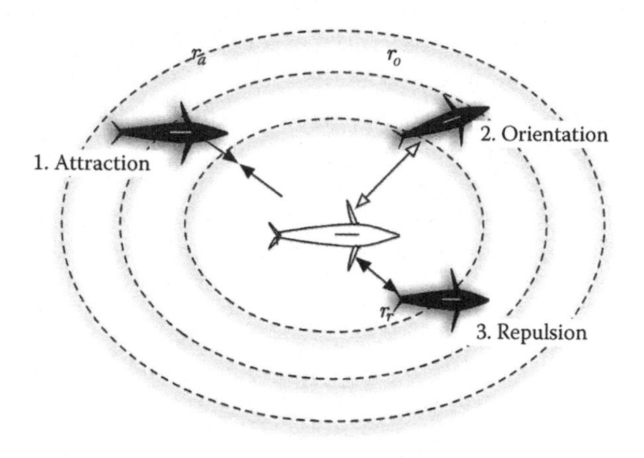

Figure 7.3 The three components are activated only at certain minimal ranges and act as counterweights to create a dynamic equilibrium.

could be many groups of fish present at the same time), denoted as S_i with $i = 1 \ldots n$, each modeling a group of fish or a group of obstacles. Each group S_i consists of a finite number of members p_j^i with $j = 1 \ldots |S_i|$.

Although the formalization and calculation is homologue for any element in any group, we will discuss the formalization of the swarm behavior rules from the perspective of the p_jth element in the group S_i. This will simplify the analysis of the algorithm.

7.3.2.1.1 Cohesion

The most prominent sub-behavior of swarming is that individuals move and stay together to form clusters and eventually form one cohesive structure. This process takes place by each member being able to sense and identify the position of other members of its group and consequently moving toward their location. This movement can be expressed in two ways: either by averaging all the movement vectors necessary to get from the individual's position to the location of each remote peer or by averaging the current position of each remote peer and then deriving a movement vector from the individual's current position to this average location. Both approaches yield the same result, the swarm's center of mass. Although the center of mass is a global property (i.e., requires knowledge about the position of all members of the group), the procedure works even if only a subset of the swarm's members are considered or only those visible to the individual. If the individual commands only over a limited sensory input range and therefore considers only a subset of the group's members (most likely its neighbors within a certain range), the same algorithm will yield an approximation of the group's center of mass and converge to the same results, if for every individual there exists at least one peer in its visibility.

Intuitively, this movement pattern creates cohesion in the following way: when the individual is already in the center of the group, it is surrounded by other peers on all sides. Therefore, its perceived center of mass and thus its intended movement of direction will be exactly or nearby its current position and its urge to move will be small. In the situation where the individual is at the outside boundary of the group, all its peers are located on one side and its resulting urge to move toward these individuals (the perceived center of mass) will keep the structure together.

$$\vec{v_c} = \frac{\sum_{k \in S_i} \overrightarrow{p_j^i p_k^i}}{|S_i| - 1} \tag{7.1}$$

Equation 7.1 formalizes this behavior. Given element p_j in group S_i, the center of mass perceived by that element can be found through vector addition of all movement vectors to members in its group. In the case that the individual has a limited sensory input range r_a, it just considers those elements $p_k^i \in S_i$ where $\left|\overrightarrow{p_j^i p_k^i}\right| < r_a$.

7.3.2.1.2 Obstacle Avoidance

It is also critical that while staying together and forming groups, individuals do not get into too close proximity to each other and stay at sufficient distance to other obstacles. Such obstacles can be members of other groups or other foreign objects to which a minimal distance must be maintained. As stated earlier, this sub-behavior acts as a counterweight, balancing the contraction as induced by the cohesion rule, and enforces that a minimal distance between objects is maintained at all times.

If a particular object comes closer than this minimal range r_r, the individual tries to restore this minimum clearance in the fastest way possible, which is denoted by a negative vector from its current position to the violating object: a straight reverse. If more than one object triggers this rule, and therefore must be avoided simultaneously, the addition of all "avoidance" vectors creates a movement that is likely to restore the minimal clearance as soon as possible.

$$\vec{v_o} = \frac{-\sum_{k \in C_s} \overrightarrow{p_j^i p_k^i} - \sum_{k \in C_o} \overrightarrow{p_j^i p_k^i}}{|C_s \cup C_o|} \tag{7.2}$$

where

$$C_s = \left\{ k \in S_i \middle| \left|\overrightarrow{p_j^i p_k^i}\right| < r_o \right\}$$

and

$$C_o = \left\{ k \in \left(\bigcup_{\forall k | k \neq i} S_j \right) \mid \left| \overrightarrow{p_j^i p_k^i} \right| < r_o \right\}$$

Equation 7.2 models this avoidance behavior. Let C_s denote the set of all members of the individual's own group that are too close and trigger the avoidance rule. Let C_o denote members of other groups or foreign obstacles that also violate the minimal clearance. For both these sets, the sum of avoidance vectors as measured from p will yield a corresponding avoidance movement.

7.3.2.1.3 Alignment

While not being mandatory to create working swarm behavior, the sub-behavior of alignment helps to maintain structure in the group, and thus implicitly supports the rules of cohesion and obstacle avoidance. Following this rule, individuals align their direction and speed to those around them, thus addressing the issue of predicatively maintaining a cohesive, but not too dense, structure. This alignment continuously adjusts the elements' velocities such that group members do not loose the connection with the group nor move into repulsion range in the first place.

$$\vec{v_a} = \frac{\sum_{k \in S_{i_n}} \vec{v_p} - \vec{v_k}}{|S_{i_n}|} \tag{7.3}$$

where $S_{i_n} = \left\{ k \in S_i | \left| \overrightarrow{p_j^i p_k^i} \right| < r_a \right\}$

To calculate the necessary alignment, which is described in Equation 7.3, each particular individual again considers only peers within a certain range, here the alignment distance r_o. For this proactive concept of work, this distance must lie between the attraction and repulsion range.

The overall resulting direction of movement for each individual is then determined through the synthesis of the movement vectors of the sub-behaviors. These may just be added as vectors, but for practical considerations should be weighted to create certain global properties (see Equation 7.5). These weights have been found to be species specific [8] and fine-tune the emergent behavioral characteristics at the global level. High weighting of cohesion (large ω_c) will, for example, urge the group to stay together in a single structure and avoid dissections as much as possible. For high weighting of obstacle avoidances (large ω_o), the group will demonstrate very timid reactions against outside obstacles (and members within collision distance) and perform rapid escape maneuvers.

$$\vec{v_p} = \vec{v_c} * w_c + \vec{v_o} * w_o + \vec{v_a} * w_a \tag{7.4}$$

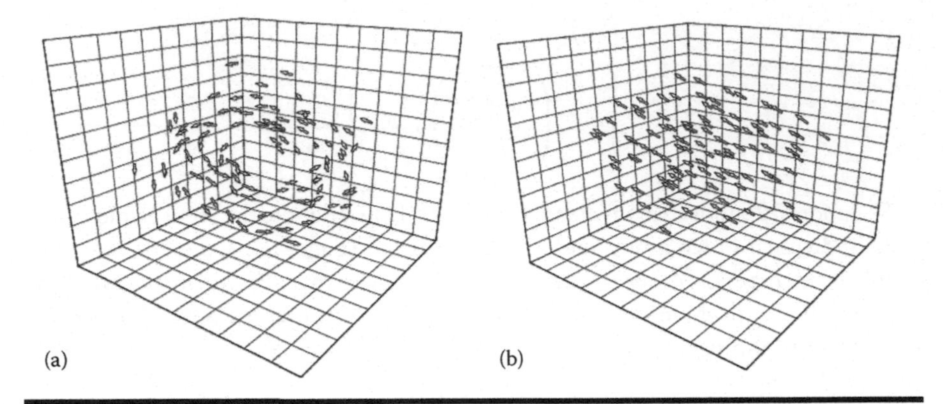

Figure 7.4 Different sizes of attraction, orientation, and alignment ranges cause the group to show different types of global behavior. (a) Group members move in a torus: small r_o and high r_a and (b) group members create high parallel orientation: mid r_o, high r_a. (From Couzin, I.D. et al., *J. Theor. Biol.*, 218, 1, 2002. With permission.)

In addition to the weighting of sub-behaviors, the overall global behavior also depends on the respective sizes of the cohesion, obstacle avoidance, and alignment window. Although in general it can always be assumed that $r_r < r_o < r_a$, the relative sizes of the three windows change the overall structuring of the group members. As shown in Figure 7.4, small alignment windows (r_o) and high ranges of attraction (r_a) let the members form a torus that spins around an empty core when individuals move with a minimum speed. At mid-size orientation ranges (r_o) and high attraction ranges (r_a), group members form a group with high parallel orientation that is able to rapidly react to outside influences.

7.3.3 How Swarm Behavior Can Be Transferred to Cognitive Radio Networks

After the biological concept of swarming has been formalized and the key parameters have been identified, it is possible to use this theoretical model toward an application in cognitive radio networks. However, before applying these mechanisms of biological coordination to communication networks, one must adjust the concepts and strategies to fit the requirements and assumptions of the target domain, that is, adjust the sensory input and resulting actions that are used in the algorithm to meaningful senses and actions in a network of cognitive radios. This section will focus on this adjustment process and outline how such an adaptation of the algorithm can be conducted.

As the mechanism of swarming behavior has evolved as a biological response strategy in various fish and bird species, it is clear that the approach is exactly tailored to the needs, senses, and actions of the organisms using this algorithm. Instead of directly applying the algorithm based on the formalization derived in the previous section and yielding a cognitive radio control algorithm that makes use of the cognitive radio's capabilities in a suboptimal way, it is necessary to make some changes to the swarm algorithm. This is due to a variety of key differences between the source and the target domain:

7.3.3.1 Controlled Variables

As the mechanism is originally used to create and maintain group structuring among animals, it adjusts the physical position of individuals as the controlled variable. Although this is a meaningful strategy in the original domain, controlling and adapting spatial location is undesired in the context of a cognitive radio network, as the devices' positions are typically predetermined and the system should use its adaptation capabilities to provide a good communication context at these given locations.

7.3.3.2 Dimensionality

In its original domain, the algorithm operates on three dimensions, the spatial separations on the x-, y-, and z-axis, which also have an identical interpretation as it is irrelevant for the definition of spatial separation between two individuals if the remote object's distance is greatest on the x-, y-, or z-axis. In contrast to this, a cognitive radio system will have a large number of parameters that it can choose to adapt to and therefore has a high number of dimensions on which the algorithm may operate. These dimensions will also have different interpretations and the distance metrics cannot be directly exchanged, that is, reducing separation along dimension "transmit power" by three units does not improve overall reception if separation along the dimension "frequency" is increased by one unit.

7.3.3.3 Variable Types and Dynamics

In addition to a different interpretation of dimensions, the typing of variable can frequently differ between the source and the target domain. While spatial positioning and distance metrics use continuous variables, many variables in a cognitive radio system are not expressed as continuous values, but rather integers that may only be defined between certain lower and upper boundaries. This makes the use of a Euclidean "distance" equivalent for swarming difficult. An additional aspect is created, as certain variables in a cognitive radio system, for example, signal strength as a function of transmit direction, may also exhibit nonlinear behavior. This must be accounted for in the "distance" function.

Although these general differences exist, it is still feasible to transfer the algorithm to be used for the control of communication networks. For this step, and to overcome the differences discussed earlier, it is necessary to adjust the algorithm to the requirements and assumptions of a wireless communication network and replace the corresponding input and output variables to allow for proper operation in the target domain.

One approach to proctor this adjustment is to transfer the algorithm semantically, that is, for the entire algorithm and for each sub-behavior contributing to the algorithm's working, an abstract function that this sub-behavior provides in nature is extracted and an equivalent abstract function meaningful to a cognitive radio but at a high level performing a similar objective is developed. This equivalent function can then be populated using the adaptation parameters and sensing capabilities that a specific cognitive radio has to obtain for a concrete implementation.

When viewed at a high level, the main objective of a group of fish to form a school is to minimize the danger of predators [2], as a group is more difficult to attack than individual members. Forming and staying in a school formation offers protection to the individual, and individuals adjust some of their available adaptation parameters (their spatial position) to meet this objective.

When viewed at a high level, the objective that a cognitive radio network has is quite similar. It also wants to protect its communications against negative outside influences and obtain the best possible connectivity between its members. Although being physically close would certainly improve the communication links of the cognitive radio network, each cognitive radio also has a variety of other parameters it can choose to adapt to achieve a good communication environment while shielding itself against negative outside influences, for example, its transmission power, frequency setting, modulation scheme, or encoding parameters.

These parameters, when combined with a statistical analysis of how they affect a given target variable (in this situation, maintain good links to peers and reduce outside interference) as obtained through the use of fractional factorial designs, are then used to adapt and transfer the algorithm to the target domain. If the objective was to improve inner-network communication and reduce the impact of interference, a cognitive radio could then, for example, increase its transmission power to achieve that objective. Similarly, if equipped with a directional antenna, steering the antenna in the direction of the receiver would achieve the same result.

Corresponding mappings can then be found for all other available parameters to which the cognitive radio can freely adapt, and for all components of the algorithm that need to be transferred. This will ensure that while the algorithm is suited to be applied in the target domain, the general functioning will still be preserved.

7.3.4 Example Adaptations for a Cognitive Radio Swarm Algorithm

The last section discusses how the process of transferring an algorithm can be conducted in general; this section will highlight this adaptation process using a set of specific cognitive radio example parameters. These mappings will be conducted using the process described earlier for the three sub-behaviors of cohesion, obstacle avoidance, and alignment using the cognitive radio parameters transmission power, antenna directionality, frequency setting, and energy expense.

7.3.4.1 Cohesion

The abstract function of the cohesion rule is to provide an action that will maximize the algorithm's overall objective, in the case of schooling fish gaining protection, or in the case of cognitive radio networks improving the inner-network communication links and reducing outside influences. For the area of schooling fish, this can be achieved by moving the individuals closer to each other, similarly wireless communication devices would correspondingly try to maximize the strength of their communication links to peers in the network.

This strengthening of links can be done in multiple ways using the adaptability of the cognitive radio. For example, as the radio can increase its transmission power, it could increase its power until a sufficient link quality is achieved. If we assume symmetrical links for the sake of simplicity and a link between nodes i and j has a link gain of g_{ij}, the cognitive radio transmitter intends to send at a high enough power so that the remote station will receive the signal at a sufficient signal strength. If P_{min} denotes the minimum required received signal strength and P_{tx} denotes the sender's transmission power, the signal strength received at the remote station given the link gain g_{ij} can be expected to be $P_{tx} * g_{ij}$. The sender intends to use sufficient power so that the signal is received with at least P_{min} at the destination; thus, if the expected received power is less than the minimum required strength, it needs to continuously increase its emission. The trivial solution would certainly be to always transmit with maximum power settings, but this strategy will unnecessarily deplete a mobile station's battery, and if all network nodes follow the same approach, this will lead to unnecessary high interference levels in the network. Therefore, this setting must be counterbalanced, a function accomplished by the obstacle avoidance rule.

A second strategy to increase the fidelity of the links would be to modify antenna patterns if the cognitive radio is equipped with directional antennas, as other interfering transmissions coming from different directions could be dampened as much as possible. Ideally, the outgoing transmission should be directed to the receiver's current position, who would also only listen in the direction of the sender. The less accurate this steering would be, the lower would be the power received at the remote station. While this directionality loss would depend on the specifics of the directional antenna at hand, the hypothetical antenna with a cosine directionality gain would highlight the concept: if you consider $\overrightarrow{\text{dir}_{rx}}$ the receiving direction of the remote station and $\overrightarrow{\text{dir}_{tx}}$ as the transmitting direction of the local station, the offset of the two directions will determine how much energy can be received at the remote node. If both stations are aligned, the full signal can be received, for orthogonal settings no signal will be received at the other end. Using the hypothetical cosine antenna, this directionality factor will be the vector product of $\overrightarrow{\text{dir}_{tx}} * \overrightarrow{\text{dir}_{rx}}$.

The strength of the received signal would finally also depend upon the frequency being used to transmit and receive the signal. As there exists variability in the channels, that is, different channels will be better suited to transmit the signal due to less interference or competing stations, the cognitive radio could also select the parameter frequency setting to maximize its communication to remote stations. In contrast to the previously discussed parameters, however, frequency is a binary variable; if we assume a discrete nonoverlapping channel system, a transmission can only be received if being transmitted on the exact same channel. If the channels are overlapping or the cognitive radio can freely adapt its transmission frequencies at a fine granularity, this effect is less drastic. To use this parameter to maximize the abstract object of cohesion, the cognitive radio adjusts its transmission and received frequency in such a way that it maximizes its received signal power.

These observations about the cognitive radio's available parameters' effect on the value of cohesion can then be summarized in a similar way as in Equation 7.1, as shown in Equation 7.5. Following the characteristics of the parameters transmission power, antenna directionality, and frequency setting as discussed earlier, the cohesion sub-behavior using Equation 7.5 now will try to align the antenna direction, increase the transmission power to a sufficient level, and use a transmission frequency that will maximize the received signal strength at the remote station, such that the maximal possible link strength is achieved. Note that these parameters might be adjusted and maximized on a per node basis, that is, the radio would use different parameters for different remote stations, maximizing link strength for each as much as possible.

$$\left\{ \overrightarrow{\begin{pmatrix} 0 \\ f_i \\ \overrightarrow{RX\text{dir}_i} \end{pmatrix} \begin{pmatrix} P_{tx} * g_{ij} - P_{\min} \\ f_p \\ \overrightarrow{TX\text{dir}_{pi}} \end{pmatrix}} \right\} \quad \forall i \in S_i \tag{7.5}$$

7.3.4.2 Obstacle Avoidance

The abstract function of the obstacle avoidance rule is to minimize the effect of outside influences and to provide a counterbalance to the cohesion rule. In the case of the schooling fish, this rule maintains a minimal distance from all other objects. In the case of the cognitive radio network, this rule would minimize the impact of outside influences and limit the growth of certain parameters if too high levels have negative side effects, as discussed for the example of transmit power. It therefore makes sense to consider two cases under the collision avoidance rule, first the case of avoiding outside influences and second its function as a counterweight limiting the cognitive radio's parameters to sufficient but reasonable levels.

7.3.4.2.1 Avoiding Outside Influences

In a later deployment, each cognitive radio will be subject to some amount of interference, either by primary users or interference by surrounding secondary users. Each of these users will contribute to some extent to the noise floor that each radio will be exposed to on any given channel, and it is the radio's objective to minimize the amount of experienced interference to maintain a good communication environment. The easiest way to avoid these interferences would be to switch to a channel that currently experiences the lowest noise floor and where the cognitive radio can still communicate with its peers.

7.3.4.2.2 Providing a Counterweight

As discussed earlier, the cohesion rule will try to improve the communication links, which may result in some parameters being continuously increased to a maximum value, which then hurts the performance of the entire network. The collision avoidance rule acts as a counterweight and limits such parameter values to the minimum possible value that still meets the underlying objective. For the example of transmission power discussed earlier, there exists a minimal required signal strength P_{min} that should be met or exceeded by a small amount. If this is carried to excess, no further improvements can be gained, it will rather result in degraded performance due to high-energy expenses and systemwide interference. The corresponding collision avoidance rule would therefore penalize the transmitter for every unit that exceeds the minimum required value (+ perhaps a safety margin) and consequently self-regulate the system so that it stays at sufficient and reasonable power levels. Similarly, other counterweighting rules must be designed if the parameters used in the cohesion rule may be subject to self-amplification or if the value function is set up in such a way that the swarm algorithm would continuously increase the parameter value.

$$\frac{-\sum_{i\in S_i}\left|\overrightarrow{\left(\frac{P_{tx} * g_{ij} - P_{min}}{ce(P_{tx})}\right)}\right| - \sum_{i\in C_o} P_{tx}}{|S_i \bigcup C_o|} \tag{7.6}$$

where $C_o = \left\{k \in \left(\bigcup_{\forall k|k\neq i} S_j\right) \mid \left|\overrightarrow{p_j^i p_k^i}\right| < r_o \text{ and } (f_k = f_j)\right\}$

Equation 7.6 models these considerations as a rule that can be directly used by the swarm algorithm's collision avoidance sub-behavior. In addition to counterweighting the transmission power parameter, this formalization also considers the expenses generated through the radio's energy consumption, thus instructing the system to minimize these expenses as well, as each consumed unit of energy penalizes the overall utility function by a certain amount specified by the function *ce*().

7.3.4.3 Alignment

The abstract function of alignment is to predicatively react to group members' changes such that a stable configuration is maintained and no change in the triggering of the other two sub-behaviors is required. The sub-behavior of alignment is not a requirement for the proper function of swarm behavior. In the case of cognitive radio control through swarm behavior, there also exists another issue that makes the use of alignment in the way it was used in the biological algorithm difficult, if not impossible, as the system parameters are frequently discrete and thus no intermediate values exist that could be taken on before a rule such as collision avoidance or cohesion is actually triggered.

To create a function similar to the biological algorithm, the cognitive radio could do a trend prognosis on the parameter subject to control with the cohesion and obstacle avoidance rules, and thus obtain the possibility to predicatively react to redundant future changes.

7.4 Cognitive Radio Networks

After the discussion of the general biological concept of swarming and a presentation of how such a biological algorithm can be transferred and adapted to the domain of communication networks, this section presents a specific cognitive radio swarm model that was implemented as a hardware deployment (simulation results are available in [5]). This section will also focus on certain aspects not present in the biological algorithm, that is, requirements and limitations that are unique to technical hardware and therefore need to be integrated into the algorithm itself.

7.4.1 A Cognitive Radio Swarm Model

Using the transferral approach presented in the last section, we implemented a cognitive radio control algorithm based on the principles of swarming to distributively manage the radios' configuration. Although such an algorithm could use a variety of different parameters for the adaptation process, this work — to explore the feasibility of the general concept and to maintain tractable performance analysis—will focus only on a frequency adaptation model through swarming and leave the configuration of the other parameters open to different components of a cognitive radio system.

The control algorithm's main objective in the framework of the swarm behavior is to achieve the lowest possible interference to the local radio and the network as a whole. This interference is measured in terms of background noise and interference generated by other stations that are not a part of the cognitive radio network at hand. While avoiding interference to allow for the best possible link fidelity, the control algorithm must also consider the remote station's spectrum availability and the current network flows that are either generated, received, or routed by the local station at hand. Thus, the swarm algorithm's objective can be summarized as follows: select the local transmitter frequency/frequencies in such a way that the interference experienced by the local station (and through the principals of the swarming approach also simultaneously maintained for all remote stations with which that the local radio is in contact) is minimal while maintaining a feasible link connection to all remote stations in the network with which the radio is currently communicating.

Although both hardware and software deployment are built on exactly the same model of swarming as discussed in the previous section and use the same objective function as described earlier, the implementation specifics are sufficiently disjoint to warrant explaining each implementation part separately.

7.4.2 Hardware Implementation Using Low-Cost Commodity Hardware

This section discusses a series of experiments that were performed on hardware to evaluate the ability of the control algorithm to configure a cognitive radio network in the presence of interference and to interoperate with other solutions in heterogeneous environments. This task of interference avoidance and frequency reuse is just one of many tasks that can be performed by a cognitive radio network. In addition to spectrum allocation and dynamic adaptability, cognitive radios can be used for geo-location-aware configuration or general dynamic spectrum access both in cooperative and in uncooperative scenarios.

7.4.2.1 Implementation Strategy

Due to economical considerations, the implementation was based on a commodity hardware platform, specifically a set of Dell Inspiron laptops using a built-in wireless network interface card. This card was internally using the Atheros wireless chipset, which provided a low-cost, limited software–defined radio due to the flexible reconfigurability that can be achieved using the wireless driver software. In the deployment at hand the wireless laptops were running the Linux operating environment, thus providing the opportunity to flexibly adapt the card's parameters using the Madwifi [10] driver.

The systems could also be equipped with limited sensing capabilities as provided by the WiSpy spectrum scanner [14], a low-cost noise floor measurement unit for the IEEE 802.11 frequency band. A custom-written driver for this device would provide the cognitive radio with a just-in-time view of the environmental noise floor and would provide measurements with a 1 MHz granularity with a maximum of 2 millisecond resolution.

These two devices, the transceiver card and the spectrum sensor for the 802.11 channels, were directly controlled by the swarm behavior control algorithm, which was implemented as a separate structure aside from the regular network protocol

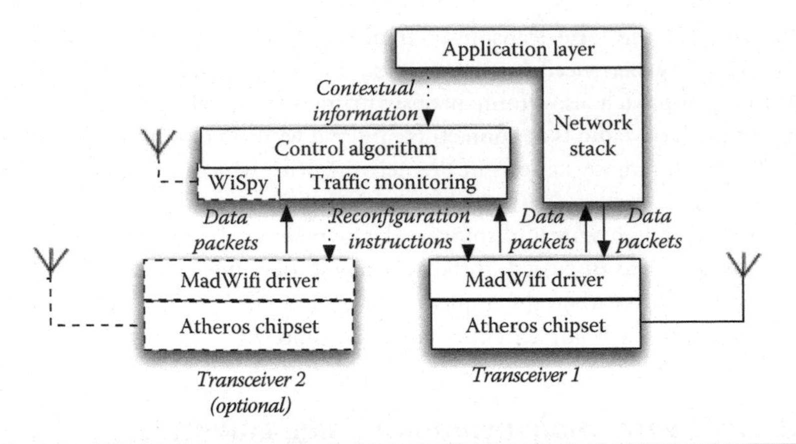

Figure 7.5 System architecture of the control algorithm's implementation on a commodity hardware platform.

stack. This architecture was deliberately chosen to minimize software maintenance expenses and to provide for maximal software portability of the system.

Figure 7.5 shows the schematic architecture of this approach. The Atheros chipset connected to an external antenna is controlled by the Madwifi driver that exports a regular network interface toward the Linux operating system. Using this wireless interface, data packets flow to and from the network stack, through which they are passed to reach the application layer. This part of the deployment uses all stock hardware and software components and does not require any modification of any parts. The Madwifi driver in this cognitive radio setup is also used to export a second interface to which all received and decodable packets (not only those that are specifically intended for the local receiver) are being sent. This so-called monitor-interface allows the cognitive radio control algorithm to overhear all data on the channel and thus get a passive view of which stations are present and transmitting on the frequency. While this interface is being used for active data communication with other network nodes, it is typically set to a specifically chosen transmission frequency during most of the time, as switching frequencies on this primary interface might result in frames being temporarily unreceiveable by the local nodes. To overcome this difficulty, the control algorithm also has the opportunity of managing and listening to a second wireless interface. With this optional network adaptor being used only for overhearing network traffic and thus passively collecting information about the current network environment, the primary interface card can be completely devoted to receiving and sending payload traffic. In addition to monitoring these devices, the control algorithm is also directly collecting data from the WiSpy, and is, in addition, able to learn other contextual information directly from the application layer about network topology issues such as, to which remote stations links should be established and maintained or which remote stations are not part of the current application. After reading all this environmental information as input parameters, the control algorithm then makes frequency selection decisions using the algorithm described in the previous section and correspondingly configures the primary (and if present the secondary) transceiver's frequency.

To save resources, the control algorithm is only invoked in certain intervals that are freely configurable, and should be varied depending on the variability of the environment and the system's application context. Even though the control algorithm is in sleep mode, most of the time it will still acquire all information generated during its idle period as the input streams to the control algorithm buffer incoming notification messages and sensor data. In order to analyze the behavior of the system at a fine granularity and to observe its adaptation decision and the resulting reactions of the other network peers, the control algorithm was invoked only in one second intervals.

As it is difficult to manage the availability of links in a hardware testbed, that is, control which node is able to see and can communicate with another peer, network topologies were virtually created through the application layer's interface with the control algorithm that allowed for the generation of different network topologies during the experiments.

Following the general architectural design considerations previously discussed, the cognitive radio control algorithm encapsulating the swarm rules was built as a separate structure parallel to the general network stack, so that all incoming data packets could be intercepted without implementing any changes to the inner workings of the operating system. The systems were placed in an ad hoc mode and initiated to send out beacons every 100 milliseconds.

7.4.2.2 Experimental Designs

At initialization, each node's wireless network adapter was tuned to a frequency chosen at random. Over the course of the experiment, it was the swarm algorithm's objective to maintain connectivity with the other nodes in the network and to avoid interference sources; to achieve that, the cognitive radio's control algorithm would be executed in one second intervals and form a decision as to which frequency it would set the wireless network adapter. This decision would be made using the rules of swarming behavior as described, and was based only on passive observation of its environment. This environmental information was collected by monitoring incoming data traffic on the channel to which the wireless network adapter was OK and by overhearing packets on adjacent wireless channels that spilled over and were possible to decode.

To evaluate the swarming approach to cognitive radio network control, two classes of experiments were conducted. The first type of experiments analyzed the swarm algorithm's ability to configure the network nodes using local information and observation only. The second type of experiments tested the interoperability between control algorithms, that is, it was determined whether the nodes controlled by the swarm approach could coordinate and cooperate with nodes following a different, incompatible protocol.

7.4.2.3 Convergence with Local Information

The objective of the first set of experiments was to determine whether local independent control exhibited by our swarming behavior control algorithm was suitable to properly configure and adapt a network of cognitive radio nodes to a common configuration.

When deployed in an environment, the cognitive radio network must perform two basic functionalities: (1) coordinate within a network to agree on a configuration that will allow all parties to communicate with each other and (2) vacate the spectrum and rendezvous on a different frequency once the band is claimed by a primary user or degraded through outside interference. It is these two properties for which the control algorithm was tested in this experiment.

These two configuration and adaptation tasks can be mapped into the following optimization problem: Let N denote the number of secondary user networks, each containing an unspecified number of network nodes; let M denote the number of primary users; and let C denote the number of channels suited for communication. Each node and each primary user resides on a particular channel, $\in [1, C]$. Let $f(i, j)$ be a binary function indicating whether a node of network $j \in [1, N]$ is present on channel i, $g(i)$ denotes a function returning the number of networks present on channel i, and $h(i)$ denotes a function returning the number of primary users on channel i. $\text{int}(i, j)$ is a binary function assessing whether any node of network i is in interference range of any node of network j.

The first functionality of the cognitive radio network, coordinate within a network to agree on a configuration suitable for all nodes to communicate, therefore requires formally that the value function "channel score,"

$$\text{CS} = \sum_{n=1}^{N} \max\left(\left(\sum_{c=1}^{C} f(c, n)\right) - 1, 0\right)$$

is minimized to reach the least possible value 0. When CS = 0, this assures that in the given channel selection problem, all networks of secondary users have converged on a frequency allocation that assures connectivity inside one's own network.

In addition, each network of secondary users should avoid outside interference as caused by other competing networks as much as possible. Stated in the optimization problem, this requires that the value function "channel utilization (sec),"

$$\text{CU}_\text{S} = \sum_{c=1}^{C} \max(g(c) - 1, 0)$$

is also minimized to reach the least possible value 0, thus preventing from two networks within interference range from selecting the same channel for communication unless it is inevitable. This situation will occur when $C < N$ and there exists at least one pair of networks that are in interference range from each other. For these situations, we can define the necessary channel overlap CU_S^* as seen from node i in network n as

$$\mathrm{CU}_\mathrm{S}^*(i,n) = \left(\sum_{j=1}^{N} \inf(n,j)\right) - C - 1.$$

Similarly, each network of secondary users must vacate frequencies claimed by primary users, thus the value function "channel utilization (pri)"

$$\mathrm{CU}_\mathrm{P} = \sum_{n=1}^{N} \max\left(\left(\sum_{c=1}^{C} f(c,n) * h(c)\right), 0\right)$$

must be minimized to reach the value 0.

The swarming algorithm present in each node i of the cognitive radio network n will independently try to minimize the overall value function

$$V = \mathrm{CS} + \left(\mathrm{CU}_\mathrm{S} - \mathrm{CU}_\mathrm{S}^*(i,n)\right) + \mathrm{CU}_\mathrm{P}$$

using only data obtained from passive observation of its environment so that the value function will reach its best possible value 0.

Given this value function that formally expresses the biological swarm behavior for the discrete domain of the channel selection problem, each node in the testbed was equipped with a cognitive radio control algorithm that would independently minimize the value function discussed earlier using the rules of swarming behavior. Following the specification of the IEEE 802.11 ad hoc mode, each node was transmitting beacons to make known its presence.

After the initialization phase and continuing in intervals afterward, a strong source of interference was placed on the channel on which the cognitive radio network had converged for communication, which interrupted communication within the network and then had to adapt and rendezvous on a different channel.

To demonstrate the ability of the swarming behavior to control an entire network and work with only a local and partial view of its environment, the deployment was set up in such a way that only one of the four cognitive radio nodes was directly affected by the interference source. After sensing the increased noise floor, the node affected by the interference source therefore vacated the channel that was now unsuitable for communication, and moved to an alternative channel.

As modeled in this test-bed experiment, events such as a primary user reclaiming spectrum or a channel becoming unavailable due to interference may only be visible to parts of a network in a deployment. Nevertheless, if a communication channel suddenly becomes unusable in a certain area and communication to nodes in this vicinity is interrupted, the rest of the cognitive radio network needs to learn about this event and reconfigure accordingly.

As the swarm control algorithm is designed to operate with local and passively collected information only, it is not necessary for the node that suddenly becomes jammed to communicate channel switch information to the rest of the network. Instead, it rather vacates the spectrum and can assume that all other nodes following the swarming behavior algorithm will detect the void and attempt to rendezvous on an alternative channel.

Figure 7.6 shows the typical response of a swarm-controlled network. The graph displays the view of the network environment of node 1, which was equipped with a WiSpy device for noise floor measurements. The background of the figure displays the measured noise floor at node 1 for a variety of frequencies depicted on the *y*-axis* and over time as depicted on the *x*-axis. The brightness of the background shows the intensity of the received noise floor, and one can easily see the impact of the interference source on channel 2 later switching to channel 1. The noise floor intensity was sampled at 100 millisecond intervals, whereas each node's control algorithm was waking up and making adaptation decisions in one second intervals. The control algorithms' long sleep cycles were chosen in the demo to achieve a fine granularity in the data, as all adaptation decisions are propagating slowly through the network and were therefore easy to observe and record without very tight distributed clock synchronization. In a deployment, much shorter sleep cycles are likely to be chosen.

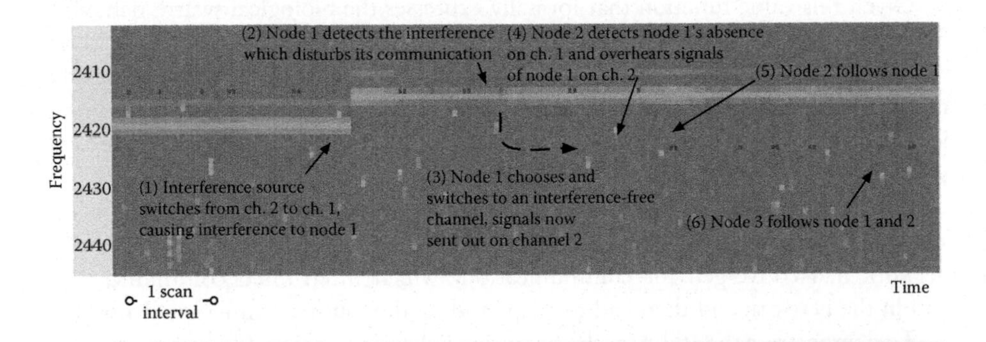

Figure 7.6 A typical response of a network controlled by the swarm algorithm: As soon as a node affected by the interference source switches to an alternative frequency, all other nodes able to observe its behavior will follow and rendezvous.

---

* The node was measuring 100 MHz of spectrum at a time; this graph is showing only the top 40 MHz of these measurements that exhibited activity during this sample experiment.

Figure 7.6 also displays node 1's passive view of the other cognitive radio nodes in the network. Each colored dot in the spectrum measurement represents a packet received on that particular frequency at that point in time. The color of the dot symbolizes the packet's sender, in this example node 2, as blue dots and node 3 as red dots.

As can be seen in the graph, in the beginning of this sample experiment the interference source was placed on channel 2 (2.417 GHz) with a 10 kHz sweep, such that communication on that particular channel would be distributed or interrupted. The three-node network had configured itself to use channel 1 (2.412 GHz) for communication, outside the influence of the interference source. Then the interference source was tuned to disturb channel 1. As node 1 was aware and affected by this jamming source, it chose and switched to an alternative frequency, channel 3 in this example. As the presence of the interference source was unknown to both nodes 2 and 3, these nodes continued to exchange data traffic on channel 1.

Due to the passive nature of the swarming approach, node 1 had not directly announced its frequency adaptation to the other nodes in the network (this might not be permitted or feasible on the now interfered channel 1), but rather started transmitting on its ad hoc mode beacons, network management frames such as ARP requests, and data packets on channel 3 as indicated in the figure.* During the next three scan intervals, node 2 detected that its neighbor had left channel 1 and moved to a new frequency. While it did not know the reason for this adaptation, it automatically rendezvoused with node 1 on channel 3. Similarly, node 3 detected the adaptation of nodes 1 and 2, and followed the switching neighbors to the new frequency.

To quantify how fast these frequency adaptation decisions would propagate within a network run by the swarm algorithm, a series of 25 experiments similar to the one described earlier was conducted. Each of the experiments started with a network of four cognitive radio nodes that had converged to a common configuration. Only one of the nodes was affected by the interference source and was able to sense the noise floor across the spectrum. All other network nodes were indifferent to the channel selection, but strived to maintain connectivity with their neighbors and achieve a fully converged network. The interference source was then tuned to the channel the network had selected for transmission. As soon as the first node (the one affected) selected and switched to an alternative frequency, it was measured how long it took until the entire network had followed the affected node and achieved stability to a new networkwide configuration.

Figure 7.7 shows the cumulative distribution of convergence time for 25 adaptation experiments. As can be seen in the graph, it takes the network in 50 percent of the experimentation runs less than 12 scanning cycles to propagate the channel switch

---

* As the graph depicts the received packets by node 1, its own transmissions are not indicated as colored dots.

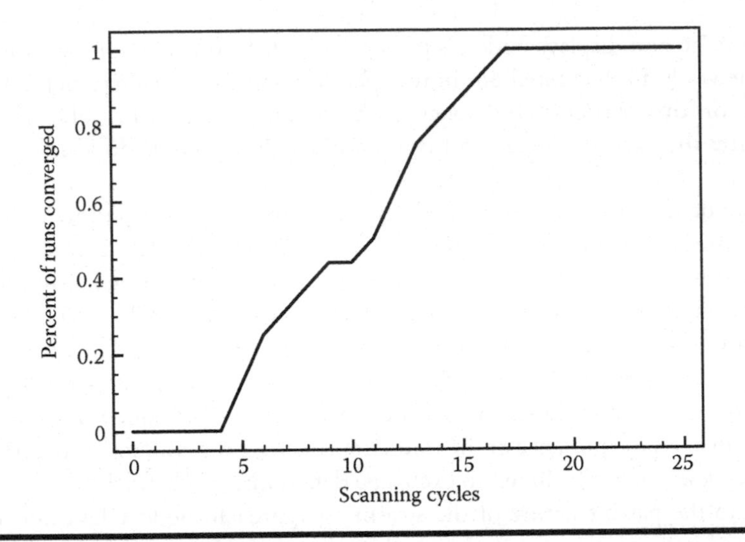

Figure 7.7 Cumulative distribution (CDF) of the network's convergence time to a stable systemwide configuration.

information starting from one node to converge to a networkwide configuration. After 17 scanning cycles, all the runs in this experimentation had converged.

7.4.2.4 Interoperability with Different Control Algorithms

The second experiment determined how well the independent local control algorithm could interoperate with other control algorithms deployed in a network that would use a potentially incompatible signaling or adaptation protocol.

When deployed in an environment, a particular cognitive radio network might only be one among many other networks of secondary users that the network at hand might need to avoid or cooperate with. In unmanaged deployments, however, one cannot assume that each network of secondary users will follow a protocol that is compatible with and understood by every other secondary user; in fact it will most likely be the situation that this is not the case. It is therefore of great importance that a certain cognitive radio control algorithm provides a degree of interoperability with other networks of secondary users and is able to either avoid or cooperate with them. In this experiment, the swarming approach was tested to meet these two requirements of being able to avoid and cooperate with incompatible networks.

An algorithm is defined to successfully avoid another network, if it migrates all its nodes to a different channel than the one that is currently being used by the competing network. If there is no vacant or suitable channel to migrate to, an algorithm should select such a channel or transmission parameter configuration that minimizes the outside interference on its nodes as much as possible. Consequently, an algorithm is defined to successfully cooperate with another network, if it selects such a configuration that all its nodes are able to communicate with the other network, that is, in these experiments the swarming approach should mirror all the frequency selection decisions the other protocol is making.

In this experiment, two cognitive radio networks were deployed, one being controlled by the swarming approach, the other one coordinated by a second, incompatible algorithm. To simplify the discussion, these networks will be referred to as network S and O, respectively. In the first part of the experiment, network S was instructed to view network O as a competing network, that is, it would try to maximize its own utility and avoid using the same channels as O, thus reducing potential interference on its own transmissions. In the second part of the experiment, network S was instructed to view the nodes of network O as friendly, thus the swarming algorithm had to configure itself in such a way that it would allow interoperability with the nodes of network O, even though they were following a different configuration protocol.

It can be expressed that each network node is trying to minimize the value function

$$V = \mathrm{CS} + (\mathrm{CU_S} - \mathrm{CU_S^*}(i, n)) + \mathrm{CU_P},$$

whereas in the first part of the experiment (avoidance), S and O are not part of the same network N, so that in the term

$$\mathrm{CS} = \sum_{n=1}^{N} \max\left(\left(\sum_{c=1}^{C} f(c, n)\right) - 1, 0\right)$$

only the coherence of S's nodes is of importance to the swarming algorithm, as well as that S is using a different channel than O as required through

$$\mathrm{CU_S} = \sum_{c=1}^{C} \max(g(c) - 1, 0).$$

In the second part of the experiment (cooperation), S and O are considered by the swarm algorithm's nodes to be part of the same logical network N, thus the algorithm will try to assemble all nodes in the same channel (condition CS) and not consider the two subnetworks as competing ($\mathrm{CU_S} = 0$ for S and O).

7.4.2.5 Avoidance

As discussed earlier, the first part of the second experiment determined to which extent the swarm algorithm can detect, avoid, and synchronize around a second, competing cognitive radio network. The second cognitive radio network was controlled through a means not known by the swarm algorithm, which could only observe the actions of the competing network.

Once the swarm algorithm had converged on a common frequency allocation, the competing network would invade that same channel and use it for its own communication. All nodes running the swarm algorithm could detect this invasion by listening to the transmitted packets on the channel and could potentially, on their own, select a new channel for transmission. As all other channels were idle in this controlled experiment, from their perspective, any other channel would have scored better and each node was allowed to make the first move. To model information that might be distributed unequally within a network, one of the swarm nodes was again equipped with a WiSpy device and could select channels based upon the traffic patterns on each frequency and its overall noise floor level while all other nodes would form their decision only on observed traffic patterns. The experiment measured the time it took after the competing network invaded the swarm network's channel until the swarm had fully converged on an alternative frequency. This experiment was replicated 25 times, both for one and two invading nodes.

Figure 7.8 shows the cumulative distribution of convergence time for one and two invading nodes, each replicated 25 times. As can be seen in the figure, in 50 percent of the experiments, the swarm algorithm converged to an alternative frequency allocation with three adaptation cycles after two nodes intruded, and

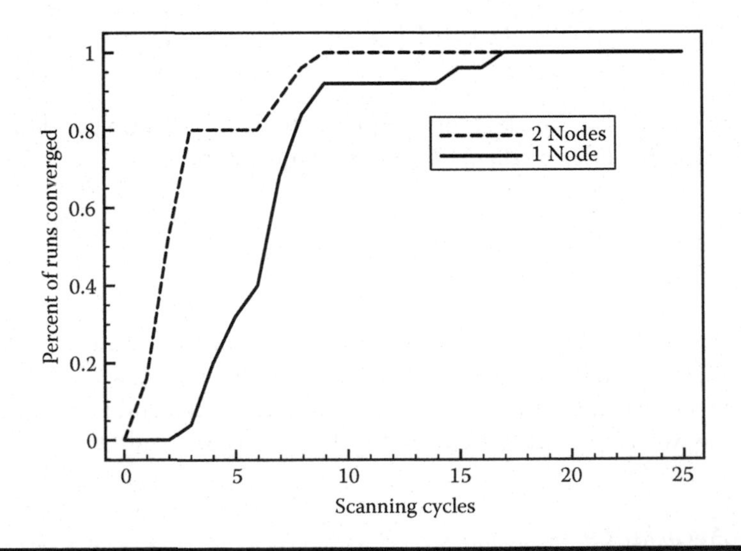

Figure 7.8 Cumulative distribution (CDF) of the network's convergence time when synchronizing with other, incompatible nodes.

seven adaptation cycles for one intruder. The entire network has converged in all runs in nine and seventeen adaptation cycles after intrusion, respectively. The higher performance in the case of two intruding nodes is due to two factors: first, the more competing nodes suddenly join the frequency, the earlier will other nodes learn about their presence from overheard data packets. Second, because for these experiments only a total of four cognitive radio nodes was used, when more nodes were controlled by the second algorithm, there were less nodes available to run the swarm algorithm.

7.4.2.6 Cooperation

The second part of the experiment tested how well the swarm algorithm could cooperate and synchronize with a second, incompatible algorithm. In this experiment, once all nodes had converged on a common frequency allocation, the nodes controlled by the second algorithm would make a collective decision to switch to a different, arbitrarily chosen, channel. As these nodes would not announce or plan this adaptation decision with the nodes of the swarm algorithm, these could only learn about this adaptation process by noticing the void where there had been packets transmitted on the current channel by the other nodes and overhearing packets on adjacent channels from these nodes if they were able to decode these packets due to spillovers. As soon as each swarm node had noticed this unannounced adaptation decision, it would rejoin the other nodes by switching to the same channel these nodes had chosen. In these setups, the time was measured that it took from the frequency adaptation of the second algorithm until all swarm nodes had caught up with the new frequency selection.

Figure 7.9 shows the cumulative distribution of convergence time for the adaptation process, replicated in 25 experiments for both one and two nodes being guided by a second control algorithm. As can be seen in the figure, in 50 percent of the scenarios, the swarm algorithm has learned the configuration decision of the second algorithm and has configured its nodes to meet up with the network nodes after six adaptation cycles when following one externally controlled node, and after four cycles when following two externally controlled nodes. In all the experiment replications, all swarm nodes have successfully mirrored and rejoined the other network after 24 and 19 adaptation cycles, respectively.

The time it takes for channel switch information to propagate through the entire network certainly depends on the amount of network nodes, the deployment density and the topology of the network, for example, how many neighbors are visible by a certain node.

As there were not enough cognitive radio systems to evaluate the algorithm's adaptation performance in larger network sizes using hardware, further experiments to analyze its performance characteristics were conducted in simulation. The results of these simulations are available in [5].

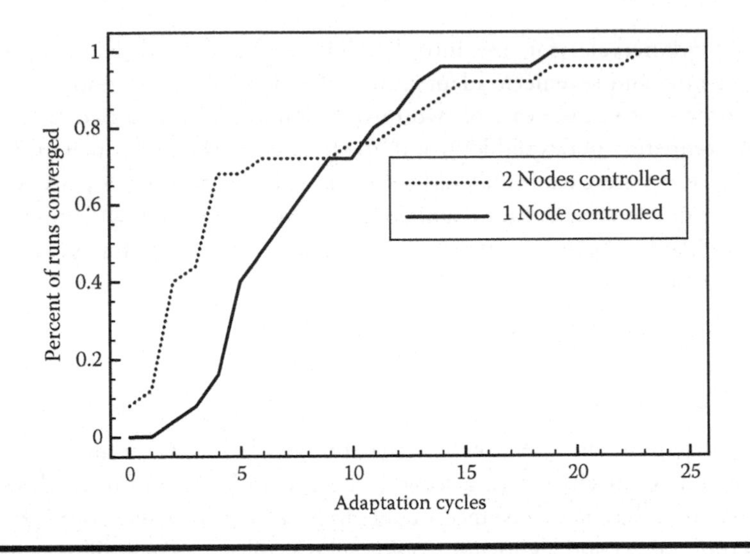

Figure 7.9 Cumulative distribution (CDF) of the network's convergence time when avoiding the presence of other, incompatible nodes.

7.4.3 Technical Limitations and Their Implications to the Biological Algorithm

Before concluding this chapter, it is worth examining the limitations of this approach. When applying the algorithm in the context of cognitive radio networks there might exist some technical properties that might limit the algorithm, for example, the radio might only be able to sense a limited number of frequencies at a time, or the transceiver can only receive and transmit a certain number of packages in any given time frame. To address these limitations, depending on the type and severity, it might potentially be necessary to modify and enhance the biological algorithm to also allow for convergence in the presence of hardware limitations. This section will discuss, using the biological analogy, how these limitations may be perceived from the perspective of swarming and a swarming-compatible solution can be crafted.

Figure 7.10 shows the impact of a cognitive radio only able to scan a limited number of frequencies at a time on swarming behavior. If we imagine running the algorithm in the 802.11 band, thus having 11 channels available at each time, technical limitations might only allow the hardware to sense and sample, for example, five frequencies at a time (current transmit frequency $\pm$ 2 channels). In such a case a certain amount of activity in the cognitive radio's neighborhood will go undetected. If these activities take place on frequencies other than those five monitored, these events will not be considered by the cognitive radio control algorithm, and convergence might therefore be slower than that if the algorithm could scan all available

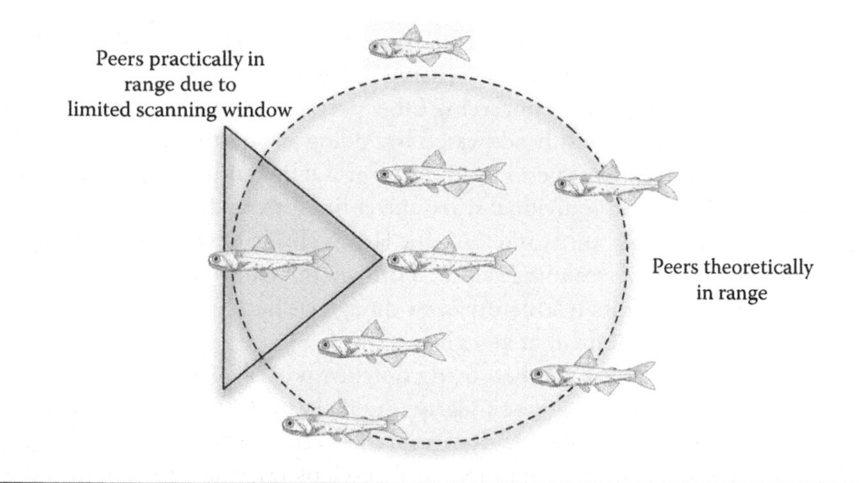

Figure 7.10 In case the hardware is not able to scan the environment at a high rate, adjustments can be made to the algorithm to allow for fast convergence.

frequencies at a time. This implication becomes visible if we translate this 5/11 ratio back to the original domain of swarming in schools of fish. A 5/11 ratio of visibility to a fish would mean that each individual is only able to see and detect other group members in a 160° viewing window as indicated in the figure. This would mean that many peers, even though present in its cohesion or obstacle avoidance range, are not visible to the individual and the individual therefore cannot react to their presence.

If we visualize this issue in the context of the biological domain as shown in Figure 7.10, it is easy to identify which problems to convergence will be created:

- There exist other peers in scanning range but not within the limited scanning window of the individual; therefore, it does not consider these hidden group members and does not align with them.
- Group members might be temporarily within the scanning window, but due to either their or the individual's movement they leave this window and are therefore also lost for the decision-making process.
- The individual might be within the scanning window of other peers, but they do not join up with the individual as initiated by the cohesion rule, because they see other items in their windows that prohibit an approach due to the triggering of the obstacle avoidance rule.

With this understanding, however it is very easy to design remedies that will address the issues of limited scanning windows, and this intuitively crafted solution for the biological domain will also address the convergence issues in a cognitive radio network. The remainder of this section will introduce three additional components to the cohesion and obstacle avoidance rules that will allow the swarm algorithm to converge even in the presence of partial sensing capabilities.

The first issue of an individual not seeing other peers, as they do not appear within its scanning window, can be easily addressed by adding an additional component to the cohesion rule. Triggered at periodic intervals or if it does not see any other peers in its scanning window, an individual scans the entire surrounding area (in the fish example performing a 360° turn) and joins the largest cluster of group members that it has identified during this scanning action. Through this additional component the group can create and maintain structure even though each individual is not capable of seeing the entire environment at any given time.

The second issue of other members being only temporarily within the individual scanning window can be solved in a similarly intuitive way. Consider the case where a peer leaves the individual's view and does not return into its view. From a biological perspective, it would make sense to briefly turn and look into the direction the other group member disappeared to, and if there exists a cluster that the other member was joining, to join as well.

That these intuitive rules would also address potential convergence problems due to limited sensing in cognitive radio networks is shown in Figure 7.11a. After a series of adaptation decisions the nodes A, B, C, and D settle on frequency 4, whereas nodes F and G detected each other on frequency 9. As neither the first nor the second group can see the others due to the limited scanning window (channel 4 ± 2 and channel 9 ± 2), both groups stay on their corresponding channels and therefore do not form a joint network structure.

Consider now node E that is in range to peer D, F, and G and therefore has links to both subnetworks. If E would execute the first additional convergence rule, that is, scan all frequencies in corresponding intervals, it would first see and announce its presence on channel 4, thus briefly joining the first subnetwork. On its way to complete the entire environmental scan, it would then come across the second

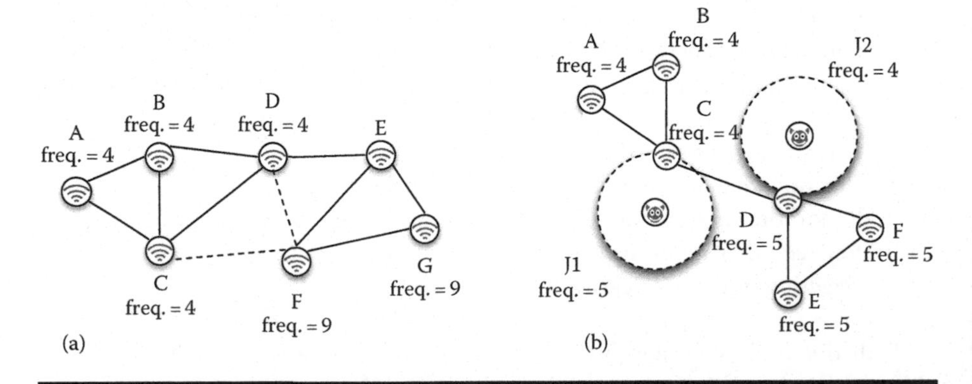

Figure 7.11 Fast convergence is maintained in the presence of limited scanning capabilities when additional rules are specified.

subnetwork on channel 9 where it would detect the presence of other nodes and be also detected by them. Because, from the perspective of node E, channel 9 is the better option (two network nodes instead of one), it will stay and permanently associate with F and G. After a while D notices that node E, which temporarily associated with it, left and did not come back, and executing the second additional convergence rule, now also scans the other frequencies to investigate where that node left to. Once it detects the other cluster it will join these nodes if possible and a corresponding scan-and-join action will propagate based on the same rule through the first subnetwork, thus creating a cohesive network structure despite the limited sensing capabilities.

The third issue of remote peers not joining up with a local node, because they see obstacles that the local node does not see, can also be solved through an intuitive additional rule. If in the school of fish the remote peer is not joining the local cluster, the local individual can infer that from the perspective of the remote group member its respective position is superior to the local node's situation and therefore does not join. While this may or may not be the case and cannot be decided for sure by an individual due to the limited scanning capabilities, if this static situation persists for a certain amount of time, either one should join the other group member on a trial basis as it can then determine if the remote situation is indeed superior and in that case stay there, which will trigger the remaining cluster to mirror that decision.

Figure 7.11b depicts how such a situation can look like in a cognitive radio network. Nodes A, B, and C have converged on frequency 4 and nodes D, E, and F have settled on frequency 5. While node C detects node D on an adjacent frequency, it does not join this remote node because it also senses the influence jammer J1 on channel 5, which from its local perspective would make the remote configuration an inferior choice than its current one and therefore does not join the other subnetwork. After a certain time in this static, suboptimal situation has passed either one of the two nodes (C or D) decides to try out the remote station's selection. If the selection turns out to be better it will stay there, otherwise it will return to its original setting. In this example, node C will not be able to create the merger, but once D tries out channel 4 it will stay there and the remainder of its subnetwork will mirror this decision, thus merging the two subnetworks into a cohesive structure.

With the help of these rules it is then possible to maintain convergence using the swarm algorithm for both the biological original and the technical adaptation of this concept in the presence of limited sensing capabilities. The reason why such rules have not evolved in the biological domain, and needed to be derived at this point, is that many species engaging in swarming have 360° vision.

7.5 Conclusion

Based on the theoretical analysis and hardware implementation as presented in this chapter, it can now be concluded that cognitive radio networks can be configured efficiently using an approach based on locally observable information and decision making. This alternative strategy of configuration management pursued by each individual can be realized on various levels. When it comes to the swarm algorithm used for the coordination of cognitive radio nodes into a dynamically adapting network structure, it became evident that a localized approach is well suited for managing such a system under dynamic environmental conditions and can cope well with limited sensing capabilities and only partial knowledge about the system. The algorithm however showed certain limitations when large network sizes came into play due to its probabilistic inner workings; however, there exist additional components that could address these issues and should to be investigated in future research. We also demonstrated that local control based on biologically inspired algorithms is well suited for the coordination of cognitive radio nodes in heterogeneous environments. In summary, it can therefore be concluded that the area of local control is a feasible and worthwhile avenue for cognitive radio configuration and coordination and shows promising prospects for other areas in wireless systems such as ad hoc networks.

References

1. S. Camazine, J.-L. Deneubourg, N. R. Franks, J. Sneyd, G. Theraula, and E. Bonabeau. *Self-Organization in Biological Systems.* Princeton University Press, Princeton, NJ, 2001.
2. C. W. Clark and R. Ducas. Balancing foraging and antipredator demands: An advantage of society. *Am. Nat.*, 144:542–548, 1994.
3. I. D. Couzin, J. Krause, R. James, G. D. Ruxton, and N. R. Franks. Collective memory and spatial sorting in animal groups. *J. Theor. Biol.*, 218:1–11, 2002.
4. C. Doerr, D. C. Sicker, and D. Grunwald. What a cognitive radio network could learn from a school of fish. *3rd International Wireless Internet Conference* (*WICON*), Austin, TX, 2007.
5. C. Doerr, D. Grunwald, and D. C. Sicker. Local independent control of cognitive radio networks. *International Conference on Cognitive Radio Oriented Wireless Networks and Communication* (*CROWNCOM*), Singapore, 2008.
6. End-to-End Reconfigurability (E2R) Integrated Project (IP). http://www.e2r.motlabs.com, 2007.
7. W. M. Hamner. Aspects of schooling in *Euphausia superba*. *J. Crustacean Biol.*, 4(1):67–74, 1984.
8. K. Kelly. *Out of Control: The New Biology of Machines, Social Systems and the Economic World.* Addison-Wesley, Reading, MA, 1994.
9. H. Kitano. Biological robustness. *Nature*, 5:826–837, 2004.
10. The MadWifi Project. www.madwifi.com, 2008.
11. D. P. O'Brien. Analysis of the internal arrangement of individuals within crustacean aggregations (euphausiacea, mysidecea). *J. Exp. Mar. Biol. Ecol*, 128:1–30, 1989.

12. C. Reynolds. Flocks, herds, and schools: A distributed behavioural model. *Comput. Graph.*, 21:25–34, 1987.
13. S. Vogel. *Cats Paws and Catapults: Mechanical Worlds of Nature and People*. W. W. Norton and Company, New York, 2005.
14. WiSpy spectrum analyzer. www.metageek.net, 2008.

Chapter 8

Quality-of-Service in Cognitive WLAN over Fiber

Haoming Li, Qixiang Pang, and Victor C. M. Leung

Contents

8.1 Introduction

Cognitive wireless local area network (WLAN) over fiber (CWLANoF) is a new architecture [1] that applies advanced cognitive radio [2] and advanced broadband radio over fiber (RoF) [3] technologies to infrastructure-based IEEE 802.11 WLAN extended service sets (ESSs) comprised of multiple access points (APs), each forming its own basic service set (BSS), to provide centralized radio resource management (RRM) and equal spectrum access through cooperative spectrum sensing. In this chapter, we examine how this architecture supports QoS provisioning in a WLAN ESS with a high degree of flexibility. We first review the architecture of CWLANoF [1] and discuss approaches on spectrum sensing and interference avoidance/mitigation in CWLANoF. After surveying existing RRM methods, most of which employ fixed channel assignment (FCA), we discuss how CWLANoF enables new dynamic channel assignment (DCA) strategies, and propose a reinforcement learning (RL) approach for QoS provisioning in CWLANoF. The QoS-provisioning problem is formulated as a Markov decision process (MDP), whose parameters are defined according to the QoS framework specified by 802.11e [4]. We present a solution framework using the Q-learning algorithm [5], which has several advantages that contribute to the suitability of this framework. The symbols and abbreviations frequently used in this chapter are listed in Table 8.1.

8.2 Cognitive WLAN over Fiber

IEEE 802.11 WLANs share the industrial, scientific, and medical (ISM) band with other devices, such as Bluetooth radios and microwave ovens. As these ISM-band devices are independently operated, it is difficult for a WLAN AP to negotiate radio frequency (RF) spectrum usage with them. Whereas cognitive radio techniques have been proposed for secondary users to exploit spectrum holes left unused in licensed frequency bands by primary users of the allocated spectrum, such techniques may also be exploited to enhance the efficient utilization of the unlicensed ISM band via spectrum sensing [6], and interference avoidance and coexistence [7]. In this case, each AP senses interference from other ISM-band users and changes its own frequency allocation when interference occurs. It then informs neighboring APs in the ESS through some inter-AP protocol, so that they can change their frequency allocations accordingly to avoid inter-AP interference. If there is a central WLAN controller in the ESS, the interfered AP can also report the interference event to the WLAN controller, which then coordinates the frequency plans across neighboring APs. However, spectrum sensing in the existing schemes is carried out at individual APs in a distributed manner, which may impair its reliability due to propagation impairments [8]. Frequency allocations at APs also have limited flexibility because commercial APs are usually equipped to utilize a single radio channel in each frequency band that an AP is equipped to operate with; for example, an 802.11a/g dual-band AP can operate over at most two radio channels simultaneously. However, existing cognitive radio techniques that classify users into primary and secondary users may not be directly applicable in the ISM band, because users of the ISM band have equal rights to access the radio spectrum and cannot be classified as primary or secondary.

8.2.1 Cognitive Radio in WLAN over Fiber

The CWLANoF technology can be applied to more efficiently utilize the ISM band in a WLAN ESS by employing cognitive radio techniques that have been suitably

Table 8.1 Frequently Used Abbreviations and Symbols

Terms	*Explanations*
ACI	Adjacent-channel interference
AI	Artificial intelligence
AP	Access point, defined in IEEE 802.11
BSS	Basic service set, defined in IEEE 802.11
CCI	Co-channel interference
CCU	Central control unit, defined in the radio over fiber architecture
CogAP	Cognitive access point, defined in the cognitive wireless local area network over fiber architecture
CSP	Constraint satisfaction problem
CWLANoF	Cognitive wireless local area network over fiber
DCA	Dynamic channel assignment
ESS	Extended service set, defined in IEEE 802.11
FAP	Frequency assignment problem
FCA	Fixed channel assignment
ILP	Integer linear programming
IM3	Third-order intermodulation
IP	Integer programming
K^n	Complete graph with n vertices
LP	Linear programming
$m(v)$	Number of colors demanded by vertex v
MDP	Markov decision process
N_c^i	Number of channels needed at the ith antenna
RAU	Remote antenna unit, defined in the radio over fiber architecture
RF	Radio frequency
RL	Reinforcement learning

Table 8.1 (continued) Frequently Used Abbreviations and Symbols

Terms	*Explanations*
RoF	Radio over fiber
RRM	Radio resource management
STA	Station, defined in IEEE 802.11
TPC	Transmission power control
v	Vertex in a graph

modified for use in the ISM band. The centralized architecture of CWLANoF systems enables cooperative sensing and, consequently, reduces the interference detection time while improving the detection accuracy. The multichannel-carrying capability of advanced broadband RoF systems [3] can significantly increase available radio resources at each WLAN AP. By implementing dynamic RRM based on accurate spectrum sensing, interference avoidance or mitigation can be easily accomplished. Effectively, the CWLANoF architecture enables the new concept of applying cognitive radio techniques for equal spectrum access in the ISM band. Before elaborating on this concept, we first give a brief introduction to CWLANoF systems.

In a conventional WLAN, each AP incorporates an 802.11 radio modem and a bridge between 802.11 and the distribution system, usually an 802.3 Ethernet. In a CWLANoF system, radio modems and bridges in the APs are moved to a centralized common control station (STA); the resulting simplified APs are now named as remote antenna units (RAUs). By centrally processing broadband RF signals received from the RAUs, the common control STA in a CWLANoF system becomes a cognitive access point (CogAP) that has a complete picture of the radio spectrum usage in the coverage area of the WLAN ESS. RAUs are connected to the CogAP via optical fibers in a logical star topology.

The largest difference between conventional WLANs and CWLANoF systems is the number of radio channels simultaneously in use at each AP. Because a broadband RAU can receive and deliver RF signals in the entire ISM band to the CogAP, a powerful *central control unit* (CCU) can be employed at the CogAP for improved spectrum sensing using sophisticated algorithms. In a conventional WLAN system, spectrum sensing at distributed APs does not allow the direction of interference to be easily identified, but this may be possible when the CogAP collects spectrum snapshots from several RAUs for processing jointly at the CCU. The CogAP essentially maps the spectrum usage within its coverage area and, based on this, the CogAP can proactively avoid interference by optimizing channel allocations to the RAUs.

CWLANoF systems can also transmit/receive over multiple channels at each RAU [9]. Powerful dynamic RRM algorithms can therefore be implemented in the CCU to organize the RAU transmissions for interference avoidance or mitigation. The increased flexibility on RRM enables equal spectrum access in the ISM band and eases the implementation of DCA compared with conventional WLAN systems.

CWLANoF also offers a potentially higher system capacity and lower total system cost than conventional WLANs. As RAUs become cheaper, it will be possible to deploy a greater number of RAUs to form a picocellular WLAN system, providing higher throughput and wider coverage at a lower total system cost than a conventional WLAN system. Picocells in CWLANoF also provide more line-of-sight propagations; therefore, the transmission power of both RAUs and WLAN STAs can be reduced. Intercell interference can consequently be greatly reduced, thus easing the task of RRM at the CogAP.

8.2.2 CWLANoF System Model

We are interested in WLANs operating in unlicensed frequency bands. To avoid wordiness at the expense of precision, we use "ISM band" to collectively refer to the 2.4 GHz ISM band (2400–2483.5 MHz), the U-NII band in North America (5150–5350 MHz, 5725–5825 MHz), and the CEPT band B in Europe (5470–5725 MHz).

The architecture of a CWLANoF system is illustrated in Figure 8.1. In this example, the CogAP connects 12 RAUs, providing an enterprise WLAN service to a three-floor building. Each RAU is equipped with one dual-band antenna for transmission over the ISM band and another one for reception [10]. These antennas are connected to the CogAP via two independent fiber links. The fiber pairs from the RAUs form a star topology with the CogAP. The star topology does not involve optical multiplexing and, thus, allows the use of inexpensive RAUs. However, additional costs are incurred due to the number of fibers and optical interfaces needed in the CogAP. The CogAP in Figure 8.1 serves the whole building and connects to other buildings via 802.3 Ethernet over fibers or cables. STAs are desktops, laptops, or handheld terminals, equipped with 802.11a/b/g network adaptors. The placement of RAUs is based on propagation measurements and the expected density of STAs in a given area. We assume that most of the RAUs cover low-density areas and only a few RAUs cover high-density areas. One general rule is first placing RAUs for high-density areas and then for low-density areas [11]. The details of RAU placement are outside the scope of this chapter.

8.2.2.1 Medium Access Issue in CWLANoF

Compared with an AP in a conventional WLAN, a CogAP in CWLANoF can cover a much wider area, manage a larger number of active channels, and offer a larger system capacity. Before discussing possible functions of the CogAP, we clarify how the

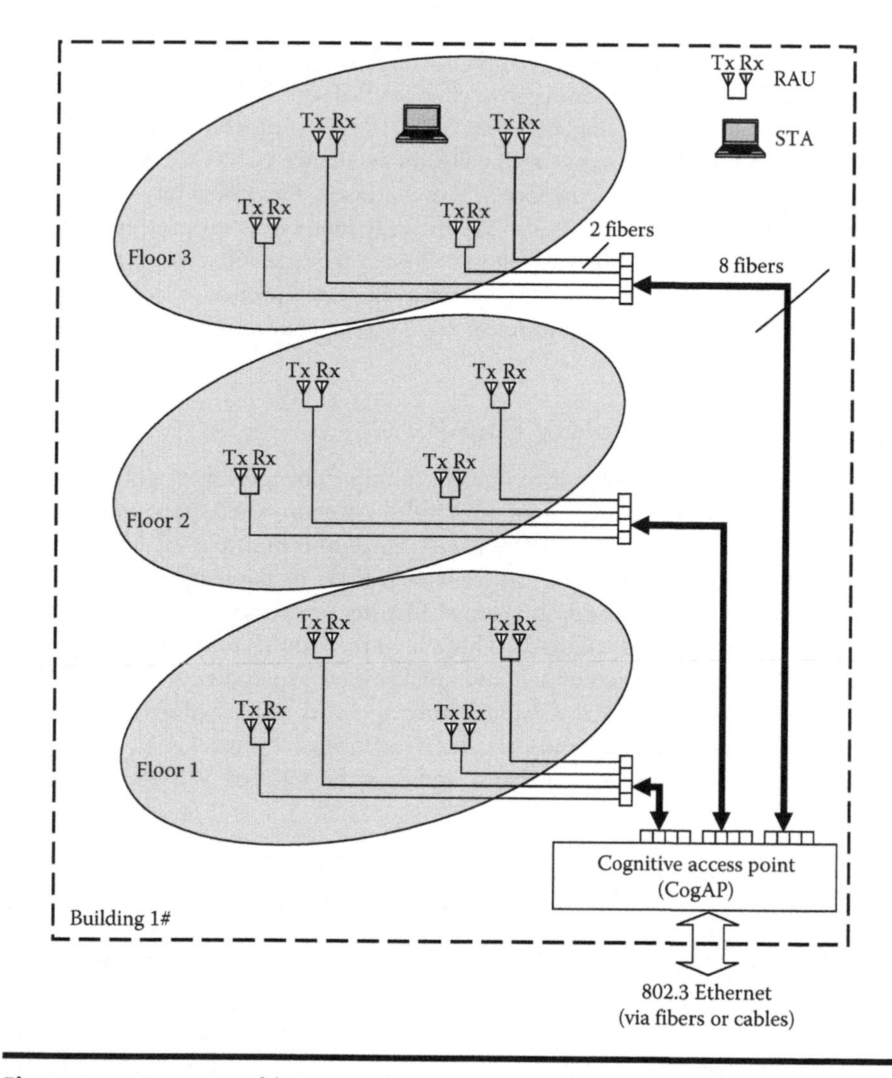

Figure 8.1 System architecture of cognitive WLAN over fiber (CWLANoF).

RoF technology impacts on the medium access control (MAC)-layer performance of a WLAN. First, protocol-processing functions in the CogAP are located farther away from the STAs than in a conventional WLAN, causing higher round-trip times (RTTs) between the CogAP and STAs. The larger the area covered by the CogAP, the lower is the total system cost, but the higher are the RTTs. These higher RTTs will reduce channel access efficiency, especially for systems employing time-division duplex. The increase in RTT can be compensated by adding double the nominal RTT to the 802.11 MAC parameters, *ACK timeout* and *CTS timeout*, such that they are larger than the distributed coordination function (DCF) interframe space

(DIFS) [12]. Simulations verified that the throughput is not noticeably affected by 1 km long fibers [13]. Another negative effect of RoF on WLAN is the lower efficiency of distributed scheduling. Since the 802.11 MAC protocol, the DCF is based on carrier-sensed multiple access with collision avoidance (CSMA/CA), when the RTTs between the STAs and the CogAP are increased, the probability of collision between the CogAP and STAs is also significantly increased; however, it does not affect the collision probability between the STAs as they are still close to each other [14]. The RTS/CTS scheme has been shown to alleviate the effect of collisions [15] by avoiding the hidden terminal problem and keeping collisions short.

8.2.2.2 Design Concepts of CogAP

Our goal in the CogAP design is to operate multiple channels over the ISM band through each RAU to increase system capacity and support equal spectrum access in the ISM band. Although the CogAP transmits signals over multiple WLAN channels to each RAU, the RF signal returned from each RAU to the CogAP contains the spectra of the entire ISM band, as channel filtering to extract the desired WLAN channels is performed at the CogAP. This allows the CogAP to detect interference within the ISM band more accurately and quickly via the spectrum usage assessment unit, as shown in Figure 8.2. Channel filtering is accomplished at the CogAP, for example, using tunable band-pass filters controlled by the CCU. Following RF/baseband conversion, the baseband signal of the selected WLAN channel is

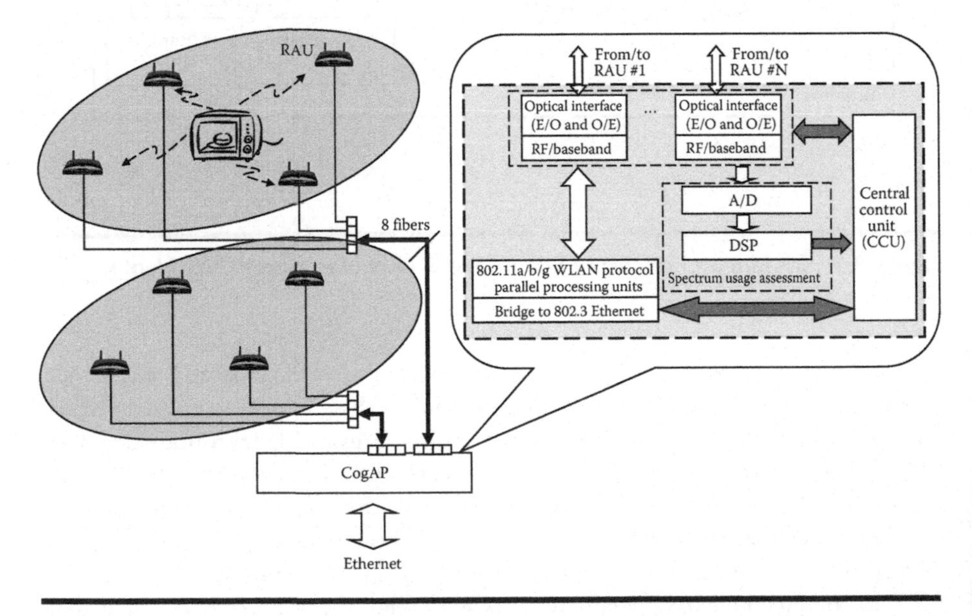

Figure 8.2 Cooperative spectrum sensing at the cognitive AP (CogAP).

then demodulated and processed by the MAC protocol processor. With large-scale integration and powerful processors, protocol processing for many WLAN channels can be performed in parallel within the CogAP.

In the physical (PHY) layer, the CogAP can exploit macro-diversity instead of micro-diversity, because signals received from widely separated RAUs have much lower correlations than those received over multiple antennas at a single location. If each RAU is also equipped with multiple antennas, we can further implement micro-diversity in conjunction with macro-diversity. However, if we keep the number of fibers between each RAU and the CogAP the same, i.e., one to transmit and one to receive, wavelength division multiplexing would then be required to deliver RF signals from/to different antennas attached to the same RAU [16]. Exploitation of macro- and micro-diversity in the proposed CWLANoF is an interesting problem left to future research. In the following, we focus on the design of the CCU to enable dynamic spectrum access in the ISM bands. This design incorporates functions including spectrum sensing, interference avoidance and mitigation, RRM, and cognitive MAC. The discussions are from the CogAP design viewpoint, such that all signals received over the ISM band that do not belong to the CWLANoF system are classified as interference.

8.2.3 ISM-Band Spectrum Sensing and Interference Avoidance/Mitigation

ISM-band spectrum sensing from the CogAP viewpoint includes detecting usage of the spectrum by external systems and monitoring WLAN co-channel interference (CCI) and adjacent-channel interference (ACI). Interference can either be avoided or mitigated. Existing spectrum sensing and interference avoidance/mitigation techniques can take advantage of centralized control in the CCU to enhance their effectiveness. In this section, we first describe potential ISM-band users, and then review existing spectrum-sensing techniques and interference avoidance/mitigation techniques. Finally, we discuss how to apply these techniques in CWLANoF systems.

8.2.3.1 External Systems Sharing ISM Band

Potential users of the 2.4 GHz band include microwave ovens, cordless phones, wireless personal area network (WPAN) devices such as Bluetooth and ZigBee, and wireless metropolitan area network (WMAN) devices such as WiMAX. Users of the 5 GHz band include cordless phones and WiMAX devices.

8.2.3.1.1 Microwave Oven

There are two types of microwave ovens: inverter based and transformer based. Their noise is modeled as continuous-wave (CW) pulse interference with a slowly drifting CW frequency around 2450 MHz [17,18]. The pulsing frequency of transformer-type ovens is AC frequency (50 or 60 Hz), and that of inverter-type ovens ranges from 10 to 30 kHz [19]. Based on WLAN experiments and power spectral density measurements of oven noise, the interference from microwave ovens covers the whole 2.4 GHz band, most intensely impacting WLAN channels 8, 9, 10, and 11 [19,20].

8.2.3.1.2 Cordless Phone

Cordless phones operate at 2400–2483.5 MHz and 5725–5850 MHz [21], using frequency-hopping spread spectrum (FHSS) or direct-sequence spread spectrum (DSSS). The bandwidth of FHSS cordless phones is 1 MHz and the 6 dB bandwidth of DSSS is less than 2 MHz [22]. The range of cordless phones is up to 30 m at 2.4 and 5.8 GHz, owing to their high transmission power. For example, some models of WDCT (worldwide digital cordless telecommunications) cordless phones set the transmission power from 23 to 29 dBm for FHSS and from 15 to 26 dBm for DSSS [23].

8.2.3.1.3 Bluetooth (802.15.1-2005)

Bluetooth devices use FHSS, hopping over 79 channels with a 1 MHz channel spacing in the 2.4 GHz band. The hopping set is reduced to 23 channels if adaptive frequency hopping is applied. The hopping frequency of data packets is up to 1600 hops per second, i.e., the minimum time staying at one frequency is 625 microseconds. The channel access is synchronized by the master in a Bluetooth piconet. The modulation is GFSK (Gaussian frequency-shift keying). We assume no communication link between Bluetooth and WLAN devices, so it is impossible to support a collaborated coexistence as defined in 802.15.2-2003.

8.2.3.1.4 ZigBee (802.15.4-2006)

ZigBee devices use 16 channels in the 2.4 GHz ISM band, with a 5 MHz channel spacing. Every 4 bits are mapped into one 32-chip sequence, and each chip is modulated by O-QPSK (offset quadrature phase-shift keying). The data rate is 250 kbps, and the chip rate is therefore 2 Mcps. ZigBee devices use CSMA/CA to access the channel. Their airtime usage is very low: under 1 percent [24]. Low costs of ZigBee devices require cheap channel filters, which determine their low transmission power (−3 to 10 dBm). Based on these properties, ZigBee devices are not likely to interfere with WLAN or other ISM-band users.

8.2.3.1.5 WiMAX (802.16)

WiMAX carriers have access to licensed bands at 2.3, 2.5, and 3.5 GHz (Europe) or a lightly licensed band at 3.65 GHz (United States). It is not likely for WiMAX

Table 8.2 Summary of ISM-Band Systems

	Frequency	*Power*	*Signal*
Microwave oven	2400–2483.5 MHz	WLAN channels 8, 9, 10, and 11 are most intensely affected	CW-pulse at 2450 MHz Pulsing frequency: 50 or 60 Hz for transformer type; 10–30 kHz for inverter type
Cordless phone	2400–2483.5 MHz 5725–5850 MHz	FHSS: 23–29 dBm DSSS: 15–26 dBm (range ≤30 m)	FHSS: BW = 1 MHz Frequency dwell time: 10 milliseconds DSSS: 6 dBr BW <2 MHz
Bluetooth	2400–2483.5 MHz	Class 1: 0–20 dBm Class 2: −6 to 4 dBm Class 3: ≤0 dBm	FHSS: BW = 1 MHz Frequency dwell time: 0.625 milliseconds, 1.875 milliseconds, 3.125 milliseconds
WLAN	2400–2483.5 MHz	≤30 dBm (United States) ≤20 dBm (Europe) ≤23.4 dBm (Japan)	802.11b: DSSS 802.11g = 802.11a
	5150–5250 MHz	≤16 dBm (IEEE)	OFDM with convolutional code
	5250–5350 MHz	≤23 dBm (IEEE)	
	5725–5825 MHz	≤29 dBm (IEEE)	
ZigBee	Not considered in this chapter		
WiMAX			

users to operate over the 2.4 or 5 GHz unlicensed bands; therefore, in this chapter, we do not consider the coexistence between WLAN and WiMAX.

A summary of the above ISM-band systems is given in Table 8.2.

8.2.3.2 *WLAN Co-Channel and Adjacent-Channel Interference*

In a WLAN ESS, the interference in the ISM band also includes CCI and ACI. The capabilities to detect CCI and ACI strongly affect RRM efficiency.

8.2.3.3 Existing Spectrum Sensing and Interference Avoidance/Mitigation Techniques

Depending on the scale of the interference area [25], local or cooperative sensing may be employed. Local sensing only involves one sensor. Cooperative sensing uses multiple sensors distributed geographically, and can be further classified as centralized or distributed. Centralized cooperative sensing requires a central control to fuse data collected from sensors; distributed cooperative sensing relies on local data exchange between neighboring sensors. The latter offers lower implementation complexity than the former [26]. Depending on available communication channels between sensors, exchanged information could be hard decisions, soft decisions, or simply analog signals.

For each individual sensor, if we know the features of the interference, feature (or coherent) detection can be performed by matched filter or cyclostationary detection. If we know nothing about the interference, energy detection can be used. If we want to learn features while detecting, machine-learning methods can be applied [27].

Feature detection is informed detection, and therefore outperforms energy detection. If the interference comes from a modulated signal with built-in periodicity, cyclostationary detection can be exploited and it outperforms matched filter detection. Feature detection is usually complex and requires perfect synchronization (for matched filter detection) or long observation intervals (for cyclostationary detection). Therefore, in practice, energy detection is widely used due to its simplicity. Energy detection over all kinds of environments has been studied, such as independent (Rayleigh) fading, independent shadowing, and spatially correlated shadowing [28]. Cooperative energy detection has been investigated for both hard cooperation [28] and analog signal cooperation, including amplify forward (AF) and decode forward (DF) [29]. The AF/DF method actually originates from signal cooperative diversity [30].

Machine learning is a new direction for spectrum sensing. It is particularly attractive for the ISM band where any type of radios satisfying certain emission constraints might be developed in the future, making it difficult to deal with some unknown systems operating in the ISM band. Even if the types of interfering systems are known, dedicating a coherent detector for each type of system would be costly. However, machine learning can learn the features of any type of system over time and update the knowledge base, which is then used by the reasoning engine to provide the sensing decision [27].

When the interference is detected but avoidance is not possible, interference mitigation techniques are needed. For this broad topic, our discussions are limited to protecting WLAN signals from interference in the ISM band.

Compared with 802.11b/g signals, microwave oven emissions are wideband interference. FHSS signals from Bluetooth or digital cordless phones can be modeled as narrow-band interference (NBI) when the WLAN packet transmission time is shorter than the hopping periods of the FHSS signals. As wideband interference mitigation is very difficult, we focus on some NBI mitigation methods specially designed for WLAN.

One interesting method of NBI mitigation in WLANs is to correct the Euclidean distance metrics used by 802.11a/g convolution decoders. Particularly, if the receiver can identify which subcarriers in an orthogonal frequency-division multiplexing (OFDM) symbol are smeared by NBI, it will think that the corresponding Euclidean distance metrics are not reliable and correct the metrics that are sent to the convolution decoder. So the NBI is mitigated by the power of error-control coding [31,32]. As the hopping period of Bluetooth FHSS is 625 microseconds, this mitigation method only works under the assumption that the WLAN data rate is high enough so the FHSS interference will not hop or only hop once during the packet transmission.

Another NBI mitigation method works in the MAC layer. When a WLAN STA senses interference, the exponential backoff mechanism would be invoked even if the packet collides with the interference instead of packets from other STAs. To avoid this unnecessary backoff, WLAN STAs in a BSS can use spectrum-sensing techniques to estimate when the interference is on and refrain from applying backoff for packet retransmissions that occur during this time. Without unnecessary backoff, packets can be sent immediately when the interference is no longer present. This MAC-based mitigation was shown to increase WLAN throughput by 30 percent in the presence of microwave oven noises [23]. Although the method works effectively for transformer-based microwave ovens due to their long pulsing periods, it does not work for inverter-type microwave ovens, where the 30~50 kHz pulsing frequency makes it ineffective to exploit the very short oven-off periods for packet transmissions.

8.2.3.4 Enhancements in CWLANoF

We now describe how to design interference detection, avoidance, and mitigation mechanisms for CWLANoF systems.

8.2.3.4.1 Detection

Energy detectors are not suitable for detection of non-WLAN interference for CWLANoF systems because they cannot detect FHSS/DSSS signals [33] and cannot distinguish between interference and normal WLAN channel usage. In fact, IEEE 802.11b gives three carrier-sensing methods: energy detection, DSSS detection, and their combination; on the other hand, 802.11a and 802.11g do not specify particular sensing methods for OFDM. WLAN STAs in the market tend to support energy detection only due to its lower cost. The sensitivity on STA cost suggests that we should focus on interference detection at the CogAP rather than at the STAs.

We consider centralized cooperative sensing to take advantage of the centralized CWLANoF architecture. Similar to the AF method, the raw spectrum data is collected from the distributed RAUs via fibers, and full diversity can be achieved. The required sensitivity at the CogAP is therefore reduced. The fiber delays between the CogAP and all RAUs are fixed; this allows for tight synchronization among signals collected from neighboring RAUs and ensures good performance of the AF method.

It is difficult to evaluate the effects of shadowing on spectrum sensing at the CogAP, because shadowing is difficult to model and varies with different environments [8]. Although theoretically RAUs are not likely to be simultaneously blocked from the interference source by the same object, it still requires experiments to verify that spatially correlated shadowing effects are small in CWLANoF systems.

To detect CCI and ACI at a RAU in a CWLANoF system, the CogAP first sends the CTS-to-self packet with a longtime reservation on all channels driven to the RAU under test. All STAs belonging to the RAU will set the network allocation vector and keep silent. The CogAP then uses multitaper method singular value decomposition (MTM-SVD) [34] or other energy detection methods to estimate CCI and ACI levels of the RAU under test.

The CogAP can also consider cyclostationary detection to detect CCI and ACI, because the CogAP knows the exact timing and beacon information contained in CCI or ACI. Notice that it is not necessary or practical to silence a frequency for a few beacon intervals. To accomplish cyclostationary detection, the CogAP could synchronize beacons of the RAU under test and the neighboring RAUs such that the beacon frame of the RAU under test is ahead of the neighboring RAUs by one beacon frame and one CTS frame. The RAU under test can then have the chance to reserve the channel for better CCI and ACI detection. Cyclostationary detection is easier to implement than MTM-SVD. And because MTM-SVD is blind detection, we believe cyclostationary detection gives better performance. The above fine-tuning required for cyclostationary detection cannot be accomplished by a WLAN controller that manages distributed APs through some inter-AP protocols.

There is one important issue worth studying for spectrum sensing in CWLANoF. As signals across the ISM band returned to the CogAP through fibers, the spectrum-sensing unit at the CogAP needs to consider inter modulation distortions caused by the nonlinearity of electrical–optical conversions.

For external interference sensing, one important topic is how many RAUs should be included in a cooperative spectrum-sensing set. For fixed-location interference, the set is static. But for cordless phones or WPAN devices, the set would vary. Choosing too many RAUs for cooperative sensing increases system complexity and might have little performance gain because of the SNR (signal-to-noise ratio) wall effect [35].

8.2.3.4.2 Avoidance

Once an interference signal is detected in a WLAN channel, the CogAP can abuse the interframe space (IFS) requirement and immediately send "disassociation" or "deauthentication" broadcast to all STAs belonging to the interfered BSS. As disassociation or deauthentication is not a request but a notification, it shall not be refused by either the STA or the AP [4]. The reason code field can be filled with type 1, "unspecified reason," or type 5, "disassociated because the AP is unable to handle all currently associated STAs." For those STAs in the power save mode, they still listen to selected beacons according to their own *ListenInterval* parameters, and then receive broadcast and multicast transmissions following certain received beacons. So the CogAP could send the beacon indicating that there are broadcast messages buffered at the CogAP. After STAs wake to fetch the buffered messages, they are disassociated or deauthenticated.

Avoiding CCI or ACI is a RRM issue and will be discussed in Section 8.4.2.

8.2.3.4.3 Mitigation

For transformer-based microwave ovens, the CogAP can either detect the interference or use a dedicated interface to directly obtain phase information of the main power, provided that a constant phase difference is maintained between the power supplies of the CogAP and the microwave oven. The CogAP can then avoid unnecessary backoff during the oven-on period—at least downlink traffic is not heavily affected by the oven. One interesting question arises: How to let STAs know the oven on/off timing without upgrading STA hardware? A possible solution is to broadcast a CTS-to-self frame right before the oven is on and reserve the channel for the oven-on period. However, the channel could be busy when the CogAP attempts to access the channel. As the CogAP can find out from the PHY-layer convergence protocol header when the channel becomes idle, it can then broadcast a CTS-to-self frame right after the channel becomes idle. The CogAP sets the modulation of the CTS frame as 1 Mbps DBPSK (differential binary phase-shift keying), hoping the STAs can understand this frame with the help of processing gain and remain silent when the oven is on. As mentioned above, it is however impossible for a BSS to coexist with an inverter-based microwave oven, because the high pulsing frequency of the oven does not leave sufficient time for packet transmissions in the WLAN.

The FHSS interference to 802.11a/g OFDM uplink signals can also be mitigated by correcting the Euclidean distance metrics corresponding to bits mapped in interfered subcarriers, as used in [31,32].

In the presence of external interference, the above mitigation methods should be combined with other PHY-layer methods, such as adaptive coding and modulation; MAC-layer methods, such as dynamic-rate switching that takes advantage of a multi-rate PHY [36,37]; and automatic fragmentation [38].

In some network scenarios, it might be possible to mitigate the uplink CCI or ACI with the help of signal processing. Supposing RAU1 sends data via Channel 1 while RAU2 receives packets from Channel 1 or Channel 6, RAU2 uplink contains CCI or ACI interference from RAU1 downlink. Because the CogAP knows the RAU1 downlink exactly, it is interesting to investigate whether an efficient way exists to cancel the RAU1 downlink signal from the RAU2 uplink signal. In the case of ACI of 802.11a/g OFDM signals, as only the edge of RAU1 uplink is affected, correcting the metrics of subcarriers on the channel edge might be helpful to the convolution decoder performance.

We emphasize that downlink traffic in a WLAN is usually heavier than uplink. The above mitigation strategies mostly focus on uplink because our interest is the CogAP design. The more practical topic would be how to mitigate interference at WLAN STAs. More importantly, interference mitigation largely complicates the CogAP design. A more efficient way to increase system throughput is dynamic RRM, which also makes QoS provisioning possible when multiple WLAN users are present. In Section 8.3, we will focus on the QoS issue and explore how to devise RRM strategies to ensure QoS in CWLANoFs.

8.3 QoS in Conventional WLANs

With Voice over Internet Protocol (VoIP) applications, such as Skype and Windows Live Messenger, and video-streaming applications, such as YouTube and web seminars, getting widely used, it is necessary to ensure their QoS in the presence of other best-effort (BE) and background (BG) traffic flows. Wireless links between the AP and STAs often suffer shadowing and signal fading, causing the drop of PHY rate and packet retransmissions. These unpredictable situations make QoS provisioning in WLAN more challenging than in wireline media. In this section, we concentrate on QoS provisioning at the CogAP.

QoS methods in 802.11 WLANs are specified in 802.11e-2005, which was later incorporated into 802.11-2007 [4]. In a QoS-aware WLAN, the AP and STAs are formally referred to as QoS AP and QoS STAs, respectively. For simplicity, we drop the "QoS" prefix and continue to refer to these as simply AP and STAs. While continuing to use a channel access structure consisting of alternating contention-free periods (CFPs) and contention periods (CPs), as specified in 802.11-1999, 802.11e also incorporates a new access element called controlled access phase (CAP) that can occur in a CP or a CFP. This will be further explained in the brief review of 802.11e channel access below to describe how QoS is enabled. Readers are referred to [4] for details on 802.11e.

8.3.1 Channel Access in 802.11e

Extending the DCF defined in 802.11-1999, 802.11e specifies the hybrid coordination function (HCF) to provide both QoS-aware contention-based access, called enhanced distributed channel access (EDCA), and contention-free access, called HCF-controlled channel access (HCCA). The previous fragment burst method in DCF is extended as transmission opportunity (TXOP). TXOPs provide protected periods for multiple frame transmissions, reducing frame exchange overhead. Two types of TXOPs are defined to facilitate QoS provisioning:

- EDCA TXOP: TXOPs located in CPs and obtained by STAs or the AP via EDCA
- HCCA TXOP: TXOPs located in CAPs and obtained by STAs or the AP via HCCA

The idea of CAP will be explained together with HCCA below.

8.3.1.1 EDCA and Priority-Based QoS

Different user priorities (UPs) are assigned different IFSs (characterized by the arbitrary IFS number, AIFSN), minimum and maximum contention window sizes (characterized by CWmin and CWmax), and maximum TXOP duration (characterized by the TXOP limit). Priority-based QoS is provided based on these differences, although still subject to the DCF mechanism based on CSMA/CA. An important feature is that the default AIFSN for the AP is 1 while the minimum AIFSN for STAs is 2. This gives the AP the highest priority to acquire the channel and set up a CAP, which is the basis of HCCA.

Traffic streams (TSs) with eight different UPs, which can be identified in 802.1D priority tags, are mapped into four access categories (ACs), corresponding to audio, video, BE, and BG traffic. The AP assigns different AIFSNs, CWmin, CWmax, and TXOP limits to different ACs so that higher-priority TSs have better opportunities to access the channel. Note that the above priority-based QoS mechanism is not capable of providing hard QoS guarantees due to the DCF mechanism.

8.3.1.2 HCCA and Parameterized QoS

Parameterized QoS provisioning aims to guarantee QoS according to the traffic specifications (TSPECs) specified by each traffic flow to be admitted and scheduled by the AP. A few important parameters in a TSPEC are nominal MAC service data unit (MSDU) size, minimum/maximum service interval, minimum/mean/peak data rate, minimum PHY rate, burst size, inactivity interval, delay bound, and surplus bandwidth allowance. The AP, functioning as the hybrid controller (HC), admits TSs according to their TSPECs and ensures the QoS of admitted TSs by using HCCA to properly schedule their packets.

Under the HCCA mechanism, the admitted TSs are scheduled for transmissions in HCCA TXOPs by the AP according to their respective TSPECs. The AP creates HCCA TXOPs by allocating CAPs for contention-free packet transmissions.

This is a much more flexible mechanism than polling in CFPs based on the point coordination function (PCF), because CAPs can be allocated in both CPs and CFPs. To allocate a CAP during a CP, the AP first acquires the channel, because of its smaller AIFSN. During CPs, CAPs always alternate with EDCA TXOPs. The use of PCF to enable CFPs will not be considered further in this chapter, because PCF has not been implemented in commercial WLAN products, as far as we know.

8.3.2 *Traffic Stream: Admission Controller and Scheduler*

To provide a parameterized QoS support, the AP is required to incorporate a TS admission controller and a TS scheduler, which admits TSs and schedules CAPs for the transmissions of MAC frames from admitted TSs based on their TSPECs. Examples of an admission controller and a scheduler, along with five typical admissible TSPECs, are given in [4, annex K].

8.4 QoS in Cognitive WLAN over Fiber

Unlike conventional WLANs, CWLANoF systems are capable of operating more than one WLAN channels via each RAU (or cell). Therefore, QoS in CWLANoFs requires not only admission control and scheduling but also DCA. Following the convention in [39], we consider DCA in the context of the generalized frequency assignment problem (FAP). The FAP in cellular systems addresses the assignment of channels to radio cells according to their different traffic demands (or loads), while avoiding excessive intracell or intercell interference. Similarly, the FAP in CWLANoF is concerned with how to satisfy the traffic loads of different BSSs within a WLAN ESS by assigning one or more WLAN channels to each BSS. If demands are relatively fixed over time, as in radio and TV broadcasting systems, the FAP can be solved by FCA; if demands vary over time, as in cellular telephone and WLAN systems, the FAP is best addressed as a DCA problem. As FCA can be considered as the foundation of DCA, we first review existing FCA methods, and then describe DCA strategies in the setting of CWLANoF. Finally, we propose a Q-learning algorithm for QoS provisioning of video and audio streams.

8.4.1 *Existing FCA Methods*

FCA problems can be solved by graph theory, integer programming (IP), constraint satisfaction methods, or heuristic methods.

8.4.1.1 FAP in Graph Theory

One general solution to the FAP is to find optimal frequency assignments that minimize the (total) interference while satisfying given demands of all cells; hence, minimum interference FAP (MI-FAP) [39–41]. A FAP is an IP problem subject to CCI, ACI, and co-site constraints [42].

In the most ideal case, we quantize CCI as either 0 or 1, ignore ACI and co-site constraints, and assume that cells have equal loads. A FAP is then mapped into the classic vertex-coloring problem in graph theory: viewing channels as colors and cells as vertices, the FAP is to color all vertices with the minimum number of colors while no adjacent vertices have the same color [42–44]. This model is, however, not suitable for unequal cell loads. We want to allow vertices to receive different number of colors to reflect different cell loads. One way is to quantize loads as integers and replicate any vertex v by its loads, $m(v)$. Edges joined to the original vertex are also replicated and connected to each of $m(v)$ vertices. This extended graph, known as split interference graph [39], certainly becomes very complex. Another way to consider unequal cell loads is to form a weighted coloring problem [45,46]. However, the above formulations still have defects: both constraints and loads are roughly quantized. These defects require a general model in IP. But graphs as still very useful to provide good visualizations of FAPs.

8.4.1.2 FAP and Integer Programming

In practice, a FAP involves integer channel numbers, continuous inter-site or co-site interference measurements, continuous transmission power, continuous throughput performance, and continuous cell loads. It is understandable that graph coloring, as a topic of discrete mathematics, is not suitable for such a problem with many continuous variables. We therefore formulate the FAP as an optimization problem, in particular, an IP problem. This IP problem can be relaxed as integer linear programming (ILP) and solved through IP solvers, such as branch-and-bound and branch-and-cut methods [39]. But the IP solvers suffer the "curse of dimensionality" [47]. The difficulty of finding optimal solutions of ILP motivates methods in machine learning and artificial intelligence (AI). Koster formulated the FAP as a constraint satisfaction problem (CSP) and solved it as a binary ILP with the help of linear programming (LP) relaxation [40].

8.4.1.3 MI-FAP and Constraint Satisfaction Problem

A CSP is represented by a triple (Z, D, C), where Z is a finite set of variables $\{x_1, x_2, \ldots, x_n\}$, D_{x_i} $(\forall i \in \{1, 2, \ldots, n\})$ are the domains of variables in Z with each D_{x_i} containing a finite number of objects of arbitrary types, and C_Z is a finite set of n-ary constraints on Z [48]. As all elements in the above CSP are defined in finite domains, this CSP is called a finite CSP (FCSP) while other CSPs are simply referred to as CSPs. Finiteness allows efficient CSP solvers.

The solution of a CSP is a tuple, i.e., a set of values of $x_i(\forall i)$, that satisfies C_Z. If such a solution tuple exists, the CSP is satisfiable. Otherwise, it is overly constrained. If a CSP has multiple solution tuples, we define an objective function, $f : (Z, D, C) \rightarrow \mathbb{R}$, to find a tuple that minimizes f. The CSP now becomes a constraint satisfaction optimization problem (CSOP). When the CSP is overly constrained, we could only satisfy some part of the constraints, and need to define an objective function f that captures the performance of this partial satisfaction. This CSP is called a partial *CSP* (PCSP), which can be viewed as a generalization of CSOP.

Koster's work used split interference graph, where the ith antenna is replaced by a complete subgraph $K^{N_c^i}$, and N_c^i is the number of channels needed at the ith antenna. Each node (transceiver) in this $K^{N_c^i}$ corresponds to x_i in the above definition; the set of allowable channels at each node corresponds to D_{x_i}. Not surprisingly, the running time of the binary ILP solver in [40] increases exponentially on $|D_{x_i}|$. When $|D_{x_i}|$ is 6, the algorithm becomes very slow. In fact, solving this PCSP is NP-hard for $|D_{x_i}| \geq 3$ [49]. With larger $|D_{x_i}|$, the gap between ILP and LP also increases. One way to improve the efficiency is to exploit particular graph structures through tree decomposition [50]. The idea was motivated by the fact that cells of a mobile telephone system are placed on a surface and interference decreases when the distance separation increases. Its graph is therefore treelike. However, WLANs do not have this property; more complicated CCI and ACI suggest that the FAP in WLAN focuses on heuristics.

8.4.1.4 Heuristic Methods in MI-FAP

Previous heuristic methods include greedy, local search (LS), tabu search, simulated annealing (SA), genetic algorithms (GA), artificial neural networks (ANN), and ant colony optimization (ACO). The greedy, LS, tabu search, SA, and GA methods have been used as heuristics to solve IP [47], while ANN and ACO come from machine-learning and AI areas.

The simplest heuristic method is greedy algorithm: we first assign channels to a node that has the largest effects on the objective. For example, the node with the maximum node degree is first picked for assignment. This is actually the first greedy algorithm in FAP, named the frequency-exhaustive or the requirement-exhaustive method, proposed by Metzger [42] and verified by Zoellner et al. [51].

Usually greedy methods are also combined with LS. LS disturbs the previous solution in a neighborhood such that the solver has a chance to escape from a local minimum and arrive at another lower minimum, sometimes at the expense of hill climbing. However, the solver might come back to the previous local minimum after hill climbing and form an endless cycle. We therefore keep the latest solutions in a tabu list and forbid any new solution that belongs to the tabu list. This is the tabu search, using the search history and, thus, performing better than LS. Tabu search methods differ on choices of neighborhood (1 exchange or 2 exchange), search-stopping rules, and the number of tabu solutions kept in the list.

LS and tabu search are deterministic searches. The SA method is a probabilistic version of tabu search. The probability of using a new solution, denoted p, depends on the objective difference produced by the new solution and the best solution so far, denoted Δ. A larger Δ produces a smaller p. The p is often defined as $e^{-\frac{\Delta}{T}}$, where T is the temperature that decreases with iterations by $T = rT$ (r is the cooling ratio). One may notice that the idea of SA is similar to the softmax method in the multiarmed bandit problem, where the probability also takes a form of an exponential function [52]. SA methods differ on initial T, r, the stopping rule, and how often T is updated by $T = rT$.

While greedy, LS, tabu search, and SA methods are focused on improving a single solution at a time, the GA method is able to improve multiple solutions (population) at one iteration (generation). New solutions (offsprings) are generated by a pair of previous solutions (parents). GAs differ on how to choose parents and how parents are combined to generate offsprings (e.g., crossover and mutation). Other heuristic methods include the ANN method from the machine-learning area and the ACO method [53]. Refer to [39] for a bibliographic annotation on heuristic methods.

There are a few reasons that most practical WLAN systems do not need optimal solvers. One reason is that the data in ILP formulations have noise—both cell loads and interferences are averaged measurements. The "optimal" solution is not really optimal. Also, DCA in CWLANoF, when viewed as a sequential solution of FCAs, requires faster FCA solvers. We now move on to DCA strategies in CWLANoF systems.

8.4.2 DCA Strategies in Cognitive WLAN over Fiber

DCA problems can be solved locally as in conventional WLAN or cooperatively as in CWLANoF. The information exchange to support DCA could be centralized or distributed. All centralized DCA schemes are cooperative; some distributed schemes are also cooperative as they exchange information between neighbors [26]. In the setting of CWLANoF, DCA involves channel allocation and load balancing. DCA is also strongly related to spectrum sensing and cognitive MAC. In this section, we focus on DCA strategies in CWLANoFs.

In CWLANoF, the CogAP assigns more channels to a busy RAU (e.g., a RAU in a conference room), while providing fewer channels to other RAUs that are located in light traffic areas. The traffic demands of each RAU can be quantized as the number of channels required, defined as the weight of the RAU; the load-balancing and channel allocation problems can therefore be formalized together as a weighted coloring problem, which is however NP-complete. Therefore, suboptimal DCA algorithms need to be developed [54–58] in conventional WLANs. These algorithms can be used to solve DCA problems in CWLANoF systems simply by replacing APs by RAUs. However, some properties of RoF systems need to be examined.

8.4.2.1 Adjacent Channels in 5 GHz Band

One distinct advantage of CWLANoF is that a single RAU can support several channels and add much more flexibility on channel allocation across RAUs. Because channel allocation alone has potentials to largely improve the system capacity, a simpler heuristic DCA scheme might be enough to satisfy dynamic traffic, without even considering the transmission power control (TPC) of RAUs. This claim is based on the assumption that 12 nonoverlapping channels in a 5 GHz band are available to any RAU (in North America). However, the actual transmission mask of an 802.11a channel is not a brick wall; spectrum leakage not only raises the interference floor of adjacent channels, but causes false deferring in these channels if they use a fixed received-signal-strength-indication threshold for clear channel assessment, which may result in free channels being wasted due to activities in adjacent channels. Experiments have shown that severe ACI exists for both 802.11a and 802.11b/g channels (channels 1, 6, and 11) when a WLAN receiver is close to a STA transmitting at an adjacent channel [56,59]. These results suggest that adjacent channels should be allocated to different RAUs and, if possible, non-neighboring RAUs.

8.4.2.2 Priority of Allocating 5 GHz Channels

The 5 GHz band is not as heavily used and has more nonoverlapping channels than the 2.4 GHz band. We could separate RRM into two steps: first allocating 5 GHz channels to each RAU based on their traffic demands and, if possible, the geographical distribution of 802.11a-capable STAs; then allocating 2.4 GHz channels to some of the RAUs. Note that some STAs may only support the lower band of 5 GHz: 5150~5350 MHz. Channel allocation and load balancing are dynamic; thus, the above historical data can be accumulated at the CogAP.

Denser BSSs operating at the 5 GHz band could provide uniform STA rates in one BSS; this helps increase the BSS capacity when a DCF is used. Sparser BSSs at the 2.4 GHz band agree with the longer range of 2.4 GHz signals and avoid a sophisticated 2.4 GHz frequency planning. To reduce CCI in the 2.4 GHz band, we only allow one 2.4 GHz channel to be operated at each RAU.

8.4.2.3 Third-Order Intermodulation

The desired-to-undesired signal power ratios (DURs) of multichannel signals from/to a single RAU may be degraded due to the nonlinearity of fiber links [9]. The DUR can be measured by composite triple beat (CTB) and third-order intermodulation (IM3) [60]. As a degraded DUR causes reduced throughput per radio channel, before adding radio channels to a RAU, the DCA algorithm should estimate the DUR degradation that this additional channel might cause. The DCA algorithm can use the analysis method developed in [9] to estimate the DURs of multi channel signals by using premeasured carrier-to-noise ratio and adjacent-channel leakage ratio of single-channel signals.

The DCA algorithm might be able to pick one channel that gives an acceptable DUR degradation. If no such channel exists for a heavily loaded RAU, some of STAs associated with this RAU can be disassociated and then reassociated to another less loaded RAU. This function would be useful for a conference scenario.

The CCU also needs to control the optical interfaces to adjust the optical modulation index of lasers. This is particularly important in controlling intermodulation when transmitting multichannel signals to a single RAU [9].

Note that the IM3 issue had been addressed in mobile telephone FAPs [61,62].

8.4.2.4 Near–Far Effects over Multiple Uplink Channels

The CogAP can easily control the power of downlink signals at all channels, while uplink signals come from the geographically dispersed STAs and have a wider dynamic range than the downlink signals if the STAs are not subject to TPC. Therefore, at each RAU, an automatic gain control (AGC) unit is placed before the laser to avoid overmodulation, as shown in Figure 8.3 [10]. This AGC unit might cause the near–far effect in CWLANoF systems. Suppose that one STA is very close to the RAU and another STA is farther. If both STAs are simultaneously transmitting packets to the RAU with the same transmission power over different uplink channels, the stronger uplink signal would trigger AGC at the RAU and the weaker signal would suffer an SNR loss at the CogAP. The most efficient solution to this near–far effect is certainly the TPC of STAs, as used in CDMA networks. However, TPC is mandatory only for STAs operating at the 5 GHz band in Europe (802.11h-2003).

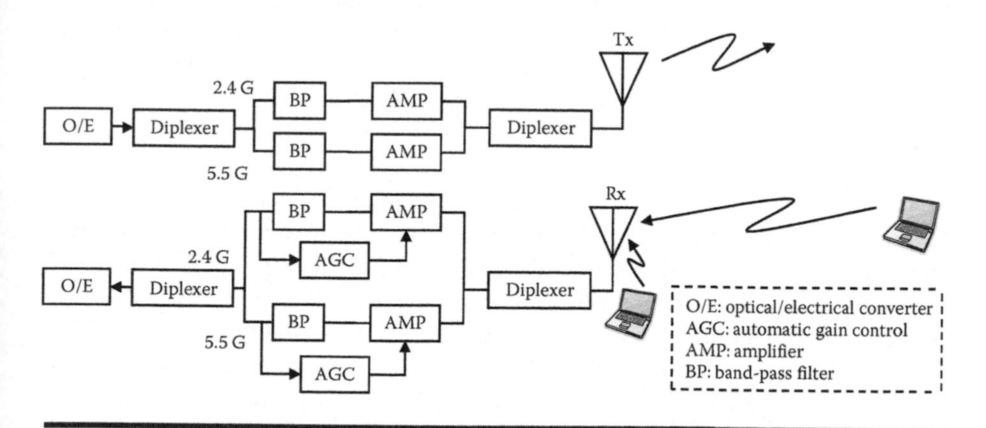

Figure 8.3 Near–far effects at remote antenna units (RAUs).

From the CogAP viewpoint, if we can limit simultaneous uplink traffic, the near–far effect could be alleviated. This scheduling constraint is useful when symmetric video or audio streams are delivered during CAPs.

8.4.3 QoS Provisioning Using Q Learning

So far, we have reviewed TS admission control and scheduling methods provided by 802.11e in Section 8.3, examined existing FCA methods in Section 8.4.1, and presented DCA strategies of CWLANoFs in Section 8.4.2. We showed that QoS provisioning in CWLANoFs requires admission control, scheduling, and DCA. As the large flexibility of CWLANoFs suggests a heuristic DCA method, our focus is to devise a low-complexity QoS-provisioning method that combines aforementioned heuristic DCA strategies in CWLANoFs and an admission controller in the framework of 802.11e. (Scheduling is an interesting research problem that will however not be discussed in this chapter.) We start with modeling the QoS-provisioning problem as a MDP, and then use the RL method, particularly, the Q-learning algorithm to solve the MDP. The advantage of using RL is to learn the system by interacting with it during its normal operations [63].

8.4.3.1 Assumptions and Abbreviations

We first state the assumptions and abbreviations to be used in the MDP model formulation. For each RAU, the number of nonoverlapping 2.4G/5G channels, N_{ch}, is considered as 15, including channels 1, 6, and 11 at 2.4 GHz and 12 channels at 5 GHz. In North America, the 5 GHz band is divided into two parts: the U-NII lower and middle bands have 8 channels and the U-NII upper band has 4 channels. The N_{ch} varies with other locations in the world. If the CogAP senses the presence of external interference (microwave or class I Bluetooth devices) around RAU1, the N_{ch} might decrease. Channels assigned to neighboring RAUs also reduce the N_{ch} of RAU1. We assume that the CogAP could sense external interference precisely. Although CCI and ACI can be estimated by the CogAP by measuring inter-RAU interference, the effects of CCI and ACI on the available admission capacity of BSSs are uncertain.

To reduce the state space, at any time, only one RAU is considered in the MDP. Note that in CWLANoF, each RAU might control multiple BSSs by operating multiple channels. All STAs in CWLANoFs are assumed to support 802.11e, 802.11a, and 802.11b/g. We further assume STAs always associate with the BSS that has the strongest signal. Therefore, when a BSS cannot accommodate a TS, we do not consider the strategy of disassociating the STA and then having the TS accepted by a neighboring BSS, because the weaker signal of this STA might reduce the capacity of the neighboring BSS. The preferable strategy is to form a new BSS in the same cell, taking advantage of the multiple-channel capability of CWLANoF.

The abbreviations and symbols shown in Table 8.3 will be used for MDP modeling.

Table 8.3 Abbreviations and Symbols in MDP Modeling

Terms	*Explanations*
AC	Access category. There are four access categories defined in 802.11e: AC_BE for best effort, AC_BG for background, AC_VI for video, and AC_VO for voice.
UP	User priority, which can be inferred by the AP from the TSPEC. There are eight UPs corresponding to the four ACs in 802.11e.
TS	Traffic stream. Each TS can be characterized by TSPEC.
TSPEC	Traffic specifications. Five admissible TSPECs are recommended in 802.11e: continuous-time QoS traffic (HCCA), controlled-access constant bit rate (CBR) traffic (HCCA), bursty traffic (HCCA), unspecified non-QoS traffic (HCCA), and contention-based CBR traffic (EDCA).
AAC	Available admission capacity in units of 32 microsecond/second. It is contained in the BSS load element and specifies the remaining amount of medium time available via explicit admission control. The AAC of a RAU is the sum of the AAC of all BSSs generated by this RAU.
U_{ch}	Channel utilization, in time percentage. A BSS load element defined in 802.11e includes AAC, U_{ch}, and the number of associated STAs in the BSS.
N_{ACI}	The number of admitted users with the access category ACI, where ACI $\in \{1,2,3,4\}$, corresponding to AC_BG, AC_BE, AC_VI, and AC_VO, respectively.
chid	The channel number defined by the CogAP designer. One possible choice is to use the channel-numbering system defined in 802.11-2007 [4, annex J]. For North America, chid $\in \{1,6,11,36,40,44,48,52,56,60,64,149,153,157,161\}$.
rid	The ID of a RAU.
bid	The ID of a BSS.
f	A traffic stream.

8.4.3.2 MDP Modeling

QoS provisioning in CWLANoF is modeled as a MDP. When an ADDTS (add traffic stream) request arrives from a STA, the agent (implemented in the CogAP) checks the traffic load of the BSS being associated by the STA, senses the ISM band, and determines available channels in the cell from where the TS comes. We call the

"ADDTS request frame" as an event, e. Only one event, ADDTS request frame, is considered, because a DELTS (delete traffic stream) request frame requires no actions from the CogAP. The states of our MDP formulation, s_t, should include the BSS load and the information of available channels that can be further assigned to the RAU. For each ADDTS request frame, the CogAP takes one action, a_t, according to the state s_t. The MDP then moves to the next state, $s_{t+1} = \delta(s_t, a_t)$, in the deterministic case; in the nondeterministic case, the probability of moving to s_{t+1} is $P^{a_t}_{s_t,s_{t+1}}$. Each action has a delayed reward, r_t, whose definition reflects the CogAP's QoS-provisioning policy $\pi : S \rightarrow A$. In the deterministic case, $a_t = \pi(s_t)$ and $r_t = r_t(s_t, a_t)$; in the nondeterministic case, the probability of a_t is taken to be $\pi(s_t, a_t)$ and the expected value of the reward r_t is $R^{a_t}_{s_t,s_{t+1}}$. As δ_t and r_t might be nondeterministic, the CogAP might not know δ_t and r_t when a_t is taken.

The MDP is a process of exploring the system and trying to choose the optimal action such that the accumulated rewards, V_t, can be maximized. There is an inherent trade-off between exploration and exploitation in the MDP. In the setting of CWLANoF, the MDP for QoS provisioning is to explore the radio propagation environment and ensure the QoS of video and audio streams by choosing optimal decisions on the TS admission and the new channel number assigned to the RAU. The goal is to maximize the AAC or U_{ch} of the CWLANoF system. The above is a brief introduction on MDP modeling in RL. Refer to [52] and [64] for more treatments. In the following, we state our formulations on event, state, action, reward, and value.

8.4.3.2.1 Event

We consider both TS admission control and DCA together; so, the state should include the event. A smaller number of events produces a smaller state space. We therefore limit the events as

- ADDTS request of an audio stream
- ADDTS request of a video stream

By assuming that all audio streams share a common TSPEC parameter and so do all video streams, we obtain a small state space. In practice, we can quantize the categories of audio or video streams in terms of their medium time, which can be derived by the CogAP through a procedure in annex K.2.2 of 802.11-2007 and set in the ADDTS response frame. A smaller number of quantization levels results in a smaller state space. In this sense, the TSs to be admitted in CWLANoF are analogous to different classes of users in a cellular telephone system.

According to 802.11-2007, the AP cannot terminate any TS unless the corresponding STA requests to do so, or the inactivity interval of the TS has elapsed.

8.4.3.2.2 State

In cellular telephone networks, states are defined as {cell ID, N_{ch}} for DCA purpose [63], or as {the number of class-1 and class-2 users, call arrival event from a class-1 or class-2 user} for call admission control purpose [65], or as the combination of these two state representations for both call admission control and DCA purposes [66]. Considering that the QoS framework in 802.11e uses medium time as the indication of BSS admissible capacity, we use quantized medium time to represent system states: {rid, N_{ch}(rid), quantized AAC(bid), quantized medium time(f), ACI(f)}, where N_{ch}(rid) specifies the number of available channels in the RAU after considering the external interferences and CCI to neighboring RAUs at the same floor, f is the TS, and ACI(f) is mapped from the UP of f and quantifies the priority of f.

To capture the characteristics of the ISM band, such as the 2.4 GHz band channelization and ISM-band interference properties (microwave oven and a 5.8 GHz cordless phone), the above state representation can be refined as {rid, chid(rid, 2.4 GHz band), N_{ch}(rid, U-NII lower and middle band), N_{ch}(rid, U-NII upper band), quantized AAC(bid), quantized medium time (f), ACI(f)}. This representation in fact captures the following strategies or properties of DCA in CWLANoF:

- Only one channel (1, 6, or 11) is driven to one RAU at the 2.4 GHz band.
- U-NII lower and middle bands have much less external interferences than other frequencies in the ISM band.
- Microwave ovens mostly affect four channels at 2.4 GHz: 8, 9, 10, and 11.

Therefore, a higher value is expected from this state formulation, at the expense of a larger state space.

8.4.3.2.3 Action

Given an ADDTS video or audio event, three actions are available:

- A1: Accept the TS in the current BSS.
- A2: Reject the TS.
- A3: Reject the TS and disassociate the STA, while channel i is assigned to the RAU. This new channel forms a new BSS, so the STA could associate with the new BSS later and have the TS accepted.

8.4.3.2.4 Reward

The reward function is defined as

- $r(A1) = ACI * AAC(bid)$, where AAC(bid) is the AAC of the BSS after the TS is admitted.
- $r(A2) = C_{UP}$.

- $r(A3) = \mathrm{ACI} * C_{mc} * \Delta\mathrm{AAC}(\mathrm{rid},i)$, where $\Delta\mathrm{AAC}(\mathrm{rid},i)$ is the AAC gain of the RAU and the RAUs possibly affected by the newly added channel, *i*. The effects of the new channel include IM3, CCI, and ACI. Severe CCI or ACI after adding channel *i* might result in a negative value of $\Delta\mathrm{AAC}(rid,i)$.

The constant C_{UP} can be set as a positive number such that when the system is heavily loaded, TSs with lower ACs will be rejected while TSs with a higher UP can be accepted. The constant C_{mc} (mc for multiple channel) is set between 0 and 1. It reflects the risk of assigning one more channel to the RAU. A larger C_{mc} encourages driving multiple channels into RAUs; however, unpredictable indoor radio propagation might incur severe CCI and ACI to neighboring RAUs and degrade the already admitted TS in these RAUs. The C_{mc} works as a discounting factor to discourage risky multiple-channel driving. The above constants, C_{UP} and C_{mc}, can be adjusted by the system administrator during operations.

8.4.3.2.5 Value

The discounted accumulated rewards are defined as the value, $V(\pi, s)$, which depends on the policy, π, and the state s. The value and the Q value are given by [52]

$$V(\pi, s) = \mathbf{E}_{\pi}\left[\sum_{i=0}^{\infty} \gamma^{i} r_{t+i} | s_t = s\right], \tag{8.1}$$

$$Q(\pi, s, a) = \mathbf{E}_{\pi}\left[\sum_{i=0}^{\infty} \gamma^{i} r_{t+i} | s_t = s, a_t = a\right], \tag{8.2}$$

where the constant γ is the discount rate: $0 \leq \gamma < 1$; $\mathbf{E}_{\pi}$ is the expectation operation when the agent sticks to the policy, π.

The Bellman equation for the value $V(\pi, s)$ reveals the recursive relationship between the value of the current state, $s_t = s$, and its succeeding state, $s_{t+1} = s'$:

$$V(\pi, s) = \sum_{a} \pi(s, a) \sum_{s'} P^{a}_{s,s'} \left[R^{a}_{s,s'} + \gamma V(\pi, s')\right], \tag{8.3}$$

where $R^{a}_{s,s'} = \mathbf{E}(r_t | s_t, a_t, s_{t+1})$. The Q value is introduced to avoid the trouble of evaluating $V(\pi, s)$ when the agent wants to choose the optimal action:

$$Q^{*}(s, a) = \mathbf{E}[r_t + \gamma V^{*}(s_{t+1}) | s_t = s, a_t = a]. \tag{8.4}$$

In fact, the optimal action can be chosen by maximizing $Q(s, a)$, which can be approximated by the Q-learning algorithm developed by Watkins [67].

In CWLANoF systems, due to the complex indoor radio propagation environment, previously defined rewards cannot be accurately estimated right after the action is taken, but have to be measured for a certain period. This nondeterminism of reward functions motivates us to use the Q-learning algorithm [5].

8.4.3.2.6 Q-learning algorithm

Given the current state, s; the taken action, a; the succeeding state, s'; and the received reward, r; the Q value is estimated as

$$\hat{Q}_n(s,a) = (1-\alpha_n)\hat{Q}_{n-1}(s,a) + \alpha_n\left[r + \gamma \max_{a'} \hat{Q}_{n-1}\left(s',a'\right)\right], \tag{8.5}$$

where the time-decaying weight, α_n, can be chosen as $\frac{1}{1+N_{\text{vst}}(s,a,n)}$. $N_{\text{vst}}(s,a,n)$ is the number of visits to the state–action pair (s, a) up to iteration n. The gradually decreasing α_n ensures the convergence of the Q value when the reward function is nondeterministic. Other choices of α_n sequence could also ensure the convergence as long as they satisfy the convergence theorem proved by Watkins and Dayan [5].

8.4.3.2.7 Q-value Table

We choose to use a lookup table to store the estimated Q values. The number of entries is determined by the number of state–action pairs. In practice, we could eliminate as many invalid state–action pairs as we can to improve the convergence rate of the Q-learning algorithm. Once the Q value lookup table is established, the optimal action is the one that maximizes the (estimated) Q value.

8.5 Conclusions and Open Issues

We introduced a new concept of cognitive WLAN over fiber, CWLANoF, featuring a CogAP that can provide a cost-effective and efficient method for devices to equally share the ISM band by taking advantage of cognitive radio capabilities. Based on the cognitive assessments, the CogAP is able to fairly allocate resources in a collaborative way. It can provide equal access in the ISM band and have the potential of alleviating the congestion in this frequency band. Based on the DCA strategies in CWLANoF and the QoS framework in 802.11e, we then devised a Q-learning based QoS-provisioning scheme for video and audio streams in CWLANoF systems. The scheme lies in an RL field. It avoids solving complex optimization problems while being able to explore the CWLANoF system during normal operations.

8.5.1 Open Issues in QoS Provisioning

The proposed Q-learning scheme is to set a new direction on QoS provisioning in cognitive radio, particularly, in CWLANoF systems. Many details are left for future research.

8.5.1.1 Semi-Markov Decision Process Modeling

MDP modeling may not be accurate because the sojourn time of each state is not equal. Semi-MDP modeling is more precise, while time-averaged Q values can be defined (instead of discounted Q values in a MDP model). Practical systems might prefer time-averaged values.

8.5.1.2 Improving Rate of Convergence

In the deployment stage of a CWLANoF system, it is important to develop suitable training methods to increase the rate of convergence of the DCA algorithm, especially when the scale of the CWLANoF system is large.

8.5.1.3 Backward Compatibility of QoS Provisioning

In practice, there exist five types of STAs: 11a/b/g/e, 11b/g/e, 11a/b/g, 11b/g, and 11b, assuming STAs supporting 802.11a always support 802.11b/g. How to provide backward compatibility with the proposed QoS-provisioning scheme remains an open issue. A more difficult situation is that there might be no STA supporting the sophisticated HCCA that is more suitable for multimedia TSs. The CogAP then has to rely on EDCA to provide QoS for these TSs. This would be more difficult to do when compared with HCCA-based QoS provisioning. The coordination of scheduling, TS admission control, and DCA at the CogAP is another practical issue.

8.5.1.4 Degraded Admission and Dynamic Source-Coding Adjustment

By extending the idea in [68] termed degraded admission, the CogAP could dynamically adjust parameters in the TSPEC such that a TS could be admitted at a lower mean data rate without assigning one more channel to the RAU. The risk of interfering other RAUs is therefore reduced when the CWLANoF system is heavily loaded. How to incorporate this idea into the proposed QoS-provisioning scheme would be an interesting research problem.

8.5.2 Open Issues in Cognitive WLAN over Fiber Architecture

QoS provisioning in CWLANoF is built on DCA or, more generally, RRM algorithms in the architecture of CWLANoF. This architecture opens new revenues for future research on RRM:

- More advanced RRM algorithms
- Cross-layer RRM algorithms with the management of individual OFDM channels
- RRM algorithms with multiple-input, multiple-output (MIMO) technology
- RRM algorithms with space division multiple access

8.5.2.1 Other 802.11 Projects and CWLANoF

RRM is closely related to dynamic frequency selection (DFS), TPC, QoS, intercell handover, and stronger control over STAs. Some of these issues have already been studied by IEEE 802.11 work groups: 802.11h-2003 for DFS and TPC, 802.11e-2005 for QoS, 802.11k for RRM enhancements, 802.11r for VoIP handover, and 802.11v for wireless network management. The centralized architecture of CWLANoF provides a realistic platform for realizing many of these 802.11 projects.

In cases of inter-building interference, the CogAP needs to communicate with other CogAPs or WLAN controllers via some inter-AP protocols to negotiate channel allocations. This communication is supported by the ongoing internet engineering task force (IETF) project, control and provisioning of wireless access points (CAPWAP) [69]. This project will output a request for comments (RFC), "CAPWAP protocol binding for IEEE 802.11," to support inter-AP protocols in 802.11 networks. By complying with this RFC, the CogAP is able to communicate with future lightweight AP (or thin AP) systems.

8.5.2.2 Standardization Activities on Radio over Fiber

To the best of our knowledge, standardization activities on RoF so far include the measurement methods from microwave and millimeter-wave to photonic converter, photo detector, and devices for RoF system [70]. The proposal is expected to be complete by the end of 2010 [70,71]. It is currently conducted by the Institute of Electronics, Information and Communication Engineers (IEICE) in Japan, subject to the subcommittee (SC) 46F of the technical committee 103 (TC103) in the International Electrotechnical Commission (IEC).

The lack of standardization may be a reason why there are few commercial products on RoF. In fact, Zinwave is the only company we know to provide RoF products for indoor wideband coverage [72]. To speed up the deployment of CWLANoF systems, further standardization activities on RoF are required.

Acknowledgment

This work was supported in part by a grant from Bell Canada through the Bell University Laboratories program and by the Canadian Natural Sciences and Engineering Research Council through grant CRDPJ 320552-04.

List of Abbreviations

ACs	access categories
APs	access points
ACI	adjacent channel interference
AF	amplify-forward
ACO	ant colony optimization
AI	artificial intelligence
ANN	artificial neural networks
AGC	automatic gain control
BG	back-ground
BSS	Basic Service Set
BE	best-effort
CSMA/CA	carrier-sensed multiple access with collision avoidance
CCU	central control unit
CCI	co-channel interference
CogAP	cognitive access point
CWLANoF	cognitive wireless local area network over fiber
CTB	composite triple beat
CSOP	constraint satisfaction optimization problem
CSP	constraint satisfaction problem
CPs	contention periods
CFPs	contention-free periods
CW	continuous-wave
CAPWAP	control and provisioning of wireless access points
CAP	controlled access phase
CBR	constant bit rate
DF	decode-forward
DELTS	delete traffic stream
DURs	desired-to-undesired signal power ratios
DSSS	direct-sequence spread spectrum
DBPSK	differential binary phase shift keying
DCF	distributed coordination function
DIFS	distributed interframe space
DCA	dynamic channel assignment
DFS	dynamic frequency selection
EDCA	enhanced distributed channel access

ESSs	Extended Service Sets
FCSP	finite CSP
FCA	fixed channel assignment
FAP	frequency assignment problem
FHSS	frequency-hopping spread spectrum
GFSK	Gaussian frequency-shift keying
GA	genetic algorithms
HCCA	HCF controlled channel access
HC	hybrid controller
HCF	hybrid coordination function
IEICE	Information and Communication Engineers
ILP	integer linear programming
IP	integer programming
IFS	interframe space
IEC	International Electrotechnical Commission
IETF	internet engineering task force
LP	linear programming
LS	local search
MSDU	MAC service data unit
MDP	Markov decision process
MAC	medium access control
MI-FAP	minimum interference FAP
MIMO	multiple-input, multiple-output
MTM-SVD	multitaper method-singular value decomposition
NBI	narrow-band interference
OFDM	orthogonal frequency-division multiplexing
PCSP	partially CSP
PHY	physical layer
PCF	point coordination function
RF	radio frequency
RoF	radio over fiber
RRM	radio resource management
RL	reinforcement learning
RAU	remote antenna unit
RTTs	round-trip times
RFC	request for comments
SNR	signal-to-noise ratio
SA	simulated annealing
SC	sub-committee
IM3	the third-order intermodulation
TSPECs	traffic specifications
TSs	traffic streams
TXOP	transmission opportunity

TPC	transmission power control
UPs	user priorities
VoIP	voice over Internet protocol
WLAN	wireless local area network
WMAN	wireless metropolitan area network
WPAN	wireless personal area network
WDCT	worldwide digital cordless telecommunications

References

1. H. Li, Q. Pang, and V. C. M. Leung, Cognitive access points for dynamic radio resource management in wireless LAN over fiber, in *Proceedings of the WWRF 20th Meeting*, Ottawa, Canada, Apr. 2008.
2. J. Mitola and G. Q. Maguire, Cognitive radio: Making software radios more personal, *IEEE Pers. Commun. Mag.*, 6(4): 13–18, Aug. 1999.
3. T.-S. Chu and M. J. Gans, Fiber optic microcellular radio, *IEEE Trans. Veh. Technol.*, 40(3): 599–606, 1991.
4. Available: http://standards.ieee.org/getieee802/download/802.11-2007.pdf (Online).
5. C. J. C. H. Watkins and P. Dayan, Technical note: Q-learning, *Mach. Learn.*, 8: 279–292, 1992.
6. G. Ganesan, Y. G. Li, B. Bing, and S. Li, Spatiotemporal sensing in cognitive radio networks, *IEEE J. Sel. Areas Commun.*, 26(1): 5–12, 2008.
7. Q. Pang and V. C. M. Leung, Channel clustering and probabilistic channel visiting techniques for WLAN interference mitigation in Bluetooth devices, *IEEE Trans. Electromagn. Comput.*, 49(4): 914–923, 2007.
8. S. Mishra, A. Sahai, and R. W. Brodersen, Cooperative sensing among cognitive radios, in *Proceedings of the IEEE ICC 2006*, Istanbul, Turkey, Jun. 2006, pp. 1658–1663.
9. T. Niiho, M. Nakaso, K. Masuda, H. Sasai, K. Utsumi, and M. Fuse, Transmission performance of multichannel wireless LAN system based on radio-over-fiber techniques, *IEEE Trans. Microw. Theory Tech.*, 54(2): 980–989, 2006.
10. M. Sauer, A. Kobyakov, and J. George, Radio over fiber for picocellular network architectures, *J. Lightwave Technol.*, 25(11): 3301–3320, 2007.
11. A. Hills, Large-scale wireless LAN design, *IEEE Commun. Mag.*, 39(11): 98–107, 2001.
12. K. K. Leung, B. McNair, L. J. Cimini Jr., and J. H. Winters, Outdoor IEEE 802.11 cellular networks: MAC protocol design and performance, in *Proceedings of the IEEE ICC 2002*, Vol. 1, New York, Apr. 2002, pp. 595–599.
13. B. L. Dang and I. Niemegeers, Analysis of IEEE 802.11 in radio over fiber home networks, in *Proceedings of the IEEE 30th Conference on Local Computer Networks (LCN) 2005*, Washington, DC, 2005, pp. 744–747.
14. M. G. Larrodé, A. M. J. Koonen, and P. F. M. Smulders, Impact of radio-over-fibre links on the wireless access protocols, in *Proceedings of the NEFERTITI Workshop*, Brussels, Belgium, Jan. 2005.

15. A. Das, M. Mjeku, A. Nkansah, and N. J. Gomes, Effects on IEEE 802.11 MAC throughput in wireless LAN over fiber systems, *J. Lightwave Technol.*, 25(11): 1–8, 2007.
16. H. Kim, J. H. Cho, S. Kim, K. U. Song, H. Lee, J. Lee, B. Kim, Y. Oh, J. Lee, and S. Hwang, Radio-over-fiber system for TDD-based OFDMA wireless communication systems, *J. Lightwave Technol.*, 25(11): 1–9, 2007.
17. K. L. Blackard, T. S. Rappaport, and C. W. Bostian, Measurements and models of radio frequency impulsive noise for indoor wireless communications, *IEEE J. Sel. Areas Commun.*, 11: 991–1001, Sept. 1993.
18. A. Kamerman and N. Erkocevic, Microwave oven interference on wireless LANs operating in the 2.4 GHz ISM band, in *Proceedings of the IEEE PIMRC 1997*, Vol. 3, Helsinki, Finland, Sept. 1997, pp. 1221–1227.
19. T. Murakami, Y. Matsumoto, K. Fujii, and A. Sugiura, Performance analysis of Bluetooth system in the presence of microwave oven noises, *Electron. Commun. Jpn. I Commun.*, 89(11): 24, 2006.
20. Available: http://www.wi-fiplanet.com/tutorials/article.php/3116531 (Online).
21. Available: http://www.ofta.gov.hk/en/ad-comm/rsac/paper/rsac6-2002.pdf (Online).
22. M.-J. Ho, M. S. Rawles, M. Vrijkorte, and L. Fei, RF challenges for 2.4 and 5 GHz WLAN deployment and design, in *Proceedings of the IEEE WCNC 2002*, Vol. 2, Orlando, FL, March 2002, pp. 783–788.
23. S. Srikanteswara, G. Li, and C. Maciocco, Cross layer interference mitigation using spectrum sensing, in *Proceedings of the IEEE GLOBECOM 2007*, Washington, DC, Nov. 2007, pp. 3553–3557.
24. Available: http://standards.ieee.org/getieee802/download/802.15.4-2006.pdf (Online).
25. S. Mishra, R. Tandra, and A. Sahai, Coexistence with primary users of different scales, in *Proceedings of the IEEE DySPAN 2007*, Dublin, Ireland, Apr. 2007, pp. 158–167.
26. I. Akyildiz, W. Lee, M. Vuran, and S. Mohanty, Next generation/dynamic spectrum access/cognitive radio wireless networks: A survey, *Comput. Netw. Int. J. Comput. Telecommun. Netw.*, 50(13): 2127–2159, 2006.
27. C. Clancy, J. Hecker, E. Stuntenbeck, and T. O'Shea, Applications of machine learning to cognitive radio networks, *IEEE Wireless Commun. Mag.*, 14(4): 47–52, Aug. 2007.
28. A. Ghasemi and E. Sousa, Collaborative spectrum sensing for opportunistic access in fading environments, in *Proceedings of IEEE DySPAN 2005*, Baltimore Harbor, MD, Nov. 2005, pp. 131–136.
29. G. Ganesan and Y. G. Li, Cooperative spectrum sensing in cognitive radio—Part I: Two user networks, *IEEE Trans. Wireless Commun.*, 6(6): 2204–2213, Jun. 2007.
30. J. N. Laneman and D. N. C. Tse, Cooperative diversity in wireless networks: Efficient protocols and outage behaviour, *IEEE Trans. Inf. Theory*, 50: 3062–3080, Dec. 2004.
31. S. Vogeler, L. Broetje, K.-D. Kammeyer, R. Rueckriem, and S. Fechtel, Blind Bluetooth interference detection and suppression for OFDM transmission in the ISM band, in *Proceedings of IEEE ACSSC 2003*, Vol. 1, Pacific Grove, CA, Nov. 2003, pp. 703–707.

32. M. Ghosh and V. Gadhi, Bluetooth interference cancellation for 802.11g WLAN receivers, in *Proceedings of IEEE ICC 2003*, Anchorage, AK, May 2003, pp. 1169–1173.
33. D. Cabric, S. M. Mishra, and R. W. Brodersen, Implementation issues in spectrum sensing for cognitive radios, in *Proceedings of IEEE ACSSC 2004*, Vol. 1, Pacific Grove, CA, Nov. 2004, pp. 772–776.
34. S. Haykin, Cognitive radio: Brain-empowered wireless communications, *IEEE J. Sel. Areas Commun.*, 23(2): 201–220, Feb. 2005.
35. R. Tandra and A. Sahai, SNR walls for signal detection, *IEEE Trans. Signal Process.*, 2(1): 4–17, Feb. 2008.
36. G. Holland, N. Vaidya, and P. Bahl, A rate-adaptive MAC protocol for multi-hop wireless networks, in *Proceedings of ACM MobiCom 2001*, Rome, Italy, July 2001, pp. 236–251.
37. A. Kamerman and L. Monteban, WaveLAN II: A high-performance wireless LAN for the unlicensed band, *Bell Labs Tech. J.*, 2(33): 118–133, Summer 1997.
38. A. C.-C. Hsu, D. S. L. Wei, and C.-C. J. Kuo, Coexistence mechanism using dynamic fragmentation for interference mitigation between Wi-Fi and Bluetooth, in *Proceedings of IEEE MILCOM 2006*, Washington, DC, Oct. 2006, pp. 1–7.
39. K. Aardal, S. P. M. van Hoesel, A. M. C. A. Koster, C. Mannino, and A. Sassano, Models and solution techniques for frequency assignment problems, *Ann. Oper. Res.*, 153(1): 79–129, 2007.
40. A. M. C. A. Koster, Frequency assignment—Models and algorithms, PhD dissertation, Maastricht University, Maastricht, the Netherlands, 1999.
41. A. Eisenblätter, Frequency assignment in GSM networks: Models, heuristics, and lower bounds, PhD dissertation, Technische Universität Berlin, Berlin, Germany, 2001 (Online). Available: http://www.zib.de/bib/diss/
42. B. H. Metzger, Spectrum management technique, in *Presentation at 38th National ORSA Meeting*, Detroit, MI, 1970.
43. W. K. Hale, Frequency assignment: Theory and applications, *Proc. of IEEE*, 68(12): 1497–1514, 1980.
44. F. S. Roberts, T-colorings of graphs: Recent results and open problems, *Discrete Math.*, 93: 229–245, 1991.
45. C. McDiarmid and B. Reed, Channel assignment and weighted coloring, *Networks*, 36(2): 114–117, 2000.
46. C. McDiarmid, Discrete mathematics and radio channel assignment, *Recent Advances in Algorithms and Combinatorics*, New York: Springer, 2003, pp. 27–63.
47. L. A. Wolsey, *Integer Programming*, New York: John Wiley & Sons, 1998.
48. E. Tsang, *Foundations of Constraint Satisfaction*, London, U.K. and San Diego, CA: Academic Press, 1993.
49. A. M. C. A. Koster, C. P. M. van Hoesel, and A. W. J. Kolen, Solving partial constraint satisfaction problems with tree decomposition, *Networks*, 40(3): 170–180, 2002.
50. N. Robertson and P. Seymour, Graph minors II. Algorithmic aspects of tree-width, *J. Algorithms*, 7: 309–322, 1986.
51. J. A. Zoellner and C. L. Beall, A breakthrough in spectrum conserving frequency assignment technology, *IEEE Trans. Electromagn. Compat.*, 19: 313–319, 1977.

52. R. S. Sutton and A. G. Barto, *Reinforcement Learning: An Introduction*, Cambridge, MA: MIT Press, 1998.
53. M. Dorigo, V. Maniezzo, and A. Colorni, Ant system: Optimization by a colony of cooperating agents, *IEEE Trans. Syst., Man, Cybern. B*, 26(1): 29–41, 1996.
54. A. Kumar and V. Kumar, Optimal association of stations and APs in an IEEE 802.11 WLAN, in *Proceedings of the National Conference on Communications (NCC) 2005*, IIT Kharagpur, India, Feb. 2005.
55. A. Mishra, V. Brik, S. Banerjee, A. Srinivasan, and W. Arbaugh, A client-driven approach for channel management in wireless LANs, in *Proceedings of the IEEE INFOCOM 2006*, Barcelona, Spain, Apr. 2006, pp. 1–12.
56. I. Broustis, K. Papagiannaki, S. V. Krishnamurthy, M. Faloutsos, and V. Mhatre, MDG: Measurement-driven guidelines for 802.11 WLAN design, in *Proceedings of ACM MobiCom 2007*, New York, 2007, pp. 254–265.
57. B. Kauffmann, F. Baccelli, A. Chainteau, V. Mhatre, K. Papagiannaki, and C. Diot, Measurement-based self organization of interfering 802.11 wireless access networks, in *Proceedings of IEEE INFOCOM 2007*, Anchorage, AK, May 2007, pp. 1451–1459.
58. V. Mhatre, K. Papagiannaki, and F. Baccelli, Interference mitigation through power control in high density 802.11 WLANs, in *Proceedings of IEEE INFOCOM 2007*, Anchorage, AK, May 2007, pp. 535–543.
59. D. Valerio, F. Ricciato, and P. Fuxjaeger, On the feasibility of IEEE 802.11 multi-channel multi-hop mesh networks, *Comput. Commun.*, 31(8): 1484–1496, 2008.
60. J. Rogers and C. Plett, *Radio Frequency Integrated Circuit Design*, Boston, MA: Artech House, 2003.
61. D. J. Castelino, S. Hurley, and N. M. Stephens, A tabu search algorithm for frequency assignment, *Ann. Oper. Res.*, 63: 301–319, 1996.
62. D. H. Smith, R. K. Taplin, and S. Hurley, Frequency assignment with complex co-site constraints, *IEEE Trans. Electromagn. Compat.*, 43(2): 210–218, 2001.
63. J. Nie and S. Haykin, A Q-learning based dynamic channel assignment technique for mobile communication systems, *IEEE Trans. Veh. Technol.*, 48(5): 1676–1687, Sept. 1999.
64. T. M. Mitchell, *Machine Learning*, New York: McGraw-Hill Companies, Inc., 1997.
65. S.-M. Senouci, A.-L. Beylot, and G. Pujolle, Call admission control in cellular networks: A reinforcement learning solution, *Int. J. Netw. Manag.*, 14(2): 89–103, 2004.
66. S.-M. Senouci and G. Pujoile, Dynamic channel assignment in cellular networks: A reinforcement learning solution, in *Proceedings of the 10th International Conference on Telecommunications (ICT) 2003*, Vol. 1, Papeete, French Polynesia, Feb. 2003, pp. 302–309.
67. C. J. C. H. Watkins, Learning from delayed rewards, PhD dissertation, Cambridge University, Cambridge, MA, 1989.
68. F. Yu, V. W. S. Wong, and V. C. M. Leung, Efficient QoS provisioning for adaptive multimedia in mobile communication networks by reinforcement learning, in *Proceedings of the First International Conference on Broadband Networks (BroadNets) 2004*, San José, CA, Oct. 2004, pp. 579–588.

69. IETF, Control and provisioning of wireless access points (capwap), 2007. Available: http://www.ietf.org/html.charters/capwap-charter.html (Online).
70. Available: http://roms.comm.eng.osaka-u.ac.jp/~komaki/RoF%20Backhaul%20NW.pdf (Online).
71. S. Komaki, H. Ogawa, and J. Ichikawa, Feasibility study and standardization activities on radio on fiber devices, in *Proceedings of International Symposium on Signals, Systems and Electronics (ISSSE) 2007*, Montreal, QC, Canada, Jul. 2007, pp. 89–94.
72. Available: http://www.zinwave.com (Online).

Chapter 9

Game Theory for Dynamic Spectrum Access

Samir Medina Perlaza, Samson Lasaulce,
Mérouane Debbah, and Jean-Marie Chaufray

Contents

In this chapter, the competitive interaction of radio devices dynamically accessing the spectrum is studied using tools from game theory. Depending on the scenario under consideration, the dynamic spectrum access (DSA) is modeled by different types of games following both a noncooperative and a cooperative approach. In the first case, each radio device aims to selfishly maximize an individual performance metric (e.g., individual data rate), while in the second case, such maximization concerns global network parameters (e.g., network sum-rate). In each case, we analyze network equilibria that allow network designers, operators, or manufactures to predict the behavior and the performance of cognitive networks or terminals.

9.1 Introduction

For many years, the management of the radio spectrum has been based on the classic property right model. In this model, operators are granted licenses that allow them to exclusively use a certain frequency band in a given area for a specific service [3]. However, results from recent studies [1] show that this model often leads to an inefficient usage of the spectrum in terms of spectral efficiency (bits/Hertz) due to two main reasons. First, the spectrum remains unused during the time that the licensed (primary) systems are idle, and second, the spectrum can be congested in one area while it remains unused in another due to a low spatial density of radio devices.

In this chapter, we consider the efficient use of the radio spectrum and analyze dynamic spectrum access (DSA) as a paradigm to improve the spectral efficiency of multiple communication systems subject to mutual interference. Here, we consider DSA as either a dynamic spectrum allocation in the case where a central controller exists or a spontaneous access where cognitive radios (CR) [2] autonomously decide to access the spectrum in a scenario where no central control is present. CRs are radio devices equipped with sensing systems that allow them to be aware of the environment they are operating in. More specifically, they are able to identify other active radios and estimate unused radio resources (frequency bands, time slots, spreading sequences, spatial directions, etc.). Additionally, such devices are able to self-adapt (in terms of coding-modulation scheme, power allocation, etc.) to compete with other devices and use the radio spectrum more efficiently.

There are at least two scenarios of high practical interest for the study of DSA [4]: hierarchical spectrum access (HSA) and open spectrum access (OSA). In HSA, CRs coexist with legacy systems if and only if the additional interference overcome by the preexisting systems is below a specific threshold. Such thresholds can be predefined by network operators, manufacturers, or regulation entities to ensure a certain quality of service (QoS) in the primary systems. In this scenario, CRs only transmit using radio resources left unused by the primary systems. Hence, CRs are often called opportunistic or secondary radio devices. Such unused radio resources are called spectrum access opportunities (SAO) or available channels. Typically, an available channel consists of non-occupied time slots in time division multiple access (TDMA), frequency bands in frequency division multiple access (FDMA), spatial directions in spatial division multiple access (SDMA), tones in orthogonal frequency division multiple access (OFDMA), spreading codes in code division multiple access (CDMA) or a combination of any of those. In HSA, once the SAOs have been identified, each CR decides whether to transmit based on their own performance criterion.

In OSA, the notion of primary and secondary systems does not exist, at least in its conventional definition. In this scenario, each terminal has the same rights to access the spectrum at any time. OSA typically includes the case of unlicensed bands (e.g., the industrial, scientific, and medical [ISM] band [2.400, 2.500] GHz). Radio devices operating in these bands include cordless telephones, wireless sensors, and devices operating under the standards of Wi-Fi (IEEE 802.11), Zig-Bee (IEEE 802.15.4), and Bluetooth (IEEE 802.15.1). Here, each technology implements different coding and modulation schemes, so there exists neither a common multiple access (MA) technique nor a signaling system to harmonize the use of these bands. Different governmental agencies, such as the European Radio Communications Office (ERO) in Europe and the Federal Communications Commission (FCC) in the United States, have defined a set of few rules either in terms of power spectral density *masks* or in terms of time duty cycles, depending on the application [5]. The power spectral density mask defines the limits on the peak-to-average power ratio (PAPR) as a function of the frequency offset around the central frequency. The time duty cycle defines the longest cumulative period a specific device is allowed to transmit within a time unit.

In this chapter, we assume that CRs can sense their environment in a sufficiently reliable manner so that SAOs can be perfectly identified. Then, if SAOs are known by each CR, the common problem with either HSA or OSA is that there is a group of terminals competing for spectral resources. The rules the terminals must follow may differ, but in each case, it is a problem of interactions between cognitive entities that must make decisions to optimize their performance metrics. Hence, game theory, a branch of mathematics that studies the interaction between several decision makers, is the dominant paradigm to analyze such problems [6,8,9]. Such analysis will be the focus of this chapter.

This chapter is organized as follows: in Section 9.2 we describe the problem of DSA and identify several scenarios depending on the network topology. More specifically, we analyze the DSA problem in both the multiple access channel (MAC) and interference channel (IC) [15,16]. In Section 9.3 we present fundamental concepts of game theory used to study DSA. Here, we model the DSA problem as games following both cooperative and noncooperative approaches. In Section 9.4 we focus on OSA games, which includes the case of unlicensed bands. Here, we study how noncooperative games often lead to suboptimal equilibrium points and present several techniques to improve the game outcomes. The two approaches presented for including a certain degree of cooperation between the players, and therefore improving the equilibrium efficiency are the repetition of the game (repeated games) and coalitions between several terminals to jointly compete against other coalitions (coalitional games). In Section 9.5, we discuss games modeling the coexistence between primary systems and opportunistic systems (hierarchical spectrum access [HSA]). Here, another technique to improve the game outcome on noncooperative games is studied. We introduce a certain degree of hierarchy among the terminals (Stackelberg games) regarding, for instance, either their nature (primary or secondary radio devices) or their decoding order at the receiver. Finally, in Section 9.6 we present open issues related to both OSA and HSA problems and state our conclusions.

9.2 Formulation of the DSA Problem

Dynamic access to the radio spectrum suggests either a dynamic allocation of SAOs (if a central controller exists) or a spontaneous access (if no central controller exists). However, two or more radio devices transmitting on the same channel (using the same SAO) might either degrade or break the communication off due to mutual interference. Hence, the formulation of the DSA problem can be summarized by the question: how does one optimally design the system to let each radio device (not necessarily a CR) use the available channels while maximizing the spectral efficiency, that is, the number of successfully transmitted bits per second over total available bandwidth?

In this chapter, we constraint the study of DSA to the choice of the best channels (SAO) and the optimal transmit powers per channel. It should be noted that other degrees of freedom, such as modulation-coding schemes, constellation size, polarization, and type of receiver, might also be considered to increase the spectral efficiency [54,55].

Regarding the topology, we focus on both the multiple access channel (MAC) and the interference channel (IC). In the former, several radio devices transmit to a single receiver. In the latter, each radio device transmits to a different receiver. This apparent simple difference between IC and MAC immediately implies different constraints regarding the channel state information (CSI), interference cancellation, and moreover signaling among the active users in the network.

9.2.1 DSA in Multiple Access Channels

The multiple access channel, also known as the many-to-one channel, consists of K transmitters aiming to communicate with a single receiver using a common channel [36]. If $N \geq 1$ channels are available, then there exists N independent or *parallel* MACs, where transmissions in different MACs do not interfere with each other. For instance, this model corresponds to the uplink channel in a single-cell multi-carrier cellular system.

Regarding notational aspects, the channel gain from transmitter i to the receiver over the channel n is denoted by $h_{i,n}$. We assume a block flat-fading channel model such that channel realizations remain constant during the transmission of M consecutive symbols. All the channel realizations, $\forall i = \{1, \ldots, K\}$ and $\forall n = \{1, \ldots, N\}$ are drawn from a Gaussian distribution with zero mean and unit variance. The power allocated by transmitter i to channel n is denoted by $p_{i,n}$. Each transmitter is power-limited, that is, for the ith transmitter, its transmit power cannot exceed $p_{i,\max}$, i.e., $\forall i \in \{1, \ldots, K\}$, $\sum_{n=1}^{N} p_{i,n} \leq p_{i,\max}$.

The symbol sent by transmitter i over channel n is represented by $x_{i,n}$. We consider that transmitted symbols $\forall i = \{1, \ldots, K\}$ and $\forall n = \{1, \ldots, N\}$ are random variables with zero mean and unit variance. The noise at the receiver is denoted by w and corresponds to an additive white Gaussian noise (AWGN) process with zero mean and variance σ^2. In matrix notation, the channel realizations are written as an $N \times N$ diagonal matrix $\boldsymbol{H}_i = \text{diag}\left(h_{i,1}, \ldots, h_{i,N}\right)$. Using a similar notation, the transmit powers, transmitted symbols, and noise are written as $\boldsymbol{P}_i = \text{diag}\left(p_{i,1}, \ldots, p_{i,N}\right)$, $\boldsymbol{x}_i = \left(x_{i,1}, \ldots, x_{i,N}\right)^T$, and $\boldsymbol{w} = (w_1, \ldots, w_N)^T$, respectively. Then, the received signal sampled at the symbol rate $\boldsymbol{r}_i = \left(r_{i,1}, \ldots, r_{i,N}\right)^T$ can be expressed as

$$\forall t \in \{1, \ldots, M\}, \quad \boldsymbol{r}_i(t) = \boldsymbol{H}_i \boldsymbol{P}_i^{\frac{1}{2}} \boldsymbol{x}_i(t) + \sum_{j \neq i}^{K} \boldsymbol{H}_j \boldsymbol{P}_j^{\frac{1}{2}} \boldsymbol{x}_j(t) + \boldsymbol{w}(t). \tag{9.1}$$

According to this signal model, the received signal to interference plus noise ratio (SINR) on channel n for transmitter i, denoted by $\gamma_{i,n}$ for all $i \in \{1, \ldots, K\}$ and for all $n \in \{1, \ldots, N\}$, is

$$\gamma_{i,n} = \frac{p_{i,n} |h_{i,n}|^2}{\sigma^2 + \sum_{j \neq i}^{K} p_{j,n} \left|h_{j,n}\right|^2}. \tag{9.2}$$

In the MAC model, the receiver knows the codebooks used by all the transmitters in the network [36]. This allows the receiver to use techniques such as multiuser detection and interference cancellation [35]. Additionally, the receiver is able to estimate the channel of all the transmitters. Conversely, without any additional signaling, each transmitter is able to estimate only its own channel. This fact implies that if the receiver is equipped with enough processing capabilities, the network performance could be optimized at the receiver and optimal transmission parameters (e.g., power allocation and modulation-coding scheme) can be fed back to the transmitters by using a control signaling protocol. The same procedure will be constrained at each transmitter due to the lack of knowledge about the other transmitter's channel gains. As described later, the availability of complete channel state information plays a key role in the DSA problem.

9.2.2 DSA in Interference Channels

The interference channel, known also as the many-to-many channel, consists of a set of K point-to-point links close enough to produce mutual interference due to the coexistence on the same channel. If $N \geq 1$ channels are available, then there exists N independent or parallel ICs, where transmissions in different ICs do not interfere with each other. This topology typically appears in self-organized networks (ad-hoc networks) where nodes communicate in pairs, possibly using the same set of SAOs.

To describe the IC model we keep the same notation and assumptions presented in the MAC case (Section 9.2.1). Only a slight modification is introduced to denote the channel realization from transmitter j to receiver i on channel n. Here, it is denoted by $h_{i,j}^{(n)}$. The channel transfer matrix from transmitter j to receiver i is denoted by the $N \times N$ diagonal matrix $\boldsymbol{H}_{i,j} = \text{diag}\left(h_{i,j}^{(1)}, \ldots, h_{i,j}^{(N)}\right)$. The noise at receiver i over channel n is denoted by $w_{i,n}$ and $\boldsymbol{w}_i = \left(w_{i,1}, \ldots, w_{i,N}\right)^T$. The received signal sampled at the symbol rate at the receiver i, denoted by the N-dimensional vector $\boldsymbol{r}_i(t) = \left(r_{i,1}(t), \ldots, r_{i,N}(t)\right)^T$ at sampling instant t, is

$$\forall t \in \{1, \ldots, M\}, \quad \boldsymbol{r}_i(t) = \boldsymbol{H}_{i,i}\boldsymbol{P}_i^{\frac{1}{2}}\boldsymbol{x}_i(t) + \sum_{j \neq i}^{K} \boldsymbol{H}_{i,j}\boldsymbol{P}_j^{\frac{1}{2}}\boldsymbol{x}_j(t) + \boldsymbol{w}_i(t), \qquad (9.3)$$

The expression of the SINR over channel n at the receiver i, denoted by $\gamma_{i,n}$ for all $(i, n) \in \{1, \ldots, K\} \times \{1, \ldots, N\}$, is

$$\gamma_i^n = \frac{p_i^n |h_{i,i}^n|^2}{\sigma^2 + \sum_{j \neq i}^{K} p_j^n \left|h_{j,i}^n\right|^2}. \qquad (9.4)$$

Often, in the IC model, each point-to-point link uses different codebooks that constraints the usage of multiuser detection techniques, and interference cancellation [35]. Moreover, each receiver is able to estimate only its own channel realization due to the usage of different code books in each link. Thus, the DSA problem turns out to be more challenging than in the MAC case due to the lack of knowledge about the channel gains of all the other transmitters.

9.2.3 General Assumptions

For both types of network topologies, we assume that all the transmitters have perfect channel state information (CSI), that is, each transmitter knows the channel realizations $\boldsymbol{h}_i = \left(h_{i,1}, \ldots, h_{i,N}\right)$ for all $i \in \{1, \ldots, K\}$ in the MAC case, and $\boldsymbol{H}_{i,j} = \text{diag}\left(h_{i,j}^{(1)}, \ldots, h_{i,j}^{(N)}\right)$ for all $(i, j) \in \{1, \ldots, K\}^2$ in the IC case. For ease of presentation but may be with loss of generality, we assume that all the radio devices are limited by the same power constraint $p_{\max}$, that is, $\forall i \in \mathcal{K}$, $p_{i,\max} = p_{\max}$. Additionally, we assume that all the transmitters are aware of such common maximum power threshold.

9.3 Dynamic Spectrum Access as a Game

In this section, we present the fundamental concepts of game theory that will be used to study the problem of DSA. First, we introduce the concept of games in normal form and we present the cooperative and noncooperative approaches. Afterward, we introduce the concept of Nash equilibrium in noncooperative games and optimality measures such as Pareto optimality and price of anarchy (PoA). Finally, a formulation of the DSA problem in game-theoretic terms is provided.

9.3.1 The Game Model

As mentioned earlier, game theory provides a natural mathematical framework to analyze strategic interactions between several decision makers. The set of rules governing such interaction is a game. The simplest representation of a game is the normal form. In normal form, a game is defined as follows:

Definition 9.1 (Normal Form Game) A game in normal form is denoted by $\left\{\mathcal{K}, \mathcal{S}, \{u_k\}_{\forall k \in \mathcal{K}}\right\}$ and is composed of three elements

- A set of players: $\mathcal{K} = \{1, \ldots, K\}$.
- A set of strategy profiles: $\mathcal{S} = \mathcal{S}_1 \times \cdots \times \mathcal{S}_K$, where $\mathcal{S}_k$ is the strategy set of the kth player.
- A set of utility functions: The kth player's utility function is $u_k : \mathcal{S} \rightarrow \mathbb{R}_+$ and is denoted by $u_k(s_k, \boldsymbol{s}_{-k})$ where $s_k \in \mathcal{S}_k$ and $\boldsymbol{s}_{-k} = \left(s_1, \ldots, s_{k-1}, s_{k+1}, \ldots, s_K\right) \in \mathcal{S}_1 \times \cdots \times \mathcal{S}_{k-1} \times \mathcal{S}_{k+1} \times \cdots \times \mathcal{S}_K$.

The set of players is a finite set $\mathcal{K} \subset \mathbb{N}$ of which each element represents a transmitter. The strategy set $\mathcal{S}_k$ contains the set of actions player k might take in the game. The utility function $u_k(s_k, \boldsymbol{s}_{-k})$ allows a player to evaluate the convenience of its strategy s_k with respect to the other players' strategies $\boldsymbol{s}_{-k}$.

Dynamic spectrum access can be modeled as a game (Section 9.4.1), when each transmitter is a player, the choice of its transmitting parameters is its strategy, and its utility function is described either in terms of its individual or in terms of the network quality of service (QoS) parameters. The choice of a utility function leads to two different kinds of games: noncooperative and cooperative games.

9.3.2 Noncooperative and Cooperative Games

In a noncooperative game, each player is selfish and unconcerned about all the other players' performance. Each terminal chooses its strategy to optimize its own performance metric under the assumption that all players are rational and adopt the same selfish behavior [20]. Thus, each player's utility function is defined in terms of local QoS targets (e.g., individual transmission rate) [7,24–26]. In a cooperative approach, each player aims at maximizing a common benefit for the set of players assuming that all the other players have adopted the same cooperative behavior. A common benefit could be interpreted, for instance, as the sum of individual benefits (social welfare problem).

Often, if the performance metric (utility function) is well chosen, noncooperative games might be played by each player using only local information (e.g., channel gains and power constraints regarding only a given player). However, cooperative games often require information regarding all the players' local information. Hence, cooperative games are often used either when there exists a central controller (e.g., a base station) that has complete information about all the players or when communication among all the players is possible (e.g., common signaling is available and affordable) [28,29]. In the next sections, we study some of the features of cooperative and noncooperative games required for our discussions on DSA in Sections 9.4 and 9.5.

9.3.3 The Nash Equilibrium Concept

An important concept in noncooperative game theory is the Nash equilibrium (NE) [49,50]. An NE corresponds to a profile of strategies $\boldsymbol{s}^* = \left(s_1^*, \ldots, s_K^*\right)$ for which each player's strategy s_k^*, $\forall k \in \mathcal{K}$ is the optimal response to all the other players' strategies $\boldsymbol{s}_{-k}^*$.

Definition 9.2 (Nash Equilibrium) In the game $\left\{\mathcal{K}, \mathcal{S}, (u_k)_{\forall k \in \mathcal{K}}\right\}$, a strategy profile $\boldsymbol{s}^* = \left(s_1^*, \ldots, s_K^*\right) \in \mathcal{S}$ is an NE if it satisfies,

$$\forall k \in \mathcal{K} \text{ and } \forall s_k \in \mathcal{S}_k, \quad u_k\left(s_k^*, \boldsymbol{s}_{-k}^*\right) \geq u_k\left(s_k, \boldsymbol{s}_{-k}^*\right). \tag{9.5}$$

That is, at the NE, any unilateral deviation from the strategy profile $\boldsymbol{s}^*$ of player k, $\forall k \in \mathcal{K}$ will not increase its utility function u_k. Hence, at the NE, it does not exist any motivation for a player to deviate from the NE strategy profile [14,49,50]. As players are selfish and decide by themselves their strategy, one question arises: does an NE lead to an efficient game outcome?

9.3.4 Optimality Measures

The NE outcome is a stable solution to unilateral deviations; however it might not be optimal [14,27]. A formal measure of any game outcome's optimality is the Pareto optimality. A Pareto optimal strategy profile can be described as follows:

Definition 9.3 (Pareto Optimality) In the game $\left\{\mathcal{K}, \mathcal{S}, \{u_k\}_{\forall k \in \mathcal{K}}\right\}$, let $\boldsymbol{s} = (s_1, \ldots, s_K)$ and $\boldsymbol{s}_i' = \left(s_1', \ldots, s_K'\right)$ be two different strategy profiles in $\mathcal{S}$. Then, if

$$\forall k \in \mathcal{K} \quad u_k\left(s_k, \boldsymbol{s}_{-k}\right) \geq u_k\left(s_k', \boldsymbol{s}_{-k}'\right), \tag{9.6}$$

with strict inequality for at least one player, the strategy profile $\boldsymbol{s}$ is Pareto-superior to the strategy profile $\boldsymbol{s}_i'$. If there exists no strategy that is Pareto superior to $\boldsymbol{s}_i$, then $\boldsymbol{s}_i$ is Pareto optimal.

Often, the NE strategy profile is not Pareto optimal and the loss of performance observed in a noncooperative game due to the lack of cooperation is a common optimality measure called the *price of anarchy* (PoA) [34].

Definition 9.4 (Price of Anarchy) Let the triplet $\left\{\mathcal{K}, \mathcal{S}, \{u_k\}_{\forall k \in \mathcal{K}}\right\}$ be a noncooperative game and let $\mathcal{S}^*$ be its set of NE strategy profiles. Then, the ratio

$$\text{PoA} = \frac{\max\limits_{\boldsymbol{s} \in \mathcal{S}} \sum_{k=1}^{K} u_k\left(s_k, \boldsymbol{s}_{-k}\right)}{\min\limits_{\boldsymbol{s} \in \mathcal{S}^*} \sum_{k=1}^{K} u_k\left(s_k, \boldsymbol{s}_{-k}\right)} \tag{9.7}$$

is the price of anarchy (PoA) of the game $\left\{\mathcal{K}, \mathcal{S}, \{u_k\}_{\forall k \in \mathcal{K}}\right\}$.

9.4 Open Spectrum Access Games

In this section, we study the DSA problem following the OSA model described in Section 9.1. As stated before, the OSA scenario is typical of non-licensed bands where there exists neither common signaling due to the use of different communications standards (Wi-Fi, Zigbee, etc.), nor uniform QoS requirements, due to the different applications (voice, video, data, etc.) radio devices are used for. We present a simple game to model such interactions. Afterward, we present existing results using this model, namely, the existence of NE points and its optimality analysis.

9.4.1 Formulation of the Game

We model the interaction between K radio devices in the OSA model sharing N orthogonal channels as a strategic game. We study the different variants of such a game considering the topology of the network (MAC or IC), as well as the goals of each player (cooperative or noncooperative). This interaction can be defined as follows:

Definition 9.5 (OSA Game in Normal Form) In normal form, the OSA game is denoted by $\left\{\mathcal{K}, \mathcal{S}, \{u_k\}_{\forall k \in \mathcal{K}}\right\}$, where each of the elements of $\mathcal{K} = \{1, \ldots, K\}$, represents one of the K transmitters. The strategy set is $\mathcal{S} = \mathcal{S}_1 \times \cdots \times \mathcal{S}_K$, where $\mathcal{S}_i$ is the strategy set of player i:

$$\mathcal{S}_i = \left\{ \boldsymbol{p}_i = \left(p_{i,1}, \ldots, p_{i,N}\right) : \forall n \in \{1, \ldots, N\}, \; p_{i,n} > 0 \quad \text{and} \quad \sum_{n=1}^{N} p_{i,n} \leq p_{\max} \right\}. \tag{9.8}$$

The utility function for the ith player is $u_i(\boldsymbol{p}_i, \boldsymbol{p}_{-i})$, and it is defined differently depending on the topology of the network (e.g., MAC or IC) and the approach (cooperative or non-cooperative) adopted by the players.

Next, we refine this definition by specifying the utility function for each particular case of study (MAC/IC and cooperative/noncooperative).

9.4.1.1 The Choice of the Utility Function

In the current literature, the utility functions are often defined as a function of either the achieved data rate or as the ratio between the successfully transmitted bits per second (goodput) and the total transmit power. In our discussion, we call the former case rate-efficient OSA and the latter energy-efficient OSA.

The utility functions in the rate-efficient OSA game following a noncooperative approach can be defined as

$$\forall k \in \mathcal{K}, \quad u_k(\boldsymbol{p}) = \sum_{n=1}^{N} \log\left(1 + \gamma_{k,n}\right). \tag{9.9}$$

which corresponds to the sum data rate achieved by player k over all its available channels [30–33,40]. Here, $\boldsymbol{p}$ is a $N\,K$-dimensional vector obtained by concatenating the vectors $\boldsymbol{p}_i$, $\forall i \in \mathcal{K}$. The SINR $\gamma_{i,n}$ is either the expression (9.2) for the MAC case or the expression (9.4) for the IC case.

The utility function in the energy-efficient OSA games is defined as

$$u_k(\boldsymbol{p}) = \frac{\sum_{n=1}^{N} R_{k,n} f\left(\gamma_{k,n}\right)}{\sum_{n=1}^{N} p_{k,n}} \tag{9.10}$$

The data rate $R_{k,n}$ is fixed by the particular modulation-coding scheme used by transmitter k on channel n. In the sequel, we assume that all transmitters are subject to the same rate R over all the channels. The function $f\left(\gamma_{i,n}\right) : \mathbb{R}_+ \rightarrow [0,1]$, known in the literature as the efficiency function, is assumed to be sigmoidal function that approximates the fraction of successfully transmitted bits per frame given the SINR $\gamma_{i,n}$ over the channel n. A typical approximation of the $f(.)$ function is

$$f(\gamma) = (1 - e^{\gamma})^M, \tag{9.11}$$

where

γ represents a given SINR
M is the framelength

The SINR $\gamma_{i,n}$ in (9.10) might be either expression (9.2) for the MAC case or expression (9.4) for the IC case. As the rates $R_{k,n}$ are constants and might be the same for all the transmitters, we ignore them and consider the utility function (9.10) as the ratio $\sum_{n=1}^{N} f\left(\gamma_{k,n}\right) / \sum_{n=1}^{N} p_{k,n}$ in the sequel.

The utility function defined in Equation 9.10 is measured in bits per second per Joule. It describes how many bits can be successfully transmitted per Joule drained from the battery. This is why the model is known as an energy-efficient model. Extensive discussions on this utility function are presented in [12].

In the cooperative game, more specifically in either the MAC or IC cooperative game, and in either the rate-efficient or energy-efficient case, the cooperative utility function is defined as

$$U(\boldsymbol{p}) = \sum_{k=1}^{K} u_k(\boldsymbol{p}), \tag{9.12}$$

where $u_k(\boldsymbol{p})$ is the utility function of the player k in the noncooperative game.

In the following sections we study the current results in OSA games using the general OSA game described in Definition 9.5 with the utility functions

(Equation 9.9) and (Equation 9.10). Other utility functions are also studied in later sections.

9.4.2 Single-Stage Games

In the current literature, the OSA problem has been analyzed considering the game defined in 9.1 for both MAC and IC channel. For the energy-efficient utility function, the MAC scenario has been studied in Refs. [10–12,44]. The IC case has been investigated in Refs. [32,33,46–48] considering the rate-efficient utility function.

9.4.2.1 MAC Single-Stage Games

Consider the definition of the OSA game (Definition 9.5) with a utility function defined by Equation 9.10 for the MAC case. A special case of this utility is the single carrier CDMA scenario, where $N = 1$. For such particular case, the utility function reduces to

$$\forall k \in \mathcal{K}, \quad u_i(p_i, \boldsymbol{p}_{-i}) = \frac{R_i f(\gamma_i)}{p_i}. \tag{9.13}$$

This utility function exhibits several properties that are carefully studied by Rodriguez in [45]. Based on this study, it is known that an NE is observed if all the transmitters achieve an SINR as close as possible to an optimal value denoted as γ^*. Such optimal SINR is the unique solution to the equation

$$\gamma^* f'(\gamma^*) - f(\gamma^*) = 0, \tag{9.14}$$

which corresponds to the unique maximum of the function $u(.)$ (Equation 9.13).

If all the players attempt to achieve the same SINR γ^*. The optimal power value is

$$p_i^* = \frac{\sigma^2}{|h_i|^2} \frac{\gamma^*}{1 - \alpha(K-1)\gamma^*}, \tag{9.15}$$

where α is the inverse of the spreading length S, $\alpha = \frac{1}{S}$ in CDMA. For all i, if the transmit power p_i^* required to achieve γ^* is higher than $p_{\max}$, then transmitting at $p_i^* = p_{\max}$ is also an NE. Thus, for the CDMA case ($N = 1$), the existence and uniqueness of the NE is always ensured [44].

When other MA technique or multi-carrier CDMA is used ($N > 1$), even though the NE always exists, it is not always unique [12,44]. First, we use the result obtained by Meshkati et al. [12], which states that

$$\forall i \in \{1,\ldots,K\}, \quad \frac{f(\gamma^*)}{\sum_{n=1}^{N} p_{i,n}} \leq \frac{f(\gamma^*)}{p_{i,L}^*}, \tag{9.16}$$

where $L = \arg\min_{n=1,\ldots,N} \{p_{i,n}^*\}$ and $p_{i,n}^*$ is obtained with

$$p_{i,n}^* = \frac{1}{|h_{i,n}|^2}\left(\frac{\gamma^*\sigma^2}{1-\alpha\Theta(n)\gamma^*}\right), \tag{9.17}$$

where $\Theta(n)$ represents the number of players simultaneously transmitting over channel n. Again, in the case where $p_{i,L}^* > p_{\max}$, an NE is achieved by transmitting at $p_{i,L}^* = p_{\max}$. The important remark here is that, at the NE each player uses only one of its available channels [12]. Nonetheless, for certain conditions over the channels gains, the NE is not always unique. To shed light on the existence and uniqueness of the NE, we consider the following example.

Example 9.1: Two-Player-Two-Channel MAC Game

Consider the particular case of a two-carrier MAC ($N = 2$) with only two active transmitters ($K = 2$). According to the discussion earlier, the optimal OSA strategy for each player is to transmit only over the subcarrier that requires the lowest transmit power level. In this respect, we can model this interaction as two players aiming at transmitting at the minimum power level required to achieve an SINR γ^* on any of the channels. Here, their strategy is transformed into the choice to transmit over channel 1 (C_1) or 2 (C_2). Considering this new definition of the game, four particular scenarios might arise. Scenarios (1) and (4): both users transmit over the first and second channel, respectively. Scenario (2): player 1 transmits over the first channel and player 2 transmits over the second channel. For scenario (3) the converse applies. For the ease of presentation, we consider single-user decoding (SUD) at the receiver.

Under the assumption that each player only decodes its own signal and treats the other player signal as noise, we calculate the required transmit power to achieve the optimal SINR γ^* depending on the channel each player decides to transmit on. In Table 9.1 we present the possible transmit power levels in this game.

Based on the power levels shown in Table 9.1, we identify the best responses each of the players might take with respect to the other player's responses. Thus, an NE might be observed on any of the four scenarios depending on the channel gains, for example,

- Equilibrium 1: (C_1, C_1)

$$\frac{|h_{12}|^2}{|h_{11}|^2} < 1-\gamma^* \quad \text{and} \quad \frac{|h_{22}|^2}{|h_{21}|^2} < 1-\gamma^*; \tag{9.18}$$

Table 9.1 Transmit Power Levels at Each Possible Equilibrium Point for the Two-Player-Two-Carrier Game with SUD

	C_1	C_2
C_1	$p_{1,1} = \frac{\sigma^2\gamma^*}{\lvert h_{11}\rvert^2(1-\gamma^*)}$ $p_{1,2} = 0$ $p_{2,1} = \frac{\sigma^2\gamma^*}{\lvert h_{21}\rvert^2(1-\gamma^*)}$ $p_{2,2} = 0$	$p_{1,1} = 0$ $p_{1,2} = \frac{\sigma^2\gamma^*}{\lvert h_{12}\rvert^2}$ $p_{2,1} = \frac{\sigma^2\gamma^*}{\lvert h_{21}\rvert^2}$ $p_{2,2} = 0$
C_2	$p_{1,1} = \frac{\sigma^2\gamma^*}{\lvert h_{11}\rvert^2}$ $p_{1,2} = 0$ $p_{2,1} = 0$ $p_{2,2} = \frac{\sigma^2\gamma^*}{\lvert h_{22}\rvert^2}$	$p_{1,1} = 0$ $p_{1,2} = \frac{\sigma^2\gamma^*}{\lvert h_{12}\rvert^2(1-\gamma^*)}$ $p_{2,1} = 0$ $p_{2,2} = \frac{\sigma^2\gamma^*}{\lvert h_{22}\rvert^2(1-\gamma^*)}$

- Equilibrium 2: (C_1, C_2)

$$\frac{|h_{12}|^2}{|h_{11}|^2} < \frac{1}{1-\gamma^*} \quad \text{and} \quad \frac{|h_{22}|^2}{|h_{21}|^2} > 1-\gamma^*; \tag{9.19}$$

- Equilibrium 3: (C_2, C_1)

$$\frac{|h_{12}|^2}{|h_{11}|^2} > 1-\gamma^* \quad \text{and} \quad \frac{|h_{22}|^2}{|h_{21}|^2} < \frac{1}{1-\gamma^*}; \tag{9.20}$$

- Equilibrium 4: (C_2, C_2)

$$\frac{|h_{12}|^2}{|h_{11}|^2} > \frac{1}{1-\gamma^*} \quad \text{and} \quad \frac{|h_{22}|^2}{|h_{21}|^2} > \frac{1}{1-\gamma^*}. \tag{9.21}$$

In Figure 9.1 we plot such conditions and the regions where the different NEs are observed. Therein, it is evident that the existence of the NE is always ensured. Nonetheless, it might not be unique. Indeed, in the region where channel gains simultaneously satisfy $1-\gamma^* \leq \frac{|h_{12}|^2}{|h_{11}|^2} \leq \frac{1}{1-\gamma^*}$ and $1-\gamma^* \leq \frac{|h_{22}|^2}{|h_{21}|^2} \leq \frac{1}{1-\gamma^*}$ the NE is not unique. In such a region, both Scenarios 2 and 3 are NEs.

Regarding this example (Example 9.1), an algorithm for achieving an NE was presented in [12]. However, such an algorithm was unable to always converge to at least one NE, even though the existence of at least one NE is always ensured. A unique NE is observed if the receiver implements interference cancellation assuming that the same decoding order is adopted on all the available channels. Nonetheless, this new game requires the knowledge of the decoding order of all the players, which is necessarily a demanding task due to the required additional signaling.

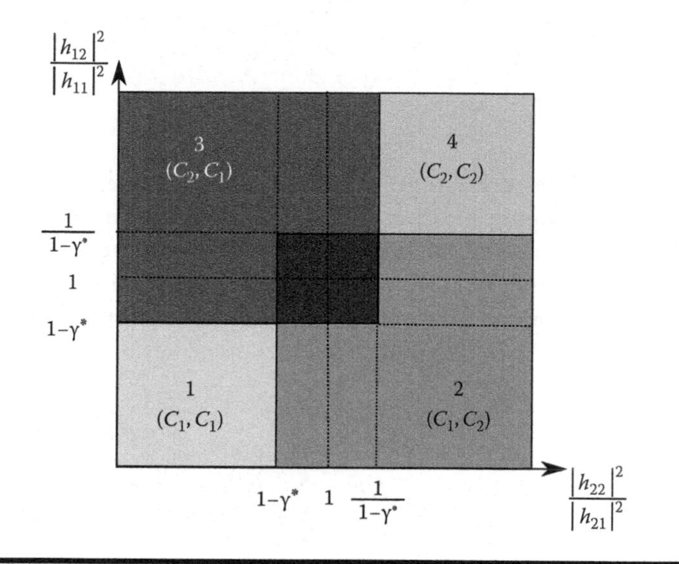

Figure 9.1 Regions of the possible NEs for the two-players two-carriers game with single-user detection (SUD).

9.4.2.2 IC Single-Stage Games

In a cooperative context, if the utility function is the sum data rate of all the players as suggested in [32], Pareto optimal strategies coincide with the border of the capacity region. Unfortunately, the capacity region for the IC is still undetermined and is a subject of intensive research [36]. For simple cases, such as the high interference regime, that is,

$$\forall n \in \{1, \ldots, N\} \text{ and } \forall (i,j) \in \mathcal{K}^2, \quad \left|h_{i,j}^{(n)}\right|^2 \left|h_{j,i}^{(n)}\right|^2 > \left|h_{i,i}^{(n)}\right|^2 \left|h_{j,j}^{(n)}\right|^2 \tag{9.22}$$

it is known that Pareto optimal sum data rates are achieved by using orthogonal channels, that is, $\langle \boldsymbol{p}_i, \boldsymbol{p}_j \rangle = 0$, $\forall (i,j) \in \mathcal{K}^2$. For instance, such optimal power profiles are obtained with frequency division multiplexing (FDM). This result is due to the fact that the condition in expression (9.22) leads to a concave optimization problem whose solution is easy to identify [37]. If the high interference regime is not met (Equation 9.22), then the Pareto optimal strategies require more complex calculations [32].

In the noncooperative context, the utility functions correspond to the individual data rate achieved by each player over all its channels (Equation 9.9). In this case, due to the selfish behavior of each player, we are interested on NE points. In [32], it has been shown that uniform power spreading among all the channels corresponds to an NE point. We denote such an optimal strategy profile as $\boldsymbol{p}_k^*$, for the player k.

Thus,

$$\forall k \in \mathcal{K}, \quad \boldsymbol{p}_k^* = \frac{p_{\max}}{N} \mathbb{1}_N \tag{9.23}$$

where $\mathbb{1}_N$ is an N-dimensional vector whose entries are all ones. This result comes from the fact that an optimal response for a given player given the strategies of all the other players is to waterfill [38] over the noise plus interference observed on its channels. This interaction leads to stable point that corresponds to spreading the total transmit power uniformly. However, the uniqueness of such NE point is not ensured. In [32], a sufficient condition for the uniqueness of such equilibrium is provided,

$$\forall n \in \{1, \ldots, N\}, \quad \sum_{j=1}^{K} \frac{\left|h_{ji}^n\right|^2}{\left|h_{ii}^n\right|^2} \leq 1. \tag{9.24}$$

This result comes from the Karush–Kuhn–Tucker conditions of the optimization problem of the set of data rates. Less-restrictive sufficient conditions for the uniqueness of the NE are provided in [38].

9.4.3 Repeated Games

When the interaction between the decision makers is modeled as a strategic game, as discussed in Section 9.3.1, it is assumed that players make their decisions simultaneously and once during the whole game. In DSA, this approach is suited for one of the following situations: (1) networks with low mobility and fixed topology; (2) the study of a network over a short period of time; (3) the users' decisions over time are taken independently, for example, if the users's utility is chosen to be its instantaneous data rate or energy efficiency (over a block). However, radio devices interact for long periods in a constantly changing environment. To take into account such practical features, we can consider a sequence of strategic games related to each other in time [43]. This model is known in game theory as repeated games.

A repeated game is a special case of dynamic game [51]. It can be seen as a strategic game that is played consecutively a finite or infinite number of times. These games are respectively referred to as infinitely repeated and finite horizon repeated games [14]. At each time or stage, the same game is played. In this section, we focus on infinitely repeated games. For such cases, the utility for each player is chosen to be a weighted sum of its own instantaneous utility (e.g., Equation 9.10 or Equation 9.9) at stage m, $u_k(m)$, that is,

$$r_k = (1 - \delta) \sum_{m=0}^{\infty} \delta^m u_k(m) \tag{9.25}$$

where δ is a normalization factor such that $0 < \delta \leq 1$, and $r_k = 1$ if $u_k = 1$. The factor δ defines if the player is more interested either on the utility obtained at the initial stages of the game or on the utility obtained in later stages.

At the end of each stage, all the players observe the outcome of the game. For a transmitter to be able to directly observe the actions of the other transmitters, additional assumptions have to be made. For example, in a two-player game, if player 1 wants to evaluate the transmit power of player 2 from the signal received from the first player, some source separation algorithm and sufficiently accurate path loss model need to be assumed. In what follows, we assume that players can observe each other player's actions.

The OSA noncooperative game described in Section 9.4.1 is modeled by an infinitely repeated game considering the utility function r_k described in Equation 9.25 where each $u_k(m)$, $\forall k \in \mathcal{K}$ and $\forall m \in \mathbb{N}$ corresponds to the individual data rate as defined in Equation 9.9. In such a game, at each stage the same analysis presented in Section 9.4.2 applies. To extend such analysis to a repeated game, we resort to the following definition.

Definition 9.6 (Perfect Sub-Game NE) A set of strategies

$$\boldsymbol{s}^* = \left(\boldsymbol{s}(t+1)^*, \ldots, \boldsymbol{s}(t+T)^*\right),$$

with $\boldsymbol{s}(t) = (s_1(t), \ldots, s_K(t))$ is a perfect sub-game (game from stage t to state $t+T$) NE if at every stage t of the sub-game, the strategy $\boldsymbol{s}^*(t)$ is an NE.

In the single-stage game discussed in Section 9.4.2 for the IC case with utility function (Equation 9.9) an NE was obtained by uniformly spreading the transmit power among all the available channels. Then, in the repeated game, uniformly spreading the total power at each stage forms a perfect sub-game NE $\forall m \in \mathbb{N}$ [32]. Exploiting the fact that the game is infinitely repeated, an improvement based on the assumption that if a Pareto optimal strategy $\boldsymbol{p}'$ (e.g., transmitting over orthogonal channels in the high interference regime) is achieved due to an initial coordination of the network, by means of a punishment policy, each transmitter is encouraged to keep the strategy $\boldsymbol{p}'_k$ at each iteration m. If a player deviates from the strategy $\boldsymbol{p}'$, the other players punish him by spreading the power over all the channels. A punishment will necessarily decrease the utility function of all the active players in the next stage [32]. Thus, no player is motivated to deviate from the initial agreement.

The strategy of punishments significantly improves the performance obtained with single-stage repeated games. In the repeated game, radio devices obtain Pareto optimal outcomes at each stage of the game. Playing a different strategy leads to a punishment that represents a lower utility function. This punishment policy can be seen as certain coordination between the players to improve the global performance (cooperative approach). It should be noted that no additional signaling is required once the players have set up their agreement.

Other application of repeated games to OSA is presented by Wendorf et al. in [21,23]. Therein, the problem of dynamic frequency selection (DFS), typical from IEEE 802.11 networks is studied. The DFS problem can be defined as in Definition 9.5, where the set of strategies of each player is redefined as

$$\forall i \in \mathcal{K}, \quad \mathcal{S}_i = \left\{p_{\max}\boldsymbol{e}_1, \ldots, p_{\max}\boldsymbol{e}_N\right\}, \tag{9.26}$$

where the N-dimensional vector $\boldsymbol{e}_i$, $\forall i \in \mathcal{K}$ is such that all its entries are zeros except the ith entry, which is a one. In DFS each radio transmits over only a single channel and always at the maximum power. The analysis in [21,23] considers only two players under the assumption that coexistence of both radio devices on the same channel leads to an outage condition due to mutual interference. This problem is modeled as a repeated game considering that at the starting point both players are using the same channel. Hence, a simplification is proposed and the strategy set for each player is reduced to the actions of either changing its channel (C) or staying (S) on the same channel. Thus $\forall k \in \mathcal{K}$, $\mathcal{S}_k = \{S, C\}$. The utility function, which is in reality a cost function (players aim to minimize it), is defined as the sum of the delay (denoted by v) induced by switching from the current channel to a new one and the delay (denoted by u) due to the multiple access interference (which increases the probability of retransmission). In Table 9.2 the game under investigation is represented in its matrix form [14,27]. The first (resp. 2) entry of the pair (x, y) corresponds to the cost for user 1 (resp. 2).

The authors of [21,23] do not link the strategies S and C to the channel being used and the channel to be used. Therefore, players randomly change to another channel. When the number of channels is large, the probability of overlapping is relatively small. Thus, the case of two players with two channels is the worse scenario

Table 9.2 Game Matrix for the DFS Game

1/2	C	S
C	(v, v)	$(v, 0)$
S	$(0, v)$	$(1 + u, 1 + u)$

Source: Wendorf, R.G. and Seidenberg, H.B., Channel-change games for highly interfering spectrum-agile wireless networks, *2nd International Symposium on Wireless Pervasive Computing,* San Juan, PR, 2007. With permission.

in terms of the probability of overlapping. Assuming a game with two channels and a certain probability (say ρ) that a user changes its channel, the authors of [21,23] derived the solution of the game in terms of such probability. They provide the optimal probability ρ^* a user should adopt during the stages of the game. The authors show that ρ^* corresponds to a unique NE. The result is compared with a cooperative approach where the utility function is defined as the sum of the channel acquisition cost (delay) for each player. The comparison shows a significant loss in performance in terms of delay due to the lack of cooperation. This work has been extended to the case of several players in [22].

This work is important in the sense that it does not consider a set of actions as the strategy of each player. It rather defines the actions and assumes the strategy set as the probability with which each user will play each action in the repeated game. In game theory, this strategy definition is known as mixed-strategies [14].

9.4.4 Bayesian Games

So far, we have considered that each player has complete information about all the other players, that is, each player knows all the other players' utilities. Regarding the DSA problem, this means that each radio has perfect CSI and perfectly knows the power limitations of all the other radio devices. However, this assumption is quite far from the situation encountered in practice. In the open spectrum access model, for example, it is more realistic to assume that a given radio transmitter only knows its own channel gains, QoS requirements, and power limitations. In this section, unless otherwise stated, each radio device only knows its own channel gain, power limitations, and QoS requirements. Game theory provides a complete framework to analyze the interaction between decision makers with incomplete information. Bayesian games are either static or dynamic games where at least one player does not know the utility function of one or more of its opponents. Thus, each user plays the game based on a probabilistic analysis. Therefore, these games are known as Bayesian games due to the Bayesian inference required to play.

Following the approach proposed by Harsanyi [39], a Bayesian game could be obtained by introducing some randomness in a strategic game. Suppose that there exists a set $\mathcal{K} = \{1, \ldots, K\}$ of players with a set of actions $\mathcal{S}_k$, $\forall k \in \mathcal{K}$ and that each player does not know the utility function of all the other players. However, each player knows that there exists a finite set of possible types $\mathcal{T}_i$ for each player $k \in \mathcal{K}$. The corresponding type for each player is a random variable that follows a probability distribution known by all the players. Each type of player has a specific utility function and strategy set, denoted u_t and $\mathcal{S}_t$, $\forall t\ in \mathcal{T}$, respectively. Thus, the incomplete information game is transformed into a game where at least the probability distributions of the unknown parameters (e.g., utility functions and strategy sets) are known.

Definition 9.7 (Bayesian Game) A Bayesian game is completely described by the following set of parameters

- A set of K players, $\mathcal{K} = \{1, \ldots, K\}$
- A finite set of T types of players $\mathcal{T} = \{t_1, \ldots, t_T\}$
- A probability density function of the different types of players: $\{f(t) \in [0,1] \mid \forall t \in \mathcal{T}\}$
- A set of T finite sets of strategies: $\mathcal{S}_1, \mathcal{S}_2, \ldots, \mathcal{S}_T$ each one for each type of player
- A set of T utility function $u_k : \mathcal{T} \times \mathcal{S} \rightarrow \mathbb{R}_+$, for each type of player

The DSA problem presented in Section 9.2 can be modeled as a Bayesian game. Here, it could be stated that radio devices do not know the exact channel realizations of all the other players in the network and, therefore, their utility function cannot be determined. Under a Bayesian formulation of the problem, we can assume that players do not know the exact channel realization, but it is known that it belongs to a known probability distribution. Under these assumptions, we can now formulate the OSA problem following the definition Definition 9.5 considering that the set of types $\mathcal{T}$ corresponds to all the possible probability distributions that can model the channel realization of each player. For simplicity, we assume that all the channel gains follow the same distribution f_h.

A more realistic model would be to consider different probability distributions for each player. This is the case, for example, if some transmitters are in line of sight with their corresponding receivers (Rice channel distribution) while the other transmitters are linked with their receivers by Raleigh-distributed channels.

The noncooperative DSA problem in the IC case is studied in [40]. The underlying assumption is that radio devices do not know any of the parameters to play the game. More specifically, CSI is not available either at the transmitter or at the receiver. Nonetheless, it is assumed that all the channel realizations are drawn from a known probability distribution f_h. The utility function for each player is chosen to be the expected value of the utility function defined in (9.9),

$$\forall k \in \mathcal{K}, \quad \hat{u}_k(\boldsymbol{p}_k, \boldsymbol{p}_{-k}) = \mathbb{E}\left[u_k(\boldsymbol{p}_k, \boldsymbol{p}_{-k})\right], \tag{9.27}$$

where, the expectation is taken over the distribution f_h.

The authors in [40] analyze the problem from a noncooperative point of view, that is, the goal of each user is to selfishly maximize its own utility function. Here, a highly desired outcome corresponds to a Nash equilibrium. When it exists, the optimal strategy profile $\boldsymbol{p}_k^* = \left\{p_{k,1}^*, \ldots, p_{k,N}^*\right\}$ is a solution to the following optimization problem

$$\forall k \in \mathcal{K}, \quad \boldsymbol{p}_k^* = \arg\max_{\boldsymbol{p}_k} \int_0^\infty \hat{u}_k(\boldsymbol{p}_k, \boldsymbol{p}_{-k}^*) f_h(\boldsymbol{h}) \mathrm{d}\boldsymbol{h}, \tag{9.28}$$

The NE for this case is the transmit power vectors $\boldsymbol{p}^* = \{\boldsymbol{p}_1^*, \ldots, \boldsymbol{p}_N^*\}$, where

$$\forall k \in \mathcal{K}, \quad \boldsymbol{p}_k^* = \frac{p_{\max}}{N} \mathbb{1}_N. \tag{9.29}$$

Thus, under the assumption of incomplete information, there is a unique NE at which all the transmitters uniformly spread all the transmittable power between all the available channels. In a further step, the authors in [40] keep the number of players limited at $K = 2$ and additionally restrict the set of actions for each player such that in this new formulation, each player might either assign all the transmittable power to a unique channel or uniformly spread it over all the N available channels. Therefore, the set $\mathcal{S}_k$ is made of $N + 1$ vectors such that

$$\forall k \in \mathcal{K}, \quad \mathcal{S}_k = \{p_{\max}\boldsymbol{e}_1, \ldots, p_{\max}\boldsymbol{e}_N, p_{\max}\mathbb{1}_N\}. \tag{9.30}$$

Moreover, it is assumed that the ith player perfectly knows the set of channels $\boldsymbol{h}_i = (h_{i,i}, h_{-i,i})$, that is, its own channel and its interfering channels. As discussed before, the NE solution corresponds to the solution of the optimization problem,

$$\forall k \in \mathcal{K}, \quad \boldsymbol{p}_k^* = \arg\max_{\boldsymbol{p}_k} \mathbb{E}\left[u_k(\boldsymbol{p}_k, \boldsymbol{p}_{-k}^*) | \boldsymbol{h}_k\right], \tag{9.31}$$

where the expectation is taken over the unknown channels $\boldsymbol{h}_j = (h_{j,j}, h_{i,j})$, for all $j \neq i$.

The authors showed again that although each player have more information (its own channel $h_{i,i}$ and its interfering channel $h_{-i,i}$) the NE is again obtained by uniformly spreading the total power over all the available channels. Moreover, the corresponding NE is not Pareto optimal.

9.4.5 Coalitional Games

In economics, for instance, rational players (e.g., companies or manufacturers) tend to create coalitions to maximize their individual or common benefits. This kind of games is called coalitional games [14,27]. In this scenario, a set of actions $\mathcal{S}_c$ is associated to each of the C coalitions or groups $\mathcal{K}_c$, $\forall c \in \{1, \ldots, C\}$. The set of actions of the kth player, $\forall k \in \mathcal{K}_c$ is then $\mathcal{S}_c$. Each coalition obtains a benefit called value, denoted v_c for the cth coalition, as a result of the actions of all the players $k \in \mathcal{K}_c$. The total benefit of the coalition is shared between all the members of the coalition. Each coalition can have a different policy $u_c(\mathcal{K}_c, v_c)$ for sharing the common benefit. Hence, the goal of each player k, $\forall k \in \mathcal{K}$ is to choose the best coalition to join. In fact, rational players will join the coalition where the highest individual benefit is obtained.

Definition 9.8 (Coalitional Game) A coalitional game is completely described by the following parameters:

- A finite set of K players, $\mathcal{K} = \{1, \ldots, K\}$
- A finite set of C coalitions $\mathcal{C} = \{1, \ldots, C\}$ such that $\forall k \in \mathcal{C}$, $\mathcal{K}_k \subseteq \mathcal{K}$ and $\forall i \neq j$, $\mathcal{K}_i \cap \mathcal{K}_j = \{\}$
- C finite sets of actions $\mathcal{S}_1, \ldots, \mathcal{S}_C$, with $\mathcal{S} = \mathcal{S}_1 \times \ldots \times \mathcal{S}_C$
- A finite set of C values, $\{v_c : \mathcal{K} \times \mathcal{S} \rightarrow \mathbb{R} : \forall c \in \mathcal{C}\}$
- A finite set of C policies, $u_c(v_c)$, $\forall c \in \mathcal{C}$

A stable outcome or equilibrium of coalitional games consists of a distribution of the players among the different coalitions where no player is interested in choosing another coalition. A common equilibrium is called the grand coalition. Here, there is only one coalition and all the players belong to it. Thus, none of the players can obtain a higher benefit by leaving the grand coalition. Coalitional games are of high importance in the DSA problem. In the OSA model, for example, terminals of the same physical layer technology, which can "understand" each other, can ally to improve its individual or common performance. Similarly, radio devices with similar power constraints or even QoS requirements might be interested in forming/joining a coalition. Nonetheless, this assumption requires radio devices to coordinate themselves to establish the available set of actions for a possible coalition. Often, coordination requires signaling among the terminals, which is not always practically appealing.

In the literature, the DSA problem has been modeled as a coalitional game in [17–19]. The idea of a coalition in the DSA problem for a unique channel $N = 1$, and a finite number of radio devices, is that receivers that belong to the same coalition jointly decode their received signals and perform interference cancellation [35]. Indeed, all the other signals from the transmitter belonging to other coalitions are treated as an AWGN. This configuration corresponds to a set of single input multiple output-multiple access channels (SIMO-MAC). Similarly, transmitters can also form coalitions, which will lead us to virtual MISO systems. If both transmitters and receivers form coalitions, the network is equivalent to a set of virtual MIMO channels.

The value v_c, $\forall c \in \{1, \ldots, C\}$ in both cases, transmitter or receiver cooperation, is then chosen as the sum of individual data rates R_k achieved by all the players $k \in \mathcal{K}_c$, that is,

$$v_k = \sum_{i \in \mathcal{K}_k} R_i. \tag{9.32}$$

Depending on its achieved individual data rate R_k, each player k decides which coalition to join. For instance, in the receiver cooperation case, depending on the coalition a receiver belongs to, it will decode a different set of interferers. The authors in [17] showed that the grand coalition maximizes the spectrum utilization. This is equivalent to the fact that terminals should jointly decode all the surrounding terminals' signals [35].

9.5 Hierarchical Spectrum Access Games

The hierarchical spectrum access model considers two different approaches, the underlaying and the overlaying approaches [3,4]. On the one hand, the opportunistic devices have to meet a certain power constraint to keep the interference level they generate on the primary systems always below their noise floor. This is the case of ultra wide band or interference alignment systems, for example [4,41,42]. Here, the opportunistic and primary transmitters can simultaneously transmit without generating harmful interference on the primary receivers.

On the other hand, the overlaying approach that targets SAOs, does not impose any limit on the transmit power. It only requires opportunistic radio devices to identify radio resources left unused by the primary network and exploit them subject to the constraint that it can be required by the primary system at any time. In this approach, the opportunistic and primary players do not transmit simultaneously. If the current power constraints and the available channels are perfectly identified, in both the overlaying and underlaying models, the influence on the primary systems can be neglected and the problem can be modeled as in Section 9.2. Hence, all the game theoretical machinery we have developed so far directly applies to this particular dynamic spectrum access scheme.

A special case of hierarchical spectrum access arises when interaction between the primary system and the opportunistic devices is allowed. Here, any action of the primary devices affects the benefit obtained by the opportunistic radio devices. For example, a primary system might offer a set of channels to the opportunistic terminals in exchange of cooperation in the form of distributed space-time coding (DSTC) [13]. This kind of interactions cannot be modeled with the tools presented earlier. Here, there are two different types of players (primary and secondary) whose priority is different when they access the radio resources. The framework provided by game theory to study this interaction is called Stackelberg games.

9.5.1 Stackelberg Games

Here we investigate situations where there is a hierarchy between players. A useful case of this kind of games are Stackelberg games, which were initially introduced by Stackelberg in 1934 [14]. In this game, there is an implicit concept of hierarchy upon the set of players. Such hierarchy naturally occurs when users are playing sequentially. For example, in a two-level Stackelberg game, the game leader moves first and the other players follow and play simultaneously. The game leader perfectly knows the set of strategies and the utilities of the followers. Similarly, it is guaranteed that the followers can observe the actions of their leader(s).

Definition 9.9 (Stackelberg Games) A Stackelberg game is a two-stage game at which one player (leader) moves at the first stage and all the other players (followers) react simultaneously at the second stage.

A Stackelberg game can be easily solved through the concept of sub-game perfect NE. In the first stage the leader, who perfectly knows all the followers' set of actions and utility functions, chooses the action that maximizes its benefit considering that each follower will react with the action that maximizes its own benefit as well. Thus, the game leader analyzes all the possible outcomes and picks up the action that maximizes its benefit considering the optimal moves for each player. In the recent literature, an interesting application of Stackelberg games in dynamic spectrum access was presented by Simeone et al. The problem is modeled as follows. There exists a unidirectional primary link from transmitter *Txp* to receiver *Rxp* and an ad hoc network with a set $\mathcal{I}$ of K point-to-point links as described in Section 9.2. The primary system divides its transmission (L bit durations) into two parts of αL bit durations and $(1-\alpha)L$ bit durations, with $0 \leq \alpha \leq 1$. The first $(1-\alpha)L$ bits are dedicated to a direct transmission from *Txp* to *Rxp*. The second αL bits are again divided in two. One part $\beta\alpha L$,with $0 \leq \beta \leq 1$, is dedicated to send information from *Txp* to *Rxp* using the ad hoc network as a means to perform distributed space time coding (DSTC). The remaining $\alpha(1-\beta)L$ bits are then granted to the ad hoc network to transmit its own data. The performance of the ad hoc network follows the one described in Section 9.2. The aim of the primary link is described depending on its available information. For the case where instantaneous CSI is available, that is, the primary system perfectly knows all the channel gains, its utility function, u_{leader}, is described in terms of the achieved data rate. A complete discussion is presented in [13]. If only the statistics of the channel realizations are available, the utility function is described in terms of the outage probability. In both cases, the goal of the leader is to maximize its own utility by deciding on the amount of bits to be coded using DSTC, the amount of bits to grant to the ad hoc network, that is, α and β, and the most convenient set of links to use in the DSTC, denoted as $\mathcal{K}_s \subseteq \mathcal{K}$. Hence, the set of actions of the leader is

$$\mathcal{S}_{\text{leader}} = \{\alpha, \beta, \mathcal{K}_s : 0 \leq \alpha \leq 1, 0 \leq \beta \leq 1, \mathcal{K}_s \subseteq \mathcal{K}\}. \tag{9.33}$$

The primary network is considered to be the leader of the game and decides the optimal parameters α^*, β^*, and $\mathcal{K}_s^*$ at the first instant of the game. The optimal values, then, correspond to the solution to

$$\max_{\alpha,\beta,\mathcal{K}_s} u_{\text{leader}}(\alpha, \beta, \mathcal{K}_s) \quad \text{s.t. } \mathcal{K}_s \subseteq \mathcal{K},\ 0 \leq \alpha, \beta \leq 1, \tag{9.34}$$

Later, the ad hoc network reacts by exploiting the $\alpha(1-\beta)L$ bit periods it has to transmit. Each transmitter in the ad hoc network plays the game described in Section 9.2. The primary player always obtains the best outcome from the game because it is privileged to play at the first moment. The followers, which follow the configuration discussed in Section 9.2, obtain a stable non-Pareto NE outcome. This NE has been already studied in Section 9.4.

Other Stackelberg formulation is presented in [52,53] for the case of MAC with $N = 1$ and the energy-efficient utility function (Equation 9.10). Therein, the interactions among the transmitters are modeled as in Definition 9.1 and a certain degree of hierarchy is introduced in two different ways: (1) assuming SUD at the receiver, one player (e.g., one primary transmitter) plays first whereas the other players (e.g., secondary transmitters) are assumed to react to the leader's decisions and (2) assuming neither leader nor followers among the players, hierarchy is introduced by assuming successive interference cancellation at the receiver. It is shown that these two hierarchical models not only improve the individual energy efficiency of all the users but can also be a way of ensuring the existence of an equilibrium and reaching a desired trade-off between the global network performance at the equilibrium and the requested amount of signaling.

In the first case, the authors consider without loss of generality (but possibly with loss of optimality) that user i is the leader of the game (and plays first). Each follower $j \neq i$ therefore plays a noncooperative game with the other followers, given what the leader plays. Interestingly, it is possible to show that, under realistic conditions, there is a unique equilibrium in this hierarchical game, which is called a Stackelberg equilibrium (SE). To define an SE, let $\boldsymbol{U}^*(p_i)$ be the set of NE for the group of followers when the leader plays strategy p_i. In other words, the leader maximizes its utility function that depends on the NE $u^* \in \boldsymbol{U}^*(p_i)$ of the followers. By denoting $(p_i^{\text{SE}}, \boldsymbol{p}_{-i}^{\text{SE}})$ the power profile at the SE, this definition translates mathematically by

$$p_i^{\text{SE}} = \arg\max_{p_i} u_i\left(p_1^{\text{SE}}(p_i), \ldots, p_{i-1}^{\text{SE}}(p_i), p_i, p_{i+1}^{\text{SE}}(p_i), \ldots, p_K^{\text{SE}}(p_i)\right), \tag{9.35}$$

where $p_j^{\text{SE}}(p_i)$, $j \neq i$, is the power at the NE of the follower j (which depends on p_i, the leader's action).

Under this formulation, the existence and uniqueness of the SE is always ensured [53]. When player i is assumed to be the leader, its optimal power allocation is

$$p_i^{\text{SE}} = \frac{\sigma^2}{|h_i|^2} \frac{\gamma^*(1+\beta^*)}{1-(K-1)\gamma^*\beta^* - (K-2)\beta^*} \tag{9.36}$$

and for each follower $j \neq i$,

$$p_j^{\text{SE}} = \frac{\sigma^2}{|h_j|^2} \frac{\beta^*(1+\gamma^*)}{1-(K-1)\gamma^*\beta^* - (K-2)\beta^*}, \tag{9.37}$$

if the following (sufficient) conditions hold: $\frac{f''(0)}{f'(0)} \geq 2\frac{(K-1)\beta^*}{1-(K-2)\beta^*}$ and

$$\phi(x) = x\left[1 - \frac{(K-1)\beta^*}{1-(K-2)\beta^*}x\right]f'(x) - f(x)$$

has a single stationary point in $]0, \gamma^*[$, where γ^* is the positive solution of the equation $\phi(x) = 0$ and β^* is the solution to (Equation 9.14).

In the second case, the receiver is assumed to implement successive interference cancellation (SIC). Under the assumption that players are perfectly decoded, the optimal power allocation at the NE is

$$\forall i \in \{1, \ldots, K\},\ p_i^{\text{SIC}} = \frac{\sigma^2}{|h_i|^2} \beta^* (1 + \beta^*)^{i-1} \tag{9.38}$$

where player denoted by i is decoded with rank $K - i + 1$ in the successive decoding procedure at the receiver. In [53], it is shown that this NE always exists and is unique.

To compare both approaches, SUD and SIC, assume a random CDMA system with $R_i = 100$ kbps for all $i \in \{1, \ldots, K\}$. An efficiency function $f(x) = (1 - e^{-x})^M$ with $M \in \{2, 5, 10, 20, 50, 100\}$. The corresponding values for $\gamma^*(M)$ are respectively 1.25, 2.66, 3.61, 4.51, 5.65, and 6.47. Figure 9.2 represents the quantity $\frac{w^{\text{SIC}}}{w^{\text{SUD}}} - 1$ in percentage as a function of the spectral efficiency $\alpha = \frac{K}{S}$ for SNR [dB] = 6 and random decoding order. Here, w^{SIC} and w^{SUD} represent the

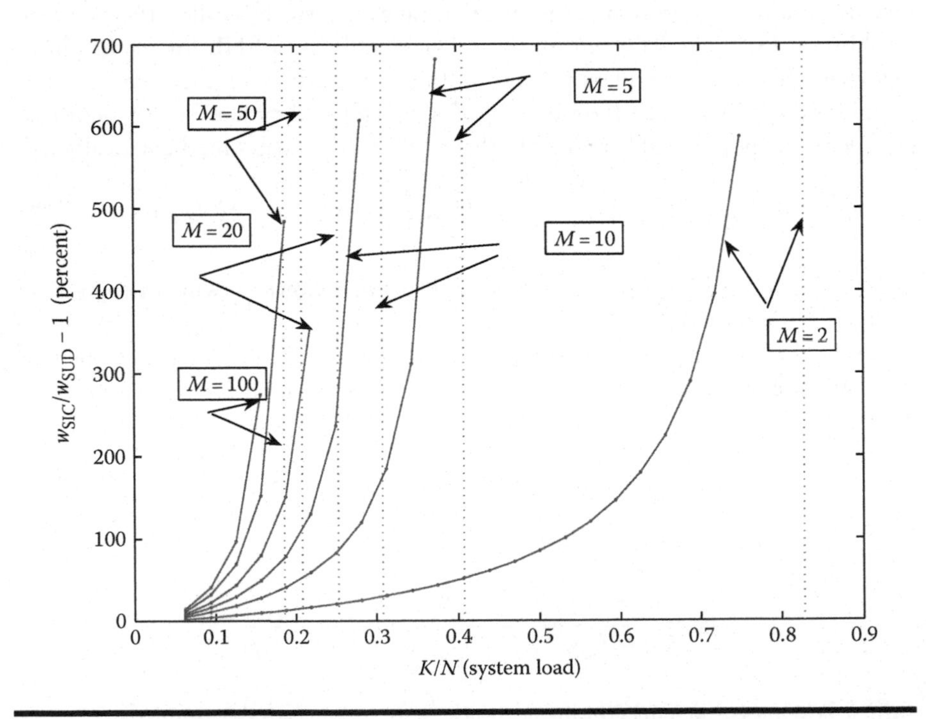

Figure 9.2 Influence of the decoding scheme (SIC/SUD) for different system loads and packet lengths.

obtained utility function in the game when SIC is implemented and the game when leader and followers exist, respectively. The asymptotes $\alpha_{\max} = \frac{1}{\beta^*(M)} + \frac{1}{S}$ are indicated by (red) dotted lines and S is the spreading length. The gains are particularly significant when the system load is relatively high, that is, when there is a significant amount of interference to be removed after despreading. In fact, when $\frac{K-1}{S}\gamma^* \rightarrow 1^+$ the noncooperative game becomes dramatically inefficient as the transmit powers at the equilibrium diverge; here, once again, we recall that we assume a nontrivial NE at which the users do not exploit all their power. Otherwise, a user who would saturate his or her power constraint would maximize his or her utility by transmitting at $P_i^{\max}$.

9.6 Concluding Remarks

We have seen that game theory is a natural paradigm to study a network where terminals are competing with each other for a common resource, namely, the spectrum. We have essentially distinguished between two kinds of games: (1) the open spectrum access game where all the transmitter–receiver pairs have the same priority; (2) the hierarchical spectrum access game where some transmitters are leaders of the game (primary transmitters) or follower (secondary transmitters). Depending on the context each of these games can be noncooperative or cooperative. In the first case, the network is totally decentralized and the terminals selfishly optimize their individual performance criteria, which can lead to an equilibrium of the network. The existence and uniqueness of a NE is an attractive feature of a decentralized network as the system owner(s) can predict the performance of the users and therefore guarantee, for example, a minimum QoS. In the case of cooperative games, the global performance of the network becomes the main target. We have presented several ways of introducing cooperation in a game: repeating the game, forming coalition, or by introducing a certain degree of hierarchy. The global network efficiency can be measured in terms of price of anarchy or by considering Pareto optimality. Another way to measure it, when transmission rates are considered for the users' utilities, is the ratio of the network sum-rate at the equilibrium to that obtained by the equivalent virtual MIMO system. Clearly, the assumed spectrum-sensing technique will play an important role in selecting the way of cooperating. Also, depending on the a priori information available to the transmitters or the used spectrum-sensing technique, the game can be played with or without complete information. In this case, we have seen that the Bayesian approach is a possible candidate to deal with this kind of (practical) situations.

We will conclude this chapter by mentioning a few open issues related to the open spectrum access problem. It is a fact that there are more and more devices using unlicensed bands. Each actor, that is, a telecom operator or a manufacturer generally designs his or her system independent of the other actors. But this assumption is likely to become less and less valid as the number of devices operating in these bands grows.

The problem of system interference naturally appears, which creates interaction between the different transmitter–receiver pairs. Obviously, there is going to be a game between many players (operators, manufacturers, etc.) who will have to design their terminals in a smart way to take into account this interaction. The problem is particularly challenging because networks will be, by nature, heterogeneous. Many questions arise. Will the different players continue to deploy their systems independently or will they try to cooperate to mitigate the interference? What kind of information can be reasonable to be assumed at a given terminal in a context where the system owners do not disclose their strategies? Is it possible to classify the different devices to apply a Bayesian approach that is robust to the uncertainty on the environment? This chapter gathers a few basic elements released in the recent literature on spectrum access games to start solving these issues but more and more contributions will be needed to provide reliable answers.

References

1. US FCC Spectrum Policy Task Force. Report of the spectrum efficiency working group. Tech. Rep., November 2002.
2. S. Haykin, Cognitive radio: Brain-empowered wireless communications. *IEEE Journal on Selected Areas in Communications*, 23(2): 201–220, 2005.
3. M. Buddhikot, Understanding dynamic spectrum access: Models, taxonomy and challenges. *IEEE DySPAN*, Dublin, Ireland, April 2007.
4. Q. Zhao and B. M. Sadler, A survey of dynamic spectrum access. *IEEE Signal Processing Magazine*, 24(3): 79–89, 2007.
5. European Radiocommunications Committee, Relating to the use of short range devices, ERC Recommendation 70-03. April 2002.
6. A. Mackenzie, L. DaSilva, *Game Theory for Wireless Engineers (Synthesis Lectures on Communications)*, 1st edn. Morgan & Claypool Publishers, San Rafael, CA, May 2006.
7. L. Berlemann, G. R. Hiertz, B. Walke, and S. Mangold, Strategies for distributed QoS support in radio spectrum sharing. *IEEE International Conference on Communications, ICC 2005*, 5: 3271–3277, 2005.
8. A. B. Mackenzie and S. B. Wicker, Game theory and the design of self-configuring, adaptive wireless networks. *IEEE Communications Magazine*, 39(11): 126–131, 2001.
9. V. Srivastava, J. Neel, A. B. Mackenzie, R. Menon, L. A. Dasilva, J. E. Hicks, J. H. Reed, and R. P. Gilles, Using game theory to analyze wireless ad hoc networks. *IEEE Communications Surveys & Tutorials*, 7(4): 46–56, 2005.
10. S. M. Perlaza, L. Cottatellucci, and M. Debbah, A game theoretic framework for decentralized power allocation in IDMA systems. *IEEE International Symposium on Personal, Indoor and Mobile Radio Communications*, Cannes, France, September 2008.
11. N. Bonneau, M. Debbah, E. Altman, and A. Hjorungnes, Non-atomic games for multi-user systems. *IEEE Journal on Selected Areas in Communications, Issue on Game Theory in Communication Systems*, 26(7): 1047–1058, 2008.

12. F. Meshkati, M. Chiang, H. V. Poor, and S. C. Schwartz, A game-theoretic approach to energy-efficient power control in multicarrier CDMA systems. *IEEE Journal on Selected Areas in Communications*, 24(6): 1115–1129, 2006.
13. O. Simeone, I. Stanojev, S. Savazzi, Y. Bar-Ness, U. Spagnolini, and R. Pickholtz, Spectrum leasing to cooperating secondary ad hoc networks. *IEEE Journal on Selected Areas in Communications*, 26(1): 203–213, 2008.
14. D. Fudenberg and J. Tirole, *Game Theory*. MIT Press, Cambridge, MA, October 1991.
15. A. Carleial, Interference channels. *IEEE Transactions on Information Theory*, 24(1): 60–70, 1978.
16. H. Sato, Two-user communication channels. *IEEE Transactions on Information Theory*, 23(3): 295–304, 1977.
17. S. Mathur, L. Sankaranarayanan, and N. B. Mandayam, Coalitional games in Gaussian interference channels. *IEEE International Symposium on Information Theory*, Seattle, WA, 2006, pp. 2210–2214.
18. S. Mathur, L. Sankaranarayanan, and N. B. Mandayam, Coalitional games in receiver cooperation for spectrum sharing. *40th Annual Conference on Information Sciences and Systems*, Princeton, NJ, 2006, pp. 949–954.
19. S. Mathur, L. Sankaranarayanan, and N. B. Mandayam, Coalitional games in cooperative radio networks. *Fortieth Asilomar Conference on Signals, Systems and Computers*, Pacific Grove, CA, 2006, pp. 1927–1931.
20. L. J. Savage, *Foundations of Statistics*, 1st edn. John Wiley & Sons, New York, 1954.
21. R. G. Wendorf and H. B. Seidenberg, Channel-change games for highly interfering spectrum-agile wireless networks. *2nd International Symposium on Wireless Pervasive Computing*, San Juan, PR, 2007.
22. R. G. Wendorf and H. Blum, A channel-change game for multiple interfering cognitive wireless networks. *Military Communications Conference*, Washington, DC, 2006, pp. 1–7.
23. R. G. Wendorf and H. Blum, Wlc38-3: Simple channel-change games for spectrum agile wireless networks. *IEEE Global Telecommunications Conference*, San Francisco, CA, 2006, pp. 1–5.
24. S. Mangold, L. Berlemann, and B. Walke, Equilibrium analysis of coexisting IEEE 802.11e wireless LANs. *14th IEEE Symposium on Personal, Indoor and Mobile Radio Communications*, Beijing, China, 2003.
25. S. Mangold, L. Berlemann, and B. Walke, Radio resource sharing model for coexisting IEEE 802.11e wireless LANs. *International Conference on Communication Technology*, Beijing, China, Vol. 2, 2003, pp. 1322–1327.
26. L. Berlemann, G. R. Hiertz, B. H. Walke, and S. Mangold, Radio resource sharing games: Enabling QoS support in unlicensed bands. *IEEE Network*, 19(4): 59–65, 2005.
27. M. J. Osborne, *An Introduction to Game Theory*. Oxford University Press, New York, August 2003.
28. F. Willems, The discrete memoryless multiple access channel with partially cooperating encoders (corresp.). *IEEE Transactions on Information Theory*, 29(3): 441–445, 1983.

29. A. Sendonaris, E. Erkip, and B. Aazhang, User cooperation diversity. Part I. System description. *IEEE Transactions on Communications*, 51(11): 1927–1938, 2003.
30. V. Belmega, S. Lasaulce, and M. Debbah, Decentralized handovers in cellular networks with cognitive terminals. *3rd IEEE International Symposium on Communications, Control and Signal Processing*, St. Julians, Malta, March 2008.
31. S. M. Perlaza, E. V. Belmega, S. Lasaulce, and M. Debbah, On the base station selection and base station sharing in self-configuring networks. *3rd ICST/ACM Intl. Workshop on Game Theory in Communication Networks*, Pisa, Italy, October 2009.
32. R. Etkin, A. Parekh, and D. Tse, Spectrum sharing for unlicensed bands. *IEEE Journal on Selected Areas in Communications*, 25(3): 517–528, 2007.
33. R. Etkin, Spectrum sharing: Fundamental limits, scaling laws, and self-enforcing protocols. PhD dissertation, EECS Department University of California, Berkeley, CA, December 2006.
34. C. H. Papadimitriou, Algorithms, games, and the internet. *28th International Colloquium on Automata, Languages and Programming*, Springer-Verlag, Crete, Greece, 2001, pp. 1–3.
35. S. Verdú, *Multiuser Detection*. Cambridge University Press, Cambridge, U.K., August 1998.
36. T. Cover and J. Thomas. *Elements of Information Theory*. John Wiley & Sons, New York, 1991.
37. S. Boyd and L. Vandenberghe, *Convex Optimization*. Cambridge University Press, Cambridge, U.K., March 2004.
38. Z.-Q. Luo and J.-S. Pang, Analysis of iterative waterfilling algorithm for multiuser power control in digital subscriber lines. *EURASIP Journal on Applied Signal Processing*, 1: 80, 2006.
39. J. C. Harsanyi, Games with incomplete information played by Bayesian players, I–III. *Management Science*, 50(12): 1804–1817, 2004.
40. S. Adlakha, R. Johari, and A. Goldsmith, Competition in wireless systems via Bayesian interference games. arXiv:0709.0516v1, September 2007.
41. L. S. Cardoso, M. Kobayashi, M. Debbah, and O. Ryan, Vandermonde frequency division multiplexing for cognitive radio. *IEEE Signal Processing Advances in Wireless Communications*, Recife, Brazil, July 2008.
42. S. M. Perlaza, M. Debbah, S. Lasaulce, and J.-M. Chaufray, Opportunistic interference alignment in MIMO interference channels. *IEEE International Symposium on Personal, Indoor and Mobile Radio Communications*, Cannes, France, September 2008.
43. R. G. Cascella, The value of reputation in peer-to-peer networks. *5th IEEE Conference on Consumer Communications and Networking*, Las Vegas, NV, 2008, pp. 516–520.
44. F. Meshkati, Game-theoretic approaches to energy-efficient resource management in wireless networks. PhD dissertation, Department of Electrical Engineering, Princeton University, Princeton, NJ, 2006.
45. V. Rodriguez, An analytical foundation for resource management in wireless communications. *IEEE Global Telecommunications Conference*, San Francisco, CA, December 1–5, 2003.

46. W. Yu, G. Ginis, and J. M. Cioffi, Distributed multiuser power control for digital subscriber lines. *IEEE Journal on Selected Areas in Communications*, 20(5): 1105–1115, 2002.
47. W. Yu, G. Ginis, and J. M. Cioffi, An adaptive multiuser power control algorithm for VDSL. *IEEE Global Telecommunications Conference*, San Antonio, TX, December 2001.
48. S. T. Chung, S. J. Kim, J. Lee, and J. M. Cioffi, A game-theoretic approach to power allocation in frequency-selective gaussian interference channels. *IEEE International Symposium on Information Theory*, Yokohama, Japan, 2003.
49. J. Nash, Non-cooperative games. *Annals of Mathematics*, 54(2): 286–295, September 1951.
50. J. Nash. Equilibrium points in *n*-person games. *Proceedings of the National Academy of Sciences*, 36(1): 48–49, 1950.
51. T. Basar and G. J. Olsder, *Dynamic Noncooperative Game Theory*. Society for Industrial & Applied Mathematics, Philadelphia, PA, 1998.
52. Y. Hayel, S. Lasaulce, R. El-Azouzi, and M. Debbah, Introducing hierarchy in energy-efficient power control games. *Second International Workshop on Game Theory in Communication Networks*, Athens, Greece, 2008.
53. S. Lasaulce, Y. Hayel, R. El-Azouzi, and M. Debbah, Introducing hierarchy in energy games. *IEEE Transactions on Wireless Communications*, 8(7): 3833–3843, 2008.
54. F. Meshkati, H. V. Poor, S. C. Schwartz, and N. B. Mandayam, An energy-efficient approach to power control and receiver design in wireless data networks. *IEEE Transactions on Communications*, 53(11): 1885–1894, 2005.
55. S. Buzzi and V. H. Poor, Joint receiver and transmitter optimization for energy-efficient CDMA communications. *IEEE Journal on Selected Areas in Communications*, 2: 459–472, 2008.

Chapter 10

Game Theory for Spectrum Sharing

Jianwei Huang and Zhu Han

Contents

Cognitive radio technology enables flexible and dynamic spectrum sharing among multiple radio networks and users and has the potential of greatly improving the spectrum utilization and network performance. This new communication paradigm, however, requires a new design and analysis framework targeting at highly flexible and distributed communication and networking. Game theory is very suitable for this task, because it is a comprehensive mathematical theory for modeling the interactions among distributed and intelligent rational decision makers. In this chapter, we discuss several game theoretical models/concepts that are highly relevant for spectrum sharing, including iterative water-filling, potential game, supermodular game, bargaining, auction, and correlated equilibrium. We also discuss several related open problems, such as the lack of proper models for dynamic and incomplete information games in this area.

10.1 Introduction

Wireless spectrum has been a tightly controlled resource worldwide since the early part of the twentieth century. The traditional way of regulating the spectrum is to assign each wireless application its own slice of spectrum at a particular location. Currently, almost all spectrum licenses belong to government identities and commercial operators. Thus, every new commercial service, from satellite broadcasting to wireless local area network, has to compete for licenses with numerous existing sources, creating a state of "spectrum drought" [1]. However, recent technology advances of smart technologies in software-defined, frequency-agile, or cognitive radios [2–4], together with reforms of the government regulation policies, may enable more flexible and efficient spectrum sharing.

In cognitive radio networks, wireless devices and networks can sense, adapt, and efficiently utilize the spectrum resource to achieve the communication targets. When end users and network operators have selfish objectives, it is natural to analyze their interactions using game theory. Even when users want to cooperate, game theory still provides a powerful mathematical framework for designing spectrum-sharing algorithms with fast convergence, robust performance, and limited information exchange requirements. In this chapter, we introduce various game theoretical models that are closely related to spectrum sharing, explain their fundamental concepts and properties, and give concrete application examples.

This chapter is organized as follows. In Section 10.2, we first review the background and trends of spectrum sharing for cognitive radios. Compared with the traditional license-based static spectrum allocation, the flexible and dynamic nature of spectrum sharing demands a new design and analysis framework. Game theory is a good candidate to fill in this theoretical gap. The background of game theory is introduced in Section 10.3, and several concrete models relevant to dynamic spectrum sharing are illustrated in Section 10.4. We finally conclude and discuss open problems in Section 10.5.

10.2 Spectrum Sharing

10.2.1 Current Spectrum Control Policy

There is a growing consensus that the current spectrum regulation policy is out of date. In the United States, the federal government established control of the electromagnetic spectrum around 90 years ago, largely as a consequence of the communication failures associated with the sinking of the Titanic [5]. The Federal Communications Commission (FCC) was established in 1934 to take the responsibility for spectrum management. From 1934 to 1990, the command-and-control model has been the core of the U.S. spectrum policy. This model is based on the assumption that simultaneous transmissions within the same spectrum at the same location will cause mutual interferences and make the transmissions useless. Thus, a highly centralized control model was adopted to assign licensees to different wireless applications to maintain their service levels. For more detailed discussions on the deficiencies with current spectrum policies, see [6,7].

There are several arguments put forward to support changing spectrum policies. First of all, there has been a rapid increase in the number of wireless users; it will be difficult to accommodate such increasing demand in the current management framework. Second, advances in communication technologies, such as error control coding and digital filtering, have made wireless receivers more immune to interference, which allow for the possibility of devices coexisting within the same spectrum. Third, many empirical studies have shown that the current spectrum usages are far from efficient—there are many spectrum holes (both in time and in space) that could be exploited if more flexible usage models are used. Fourth, the rapid development of cognitive radio technology, which enables the radio devices to detect the spectrum environment, find the spectrum holes, and tune the working frequency to exploit these spectrum holes, has made dynamic spectrum sharing feasible.

10.2.2 New Spectrum-Sharing Approaches

Several approaches have been taken to achieve more efficient use of the spectrum resource during the past decade. The FCC has reclaimed spectrum from the U.S. military and the TV industry to reallocate these spectrum resources to other (higher-valued) wireless applications, such as third-generation mobile services [1]. Another approach is applying the auction mechanism in license allocations (e.g., [8–11]).

Two potential new spectrum assignment policies are described by the FCC Spectrum Policy Force Report [5]: the exclusive use model and the commons model. The exclusive use model urges the relaxation of the current technical and commercial limitations on the existing spectrum licenses. For example, the current licensee may still have exclusive rights to the spectrum, but could allow other secondary users to access the spectrum in several flexible ways. The transmissions of the primary and secondary users could coexist, provided that a maximum interference temperature constraint is not violated at the primary user's receiver(s). Or the primary user could temporarily lease the whole spectrum to secondary users when the primary service is not in operation. Several discussions on how such secondary markets could be operated can be found in [2,12]. The commons model allows unlimited numbers of unlicensed users to share frequencies, with usage rights that are governed by technical standards or etiquettes, but with no right to protection from interference. The commons model is closely related to the open spectrum access model. In either model, the FCC wants to give spectrum users maximum autonomy in the following areas: choice of uses or services that are provided on the spectrum, choice of technology that is most appropriate to the spectrum environment, and the right to transfer, lease, or subdivide spectrum rights [5].

The flexible and dynamic nature of spectrum sharing demands a new design and analysis framework. Because the secondary users typically have selfish interests and make distributed decisions, the traditional network-oriented, centralized optimization and control methods are no longer applicable. Game theory turns out to be an ideal analysis framework for spectrum-sharing applications, as explained next.

10.3 Game Theory

A good mathematical theory for modeling the interactions among distributed entities in a network is game theory [13,14], which aims at studying interactive decision problems among intelligent, rational decision makers. In this section, we briefly introduce the necessary definitions and solution concepts that are relevant to this chapter, mainly based on discussions in [15]. Other good introductions of game theory are included in [16–18].

10.3.1 Basic Definitions

The essential elements of a game are the players, the actions, the payoffs, and the information, known collectively as the rules of the game.

Players are the individuals who make decisions, denoted by a set $\mathcal{M} = \{1, \ldots, M\}$. An action, a_i, is a choice player i can make. Player i's action set, $\mathcal{A}_i$, is the set of all the choices he can make. An action profile, $\boldsymbol{a} = \{a_i\}_{i \in \mathcal{M}}$, is a sequence of all players' actions, one from each player. For example, in an auction setting, players are the bidders and actions are the bids submitted by the bidders. A common action set for a bidder is the interval of $[0, \infty)$, i.e., he can submit any nonnegative bid. The action sets of all users other than user i are denoted as $\mathcal{A}_{-i}$.

Player i's payoff, $s_i(\boldsymbol{a})$, is a function of the action profile, $\boldsymbol{a}$, and describes how much the player gains from the game for each possible action profile. In the games we consider, a player's payoff typically equals his utility, $u_i(\boldsymbol{a})$, minus his payment, $c_i(\boldsymbol{a})$, i.e., $s_i(\boldsymbol{a}) = u_i(\boldsymbol{a}) - c_i(\boldsymbol{a})$.*,† Note that in general, we allow both a player's utility and payment to depend on the action profile. One important assumption of game theory is that all players are rational, i.e., they want to choose actions to maximize their payoffs.

Given players, actions, and payoffs, we can represent a game as $G = \{\mathcal{M}, \{\mathcal{A}_i\}_{i \in \mathcal{M}}, \{s_i\}_{i \in \mathcal{M}}\}$. There could be some exceptions, e.g., players' action sets are not independent. We will discuss such a case in Section 10.4.2.

A player's information can be characterized by an information set, which tells what kind of knowledge the player has at the decision instances. To maximize their payoffs, the players would design contingent plans known as strategies, which are mappings from one player's information set to his actions. A strategy could be pure, in which case it only contains one deterministic action for each information set; or it could be mixed, in which case it specifies a set of actions, each chosen according to a probability vector for each information set. A strategy profile is a sequence of the players' strategies, one from each player.

In static games, players choose their strategies simultaneously and only once. Each player has only one information set, which is what he knows at the beginning of the game. In this case, each pure strategy corresponds to one action, and we loosely use them interchangeably with the same notation. We mainly focus on static games in this chapter, unless otherwise stated.

A reasonable prediction of the outcome of a game is an equilibrium, which is a strategy profile where each player chooses the best strategy to maximize his payoff. Among several available equilibrium concepts, we focus mainly on the Nash equilibrium (NE). In a static game, an NE is a strategy profile, $\boldsymbol{a}^*$, where no player can increase his payoff by deviating unilaterally, i.e.,

$$s_i\left(a_i^*, a_{-i}^*\right) \geq s_i\left(a_i', a_{-i}^*\right), \quad \forall a_i' \neq a_i^*, a_i' \in \mathcal{A}_i, a_i^* \in \mathcal{A}_i.$$

---

* Sometimes we also call a player's payoff the player's surplus. Here, we assume that players' utilities are quasilinear with respect to numeraire commodity, i.e., the utilities can be measured in terms of money [19].

† Our notations are consistent with the conventions used in the communication literature, but a little different from the traditional economics literature, where the utility used here is called valuation and payoff used here actually refers to the (expected) utility. See, for example, [18].

Here, we use the notation $\boldsymbol{a} = (a_i, a_{-i})$, where a_{-i} represents the actions of all players except player i. One concept closely related to the NE is a player i's best response, which can be defined as follows:

$$\mathcal{B}_i (a_{-i}) = \arg \max_{a_i \in \mathcal{A}_i} s_i (a_i, a_{-i}) .$$

When $\mathcal{B}_i (a_{-i})$ is a singleton set, it is called the best-response function of the actions a_{-i}. It is easy to see that an NE is a fixed point of all the players' best responses, i.e., for an NE $\boldsymbol{a}^*$,

$$a_i^* \in \mathcal{B}_i \left(a_{-i}^*\right), \quad \forall i \in \mathcal{M}.$$

Note that there may be no NE or multiple NEs in a given game.

10.3.2 Bounded Rationality and Myopic Best-Response Updates

In many problems that we encounter in practice, the players only know their own payoff functions (private information assumption) at the beginning of the game. This makes it difficult for the players to determine the NE, because they cannot calculate the other players' best responses, and thus are not able to find the NE by solving for a fixed point. In game theory, a traditional way of dealing with this private information assumption is to assume that players know the distributions of other players' payoff functions and choose strategies to maximize their expected payoffs. This leads to the concept of a Bayesian NE. Most of the classical analysis of auction theory is built on this solution concept. However, assuming a publicly known distribution assumption is questionable for many spectrum-sharing scenarios.

An alternative approach is to consider a repeated game where the players play the same static game repeatedly, and choose their actions in each round to maximize their payoffs, based on the history of the other players' actions [16]. This is effectively an incremental information revelation process, where the players' actions in each round gradually reveal their underlying payoff functions. One difficulty in this approach is that there are typically an infinite number of NEs of this new repeated game if players are able to consider both the entire history and the future of the game when making decisions.

In fact, the intelligence assumption, or so-called perfect rationality, is central to classical game theory. This means that if a player knows everything that we know about the game, he can make any inferences about the situation that we can [18]. This effectively assumes that players act as a supercomputer with infinite (and free) computational capacity and can always find their best responses, no matter how complex the game is. On the other hand, in a practical game, where players are people or computer agents, perfect rationality is a problematic assumption because the computation capacities are typically limited, and the time and effort of computing the best responses could be very expensive. Thus, sometimes players can be better modeled with bounded rationality [20], especially in a complex situation, such as a repeated game where a rational choice would typically be based on a perfect recall of the entire history and an accurate forecast into the infinite future.*

In the context of our problems, we consider a specific type of bounded rationality where each player does the following: in round T of the repeated game, he chooses an action, $a_i^{(T)}$, according to his best response, $\mathcal{B}_i\big(a_{-i}^{(T-1)}\big)$, i.e., he tries to maximize his payoff assuming that other players will choose the same actions as in the immediate previous round. If every player follows the same updating rule and the action profile finally converges (i.e., everyone's action does not change from round to round), then an NE of the original static game with full information (i.e., the game in which everyone knows everyone's payoff) is found. This type of update strategy is called the myopic best-response (MBR) update [23,24].

The MBR updates could be thought as a "limited memory" interpretation [20,25] of bounded rationality, where players only remember situations in the previous round. These updates can be traced back to the early studies of oligopoly. The MBR update is one of the simplest learning mechanisms for the players in a game theoretical environment. In some interesting auction applications [26,27], the MBR update has been proved to be an ex post NE, in which there is no better strategy for a player whatever be the payoffs of other agents, as long as the other players also follow the MBR updates [28]. In some special game theoretical models, such as supermodular game [29] or potential game [30], the MBR updates have very good convergence properties, though, in more general settings, these updates can lead to cyclic behaviors that do not converge. Interested readers are referred to [31–33] for more general discussions on learning in games.

A shortcoming of the MBR update is the restricted assumption on the players' intelligence. However, we emphasize that in the cases considered here, we are often dealing with engineered systems. In these cases, this assumption can simply be reviewed as a design choice. The reason for modeling this choice is that it leads to simple systems with a desirable behavior.

10.4 Game Theoretical Models for Dynamic Spectrum Sharing

In this section, we investigate several useful, popular game theoretical models for dynamic spectrum-sharing applications. Specifically, we study iterative water-filling, potential game, supermodular game, bargaining, auction, and correlated equilibrium in the following sections, respectively.

---

* The concern of limited computational capacity and bounded rationality has been the main motivation for a new research area called computational mechanism design [21] or algorithmic mechanism design [22].

10.4.1 Iterative Water-Filling

In a multiuser environment with multiple interference channels, it is important to efficiently allocate the transmission power over the different channels to maximize the transmission rate and minimize the interferences. This is a very common scenario for dynamic spectrum sharing, where multiple secondary users want to access some common open channels. In this case, a centralized approach can achieve a global optimal solution but with very high complexity and communication overhead. One of the low-complexity distributed algorithms proposed in this context is iterative water-filling.

As an example, let us consider a multi-carrier system with M users and K sub-channels. The signal-to-interference-plus-noise ratio (SINR) of user i on the sub-channel k is given by

$$\gamma_i^k = \frac{p_i^k h_{ii}^k}{n_0 + \sum_{j \neq i} p_j^k h_{ji}^k}, \tag{10.1}$$

where

h_{ji}^k is the channel gain from user j's transmitter to user i's receiver at the sub-channel k

p_i^k is the transmitting power of user i at the sub-channel k

n_0 is the thermal noise level

The rate for user i and the sub-channel k (in bits/second/Hz) is given by

$$R_i^k = \log_2\left(1 + \gamma_i^k\right). \tag{10.2}$$

Notice that here interference is treated as noise and no multiuser detection is considered. The overall rate achieved over different sub-channels for the user i is given by $R_i = \sum_{k=1}^{K} R_i^k$. User i needs to decide its transmission power vector over all sub-channels, $\boldsymbol{p} = \left\{p_i^k, \forall k\right\}$, to maximize its data rate such that the total power is no larger than $P_{\max}$.

Next, we define the noncooperative rate maximization game as follows.

Definition 10.1 (Noncooperative Rate Maximization Game) In a noncooperative rate maximization game, $G = \left[\mathcal{M}, \{\mathcal{A}_i\}_{i \in \mathcal{M}}, \{R_i\}_{i \in \mathcal{M}}\right]$, each user i's action set is

$$\mathcal{A}_i = \left\{\boldsymbol{p}_i : p_i^k \geq 0, \sum_k p_i^k \leq P_{\max}\right\}$$

and the payoff is its total rate over all sub-bands.

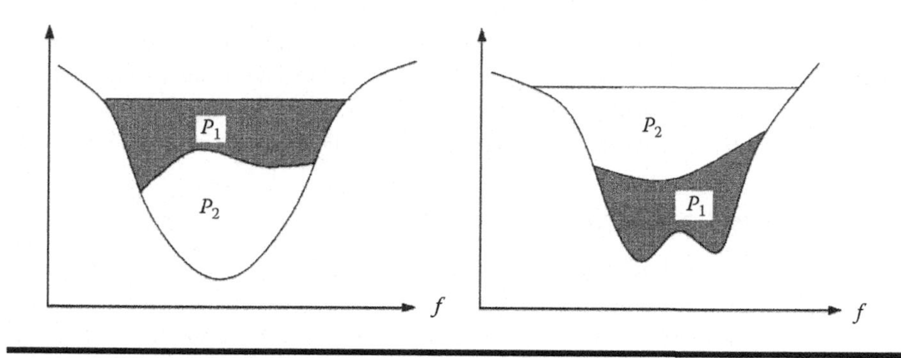

Figure 10.1 Illustration of iterative water-filling. (From Yu, W. et al., *IEEE J. Sel. Areas Commun.*, 20(5), 1105, 2002. With permission.)

The NE of the game can be found through the iterative water-filling algorithm [34]. The basic idea is to treat the interferences from the other users as the noise. Then each user employs a single-user water-filling solution iteratively, based on the changes of power levels of other users. In Figure 10.1, we illustrate a two-user example, in which the interference from the other user is treated as the noise level over a different frequency. The properties of iterative water-filling are given by the following theorems.

THEOREM 10.1 A NE always exists in the noncooperative rate maximization game.

THEOREM 10.2 Assume

$$\max_k \frac{h_{ij}^k}{h_{ii}^k} < \frac{1}{M-1}, \quad i \neq j, \tag{10.3}$$

then the iterative water-filling scheme globally converges to a NE.

Intuitively, condition (10.3) means that the interferences among users are not large enough. In particular, when the number of users, M, increases, the convergence condition becomes more stringent.

The iterative water-filling scheme is an efficient method for distributed resource allocation using only local information. However, when the interference is large, the convergence speed can be slow. Moreover, the NE is typically not optimal from the system design point of view (i.e., not maximizing total rate).

To overcome the shortcoming of iterative water-filling, we can use a referee-based game to improve the performance [68,69]. The basic idea is to introduce a referee for the noncooperative game. Pure iterative water-filling may have multiple NEs. A referee is in charge of detecting these less efficient NEs and changing the game rule to prevent the players from falling into undesirable game outcomes. It is worth mentioning that the noncooperative game is still played in a distributive way and the referee intervenes only when it is necessary. In [68,69], the above idea is employed to a multicell OFDMA network to achieve an efficient distributed resource allocation.

10.4.2 Potential Game

A potential game is a class of game that has nice convergence properties to the NEs. In potential games, although players act in a noncooperative fashion, they actually implicitly work toward a common system goal characterized as potential. In other words, the potential serves as the mathematical bridge between noncooperative and cooperative behaviors of the players.

We first introduce some useful definitions. More general definitions related to potential games can be found in [30].

Definition 10.2 (Ordinal Potential) In a game $G = \left[\mathcal{M}, \{\mathcal{A}_i\}_{i\in\mathcal{M}}, \{s_i\}_{i\in\mathcal{M}}\right]$, a function $Z : \mathcal{A}_1 \times \cdots \times \mathcal{A}_M \rightarrow R$ is an ordinal potential for G, if for every $i \in \mathcal{M}$ and every $a_{-i} \in \mathcal{A}_{-i}$, we have

$$s_i(x, a_{-i}) - s_i(y, a_{-i}) > 0 \quad \text{if and only if} \quad Z(x, a_{-i}) - Z(y, a_{-i}) > 0, \quad \forall x, y \in \mathcal{A}_i.$$

Definition 10.3 (Ordinal Potential Game) A game $G = \left[\mathcal{M}, \{\mathcal{A}_i\}_{i\in\mathcal{M}}, \{s_i\}_{i\in\mathcal{M}}\right]$ is an ordinal potential game if it admits an ordinal potential.

The following theorem summarizes several important properties of a potential game.

THEOREM 10.3 In an ordinal potential game $G = \left[\mathcal{M}, \{\mathcal{A}_i\}_{i\in\mathcal{M}}, \{s_i\}_{i\in\mathcal{M}}\right]$,

(a) Optimizers of the ordinal potential function, $Z(\boldsymbol{a})$, are NEs of game G.
(b) If the game is finite, i.e., the number of users is finite and the strategy set, $\mathcal{A}_i$, is finite for each user, then all improvement paths are finite and terminate at an NE.

There are other variations of potential games, including exact, weighted, generalized ordinal, and best-response potential games. More discussions can be found in [53].

Potential games have been extensively used in studying wireless power control. For example, the authors in [54,55] considered the case where each player chooses a scalar power level and all players' strategy sets are decoupled. In [56], the authors generalized the results to the case of vector power choice with coupled strategy sets. Furthermore, it is pointed out in [56] that there is an interesting and general relationship existing between the NEs of potential games and the equilibria of proper autonomous dynamic systems: a potential game can be interpreted as an autonomous gradient dynamic system whose Lyapunov function is just the potential of the game. This explains the convergence results in Theorem 10.3.

Next, we give an example based on the discussions in [56]. Consider a single-cell CDMA network with $\mathcal{M} = \{1, \ldots, M\}$ users. The received SINR of user $i \in \mathcal{M}$ is

$$\gamma_i(\boldsymbol{p}) = \frac{p_i h_i}{n_0 + \sum_{j \neq i} p_j h_j},$$

where h_i is the channel gain from user i's transmitter to the base station. Each user $i \in \mathcal{M}$ wants to solve the following power minimization problem:

$$\begin{aligned} \text{minimize} \quad & p_i \\ \text{subject to} \quad & f_i(\gamma_i(\boldsymbol{p})) \geq \gamma_i^{\text{thresh}}, \\ \text{variables} \quad & p_i \in [0, P_i^{\max}]. \end{aligned} \tag{10.4}$$

Here,

$P_i^{\max}$ is the maximum power
f_i is the QoS function
and γ_i^{thresh} is the QoS threshold

It is shown in [56] that solving Problem 10.4 for all users is equivalent to finding the NE of a noncooperative game, $G = \left[\mathcal{M}, \mathcal{A}, \left\{\log\left(P_i^{\max} - p_i\right)\right\}_{i \in \mathcal{M}}\right]$, where the coupled action set is

$$\mathcal{A} = \left\{\boldsymbol{p} : f_i(\gamma_i(\boldsymbol{p})) \geq \gamma_i^{\text{thresh}}, p_i \in \left[0, P_i^{\max}\right], \forall i \in \mathcal{M}\right\}.$$

Furthermore, the game G admits a potential function,

$$Z(\boldsymbol{p}) = \sum_{i \in \mathcal{M}} \log\left(P_i^{\max} - p_i\right).$$

We can then maximize function $Z(\boldsymbol{p})$ over set $\mathcal{A}$, and the corresponding maximizer(s) will be the NE(s) of game G, and thus the optimal solution(s) of Problem 10.4 for all users.

10.4.3 Supermodular Game

A supermodular game has many practical applications in economics. A key feature of the supermodular game is the "strategic complementarities"—if a player chooses a higher action, the others want to do the same thing. Supermodular games have nice properties in terms of the existence and achievability of NE. We first introduce some useful definitions, whereas more general discussions can be found in [29].

Definition 10.4 (Sublattice) A real i-dimensional set $\mathcal{V}$ is a sublattice of $\Re^i$ if for any two elements $a, b \in \mathcal{V}$, the componentwise minimum (i.e., $a \wedge b$) and the componentwise maximum (i.e., $a \vee b$) are also in $\mathcal{V}$. In particular, a compact sublattice has (componentwise) smallest and largest elements. Any compact (one-dimensional) interval is a sublattice of $\mathbb{R}$.

Definition 10.5 (Function with Increasing Differences) A twice-differentiable function f has increasing differences in variables (x, t) if $\partial^2 f/\partial x \partial t \geq 0$ for any feasible x and t.*

Definition 10.6 (Supermodular Function) A function f is supermodular in $\boldsymbol{x} = (x_1, \ldots, x_i)$ if it has increasing differences in (x_i, x_j) for all $i \neq j$.

Definition 10.7 (Supermodular Game) A game $G = [\mathcal{M}, \{\mathcal{A}_i\}_{i \in \mathcal{M}}, \{s_i\}_{i \in \mathcal{M}}]$ is supermodular if, for each player $i \in \rangle$, (a) the strategy space $\mathcal{P}_i$ is a nonempty and compact sublattice, and (b) the payoff function s_i is continuous in all players' strategies, is supermodular in player i's own strategy, and has increasing differences between any component of player i's strategy and any component of any other player's strategy.

The following theorem summarizes several important properties of a supermodular game.

THEOREM 10.4 In a supermodular game $G = [\mathcal{M}, \{\mathcal{A}_i\}_{i \in \mathcal{M}}, \{s_i\}_{i \in \mathcal{M}}]$,

(a) The set of NEs is a nonempty and compact sublattice, and so there are componentwise smallest and largest NEs.
(b) If the users' best responses are single valued, and each user uses MBR updates starting from the smallest (largest) element of its strategy space, then the strategies monotonically converge to the smallest (largest) NE.
(c) If each user starts from any feasible strategy and uses MBR updates, the strategies will eventually lie in the set bounded componentwise by the smallest and largest NEs. If the NE is unique, the MBR updates globally converge to that NE from any initial strategies.

---

* If we choose x to maximize a twice-differentiable function $f(x, t)$, then the first-order condition gives $\partial f(x, t)/\partial x|_{x=x^*} = 0$ and the optimal value x^* increases with t if $\partial^2 f/\partial x \partial t > 0$.

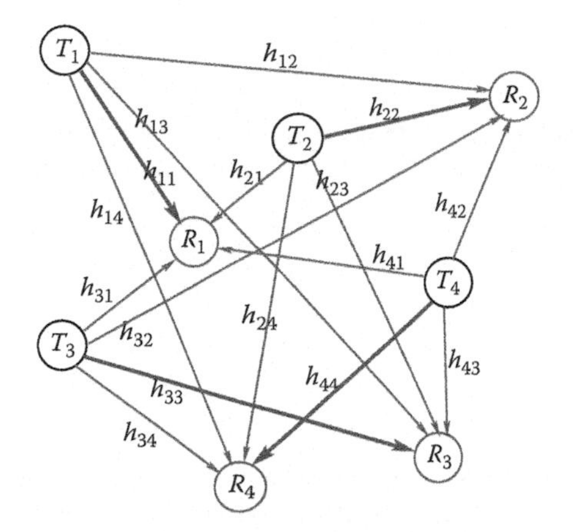

Figure 10.2 An example wireless network with four users (pairs of nodes). T_i and R_i denote the transmitter and the receiver of user i, respectively. (From Huang, J. et al., *IEEE J. Sel. Areas Commun.*, 24(5), 1074, 2006. With permission.)

In wireless communications, the supermodular game has been used to design various power control algorithms, e.g., in [34–36]. Next, we give an example based on [36], showing how supermodular game theory can help to analyze the properties of a power control algorithm.

Consider an ad hoc network with a set $\mathcal{M} = \{1, \ldots, M\}$ of distinct node pairs. As shown in Figure 10.2, each pair consists of one dedicated transmitter and one dedicated receiver.* The motivating example for this model is multiple secondary users sharing the same common open channel in a distributed fashion.

The channel gain between user i's transmitter and user j's receiver is denoted by h_{ij}. Note that in general $h_{ij} \neq h_{ji}$, as the latter represents the gain between user j's transmitter and user i's receiver. Each user i's quality of service is characterized by a utility function, $u_i(\gamma_i)$, which is an increasing and a strictly concave function of the received SINR,

$$\gamma_i(\boldsymbol{p}) = \frac{p_i h_{ii}}{n_0 + \sum_{j \neq i} p_j h_{ji}}, \qquad (10.5)$$

where

$\boldsymbol{p} = (p_1, \ldots, p_i)$ is a vector of the users' transmission powers
n_0 is the background noise power

The users' utility functions are coupled due to mutual interference. An example utility function is a logarithmic utility function, $u_i(\gamma_i) = \theta_i \log(\gamma_i)$, where θ_i is a

* For example, this could represent a particular schedule of transmissions determined by a routing and MAC protocol.

user-dependent priority parameter (e.g., related to the long-term achievable rate or queue length [76]).*

The problem we consider is to specify $\boldsymbol{p}$ to maximize the utility summed over all users, where each user i must satisfy a transmission power constraint $p_i \in \mathcal{P}_i = [P_i^{\min}, P_i^{\max}]$:

$$\max_{\{\boldsymbol{p}: p_i \in \mathcal{P}_i, \forall i\}} \sum_{i=1}^{i} u_i\left(\gamma_i(\boldsymbol{p})\right). \tag{P.1}$$

Note that a special case is $P_i^{\min} = 0$, i.e., the user may choose not to transmit.

We propose an asynchronous distributed pricing (ADP) algorithm to solve Problem P.1. We first describe the algorithm, and then show how we can interpret the algorithm as a fictitious supermodular game. This enables us to easily characterize the convergence behavior of the algorithm.

In the ADP algorithm, each user announces a single price, π_i, and sets its transmission power, p_i, based on the prices announced by other users. Prices and powers are asynchronously updated. For $i \in \mathcal{M}$, let $T_{i,p}$ and $T_{i,\pi}$ be two unbounded sets of positive time instances at which user i updates its power and price, respectively. The complete algorithm is given in Algorithm 10.1. Note that in addition to being asynchronous across users, each user also need not update its power and price at the same time.

Algorithm 10.1: The ADP Algorithm

---

(1) INITIALIZATION: For each user $i \in \mathcal{M}$, choose some power $p_i(0) \in \mathcal{P}_i$ and price $\pi_i(0) \geq 0$.

(2) PRICE UPDATE: At each $t \in T_{i,\pi}$, user i updates its price according to

$$\pi_i(t) = -\frac{\partial u_i\left(\gamma_i\left(\boldsymbol{p}(t)\right)\right)}{\partial\left(\sum_{j \neq i} p_j(t) h_{ji}\right)}.$$

(3) POWER UPDATE: At each $t \in T_{i,p}$, user i updates its power according to

$$p_i(t) = \arg\max_{\hat{p}_i \in \mathcal{P}_i} \left(u_i\left(\gamma_i\left(\hat{p}_i, p_{-i}(t)\right)\right) - \hat{p}_i \sum_{j \neq i} \pi_j(t) h_{ij} \right).$$

---

* In the high SINR regime, logarithmic utility approximates the Shannon capacity $\log(1 + \gamma_i)$ weighted by θ_i. For a low SINR, a user's rate is approximately linear in the SINR, and so this utility is proportional to the logarithm of the rate.

We next characterize the convergence of the ADP algorithm by considering the following fictitious power–price (FPP) control game:

$$G_{\text{FPP}} = \left[\mathcal{FW} \cup \mathcal{FC}, \left\{\mathcal{P}_i^{\mathcal{FW}}, \mathcal{P}_i^{\mathcal{FC}}\right\}_{i\in\mathcal{M}}, \left\{s_i^{\mathcal{FW}}, s_i^{\mathcal{FC}}\right\}_{i\in\mathcal{M}}\right].$$

Here, the players are from the union of the sets $\mathcal{FW}$ and $\mathcal{FC}$, which are both copies of $\mathcal{M}$. $\mathcal{FW}$ is a fictitious power player set; each player $i \in \mathcal{FW}$ chooses a power p_i from the strategy set $\mathcal{P}_i^{\mathcal{FW}} = \mathcal{P}_i$ and receives the payoff

$$s_i^{\mathcal{FW}}\left(p_i; p_{-i}, \pi_{-i}\right) = u_i\left(\gamma_i\left(\boldsymbol{p}\right)\right) - \sum_{j\neq i} \pi_j h_{ij} p_i. \tag{10.6}$$

$\mathcal{FC}$ is a fictitious price player set; each player $i \in \mathcal{FC}$ chooses a price π_i from the strategy set $\mathcal{P}_i^{\mathcal{FC}} = [0, \bar{\pi}_i]$ and receives the payoff

$$s_i^{\mathcal{FC}}\left(\pi_i; \boldsymbol{p}\right) = -\left(\pi_i + \frac{\partial u_i\left(\gamma_i\left(\boldsymbol{p}\right)\right)}{\partial\left(\sum_{j\neq i} p_j h_{ji}\right)}\right)^2. \tag{10.7}$$

Here, $\bar{\pi}_i = \sup_{\boldsymbol{p}} \left|\frac{\partial u_i(\gamma_i(\boldsymbol{p}))}{\partial\left(\sum_{j\neq i} p_j h_{ji}\right)}\right|$, which could be infinite for some utility functions.

In game G_{FPP}, each user in the ad hoc network is split into two fictitious players, one in $\mathcal{FW}$ who controls power p_i and the other one in $\mathcal{FC}$ who controls price π_i. Although users in the real network cooperate with each other by exchanging interference information (instead of choosing prices to maximize their surplus), each fictitious player in G_{FPP} is selfish and maximizes its own payoff function.

Denote $CR_i\left(\gamma_i\right) = -\frac{\gamma_i u_i''(\gamma_i)}{u_i'(\gamma_i)}$, and let $\gamma_i^{\min} = \min\{\gamma_i(\boldsymbol{p}) : p_i \in \mathcal{P}_i \forall i\}$ and $\gamma_i^{\max} = \max\{\gamma_i(\boldsymbol{p}) : p_i \in \mathcal{P}_i \forall i\}$. If for each user $i \in \mathcal{M}$, $P_i^{\min} > 0$ and $CR_i\left(\gamma_i\right) \in [1, 2]$ for all $\gamma_i \in \left[\gamma_i^{\min}, \gamma_i^{\max}\right]$, then we can show that G_{FPP} is a supermodular game, which means that Problem P.1 has a unique optimal solution to which the ADP algorithm globally converges.

10.4.4 Bargaining

Bargaining games refer to situations where two or more players must reach agreement regarding how to distribute a good or monetary amount. Each player prefers to reach an agreement in these games rather than abstain from doing so; however, each prefers the agreement that maximizes his own interests. Bargaining can be analyzed using cooperative game theory as follows [45–47]:

Definition 10.8 (M-Person Bargaining Problem) Let $\mathcal{M} = \{1, 2, \ldots, M\}$ be the set of players. Let $\mathbf{S}$ be a closed and convex subset of $\Re^M$ to represent the set of feasible payoff allocations that the players can get if they all work together. Let $u^i_{\min}$ be the minimal payoff that the ith player will expect; otherwise, he will not cooperate. Suppose that $\{u_i \in \mathbf{S} \mid u_i \geq u^i_{\min}, \forall i \in \mathcal{M}\}$ is a nonempty bounded set. Define $\mathbf{u}_{\min} = (u^1_{\min}, \ldots, u^M_{\min})$, then the pair $(\mathbf{S}, \mathbf{u}_{\min})$ is called an M-person bargaining problem.

Within the feasible set **S**, we define the notion of Pareto optimal as a selection criterion for the bargaining solutions.

Definition 10.9 (Pareto Optimality) An allocation $\boldsymbol{u} = (u_1, \ldots, u_M)$ is Pareto optimal if and only if there does not exist an allocation $\boldsymbol{u}' = (u'_1, \ldots, u'_M)$, such that $u'_i \geq u_i$ for all i and $u'_j > u_j$ for at least one j. In other words, there exists no other allocation that increases the performance for some users without decreasing the performance for some other users.

Next, we discuss some possible bargaining solutions. In general, there might be an infinite number of Pareto optimal points, and we need further criteria to select a bargaining result. A possible criterion is the fairness. One commonly used fairness criterion for wireless resource allocation is max–min [62], where the performance of the user with the worst channel conditions is maximized. This criterion penalizes the users with good channels and, as a result, generates inferior overall system performance. Another more reasonable solution concept is the Nash bargaining solution (NBS) [61]. In an NBS, after the minimal requirements are satisfied for all users, the rest of the resources are allocated proportionally to users according to their conditions. As a result, an NBS provides a unique and fair Pareto optimal operation point under the following conditions.

Definition 10.10 (Nash Bargaining Solution) $\bar{\mathbf{u}}$ is said to be a NBS in **S** for $\mathbf{u}_{\min}$, i.e., $\bar{\mathbf{u}} = \phi(\mathbf{S}, \mathbf{u}_{\min})$, if the following axioms are satisfied:

1. Individual rationality: $\bar{u}_i \geq u^i_{\min}, \forall i$.
2. Feasibility: $\bar{\mathbf{u}} \in \mathbf{S}$.
3. Pareto optimality: For every $\hat{\mathbf{u}} \in \mathbf{S}$, if $u_i \geq \bar{u}_i, \forall i$, then $\hat{u}_i = \bar{u}_i, \forall i$.
4. Independence of irrelevant alternatives: If $\bar{\mathbf{u}} \in \mathbf{S}' \subset \mathbf{S}$, $\bar{\mathbf{u}} = \phi(\mathbf{S}, \mathbf{u}_{\min})$, then $\bar{\mathbf{u}} = \phi(\mathbf{S}', \mathbf{u}_{\min})$.
5. Independence of linear transformations: For any linear scale transformation ψ, $\psi(\phi(\mathbf{S}, \mathbf{u}_{\min})) = \phi(\psi(\mathbf{S}), \psi(\mathbf{u}_{\min}))$.
6. Symmetry: If **S** is invariant under all exchanges of agents, $\phi_j(\mathbf{S}, \mathbf{u}_{\min}) = \phi_{j'}(\mathbf{S}, \mathbf{u}_{\min}), \forall j, j'$.

Axioms 4–6 are called axioms of fairness. The irrelevant alternative axiom asserts that eliminating the feasible solutions that would not have been chosen shall not affect the NBS. Axiom 5 asserts that the bargaining solution is scale invariant. The symmetry axiom asserts that if the feasible ranges for all users are completely symmetric, then all users have the same solution.

The NBS satisfying the above axioms can be obtained by the following optimization [47].

THEOREM 10.5 (Existence of NBS) There is a solution function $\phi(\mathbf{S}, \mathbf{u}_{\text{min}})$ that satisfies all six axioms in Definition 10.10 and can be computed as

$$\phi(\mathbf{S}, \mathbf{u}_{\text{min}}) \in \arg \max_{\bar{\mathbf{u}} \in \mathbf{S}, \bar{u}_i \geq u^i_{\text{min}}, \forall i} \prod_{i=1}^{M} \left(\bar{u}_i - u^i_{\text{min}}\right). \tag{10.8}$$

Two other bargaining solutions that have been proposed as alternatives to the NBS are the Kalai–Smorodinsky solution (KSS) and the Egalitarian solution (ES). Details can be found in [16].

There are many applications in the wireless networks using bargaining solutions, e.g., OFDMA resource allocation [63], ad hoc networks [65], mesh networks [66], and multimedia transmission [67]. In [64], the authors considered using bargaining to achieve efficient dynamic spectrum sharing. The authors defined a general framework for the spectrum access problem based on several definitions of system utilities. By reducing the spectrum allocation problem to a variant of the graph-coloring problem, the global optimization problem is shown to be NP-hard. A general approximation methodology is provided through vertex labeling. The paper investigated two strategies: a centralized strategy, where a central server calculates an allocation assignment based on global knowledge, and a distributed approach, where devices collaborate to bargain over local channel assignments toward global optimization. The experimental results show that the bargaining-based allocation algorithms can dramatically reduce interferences and lead to an order of magnitude throughput improvement compared with a naive approach.

10.4.5 Auction

Auctions are suitable to model markets where the seller(s) and buyer(s) have asymmetric information. For example, consider an exclusive use spectrum-sharing model, where the spectrum broker (i.e., the representative of the spectrum owner or the licensee) has a piece of spectrum for sale in a secondary market. However, the broker himself may not have an accurate estimate of the secondary users' value of the spectrum, as the utility functions of secondary users are typically private information. One way for the broker to extract information from the secondary users is through an auction process.

The theory of auctioning indivisible goods (single unit or multiunit) has been relatively well developed [38], but the related results cannot be directly applied to spectrum sharing where the resource should typically be allocated to more than one user (precluding single-unit indivisible auctions), and it is often difficult to divide the resources into well-defined bundles (precluding multiunit indivisible auctions). Next, we will focus our discussion on the share auction, or divisible auction.

10.4.5.1 Share Auction

A share auction [39–41] is concerned with allocating a perfect divisible good among a set of bidders. The most commonly used example in the literature comes from the financial market (such as the auction of treasury notes) [42–44]. There are two basic pricing structures in a share auction. In a uniform-price auction, all the winners (typically more than one) get some portions of the good and pay the same unit price. In a discriminatory pricing auction (sometimes called a pay-you-bid auction [40]), winning bids are filled at the bid price. Much of the results mentioned above focus on examining how different pricing and information structures affect the auction results, such as the final price, the seller's revenue, and the allocations of the divisible good.

Compared with the well-studied single-unit good auction, where bidders typically submit one-dimensional bids in the auction, some share auctions allow bidders to submit multiple combinations of price and quantity as the bids (e.g., [42,43]. This significantly complicates the auction design because the bidders have large strategy spaces. When using a share auction to allocate resources such as bandwidth in communication networks, researchers typically adopt simple one-dimensional bidding rules as in [46–51]. In these cases, the allocations to the users are proportional to the bids.

Let us consider how the share auction can be used in spectrum sharing. The following discussions are based on [52]. Consider a wireless network model that is similar to the one described in Section 10.4.3. The definitions on users (node pairs), channel gains, and utility functions are the same. The key difference here is that we do not assume that each user has a separate transmission constraint. Instead, we consider the case where there is a measurement point in the network. The aggregated interference generated by all users at the measurement point should be no larger than a threshold P, i.e.,

$$\sum_{i=1}^{i} p_i h_{i0} \leq P. \tag{10.9}$$

Here, h_{i0} is the channel gain from user i's transmitter to the measurement point. The system model is shown in Figure 10.3.

We consider two simple one-dimensional share auction mechanisms (SINR based and power based). In both share auctions, users submit one-dimensional bids representing their willingness to pay, and the manager simply allocates the received power

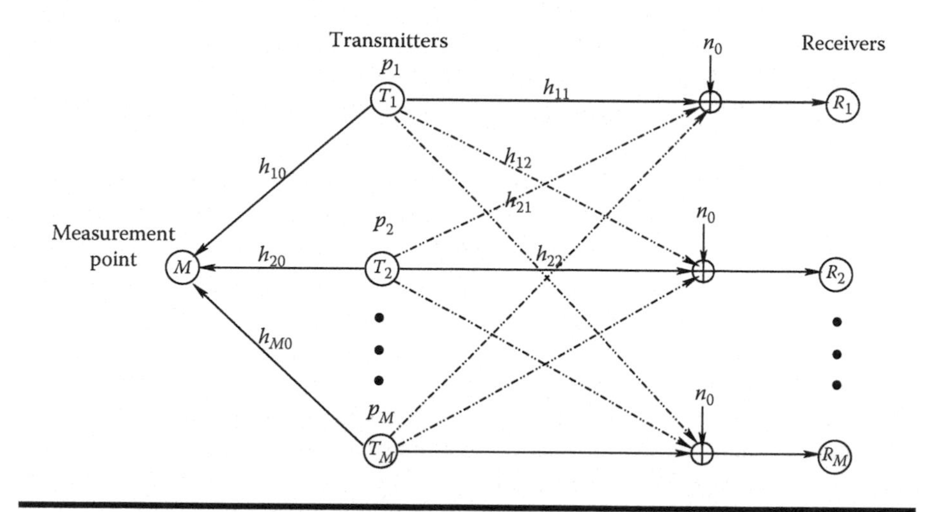

Figure 10.3 Spectrum sharing with one measurement point. (From Huang, J. et al., *Mobile Netw. Appl. J.*, 11(3), 405, 2006. With permission.)

in proportion to the bids. The users then pay an amount proportional to their SINR (or power). The manager announces a nonnegative reserve bid, β, to ensure the uniqueness of the auction result. We assume that it is a complete information game, i.e., all users' utilities and all channel gains are known to all users.

10.4.5.2 Share Auction Mechanisms

1. The manager announces a reserve bid $\beta \geq 0$ and a price $\pi^s > 0$ (in an SINR auction) or $\pi^p > 0$ (in a power auction).
2. After observing β, π^s (or π^p), user $i \in \{1, \ldots, i\}$ submits a bid $b_i \geq 0$.
3. The manager keeps a reserve power p_0 and allocates to each user i a transmission power p_i, so that the received power at the measurement point is proportional to the bids, i.e.,

$$p_i h_{i0} = \frac{b_i}{\sum_{j=1}^{i} b_i + \beta} P \quad \text{and} \quad p_0 = \frac{\beta}{\sum_{j=1}^{i} b_i + \beta} P. \tag{10.10}$$

The resulting SINR for user i is

$$\gamma_i(\boldsymbol{p}) = \frac{p_i h_{ii}}{n_0 + \sum_{j \neq i} p_j h_{ji} + p_0 h_{0i}}, \tag{10.11}$$

where h_{0i} is the channel gain from the manager (measurement point) to user i's receiver.* If $\sum_{i=1}^{i} b_i + \beta = 0$, then $p_i = 0$.
4. In an SINR (power) auction, user i pays $C_i = \pi^s \gamma_i$ $\left(C_i = \pi^p p_i h_{i0}\right)$.

These auction mechanisms differ from some previously proposed auction-based network resource allocation schemes (e.g., [46,49]) in that the bids here are not the same as the payments. Instead, the bids are signals of willingness to pay. The manager can therefore influence the NE by choosing β and π^s (or π^p). This alleviates the typical inefficiency of the NE and, in some cases, allows us to achieve socially optimal solutions.

It can be shown that under a properly chosen price π^s (π^p, respectively), the SINR auction (power auction, respectively) can achieve fair (efficient, respectively) allocation. In a fair allocation, users achieve the same SINR if they have the same utility function regardless of the network topology and channel conditions. In an efficient allocation, the total utility of the network is maximized.

The above analysis assumes that all users' utilities and all channel gains are known to all users (i.e., complete information game). As a result, the NE can be calculated in one shot (with fixed price). In practice, however, users may only know limited local information. In that case, it can be shown that users can still achieve the NE in a distributed fashion by following best-response dynamics.

10.4.6 Correlated Equilibrium

In this section, we investigate a special kind of equilibrium, correlate equilibrium. In 2006, the Nobel Prize was awarded to Robert J. Aumann for his contribution to proposing the concept of correlated equilibrium [70,71]. Unlike the NE, in which each user only considers its own strategy, correlated equilibrium achieves better performance by allowing each user to consider the joint distribution of users' actions. In other words, each user needs to consider others' behaviors to see if there are mutual benefits to explore. It has been shown that the correlated equilibrium can be better than the convex hull of the NEs [70,71].

If a user follows a single action in every possible attainable situation (i.e., information set) in a game, the action is called pure strategy. In the case of mixed strategies, the user will follow a probability distribution over different possible actions. In Table 10.1, we illustrate an example of two secondary users (row player or column player) with different actions, 0 or 1. The payoffs for both users are shown in parentheses. In Table 10.1a, we list the payoffs for two users taking actions of 0 and 1. 0 means transmitting less aggressively, while 1 means transmitting more aggressively. We can see that when both users choose action 0, they have the best overall payoff. But if

* If $h_{0i} = 0$ for all $i \in \{1, \ldots, M\}$, then the manager does not interfere with the users, and many of the results in the following section still hold. However, in the colocated case, we have $h_{0i} = 1$ for all i.

Table 10.1 Two Secondary Users' Game (from Left to Right): (a) Payoff Table, (b) NE, (c) Mixed NE, and (d) Correlated Equilibrium

	0	1
0	(5,5)	(6,3)
1	(3,6)	(0,0)

	0	1
0	0	(0 or 1)
1	(1 or 0)	0

	0	1
0	9/16	3/16
1	3/16	1/16

	0	1
0	0.6	0.2
1	0.2	0

one user transmits more aggressively using action 1 while the other still plays action 0, the aggressive user achieves a better payoff while the other user has a lower payoff, and the overall benefit is reduced. However, if both users transmit aggressively using action 1, both users obtain the lowest payoff. In Table 10.1b, we show two NEs, where one of the user dominates the other. The dominating user has the payoff of 6 and the dominated user has the payoff of 3, which is unfair. In Table 10.1c, we show the mixed NE, where two users have the probability of 0.75 for action 0 and 0.25 for action 1, respectively. The payoff for each user is 4.5.

In the case of correlated equilibrium, a strategy profile is chosen randomly according to a certain distribution. Given the recommended strategy, it is to the players' best interests to conform to this strategy. In Table 10.1b and c, the NEs and mixed NEs are all within the set of correlated equilibria. In Table 10.1d, we show an example where the correlated equilibrium is outside the convex hull of the NE. Notice that the joint distribution is not the product of two users' probability distributions, i.e., two users' actions are not independent. Moreover, the payoff for each user is 4.8, which is higher than that of the mixed strategy.

We define the correlated equilibrium in a formal way.

Definition 10.11 (Correlated Equilibrium) A probability distribution p is a correlated equilibrium of game $G = \left[\mathcal{M}, \{\mathcal{A}_i\}_{i\in\mathcal{M}}, \{R_i\}_{i\in\mathcal{M}}\right]$, if and only if, for all $i \in \mathcal{M}$, $a_i \in \mathcal{A}_i$, and $a_{-i} \in \mathcal{A}_{-i}$,

$$\sum_{a_{-i}\in\mathcal{A}_{-i}} p(a_i, a_{-i}) \left[s_i\left(a_i', a_{-i}\right) - s_i(a_i, a_{-i})\right] \leq 0, \quad \forall a_i' \in \mathcal{A}_i. \tag{10.12}$$

By dividing the inequality (10.12) with $p(a_i) = \sum_{a_{-i}\in\mathcal{A}_{-i}} p(a_i, a_{-i})$, we have

$$\sum_{a_{-i}\in\mathcal{A}_{-i}} p(a_{-i}|a_i) \left[u_i\left(a_i', a_{-i}\right) - s_i(a_i, a_{-i})\right] \leq 0, \quad \forall a_i' \in \mathcal{A}_i. \tag{10.13}$$

The inequality (10.13) means that when the recommendation to user i is to choose action a_i, then choosing action a_i' instead of a_i cannot obtain a higher expected payoff to i.

We note that the set of correlated equilibria is nonempty, closed, and convex in every finite game. Moreover, it may include the distribution that is not in the convex hull of the NE distributions. In fact, every NE is a correlated equilibrium and NEs correspond to the special case where $p(a_i, a_{-i})$ is a product of each individual user's probability for different actions, i.e., the strategies of the different players are independent [70–72]. The correlated equilibrium can be calculated via linear programming. If only local information is available, some learning algorithms such as non-regret learning can achieve the correlated equilibrium with probability 1 [72].

Using the correlated equilibrium concept, there are many possible applications for cognitive radios such as power control and spectrum access [73–75]. For example, the distributive users adjust their transmission probabilities over the available channels, so that the collisions are avoided and the users' benefits are optimized. As a result, the spectrum utilization efficiency and fairness among the distributive users can be improved. To learn the correlated equilibrium in a distributed manner, the adaptive no-regret algorithm is proposed in [73] using past history. The proposed learning algorithm converges to a set of correlated equilibria with probability 1. Learning schemes can achieve better equilibria using only the past history and without requiring more signaling and overhead. The complexity of learning algorithms can be relatively high. Moreover, there is a trade-off between convergence speed and complexity. To achieve fast convergence speed, the complexity of the learning algorithms should be high. Some simple learning algorithms have been proven to converge to the optimal solution with sufficiently long learning time. However, long learning time causes a problem in a network with high mobility, where network topologies and channel conditions have changed before the learning converges. Moreover, if the noncooperative competition is severe, the learning algorithms might converge slowly, fluctuate, or become very sensitive to randomness. So, the learning schemes based on correlated equilibrium can only achieve good performance in the situations where the noncooperative competition is not severe, there is an achievable gap between the NEs and the optimums, and the network mobility is sufficiently low.

10.5 Conclusions and Open Problems

Cognitive radio is a revolutionary wireless communication paradigm that can achieve much higher spectrum efficiency than the existing systems. Many technical challenges, however, still remain to be solved to make this vision a reality. In particular, the distributed and dynamic nature of spectrum sharing requires a new design and analysis framework. Game theory provides a natural solution for this challenging task. In this chapter, we describe several game theoretical models that have been successfully used to solve various spectrum-sharing problems.

We want to mention that the most discussed models rely on the concept of NE in a static game with complete information. Although mathematically convenient, this may not be the most suitable game theoretical model in practice. For example, instead of participating in the game only once as in a static game, secondary and primary users might interact repeatedly within a reasonably long time frame. In this case, the users make decisions based not only on the current network conditions but also on the past interaction history. A model of repeated games (either finite or infinite) will be more suitable. Moreover, the complete information assumption is difficult to be satisfied in practice, due to the fast-changing nature of the wireless channels and the bandwidth requirement to exchange channel measurements. Without complete channel information, the best users can do is to maximize their expected payoff based on their own beliefs of the unknown network information. How these beliefs are initialized and updated will affect the results of the game. Some preliminary works have been reported along these directions [77,78], and definitely more are needed.

List of Abbreviations

FCC Federal Communications Commission
NE Nash Equilibrium
MBR Myopic Best Response
SINR signal to interference plus noise ratio
ADP Asynchronous Distributed Pricing
FPP Fictitious Power-Price
NBS Nash Bargaining Solution
KSS Kalai-Smorodinsky Solution
ES Egalitarian Solution

References

1. G. Staple and K. Werbach, The end of spectrum scarcity, *IEEE Spectrum*, 41(3): 48–52, March 2004.
2. Order and further notice of proposed rule making on smart radios, Federal Communications Commission Report 03-322, 2003.
3. FCC: Workshop on cognitive radio technologies, http://www.fcc.gov/oet/cognitiveradio/, May 2003.
4. J. Mitola III, Cognitive radio for flexible mobile multimedia communications, in *Proceedings of IEEE International Workshop on Mobile Multimedia Communications*, San Diego, CA, 1999, pp. 3–10.
5. Spectrum policy task force report, Federal Communications Commission, Washington, DC, November 2002.
6. T. W. Hazlett, The wireless craze, the unlimited bandwidth myth, the spectrum auction faux pas, and the punchline to ronald coase's 'big joke': An essay on airwave allocation policy. AEI-Brookings Joint Center for Regulatory Studies Working Paper 01-02, January 2001.
7. R. P. Margie, Efficiency, predictability, and the need for an improved interference standard at the FCC, *Telecommunications Policy Research Conference*, Arlington, VA, September 2003.

8. Auctions home page, Federal Communications Commission, http://www.fcc.gov/wtb/auctions.html.
9. P. Klemperer, How (not) to run auctions: The European 3G telecom auctions, *European Economic Review*, 46(4–5): 829–845, May 2002.
10. K. Binmore and P. Klemperer, The biggest auction ever: The sale of the British 3G telecom licenses, *The Economic Journal*, 112: C74–C96, March 2002.
11. J. Cable, A. Henley, and K. Holland, Pot of gold or winner's curse? An event study of the auctions of 3G mobile telephone licences in the UK, *Fiscal Studies*, 23(4): 447–462, 2002.
12. The development of secondary markets—Report and order and further notice of proposed rulemaking, Federal Communications Commission Report 03-113, 2003.
13. J. von Neumann and O. Morgenstern, *Theory of Games and Economic Behavior*. Princeton University Press, Princeton, NJ, 1947.
14. J. Nash, Equilibrium points in *n*-person games, in *Proceedings of the National Academy of Sciences*, 36: 48–49, 1950.
15. E. Rasmusen, *Games and Information: An Introduction to Game Theory*, 3rd edn. Blackwell Publishers, Malden, MA, 2001.
16. D. Fudenberg and J. Tirole, *Game Theory*. MIT Press, Cambridge, MA, 1991.
17. R. Gibbons, *Game Theory for Applied Economists*. Princeton University Press, Princeton, NJ, 1992.
18. R. B. Myerson, *Game Theory: Analysis of Conflicts*. Harvard University Press, Cambridge, MA, 1991.
19. A. Mas-Colell, M. D. Whinston, and J. R. Green, *Microeconomic Theory*. Oxford University Press, New York, 1995.
20. A. Rubinstein, *Modeling Bounded Rationality*. The MIT Press, Cambridge, MA, 1998.
21. R. Dash, N. Jennings, and D. Parkes, Computational-mechanism design: A call to arms, *IEEE Intelligent Systems*, 18(6): 40–47, November 2003.
22. N. Nisan and A. Ronen, Algorithmic mechanism design, *Games and Economic Behavior*, 35(1): 166–196, 2001.
23. I. Gilboa and A. Matsui, Social stability and equilibrium, *Econometrica*, 59(3): 859–867, May 1991.
24. M. Kandori and R. Rob, Evolution of equilibria in the long run: A general theory and applications, *Journal of Economic Theory*, 65(2): 383–414, April 1995.
25. H. L. Cole and N. Kocherlakota, Finite memory and imperfect monitoring, Federal Reserve Bank of Minneapolis Working Paper, 2000.
26. F. Gul and E. Stacchetti, The English auction with differentiated commodities, *Journal of Economic Theory*, 92: 66–95, May 2000.
27. D. C. Parkes and L. H. Ungar, An ascending-price generalized Vickrey auction, Technical Report, Harvard University, Cambridge, MA, 2002.
28. J. Kalagnanam and D. C. Parkes, *Supply Chain Analysis in the E-Business Era*. Kluwer, Boston, MA, 2002.
29. D. M. Topkis, *Supermodularity and Complementarity*. Princeton University Press, Princeton, NJ, 1998.
30. D. Monderer and L. Shapley, Potential games, *Games and Economic Behavior*, 14: 124–143, 1996.

31. P. Milgrom and J. Roberts, Rationalizability, learning and equilibrium in games with strategic complementarities, *Econometrica*, 58(6): 1255–1277, November 1990.
32. P. Milgrom and J. Roberts, Adaptive and sophisticated learning in repeated normal form games, *Games and Economic Behavior*, 55(2): 340–371, 1991.
33. D. Funderberg and D. M. Kreps, *A Theory of Learning in Games*. MIT Press, Cambridge, MA, 1998.
34. C. U. Saraydar, N. B. Mandayam, and D. J. Goodman, Efficient power control via pricing in wireless data networks, *IEEE Transactions on Communications*, 50(2): 291–303, February 2002.
35. E. Altman and Z. Altman, S-modular games and power control in wireless networks, *IEEE Transactions on Automatic Control*, 48(5): 839–842, May 2003.
36. J. Huang, R. Berry, and M. L. Honig, Distributed interference compensation in wireless networks, *IEEE Journal on Selected Areas in Communications*, 24(5): 1074–1084, May 2006.
37. D. C. Parkes, Iterative combinatorial auctions: Achieving economic and computational efficiency, PhD dissertation, University of Pennsylvania, Philadelphia, PA, 2001.
38. V. Krishna, *Auction Theory*. Academic Press, San Diego, CA, 2002.
39. R. Wilson, Auctions of shares, *Quarterly Journal of Economics*, 93(4): 675–698, 1979.
40. L. Ausubel and P. Cramton, Demand reduction and inefficiency in multi-unit auctions, manuscript, University of Maryland, College Park, MD, 1998.
41. C. Maxwell, Auctioning divisible commodities: A study of price determination, PhD dissertation, Harvard University, Cambridge, MA, 1983.
42. K. Back and J. F. Zender, Auctions of divisible goods: On the rationale for the treasury experiment, *Review of Financial Studies*, 6(4): 733–764, 1993.
43. J. J. D. Wang and J. F. Zender, Auctioning divisible goods, *Economic Theory*, 19(4): 673–705, 2002.
44. A. Hortasu, Mechanism choice and strategic bidding in divisible good auctions: An empirical analysis of the Turkish treasury auction market, manuscript, Stanford University, Stanford, CA, 2000.
45. G. Federico and D. Rahman, Bidding in an electricity pay-as-bid auction, *Journal of Regulatory Economics*, 24(2): 175–211, 2003.
46. R. Johari and J. N. Tsitsiklis, Efficiency loss in a network resource allocation game, *Mathematics of Operations Research*, 29(3): 407–435, August 2004.
47. S. Yang and B. Hajek, An efficient mechanism for allocation of a divisible good, submitted to *Math Operation Research*, 2005.
48. S. Sanghavi and B. Hajek, Optimal allocation of a divisible good to strategic buyers, Working Paper UIUC, 2004.
49. R. Maheswaran and T. Basar, Nash equilibrium and decentralized negotiation in auctioning divisible resources, *Group Decision and Negotiation*, 12(5): 361–395, 2003.
50. R. T. Maheswaran and T. Basar, Coalition formation in proportionally fair divisible auctions, in *Autonomous Agents and Multi-Agent Systems*, Melbourne, VIC, Australia, 2003.
51. R. T. Maheswaran and T. Basar, Decentralized network resource allocation as a repeated noncooperative market game, in *Proceedings of the 40th IEEE Conference on Decision and Control*, Orlando, FL, December 2001, pp. 4565–4570.

52. J. Huang, R. Berry, and M. L. Honig, Auction-based spectrum sharing, *Mobile Networks and Applications Journal*, 11(3): 405–418, June 2006.
53. M. Voorneveld, Best-response potential games, *Economics Letters*, 66(3): 289–295, March 2000.
54. J. Neel, J. Reed, and R. Gilles, Convergence of cognitive radio networks, in *Proceedings of IEEE Wireless Communications and Networking Conference*, Atlanta, GA, 2004.
55. J. Hicks, A. MacKenzie, J. Neel, and J. Reed, A game theory perspective on interference avoidance, in *Proceedings of IEEE Global Telecommunications Conference*, Dallas, TX, 2004.
56. G. Scutari, S. Barbarossa, and D. Palomar, Potential games: A framework for vector power control problems with coupled constraints, in *Proceedings of IEEE ICASSP 2006*, Toulouse, France, 2006.
57. P. Milgrom, *Putting Auction Theory to Work*. Cambridge University Press, Cambridge, MA, 2004.
58. W. Yu, G. Ginis, and J. M. Cioffi, Distributed multiuser power control for digital subscriber lines, *IEEE Journal on Selected Areas in Communications*, 20(5): 1105–1114, June 2002.
59. H. Yaiche, R. R. Mazumdar, and C. Rosenberg, A game theoretic framework for bandwidth allocation and pricing in broadband networks, *IEEE/ACM Transactions on Networking*, 8(5): 667–678, October 2000.
60. D. Grosu, A. T. Chronopoulos, and M. Y. Leung, Load balancing in distributed systems: An approach using cooperative games, in *Proceedings of IPDPS 2002*, Washington, DC, 2002, pp. 52–61.
61. G. Owen, *Game Theory*, 3rd edn. Academic Press, San Diego, CA, 2001.
62. W. Rhee and J. M. Cioffi, Increase in capacity of multiuser OFDM system using dynamic subchannel allocation, in *Proceedings of IEEE VTC'00*, Tokyo, Japan, 2000, pp. 1085–1089.
63. Z. Han, Z. Ji, and K. J. R. Liu, Fair multiuser channel allocation for OFDMA networks using Nash bargaining and coalitions, *IEEE Transactions on Communications*, 53(8): 1366–1376, August 2005.
64. C. Peng, H. Zheng, and B. Y. Zhao, Utilization and fairness in spectrum assignment for opportunistic spectrum access, *ACM Mobile Network Applications*, 11(4): 555–576, August 2006.
65. J. E. Suris, L. DaSilva, Z. Han, and A. MacKenzie, Cooperative game theory approach for distributed spectrum sharing, in *Proceedings of IEEE International Conference on Communications*, Glasgow, Scotland, June 2007.
66. K. D. Lee, and V. C. M. Leung, Fair allocation of subcarrier and power in an OFDMA wireless mesh network, *IEEE Journal on Selected Areas in Communications*, 24(11): 2051–2060, November 2006.
67. H. Park and M. van der Schaar, Bargaining strategies for networked multimedia resource management, *IEEE Transactions on Signal Processing*, 55(7): 3496–3511, July 2007.
68. Z. Han, Z. Ji, and K. J. Ray Liu, Power minimization for multi-cell OFDM networks using distributed non-cooperative game approach, *IEEE Global Telecommunications Conference*, Dallas, TX, 2004.

69. Z. Han, Z. Ji, and K. J. Ray Liu, Non-cooperative resource competition game by virtual referee in multi-cell OFDMA networks, *IEEE Journal on Selected Areas in Communications*, 25(6): 1079–1090, August 2007.
70. R. J. Aumann, Subjectivity and correlation in randomized strategy, *Journal of Mathematical Economics*, 1(1): 67–96, 1974.
71. R. J. Aumann, Correlated equilibrium as an expression of Bayesian rationality, *Econometrica*, 55(1): 1–18, January 1987.
72. S. Hart and A. Mas-Colell, A simple adaptive procedure leading to correlated equilibrium, *Econometrica*, 68(5): 1127–1150, September 2000.
73. Z. Han, C. Pandana, and K. J. R. Liu, Distributive opportunistic spectrum access for cognitive radio using correlated equilibrium and no-regret learning, in *Proceedings of IEEE Wireless Communications and Networking Conference*, Hong Kong, China, March 2007.
74. E. Altman, N. Bonneau, and M. Debbah, Correlated equilibrium in access control for wireless communications, in *Proceedings of Networking 2006*, Coimbra, Portugal, *Lecture Notes in Computer Science*, Vol. 3976, Springer-Verlag, Berlin, Germany, 2006, pp. 173–183.
75. V. Krishnamurthy, G. Yin, and M. Maskery, Stochastic approximation based tracking of correlated equilibria for game-theoretic reconfigurable sensor network deployment, in *Proceedings of IEEE Conference on Decision and Control*, San Diego, CA, December 2006, pp. 2051–2056.
76. J. Huang, V. G. Subramanian, R. Agrawal, and R. Berry, Downlink scheduling and resource allocation for OFDM systems, *IEEE Transactions on Wireless*, 8(1): 288–296, January 2009.
77. Y. Wu, B. Wang, K. J. R. Liu, and T. C. Clancy, Repeated open spectrum sharing game with cheat-proof strategies, *IEEE Transactions on Wireless Communications*, 8(4): 1922–1933, April 2009.
78. S. Adlakha, R. Johari, and A. Goldsmith, Competition in wireless systems via Bayesian interference games, Arxiv preprint, arXiv:0709.0516, 2007.

Chapter 11

Pricing for Security and QoS in Cognitive Radio Networks

S. Sengupta, S. Anand, and R. Chandramouli

Contents

Dynamic spectrum access along with dynamic service offering and profiles are anticipated to increase hugely in the near future, as users move from long-term service provider agreements to more opportunistic service models with the help of cognitive radio (CR) networks. With radio spectrum itself traded as a commodity in a dynamic market-based scenario, wireless service providers (WSPs) will require new strategies to deploy services, define service profiles, and price them. Currently, pricing for dynamic spectrum access in "CR networks" is an open research issue as there is little understanding on how such a dynamic trading system will function so as to make the system feasible under both economic terms and performance.

In this chapter, it is carefully explained how the conflicts in the cognitive radio network system can be modeled by making use of pricing and game theory. The problem is motivated by considering typical cognitive network scenarios and by identifying characteristics that require new solutions.

11.1 Introduction

With recent proliferation in wireless networks and services, enhancements in technology, and amplification in the number of subscribers, the wireless industry is briskly moving toward a dynamic market scenario, never anticipated before. The presence of numerous wireless service providers (WSPs) in any geographic region is forcing a competitive environment where each WSP must try to maximize its profit. Essentially, a wireless service provider buys spectrum from the spectrum owner (e.g., Federal Communications Commission [FCC] in the United States) with a certain price and then sells the spectrum to the end users in the form of services (bandwidth). In such a scenario, the goal of each service provider is twofold: get a large share of users and the necessary spectrum to fulfill the demands of these users. In most countries, chunks of spectrum were allocated to the WSPs earlier statically [10]. However, with the technological advancements, economic changes, and rapid increase in subscribers, WSPs find it difficult to satisfy users and increase revenue with just the spectrum statically allocated. Moreover, with spectrum usage being both space and time dependent, a fixed, static allocation often leads to low spectrum utilization. It has been demonstrated through experimental studies [40] that there is a great amount of "white space" (unused bands) available sparsely that can potentially be used for both licensed and unlicensed services.

With the experiments proving the disadvantages of static allocation of spectra, dynamic spectrum access (DSA) [6] is thought as one of the best alternatives that can eliminate the "artificial scarcity of spectrum" introduced by the inefficient static spectrum allocation. In DSA, the spectrum owner (e.g., FCC) will create a common pool of open spectrum, and WSPs can access the spectrum bands dynamically from this common pool. Although this common pool may be created by withdrawing all the previously (statically) allocated spectrum licenses (physical merging), it is not an option because of monies already invested by the license holders. However, portions of the spectrum bands that are either unused or underutilized can be made open for the common pool (virtual merging) as a very first step. These unused parts in the spectrum that are open to all are known as the coordinated access band (CAB) [7]. Examples of such bands include, but are not limited to, the public safety bands (764–776, 794–806 MHz) and unused broadcast UHF TV channels (450–470, 470–512, 512–698, 698–806 MHz). WSPs access these additional spectrum bands dynamically for a certain duration by paying the spectrum owner and sell that spectrum in the form of services to the end users depending on increased users' requests, who, in turn, again pay for these services.

With such requirements for DSA, cognitive radio (CR) networks are seen as a key solution to meet FCC's policy and to build the future generation of wireless networks [2,19]. A cognitive radio must periodically perform spectrum sensing and operate at any unused frequency in the licensed or unlicensed band, regardless of whether the frequency is assigned to licensed services or not. But the most important regulatory aspect is that cognitive radios must not interfere with the operation in some licensed band and must identify and avoid such bands in a timely manner. DSA, based on cognitive radio network, can thus harness the idle, unutilized, or underutilized spectrum to increase the spectrum usage efficiency.

11.1.1 Dynamic WSP Switching with Cognitive Radio

As far as the end users are concerned, there is still a strong association with a single WSP, that is, a user usually gets the services from one provider for a period of time as per the contractual agreement (e.g., one or two years). However, it is anticipated that in the near future, the concept of service brokers, technically known as mobile virtual network operators (MVNO) [37], will evolve that will act as an interface between the providers and the users [24]. These service brokers will allow more flexibility to the end users to choose and connect to any WSP almost on a session-by-session basis depending on what the users' preferences are—quality of service (QoS), coverage, price, etc. The emerging wireless technology, cognitive radio [19], is anticipated to make dynamic user–WSP association a reality. The basic operating principle relies on a radio being able to sense whether a particular band is being used and utilize the spectrum band if unused. Cognitive radios can be viewed as an electromagnetic spectrum detector, which can find an unoccupied band and adapt the carrier to that band. The layer functionalities of cognitive radios can be separated into physical and medium access control layers. The physical layer includes sensing (scanning the frequency spectrum and process wideband signal), cognition (detecting the signal through energy detector), and adaptation (optimizing the frequency spectrum usage such as power, band, and modulation). The medium access layer cooperates with the sensing measurement and coordinates in allocating spectrum. Cognitive radio systems continuously perform spectrum sensing and dynamically identify unused spectrum bands. With such potential dynamic association between the users and the WSPs due to the cognitive radio networks, the most important questions that arise are as follows:

- In such cognitive radio network systems, how or which wireless service provider should a user select for a particular service?
- What price must the WSPs charge such that they are able to attract the users and increase their profits?

11.1.2 Secrecy Capacity

The admission of additional users through DSA by multiple WSPs raise an important issue of security in cognitive radio networks. As a WSP owning a particular network admits users subscribed to other networks or other WSPs, some of these admitted users could turn eavesdroppers and obtain vital information about a particular network. Although this can be countered by using ciphering mechanisms, there are still many broadcast and unicast messages transmitted in clear text [36,42]. The paradigm then shifted to using error control codes in such a way that the eavesdroppers would be unable to decode the code whereas the intended receivers would be able to successfully decode the message [34]. It then becomes interesting to determine if there exists a rate of transmission such that the bit error rate (BER) at the intended receiver is close to zero whereas that at the eavesdropper becomes close to 0.5. This rate can be determined by measuring the "secrecy capacity" of the system.

Secrecy capacity was studied for systems with key less security. A system typically consists a source (or a transmitter), a destination (or a receiver), and an eavesdropper as shown in Figure 11.1. The source transmits information that is received both by the receiver and the eavesdropper. Secrecy capacity is roughly the maximum rate at which the source can transmit such that the BER at the destination approaches zero while that at the eavesdropper approaches 0.5. For some cases, the secrecy capacity is the difference between the Shannon capacity of the channel between the source and the destination and that between the source and the eavesdropper. For example, in the system shown in Figure 11.1, let the signal-to-noise ratio (SNR) in the channel between the source and the destination be γ_d and let the SNR in the

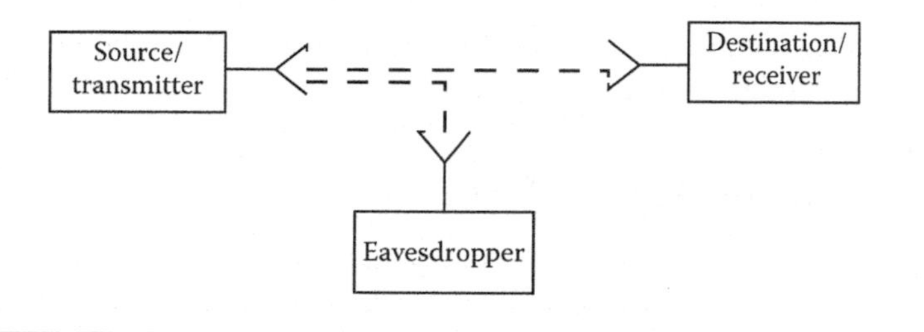

Figure 11.1 A wired/wireless channel with an eavesdropper.

channel between the source and the eavesdropper be γ_e. If the channel bandwidth is W, then the secrecy capacity, S, is given by [34]

$$S = \left[W \log_2 \left(\frac{1 + \gamma_d}{1 + \gamma_e} \right) \right]^+, \tag{11.1}$$

where $[y]^+ = \max(y, 0)$.

In CR networks, secrecy capacity provides an additional criterion for admitting secondary users. For example, consider a pair of secondary users—one acting as a transmitter and the other as its receiver. This pair of users would normally be admitted in a cognitive radio network if there is sufficient white space available and if the transmit power of the transmitter is such that any possible interference caused to primary receivers is below a specified level. However, it may be possible that the secrecy capacity of the channel between this pair of users is very low, thus making the channel insecure. Any other eavesdropper can potentially eavesdrop information due to the admission of this pair of users. However, if users are admitted based on satisfying a secrecy capacity constraint in addition to the other usual constraints, then this pair of users would be blocked and the security of the system could be improved. A complimentary example would be a case of a transmit–receive pair that possibly causes larger interference to a primary receiver and would potentially not be preferred, but can be admitted due to the larger secrecy capacity obtained because the admitted transmitter causes larger interference at the eavesdropper and hence improves the secrecy capacity.

Although secrecy capacity was studied from an information theoretic point of view, in the presence of multiple transmitters and receivers, maximizing secrecy capacity can be looked upon as a noncooperative game in which the players are the transmitters, the strategies are the transmit powers, and the utilities being the secrecy capacities of the channels between the transmitters and their corresponding receivers. To limit the transmitters from transmitting at exorbitantly high powers, it is essential to pose some kind of penalty on those transmitters transmitting at higher powers. Pricing can be introduced as a penalty on transmitters that prohibits them from transmitting at exorbitantly high powers. The concept of transmit power allocation with pricing and its effect on the secrecy capacity of cognitive radio networks is an interesting problem to investigate.

11.1.3 Pricing Interdependency

It is important to investigate the economic issues that arise due to the presence of multiple competing WSPs and that has a profound impact on the differentiated service qualities and the prices paid by the end users. Economic models have been shown to be useful in the design of such real-world market-based systems, as it is intuitive that the actions, or reactions, in response to others' actions, of the self-interested agents (spectrum owner or WSPs or end users) would be always focused on maximizing their own "profit." In this respect, concepts from economic theory can be used to guide the design process of the agents' strategies and the framework of the cognitive radio network in a distributed manner. By introducing the providers and users in an oligopoly* market-like environment, it best suites to study the concept of prices to regulate the demands of users who consume resources (bandwidth). In such a cognitive radio network trading system, the dynamic spectrum allocation to the WSPs is controlled in a time- and space-variant manner by the spectrum owner [6]. On the other hand, WSPs use the spectrum to offer services to the end users and make profit. Although these two problems, that is, dynamic spectrum allocation and dynamic service provisioning seem to be two separate problems, there is a strong correlation and pricing interdependency between them.

The most important factors that the WSPs need to consider are the estimate of amount of spectrum that they need and what *price* are they willing to pay to the spectrum owner—both of which are solely determined by the demands of the users and the revenue generated from these users. In effect, the estimation of the demand for bandwidth and revenue earned will drive the provider's strategies. The pricing offered by the providers, in turn, will affect the demand for the services by the users, thus resulting in a cyclic interdependency in a typical supply–demand scenario as depicted in Figure 11.2.

11.1.4 Paradigm Shift in CR Network System

Currently, each provider gets a fixed chunk of the spectrum and has a unique user pool that they cater to as shown in Figure 11.3a. In future, a paradigm shift as depicted in Figure 11.3b, is very likely to occur where each provider will get a part of the spectrum from the common spectrum pool as and when they need through a spectrum owner. The users will also be able to select their service provider as per their requirements through a service broker with the help of cognitive radio. The service brokers provide authentication credentials, billing account, and access

* An oligopoly is a market form in which a market is dominated by a small number of sellers (oligopolists).

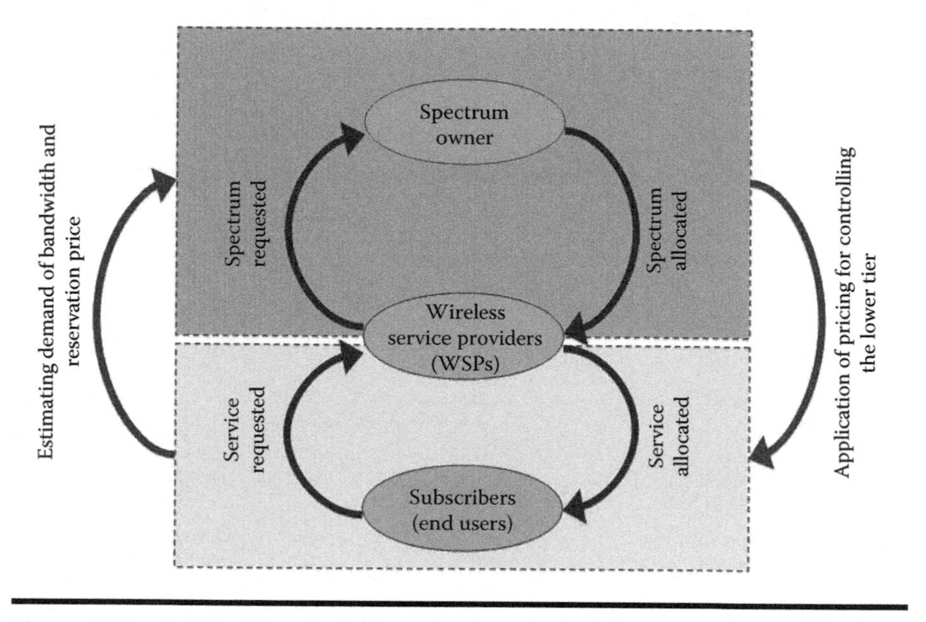

Figure 11.2 Pricing interdependency.

information of any of the service providers to the users. A CR will make the users able to sense whether a particular band is being used and, if not, to utilize the spectrum without interfering with the transmission of other users. With regard to these new developments, it is important to investigate how the pricing policies will play a very important role and control dynamic spectrum allocation and dynamic service offering in this trading system.

The dynamic spectrum allocation scenario (top part of the hierarchy in the paradigm shift scenario) with respect to a spectrum owner represents a market place with one seller (spectrum owner) and multiple buyers (WSPs). The natural question that arises in this case is "how the spectrum will be allocated" and "how much price the WSPs will be willing to pay to the spectrum owner." With exact price for the spectrum bands undetermined, this market can be modeled best using the auction-theoretic approach. The WSPs submit their requests (and bids) and the spectrum owner tries to allocate the spectrum in such a manner so as to maximize its revenue. On the other hand, WSP–user interaction (the bottom part of the hierarchy in the paradigm shift) results in conflicting objectives where both WSP and users want to maximize their individual benefits. Each service provider decides the price based on their current load and the service requested by the users. Obviously, the decision cannot be made unilaterally and the user must be involved. Through QoS and price preferences, a user selects a provider that best characterizes his preferences, usually given by the utility function. The notion of perservice static prices [18,21] are no longer followed here and the providers are allowed to set

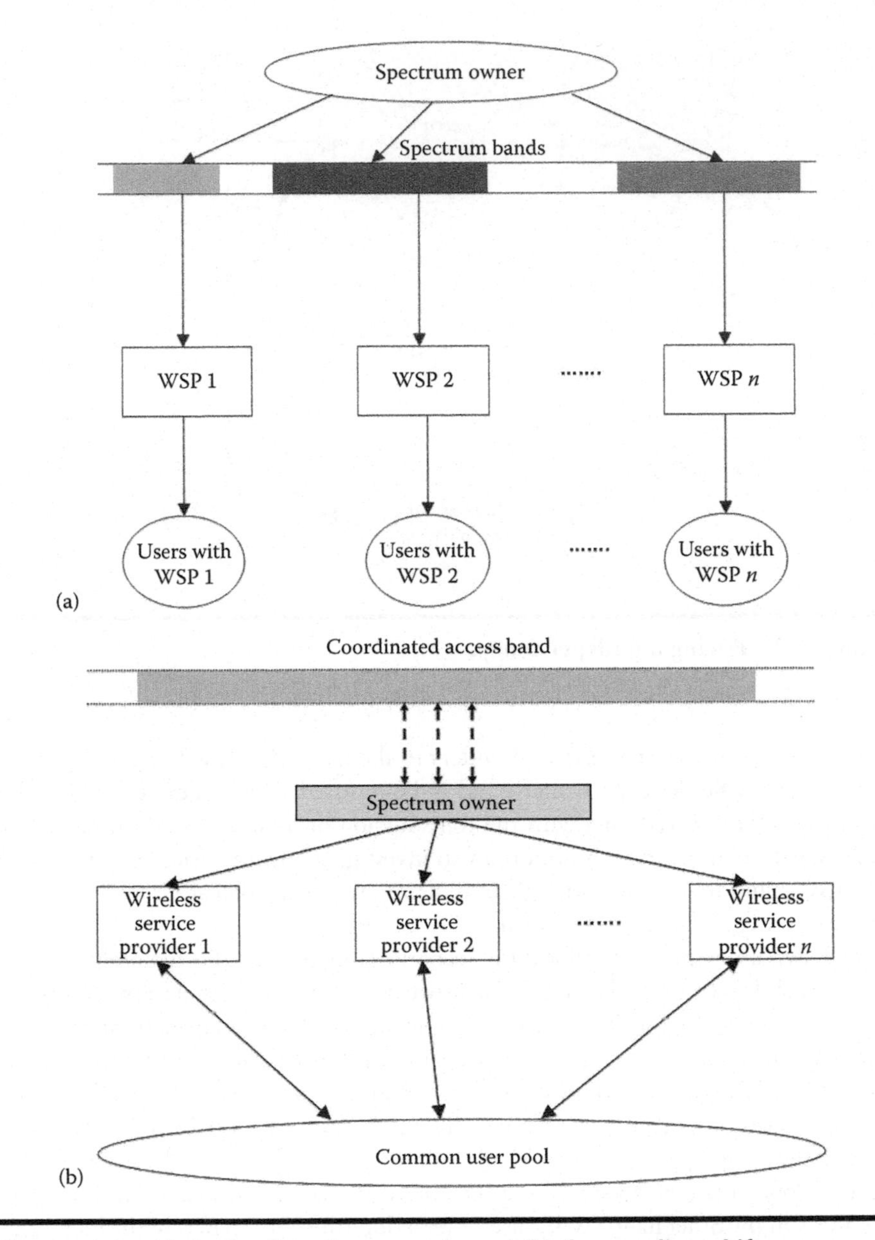

Figure 11.3 (a) Static allocation scenario and (b) the paradigm shift.

the prices dynamically to maximize their profit and minimize resource wastage. Such a market mechanism is more flexible and realistic, as there does not exist any centralized authority to determine the price of a service [28]. Pricing-based game-theoretic model thus best suits to model such conflicts in analyzing WSP–user interactions.

11.2 Price Motivation in Cognitive Radio Network

Pricing of spectrum (for dynamic access) and pricing of network services (for dynamic service provisioning) in cognitive radio networks have the overwhelming advantage that permits the spectrum owner, WSPs and users, acting individually or as organizations, to express the value they are willing to pay or receive. The aim for adapting such a dynamic market scenario is that the resources should go to those who do value it most [11,28]. The critical challenge in designing such a market is to enable real-time transactions and to eliminate the fear of market manipulation.

Pricing can also be added as a means to limit the usage of resources by the users. In other words, by imposing pricing in the form of a penalty on the usage of resources, users will be regulated against excessive usage of resources. This is slightly different from the paradigm of using pricing to generate revenue. For example, consider battery power to be one of the resources used by a wireless terminal. Excessive transmission of power reduces the battery lifetime of the terminal. Excessive transmission of power also causes interference to other users in the system. Thus, pricing can be used as a penalty to limit the transmit power. The penalty could be based on the usage (i.e., the transmit power itself) or based on the penalty caused to others (i.e., based on the interference caused to other users, which, in turn, is a function of the transmit power). A source can increase the secrecy capacity to its intended destination by increasing its transmit power indefinitely. However, this could cause large drain out of the battery or large interference on other destinations in a multiterminal network. Thus, pricing can be introduced as a means to limit the transmit power of the sources and yet maximize the secrecy capacity.

Auction theory and game theory have been recognized as the cornerstones of microeconomics, as they are used to analyze problems with conflicting objectives among interacting decision makers. These theories have been extensively used in various industries including the competitive energy market, airlines industry, and Internet services. They have been proved to be very powerful tools to deal with problems in networking and communications from an economic point of view. This is because the benefit that each entity receives in a competitive environment is often affected by the action of other entities who also try to contend for the same pool of resources. A broad overview of game theory and its application to different problems in networking and communications can be found in [31] and the references therein. Network services, including pricing issues, have been studied with the help of auctions in [12,16,17,25].

Auction theory has been used to understand markets, especially to model auction participants who bid to win and maximize profit [22]. Currently, most auction sites (e.g., eBay [38]) support a basic bidding strategy through a proxy service for a single-unit auction where bidding continues till a winner evolves. In a single-unit auction, Vickrey proved that “English-” and “Dutch-” type auctions yield the same expected revenue under the assumptions of risk-neutral participants and privately known value drawn from a common distribution [29]. Vickrey’s result is embodied in the "revenue equivalence theorem" (RET) [13]. However, with emerging markets like spectrum trading, single-unit auctions fail to address the issues where bidders bid for multiunits and multiple winners emerge [1,32]. As bidders compete for a part of the available resource and are willing to pay a price for that part only, the kind of auction model needed must be more generalized and is under investigation [4,33].

As far as the game theory is concerned, there is an emerging body of work that deals with decision making in a multi-provider setting. In [8], a market in the form of a "bazaar" was introduced where infrastructure-based wide-area wireless services are traded in a flexible manner and at any timescale. The mobile bazaar architecture allows fine-grained service through cooperative interactions based on user needs. The problem of dynamically selecting ISPs for forwarding and receiving packets has been studied in [30]. In [15], an integrated admission and rate control framework for CDMA-based wireless data networks is proposed. The providers define the admission criteria as the outcome of the game and the Nash equilibrium is reached using pure strategy. Users are categorized into multiple classes and are offered differentiated services based on the price they pay and the service degradation they can tolerate. However, dynamic pricing was not explored in [15]. The DIMSUMNet project [6] takes a novel approach for spectrum allocation in which the providers are no longer restricted to use the statically partitioned spectrum. Instead, a new paradigm of DSA is enabled through real-time spectrum allocations to different networks based on traffic demands. A spectrum broker coordinates the allocation, usage, and prices the portions of the spectrum used by different networks or providers. Similar to any market, cognitive radio network market also has sellers and buyers. Thus, the determination of efficient pricing framework with regard to spectrum and services becomes a pivotal element in a CR network from an economic perspective and performance issues.

11.3 Pricing-Driven Dynamic Spectrum Allocation

The market model of the spectrum owner and the wireless service providers is discussed formally using auction-driven marketplace to support dynamic multiparty spectrum trading. A natural question that may arise in this regard is why choose auction to model this dynamic interaction. The justification is two fold. Spectrum being a scarce resource and with exact price for the spectrum bands undetermined, auction is among the best-known market-based mechanisms to distribute scarce resources because of its perceived fairness and allocation efficiency [14]. Moreover, with the market model consisting of one seller (spectrum owner) and many buyers (WSPs), auction is an efficient method to increase revenue and spectrum usage efficiency and to reduce monopoly in the economic framework.

11.3.1 Types of Auctions

An auction is the process of buying and selling goods by offering them up for bid (i.e., an offered price), taking bids, and then selling the item to the highest bidder. In economic theory, an auction is a method for determining the value of a commodity that has an undetermined or variable price. There are several kinds of existing auction strategies. Depending on whether the bidding strategies of each of the bidders are disclosed in front of the other bidders, open and closed bid auctions are designed. In open auctions [5,10], bids are open to everybody so that a players strategy is known to other players and players usually take their turns one by one until winner(s) evolve. Bids generated by players in open bid auction can be either in increasing (e.g., English and Yankee auction) or in decreasing order (Dutch auction). An important perspective of increasing auction is that it is more in the favor of bidders than the auctioneers. Moreover, increasing open bid auction helps bidders in the early round to recognize each other and thus act collusively. Increasing auction also detracts low-potential bidders (bidders with low amount of spectrum request or low-value bid) because they know a bidder with higher bid will always exceed their bids. Closed-bid auctions are opposite to open-bid auctions and bids/strategies are not known to everybody. Only the organizer (spectrum owner in this case) of the auction will know about the bids submitted by the bidders and will act accordingly. Closed-bid auctions thus do not promote collusion. Spectrum auction is more close to the multiunit auctions. Multiple bidders present their bids for a part of the spectrum band, where the sum of all these requests exceeds the total spectrum band capacity thus causing the auction to take place. Moreover, unlike classic single-unit auction, multiple winners evolve in this auction model constituting a winner set. The determination of winner set often depends on the auction strategy taken by the spectrum owner in this case.

11.3.2 Auction Issues

A good auction design is important for any type of successful auction and often varies depending on the item on which the auction is held. Thus, auctions held on eBay [38] are quite different from the auctions applicable to the spectrum. eBay auctions are typically used to sell an art object or a valuable item. Bidding starts at a certain price defined by the auctioneer and the competing bidders increase their bids. If a bid provided by a bidder is not exceeded by any other bidder, then the auction on that object stops and the final bidder is the winner. Unlike classical single-unit auctions, spectrum auctions are multiunit where bidders bid for a part of the spectrum band, that is, the bids are for different amounts of bandwidth. Also, multiple winners evolve constituting a winner set [26]. The determination of winner set depends on the auction strategy adopted by the spectrum owner. In the dynamic spectrum auction model, the spectrum owner is the seller who owns the coordinated access band and service providers are the buyers/bidders. Recall that the service providers already have some spectrum that was statically allocated. It is the additional spectrum that is sought from the CAB. Although the objective of the spectrum owner is to sell the CAB and earn revenue, it is not at all intended that only big companies with higher spectrum demand are given additional spectra. The goal here is to increase competition and bring new ideas and services at the same time. As a result, it is necessary to make the small companies, who also have a demand of spectrum, interested in taking part in the auction. In this regard, three important issues for designing efficient spectrum auction must be considered:

1. Maximize revenue generated from bidders
2. Entice bidders by increasing their probability of winning
3. Prevent collusion

11.3.3 Auction Formulation

The problem described here has a very close connection to the classical knapsack problem, where the goal is to fill a sack of finite capacity with several items such that the total valuation of the items in the sack is maximized. Here, the sack represents the finite capacity of spectrum in the CAB that is to be allocated to the WSPs in such a manner that the revenue generated from these WSPs is maximized. In this regard, "winner determining sealed bid knapsack auction" is presented. In the proposed auction model, L WSPs (bidders) are considered who compete to acquire a total available spectrum of W. All the service providers submit their demands at the same time in a sealed-bid manner. Sealed-bid auction strategy is followed, because sealed-bid auction has shown to perform well in all-at-a-time auction bidding and has a tendency to prevent collusion [23]. Each service provider has knowledge about its own bidding quantity and bidding price but do not have knowledge about other's quantity and price.

The auction is then formulated as follows. The strategy adopted by service provider i is denoted by a tuple $q_i = \{w_i, x_i\}$, where w_i denotes the amount of spectrum requested and x_i denotes the corresponding price that the service provider is willing to pay. If the sum of the bidding quantities do not exceed the spectrum available, W, then the requested quantities are allocated. Otherwise, auction is initiated when

$$\sum_{i=1}^{L} w_i > W \tag{11.2}$$

The goal is to solve the winner-determination problem in such a way so that the spectrum owner maximizes revenue by choosing a bundle of bidders (q_i), subject to the condition that the total amount of spectrum allocated does not exceed W. More formally, the allocation policy of the spectrum owner is

$$\text{maximize} \sum_i x_i \quad \text{such that,} \sum_i w_i \leq W \tag{11.3}$$

11.3.4 Bidders' Strategies

In knapsack auction, bidders' strategies are investigated for both first- and second-price bidding schemes. In first-price auction, bidder(s) with the winning bid(s) pay their winning bid(s). In contrast, in second-price auction, bidder(s) with the winning bid(s) do not pay their winning bid but pay the second winning bid.

Let each bidder i submit its demand tuple q_i. Then the optimal allocation of the spectrum to the service providers (bidders) is done by the spectrum broker considering all the demand tuples. Let this optimal spectrum allocation be denoted as M, where M incorporates all the demand tuples q_i and is subject to conditions as given in (11.3). The allocation of M can be found through dynamic programming assuming the bids can take only integer values. The revenue generated by spectrum owner can be obtained by summing all the bids from bidder, that is, $\sum_{i\in M} x_i$. Let a particular bidder j be allocated spectrum who thus belongs to M. Then the revenue generated from the optimal allocation M minus the bid of bidder j is $\sum_{i\neq j, i\in M, j\in M} x_i$. Now it is assumed that bidder j does not exist and the auction is among the remaining $L-1$ bidders. Let the optimal allocation be denoted by M^*. The revenue generated in this case is $\sum_{i\neq j, i\in M^*, j\notin M^*} x_i$. Therefore, minimum winning bid of bidder j must be at least greater than

$$X_j = \sum_{i\neq j, i\in M^* j\notin M^*} x_i - \sum_{i\neq j, i\in M j\in M^*} x_i \tag{11.4}$$

Thus, bidder j's request is granted if $x_j > X_j$ and not granted if $x_j < X_j$. If $x_j = X_j$, bidder j is indifferent between winning and loosing. Although (11.4) gives the winning bid for bidder j, it is not necessary that bidder j will be able to afford it. There exists a reservation price beyond which a bidder is simply a passive price taker in the market.

11.3.4.1 Bidder's Reservation Price

Bidder's reservation price, commonly known as the true evaluation price, is defined as the most a bidder would be willing to pay. When a service provider buys a spectrum from the spectrum broker, the service provider needs to sell that spectrum in the form of services to the end users who pay for these services. The revenue thus generated helps the provider to pay for the fixed (static) cost for the statically assigned spectrum and the extra spectrum that the provider might need from the CAB. If the total revenue generated from the users is R and R_{static} goes toward the fixed cost, then the difference, R_{dynamic}, is the maximum amount that the provider can afford for the extra spectrum from CAB, that is,

$$R_{\text{dynamic}} = R - R_{\text{static}} \tag{11.5}$$

Note that $R_{dynamic}$ is not the reservation price but is a prime factor that governs the reservation price. A provider can bid the reservation price or a price below the reservation price depending on the kind of auction used, that is, first price or second price.

THEOREM 11.1* In knapsack second price bidding auction, dominant strategy of the bidder is to bid their reservation price.

THEOREM 11.2 In knapsack first-price bidding auction, reservation price is the upper bidding threshold.

Comments: The result shown for the winner determining knapsack auction for dynamic spectrum auction corroborates with the result shown in other contexts in the economics literature, for example, in Clarkes tax [9]. Thus it is clear that bidders have no option of manipulating this spectrum auction.

11.3.5 Spectrum Auctioning

The main factors that need to be considered for designing efficient knapsack sealed-bid auction are the increased revenue generated by spectrum broker, increased spectrum usage, and increased probability of winning for bidders. Keeping these factors in mind, a proof of concept model using knapsack auction is presented here with the following:

- *Estimation of bid tuple*: The bid tuple q_i generated by bidder i consists of (1) the amount of spectrum requested, w_i and (2) the price the bidder is willing to pay, x_i. Each bidder has a reservation or evaluation price for the amount of spectrum requested and the bid is governed by this reservation price. As the auction is of type sealed-bid, the reservation price of one bidder is not known to others.
- *Bidders' strategies*: The second-price sealed-bid mechanism is followed. Note that the first-price bidding policy could also have been chosen; the only reason for choosing second price policy is that it has more properties than first price in terms of uncertainty [29]. In the second price bidding, the bidders are honest in revealing their reservation price (refer Theorem 11.1). After each round of auction, the only information bidders know is whether their request is granted, i.e., whether they belong to the optimal allocation M or not. It is also assumed that all the bidders are present for all the auction rounds;

---

* Detailed proofs and the significance of Theorems 11.1 and 11.2 can be found in [27].

Table 11.1 Simulation Parameters

Parameter Type	*Parameter Value*
Total amount of spectrum	100
Min. amount of spectrum requested	11
Max. amount of spectrum requested	50
Min. bid for per unit of spectrum	25

bidders take feedback from previous rounds and generate the bid tuple for next round.

- *Spectrum owner's strategies*: Spectrum owner tries to maximize the revenue generated from the bidders. At the beginning of each auction round, spectrum owner collects the bid tuples and executes the dynamic programming knapsack solver and determines the winner(s). The assigned band from CAB is taken back at the end of each lease period and reused for the next round.

For better insight into the results, the proposed sealed-bid knapsack auction is compared with the classical highest bid winner strategy under the second-price bidding policy, that is, bidder(s) with the winning bid(s) do not pay their winning bid but pay the second winning bid. In the classical strategy, a bidder with highest bid is always given preference over other bidders if spectrum amount requested by this bidder is less than or equal to the total available. Other parameters considered for the illustration purpose are shown in Table 11.1.

- *Revenue and spectrum usage*: Figure 11.4a and b compare revenue and spectrum usage for two strategies for each auction round. The number of bidders considered is ten. Note that both revenue and usage are low in the beginning and subsequently increases with rounds. This is because in the initial rounds, bidders are dubious and make low bids thus generating low revenue. With increase in rounds, potential bidders emerged as expected and raised the generated revenue. It is observed that the knapsack auction generates 10 percent to 15 percent more revenue compared to the classical model and also reaches steady state faster.
- *Collusion prevention*: The occurrence of collusion must be prevented in any good auction so that a subset of bidders cannot control the auction that might decrease the spectrum broker's revenue. Thus two cases are considered: (1) when bidders collude and (2) when bidders do not collude. The bidders who collude in pair are chosen randomly. In Figure 11.5a, the average revenue generated by the spectrum owner is shown with increase in the number of competing bidders both in the presence and the absence of collusion. Although at the beginning with less number of bidders, the presence of

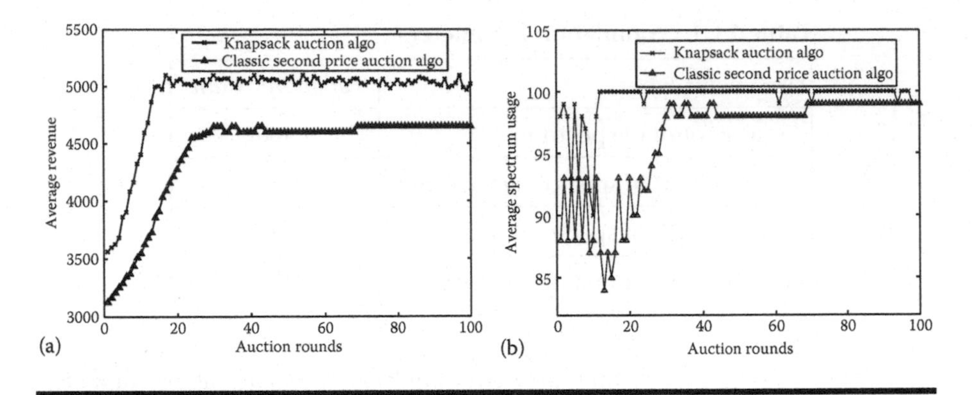

Figure 11.4 (a) Revenue maximization with auction rounds and (b) usage maximization with auction rounds.

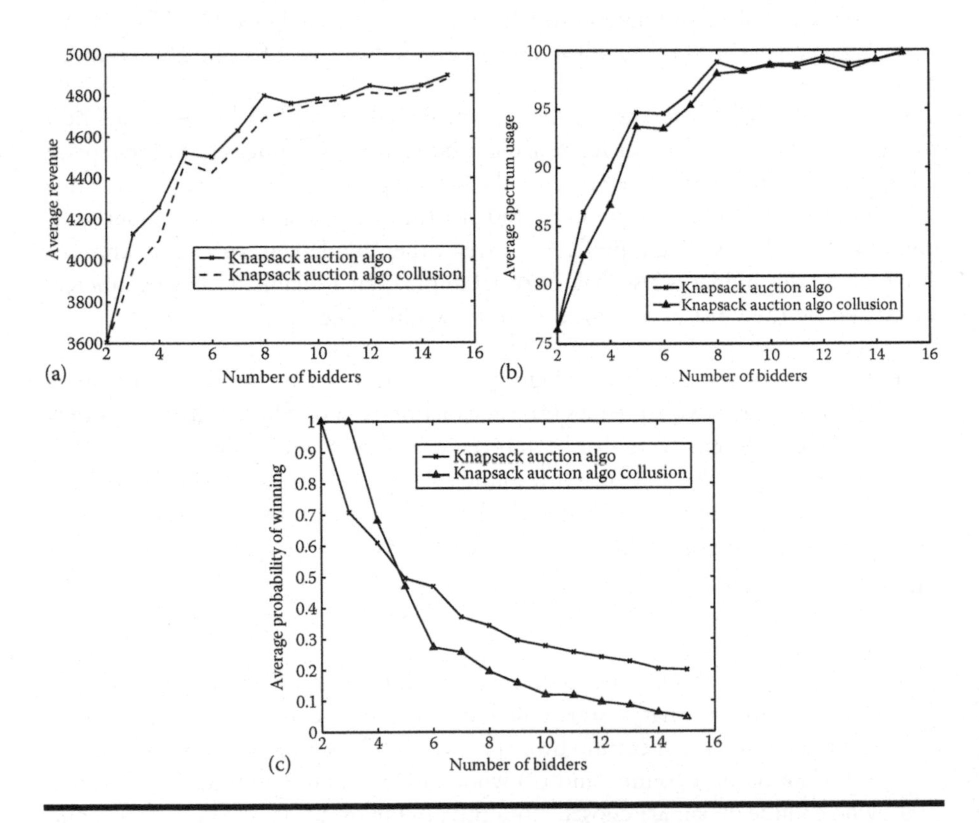

Figure 11.5 (a) Average revenue with and without collusion; (b) average usage with and without collusion; and (c) average probability of winning with and without collusion.

collusion reduces the average revenue slightly, but with increase in the number of bidders the effect due to collusion decreases. Thus with increase in the number of bidders, that is, with increase in (perfect) competition, revenue generated even in the presence of collusion reaches almost the same value as that of without collusion. Figure 11.5b similarly presents the usage of spectrum in the presence and absence of collusion. The most interesting result from bidders' perspective is shown in Figure 11.5c. The probabilities of winning, with or without collusion, increases with the number of bidders. When the number of bidders is low (less than or equal to four in our case) collusion provides a better probability of winning but as the number of bidders increases, the probability of winning with the help of collusion decreases, thus discouraging bidders to collude.

11.4 Service Provisioning Using Pricing

With the pricing interaction formally presented for spectrum owner–WSPs through auction model, the bottom tier of the hierarchy (WSPs–end users) needs to be investigated next. The market constitutes of wireless services that are sold by the WSPs and bought by the cognitive radio–enabled end users. Thus, the determination of the correct prices becomes a pivotal element for admission control and QoS provisioning. As far as the end users are concerned, with the advent of CR networks, it is anticipated that end users will have more freedom to move from long-term service provider agreements (e.g., one or two year contracts) to more opportunistic service models (session by session or perservice basis). The traditional concept of perservice static pricing is no longer considered from WSPs' perspective. In contrast, it is assumed that service providers will have more freedom in terms of choosing the price to be charged to the end users. In this new model, a WSP has freedom to change (increase or decrease) the price of a service depending on changing, load, revenue, etc. Thus, the most generic abstraction of "always greedy and profit seeking" model that exists between WSPs and end users is considered here.

11.4.1 Conflict Model

The users are the potential buyers who buy services from the WSPs. The selection of a WSP is done on a dynamic basis, that is, a user compares the offerings both in terms of QoS and price for a particular service. Once a service is completed, the user relinquishes the radio resources. As the prices offered are not static, the users do not have any information about other users' strategies, that is, the price the other users are willing to pay for a service. The users also do not know the demand for resources from other users. In such an incomplete information scenario, the benefit of a user depends not only on its own strategy but also on what others do. As it is assumed that every user is selfish (all trying to pay the least for the best service), the problem is modeled as a noncooperative game.

Service providers, very much like the users, also act in their self-interest. As a seller of the services, they determine the price for its services depending on the amount of spectrum acquired and the price paid. Similar to the noncooperative incomplete information game among the users, the service providers also do not have any exact information about other providers' strategies, such as price assigned for services, allotted resource, remaining resource, and existing load. The decisions need to be made based on this conflict model.

11.4.2 Decision Model

As a user (potential buyer), the decision problem is to select the best service provider for the session requested. Now the question arises, how to select the best service provider or rather what criteria determines the best. One answer to this question might be to select the service provider that offers the lowest price. In that case, it might be assumed that for any service that a user needs, he sends out a request to each service provider. The service providers respond by advertising a price (either price for the whole service or price per unit of resource)—the lowest of which is selected by the user. However, the lowest price does not necessarily consider the QoS that is expected. It might so happen that the price advertised by a provider is low but QoS is not met thus making the service useless for the user. The QoS perceived by a user in a network will depend on the traffic load conditions. Therefore, the load must be considered in the decision making process and a cost benefit analysis needs to be made to find the best service provider.

As a service provider (or potential seller), the decision problem is to advertise a price for a service without knowing what exact prices are being advertised by its competitors. The optimization is to find a price such that the provider is able to sustain profit in spite of offering a low price, that is, is there any price threshold to reach Nash equilibrium [20]? For finding the existence of Nash equilibrium, the preference of the providers and users—usually given by their utility functions—need to be defined first.

11.4.3 Utility Function

An utility function is a mathematical characterization that represents the benefits and costs incurred. Here, the utility functions are defined for both WSP and users. L service providers are assumed that cater to a common pool of $\mathcal{N}$ users. Let the price per unit of resource advertised by the service provider j, $1 \leq j \leq L$, at time t be $p_j(t)$. Let $b_{ij}(t)$ be the resource consumed by user i, $1 \leq i \leq \mathcal{N}$ who is served by provider j. Further assumption is that the total resource (capacity) of provider j is C_j. Then the empirical utility obtained by a user i under the provider j can be given by $a_{ij} \log(1 + b_{ij}(t))$ [31], where the coefficient a_{ij} is a positive parameter that indicates the relative importance of empirical benefit and acts as a weightage factor. Note that any other form for the empirical utility could have been chosen that increases with $b_{ij}(t)$. But the log function is chosen because the empirical benefit increases quickly from zero as the total throughput increases from zero and then increases slowly. This reflects the intuition that the initial increase in the perceived throughput is more important to a user. Moreover, log function is analytically convenient, increasing, strictly concave, and continuously differentiable.

Next, the cost components are considered. The first cost component is the direct cost paid to the provider for obtaining $b_{ij}(t)$ amount of resource. If $p_j(t)$ is the price per unit of resource, then the direct cost paid to the jth provider is given by $p_j(t)b_{ij}(t)$. This direct cost component decreases user i's empirical utility. The second cost component incurred by the user is the perceived QoS, one of the manifestations of which is the queuing delay that again depends on the resources consumed by the other users. It is assumed the queuing process to be $M/M/1$ at the links. Thus, the delay cost component can be written as

$$\begin{cases} \xi\left(\dfrac{1}{C_j - \sum_i^{N_j} b_{ij}(t)}\right) & \text{if} \sum_i^{N_j} b_{ij}(t) < C_j \\ \infty & \text{if} \sum_i^{N_j} b_{ij}(t) \geq C_j \end{cases} \tag{11.6}$$

where

N_j is the number of users currently served by provider j

$\psi(\cdot)$ is a mapping cost function of delay

Combining the utility and all the cost components, the net utility can be written as

$$U_{ij}(t) = a_{ij}\log(1 + b_{ij}(t)) - p_j(t)b_{ij}(t) - \xi\left(\frac{1}{C_j - \sum_i^{N_j} b_{ij}(t)}\right) \tag{11.7}$$

The utility of service provider j at time t is given by

$$V_j(t) = p_j(t)\sum_i^{N_j} b_{ij}(t) - K_j \tag{11.8}$$

where K_j is the cost incurred to provider j for maintaining network resources. For the sake of simplicity, this cost is assumed to be constant regardless of the amount of resources handled by provider j.

11.4.4 Price Threshold

With users' and WSPs' benefits expressed using utility functions, it is interesting to find out if there is any strategy constraints from the users' and providers' perspective and still reach the equilibrium. Consider user i has a certain resource demand and wants to connect to a provider at time t. All the providers advertise their price per unit of resource amount and the existing load. As user i wants to maximize his net utility (potential benefit minus cost incurred), he computes the resource vector that would maximize utilities from all the providers and the corresponding maximized utility vector. User i would then connect to provider j if $U_{ij}(t)$ gives the maximum value in the maximized utility vector, $[U_{i1}(t) \quad U_{i2}(t) \quad \cdots \quad U_{iL}(t)]$ and $b_{ij}(t)$ is the requested resource amount from the optimal resource vector, $[b_{i1}(t) \quad b_{i2}(t) \quad \cdots \quad b_{iL}(t)]$. This $b_{ij}(t)$ is the optimal amount of resource to be consumed by user i from provider j for advertised price $p_j(t)$. It then becomes interesting to investigate if there exists any optimal resource amount for the users and any pricing bound from the providers that will maximize respective utilities and still can reach the Nash equilibrium. To do so it is needed to be found out first whether the net utility given in (11.7) can be maximized with respect to the resource amount. If so, then a unique maximization point exists for $U_{ij}(t)$ with respect to $b_{ij}(t)$. Differentiating (11.7) with respect to $b_{ij}(t)$,

$$U'_{ij}(t) = \frac{a_{ij}}{1 + b_{ij}(t)} - p_j(t) - \xi'\left(\frac{1}{C_j - \sum_i^{N_j} b_{ij}(t)}\right). \tag{11.9}$$

It can be shown that $U''_{ij}(t) < 0$ if $\xi''\left(\frac{1}{C_j - \sum_i^{N_j} b_{ij}(t)}\right) > 0$. Then, the maximization point can be obtained by equating $U'_{ij}(t)$ in (11.9) to zero. If the users follow this strategy of demanding this optimal amount of resources for a certain advertised price, then Nash equilibrium will be achieved, where changing this strategy unilaterally by a single user will always give him the utility lesser than the maximum possible value.

From the reverse point of view, it is also clear that there exists a maximum threshold for the price $p_j(t)$ that is in the region bounded by $\frac{a_{ij}}{1+b_{ij}(t)} - \xi'\left(\frac{1}{C_j - \sum_i^{N_j} b_{ij}(t)}\right)$. For a certain demand $b_{ij}(t)$ from a user, provider j must obey this pricing constraint to maximize net utility for user i. Otherwise the user will not connect to this provider, thus leading to a loss of user and thus loss of profit for the provider. A more compact form of pricing constraint, with the help of detailed algebraic analysis and identity equations, can be found out, where it is shown that the pricing constraint $p_j(t)$ is upper bounded by,

$$\frac{a_{Ij}}{m_{Ij}(t)} - \xi'\left(\frac{1}{C_j + N_j - m_{Ij}(t)}\right) \tag{11.10}$$

where $m_{Ij}(t) = \sum_i^{N_j} m_{ij}(t) = \sum_i^{N_j}(1 + b_{ij}(t))$ and $a_{Ij} = \sum_i^{N_j} a_{ij}$. This pricing upper bound helps the provider to reach the Nash equilibrium. If all of the other providers and users keep their strategies unchanged, and a provider changes its strategy unilaterally and decides not to maintain its pricing upper bound, then that provider will not be able to maximize its users' utility and thus users will not connect to this provider. Thus, unilaterally changing the strategy will not increase the profit of the provider and Nash equilibrium will not be achieved.

11.5 Estimating the Demand for Bandwidth

The amount of extra (dynamic) spectrum that a provider needs depends on the demand for services by the users it supports. Therefore it is essential to estimate the resources consumed by the users and the price that is recovered from them. These estimates will help a provider determine the tuple $q_i = \{w_i, x_i\}$. The objective is to maximize provider's net utility, $V_j(t)$, subject to the constraint given by (11.10). Replacing $\sum_i^{N_j} b_{ij}(t)$ by $m_{Ij}(t) - N_j$, $V_j(t)$ can be obtained as

$$V_j(t) = \left(\frac{a_{Ij}}{m_{Ij}(t)} - \xi'\left(\frac{1}{C_j + N_j - m_{Ij}(t)}\right)\right)(m_{Ij}(t) - N_j) - K_j \tag{11.11}$$

Solving for first and second differentiation of $V_j(t)$ with respect to $m_{Ij}(t)$, it is found that $V_j''(t) < 0$; implying the constraint on the price will maximize the utility for the providers. Equating first differentiation of $V_j(t)$ to 0 gives the optimal value of $m_{Ij}(t)$. Let the solution of Equation 11.11 to be $m_{Ij(\text{opt})}(t)$. The optimal price that will maximize provider j's utility can be then obtained by substituting $m_{Ij(\text{opt})}(t)$ in (11.10). Thus, the optimal price is obtained as

$$p_{j(\text{opt})}(t) = \frac{a_{Ij}}{m_{Ij(\text{opt})}(t)} - \xi'\left(\frac{1}{C_j + N_j - m_{Ij(\text{opt})}(t)}\right) \tag{11.12}$$

To have a better insight into the analysis, a simple closed form of $\xi\left(\frac{1}{C_j+N_j-m_{Ij}(t)}\right)$ is assumed as $\frac{1}{(C_j+N_j-m_{Ij}(t))}$. Simplifying and equating first differentiation of $V_j(t)$ to 0 and assuming $N_j = C_j$, $m_{Ij(\text{opt})}(t)$ can be obtained in closed form as

$$m_{Ij(\text{opt})}(t) = \frac{2C_j\theta}{1+\theta} \quad \text{where } \theta = \sqrt[3]{a_{Ij}C_j} \tag{11.13}$$

Using the optimal value of $m_{Ij}(t)$, the optimal value of $p_j(t)$ is obtained as

$$p_{j(\text{opt})}(t) = \frac{a_{Ij}}{2C_j}\left(1 + \frac{1}{\theta}\right) - \left(\frac{1+\theta}{2C_j}\right)^2 \tag{11.14}$$

Thus, it is found that the providers can achieve Nash equilibrium under the given pricing constraint and at the same time they can maximize their utility if the price is set as given by (11.14). Using the relations between $m_{Ij}(t)$, $m_{ij}(t)$, and $b_{ij}(t)$ as mentioned earlier, the optimal resource consumed by user i under provider j can be given by

$$b_{ij(\text{opt})}(t) = \begin{cases} \frac{a_{ij}}{a_{Ij}}\left(\frac{2C_j}{1+\frac{1}{\theta}}\right) - 1 & \text{if } b_{ij(\text{opt})}(t) > 0 \\ 0 & \text{if } b_{ij(\text{opt})}(t) \leq 0 \end{cases} \tag{11.15}$$

11.6 Secrecy Capacity with Pricing

So far, it has been shown how pricing can be used as a means for efficient usage of resources. However, some users belonging to one WSP (e.g., WSP 1) could turn into eavesdroppers and obtain vital information from the network owned by WSP 2. Ciphering is one way to counter this, but there are many messages that are sent in clear text (i.e., without being ciphered). The paradigm of key less security [34] can be used as a method to counter such eavesdropping. Secrecy capacity provides a measure of the maximum rate of transmission upto which one can obtain key less security.

In this section, it is described how pricing can be used to limit the transmit power of the user terminals and yet maximize the secrecy capacity as discussed in Section 11.1.2. Consider a multi user system with M transmitters/sources and M receivers/destinations. Transmitter i transmits at power P_i. The objective is to obtain the optimal value of P_i that maximizes the secrecy capacity of the channel between the ith transmitter–receiver pair in the presence of an eavesdropper. It is also of interest to determine the optimal P_i's that maximize the total secrecy capacity of the system. With no knowledge of the location of the eavesdropper, the secrecy capacity maximization problem can be viewed as a Shannon capacity maximization problem [3], and can be formulated as

$$\max_{\mathbf{p}} C_i = W \log_2(1 + x_i) \quad \forall i, \tag{11.16}$$

subject to the constraints

$$0 \leq P_i \leq P_{\max} \quad \forall i, \tag{11.17}$$

where

x_i is the signal-to-interference ratio (SIR) obtained by the ith receiver, $\mathbf{p} = [P_1 \quad P_2 \quad P_3 \quad \ldots \quad P_M]$

$P_{\max}$ is the maximum power that can be transmitted by any transmitter

W is the system bandwidth

C_i is the Shannon capacity of the channel between the ith transmitter and receiver

Consider an additive white Gaussian noise (AWGN) channel with noise power spectral density N_0. Let the channel gain from transmitter i to receiver j be h_{ij}. The channel gain matrix, $\mathbf{H}$, can then be written as $\mathbf{H} = \left[h_{ij}\right]_{\substack{1 \le i \le M \\ 1 \le j \le M}}$. Let transmitter i transmit at rate r_i. Let receiver i have a gain G_i. This gain could be obtained by error-control coding or by spectrum spreading or by any other means. The gain G_i enables the reduction of the BER at the receiver i. If the gain is due to spectrum spreading, then $G_i = W/r_i$. The SIR obtained by the ith receiver, x_i, is given by

$$x_i = \frac{P_i h_{ii} G_i}{\sum_{j \neq i} P_j h_{ji} + N_0 W}. \tag{11.18}$$

It is observed from (11.16) and (11.18) that the capacity C_i depends on the vector $\mathbf{p}$ and not just P_i. The capacity maximization problem can therefore be formulated as an M-person non cooperative game in which the players are the transmitters, the strategy for player (transmitter) i is the transmit power P_i and utility function for transmitter i is the capacity C_i.

THEOREM 11.3* A power vector $\mathbf{p} = \left[P_1 \quad P_2 \quad P_3 \quad \cdots \quad P_M\right]$ is Pareto optimal if and only if $P_i = P_{\max}$ for some i.

THEOREM 11.4 The Nash equilibrium of the Shannon capacity maximization occurs when $P_i = P_{\max}$, $\forall\, i$. This equilibrium point is also Pareto optimal.

Thus, the optimal power allocation strategy would be $P_i = P_{\max}\ \forall\, i$, thus making the system inefficient in terms of energy consumption as all transmitters transmit at maximum power all the time.

Pricing can be introduced to penalize transmitters transmitting at higher powers. However, it is noted from (11.16) that the actual quantity affecting the capacity is the SIR, x_i and hence an alternative pricing strategy could be to penalize those transmitters whose corresponding receivers obtain larger SIR. Let $f_i(x_i)$ denote the price posed on transmitter i when the SIR experienced at receiver i is x_i. The pricing, $f_i(x_i)$, with respect to the SIR, x_i, could be a linear function of x_i or a nonlinear function of x_i. The proposed nonlinear pricing [3] is

$$f_i(x_i) = \lambda \frac{r_i x_i}{x_i + G_i}. \tag{11.19}$$

---

* Detailed proofs and the significance of Theorems 11.3 through 11.5 can be found in [3].

The linear pricing function is

$$f_i(x_i) = \lambda r_i x_i. \tag{11.20}$$

In (11.19) and (11.20), λ is the "pricing parameter." The capacity maximization problem with pricing can then be formulated as

$$\max_{\mathbf{p}} \hat{u}_i(P_i) = C_i(x_i) - f_i(x_i) \quad \forall i, \tag{11.21}$$

subject to the constraints in (11.17), which, in turn, can be rewritten as

$$\max_{x_i} \hat{u}_i(x_i) = C_i(x_i) - \lambda \frac{r_i x_i}{x_i + G_i} \quad \forall i. \tag{11.22}$$

This optimization problem can be solved as M independent optimization problems in each x_i. Let $\mathbf{x}^* = \begin{bmatrix} x_1^* & x_2^* & x_3^* & \cdots & x_M^* \end{bmatrix}$ be the vector of SIRs that maximize the M objective functions in (11.22). The corresponding power vector $\mathbf{p}^* = \begin{bmatrix} P_1^* & P_2^* & P_3^* & \cdots & P_M^* \end{bmatrix}$ can be obtained from the following matrix equation.

$$\mathbf{p}^* = N_0 W \left(\mathbf{I}_M - \mathbf{D}_1^{-1} \mathbf{A} \right)^{-1} \mathbf{D}_1^{-1} \mathbf{D}_2^{-1} \mathbf{1}, \tag{11.23}$$

where

$\mathbf{1}$ is the column vector of length M with all entries being unity

$\mathbf{I}_M$ is the $M \times M$ identity matrix

$\mathbf{A}$, $\mathbf{D}_1$, and $\mathbf{D}_2$ are given by

$$\mathbf{A} = \begin{bmatrix} 0 & \frac{h_{21}}{h_{11}} & \frac{h_{31}}{h_{11}} & \cdots & \frac{h_{M1}}{h_{11}} \\ \frac{h_{12}}{h_{22}} & 0 & \frac{h_{32}}{h_{22}} & \cdots & \frac{h_{M2}}{h_{22}} \\ \frac{h_{13}}{h_{33}} & \frac{h_{23}}{h_{33}} & 0 & \cdots & \frac{h_{M3}}{h_{33}} \\ \vdots & \vdots & \vdots & \ddots & \vdots \\ \frac{h_{1M}}{h_{MM}} & \frac{h_{2M}}{h_{MM}} & \frac{h_{3M}}{h_{MM}} & \cdots & 0 \end{bmatrix}, \tag{11.24}$$

$\mathbf{D}_1 = \operatorname{diag}\left(\frac{G_1}{x_1} \quad \frac{G_2}{x_2} \quad \frac{G_3}{x_3} \quad \cdots \quad \frac{G_M}{x_M} \right)$, and $\mathbf{D}_2 = \operatorname{diag}\left(h_{11} \quad h_{22} \quad h_{33} \quad \cdots \quad h_{MM} \right)$.

THEOREM 11.5 $\exists\, \lambda_{\min}, \lambda_{\max}$ such that $\forall\, \lambda \in (\lambda_{\min}, \lambda_{\max})$, the optimization problem in (11.21) subject to constraints (11.17) has a feasible solution.

Theorem 11.5 is applicable both for the case when the pricing is nonlinear [i.e., $f_i(x_i)$ given by (11.19)] and for the case when the pricing is linear [i.e., $f_i(x_i)$ given by (11.20)].

Let there be an eavesdropper in the system whose channel gain from the ith transmitter is h_{ie}. Hence, it is possible to define a vector $\mathbf{h}_e = [h_{ie}]_{1 \leq i \leq M}$. The SIR of the signal received by the eavesdropper from transmitter i, x_{ie}, can then be defined as

$$x_{ie} = \frac{P_i h_{ie} G_i}{\sum_{j \neq i} P_j h_{je} + N_0 W}. \tag{11.25}$$

The secrecy capacity of transmit receive pair i, S_i, is given by

$$S_i = \left[W \log_2 \left(\frac{1 + x_i}{1 + x_{ie}} \right) \right]^+ . \tag{11.26}$$

When the problem of allocation of powers is translated to that of allocation of optimal SIR's to each receiver, then maximizing S_i in (11.26) is the same as that of allocating SIRs to each receiver i to maximize C_i. This is because, when allocating SIR to each receiver i, the x_i is independent of x_{ie}. Similarly when pricing functions specified in (11.19) or (11.20) are applied, the allocation of SIRs to maximize $S_i - f_i(x_i)$ is the same as that of the allocation of SIRs to maximize the function $C_i - f_i(x_i)$. Hence, Theorem 11.5 is applicable to maximize the secrecy capacity for all the transmitter–receiver pairs. The sum secrecy capacity of the system, S, is defined as

$$S = \sum_{i=1}^{M} S_i. \tag{11.27}$$

Maximizing S or $S - \sum_i f_i(x_i)$ with respect to $\mathbf{x}$ is same as maximizing S_i or $S_i - f_i(x_i)$ as the objective functions S and $S - \sum_i f_i(x_i)$ can be written as a sum of M independent objective functions each depending only on one variable. Hence, maximizing the sum secrecy capacity is same as maximizing the individual secrecy capacity of each transmitter–receiver pair.

Secondary users in a cognitive radio network are admitted based on the availability of bandwidth and based on the interference caused to a primary receiver. In a cognitive radio network, typically, interference temperature constraints are used to limit the interference on primary users [35]. However, the FCC later abandoned this metric [41]. The analysis described to maximize the secrecy capacity can be used as a means to admit users by imposing a constraint on the minimum secrecy capacity experienced at each receiver. The secondary transmitter–receiver pairs that do not satisfy the secrecy capacity constraints may not be admitted into the system.

For illustration purpose, consider a cognitive radio network with primary and secondary transmitters and receivers as shown in Figure 11.6. Consider a system

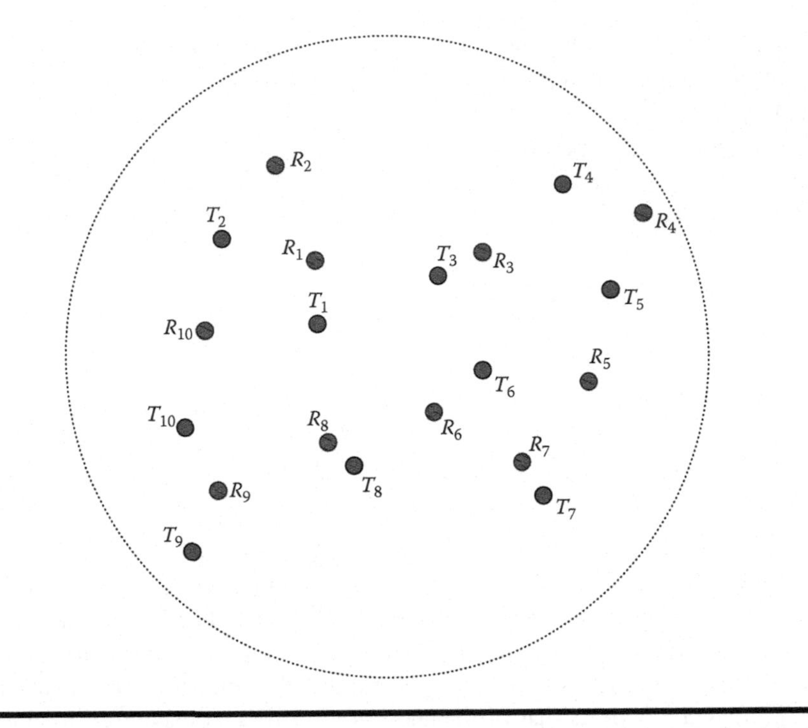

Figure 11.6 A typical scenario with transmitters and receivers. Nodes labeled $T_1, T_2, \ldots$ are the transmitters and nodes labeled $R_1, R_2, \ldots$ are the receivers.

with $M = 10$ transmitter–receiver pairs and a bandwidth of 5 MHz. Figure 11.7a shows the secrecy capacity obtained by each transmitter–receiver pair. Note that the pricing decreases the secrecy capacity for some transmitter–receiver pairs while improving it for other pairs. In particular, some users who obtain zero secrecy capacity in the absence of pricing obtain positive secrecy capacity when the nonlinear pricing is posed. This is a useful result particularly when dealing with "bottleneck" transmitter–receiver pairs, that is, transmitter–receiver pairs with the least secrecy capacity. If the transmitter–receiver pairs with zero secrecy capacity correspond to secondary users, then a cognitive radio system that admits these secondary users would suffer from poor secrecy whereas, in the presence of pricing, the overall secrecy could improve as the secrecy of the bottleneck transmitter–receiver pair improves. However, it is noted that the linear pricing does not improve the secrecy capacity. However, it cannot be concluded that the linear pricing is ineffective as shown in Figure 11.7b for a system with bandwidth 20 MHz. In this system, the system with linear pricing provides larger secrecy capacity than the one with nonlinear pricing, unlike what was observed for the 5 MHz bandwidth system. This is because, when gains are smaller, the resultant optimum SIR in the system with nonlinear pricing is larger than that in the system with linear pricing. This results in larger transmit

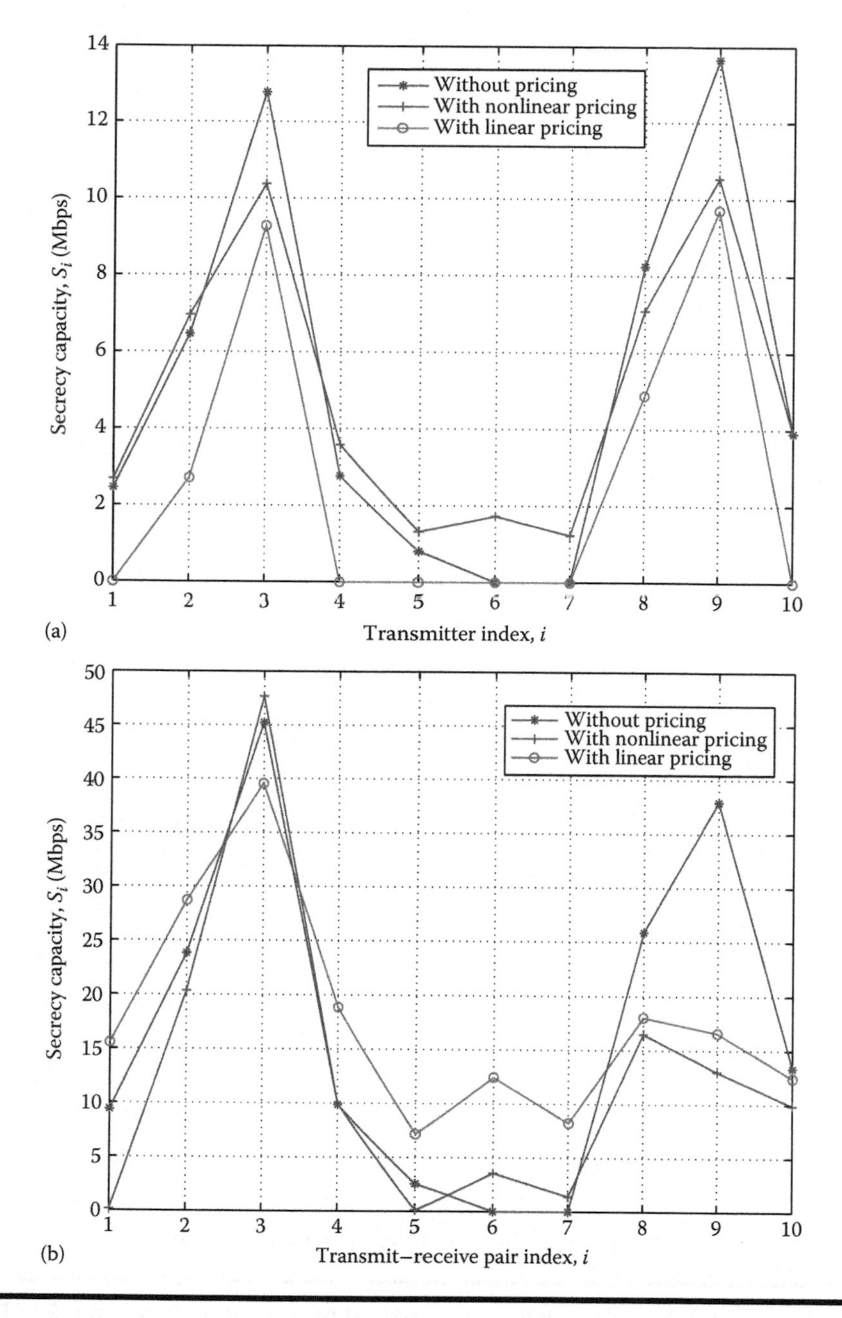

Figure 11.7 (a) Secrecy capacity of users in a system with 5 MHz bandwidth and (b) secrecy capacity of users in a system with 20 MHz bandwidth.

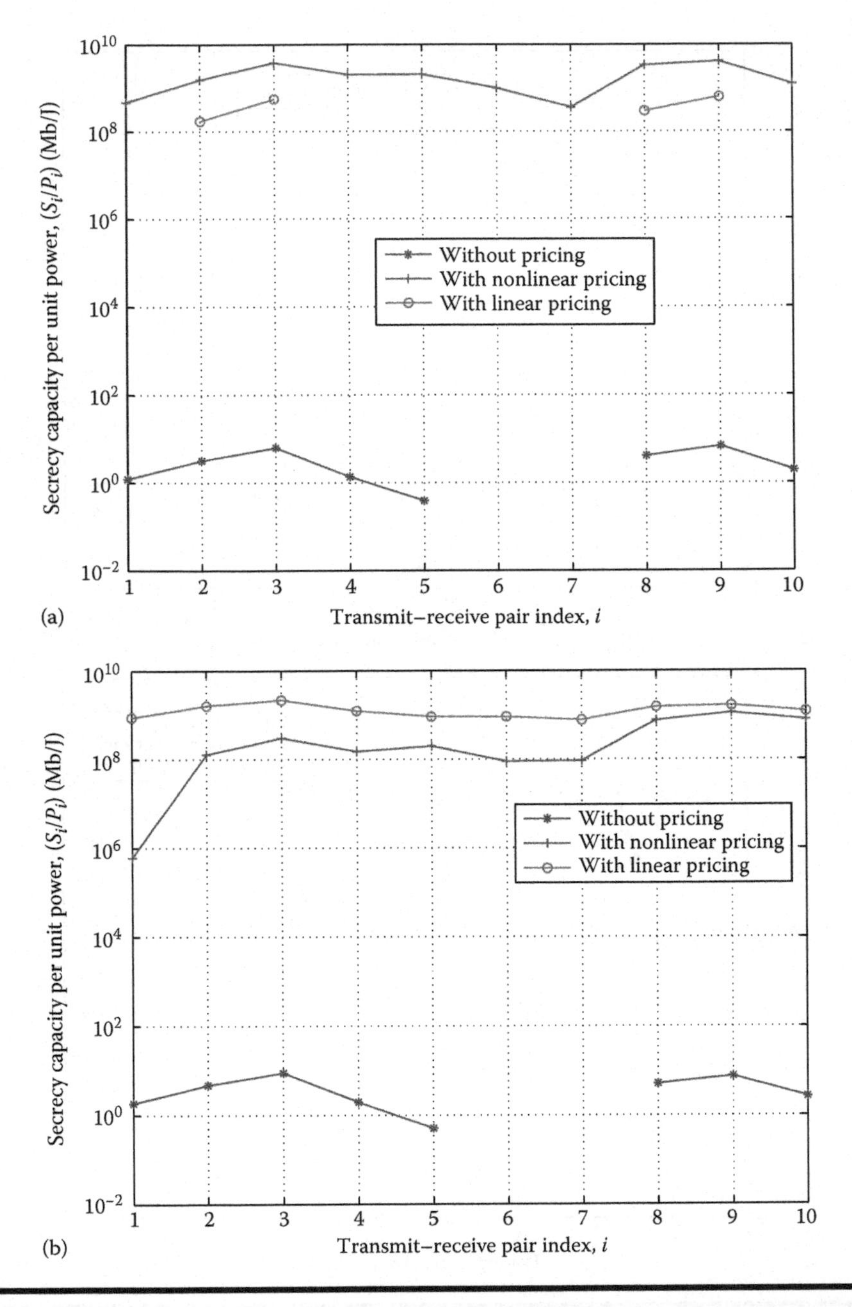

Figure 11.8 (a) Secrecy capacity per unit power in a system with 5 MHz bandwidth and (b) secrecy capacity per unit power in a system with 20 MHz bandwidth.

powers for the system with nonlinear pricing and hence results in better secrecy capacity.

Figure 11.8a and b presents the secrecy capacity per unit power for the systems with 5 MHz bandwidth and 20 MHz bandwidth, respectively.* It is observed that although pricing reduces the secrecy capacity for some users, it improves the secrecy capacity per unit power for all the users. This is because, both the nonlinear and the linear pricing function limit the transmit powers of all the transmitters. The transmit power without pricing is $P_{max} = 2$ W for all the transmitters and that with pricing is of the order of mW. Pricing thus provides an energy-efficient mechanism to improve the secrecy capacity.

11.7 Conclusions

Dynamic spectrum access coupled with cognitive radio–enabled end users will engender a flexible and competitive market for trading spectrum and wireless services. With the help of cognitive radio networks, the accessibility to different wireless networks serviced by different wireless service providers will make this dynamic trading at a finer granularity. An economic framework based on dynamic pricing-driven auction and game theory is explained that captures the interaction among spectrum owner, service providers, and end users in a multi-provider setting. A sealed-bid knapsack auction method is presented that dynamically allocates spectrum from coordinated access band and at the same time maximizes the revenue generated from WSPs. It is also shown how pricing can be used as a means to limit the usage of resources and the transmit power of the user terminals and yet maximize the utility and secrecy capacity in cognitive radio networks.

References

1. B. Aazhang, J. Lilleberg, and G. Middleton, Spectrum sharing in a cellular system, *IEEE 8th International Symposium on Spread Spectrum Techniques and Applications*, Sydney, New South Wales, Australia, 2004, pp. 355–359.
2. I. F. Akyildiz, W.-Y. Lee, M. C. Vuran, and S. Mohanty, NeXt generation/dynamic spectrum access/cognitive radio wireless networks: A survey, *Computer Networks Journal*, 50: 2127–2159, September 2006.
3. S. Anand and R. Chandramouli, Secrecy capacity of multi-terminal networks with pricing, Technical Report, Newark, NJ, 2008, http://www.ece.stevens-tech.edu/~mouli/res.html.
4. K. Back and J. Zender, Auctions of divisible goods: On the rationale for the treasury, *Review of Financial Studies*, 6(4): 733–764, Winter 1993.

* The discontinuities in Figure 11.8a and b correspond to users obtaining zero secrecy capacity.

5. R. Bapna, P. Goes, and A. Gupta, Simulating online Yankee auctions to optimize sellers revenue, System Sciences, 2001. *Proceedings of the 34th Annual Hawaii International Conference*, Maui, HI, January 3–6, 2001, p. 10.
6. M. Buddhikot, P. Kolodzy, S. Miller, K. Ryan, and J. Evans, DIMSUMnet: New directions in wireless networking using coordinated dynamic spectrum access, *IEEE International Symposium on a World of Wireless, Mobile and Multimedia Networks* (*WoWMoM*), Taormina, Italy, 2005, pp. 78–85.
7. M. Buddhikot and K. Ryan, Spectrum management in coordinated dynamic spectrum access based cellular networks, *Proceedings of the First IEEE International Symposium on New Directions in Dynamic Spectrum Access Networks*, Baltimore, MD, 2005, pp. 299–307.
8. R. Chakravorty, S. Banerjee, S. Agarwal, and I. Pratt, MoB: A mobile bazaar for wide-area wireless services, *Proceedings of the 11th Annual International Conference on Mobile Computing and Networking* (*MobiCom*), Cologne, Germany, 2005, pp. 228–242.
9. E. H. Clarke, Multipart pricing of public goods, *Public Choice*, 11(1): 17–33, 1971.
10. G. Illing and U. Kluh, *Spectrum Auctions and Competition in Telecommunications*, The MIT Press, London, U.K., 2003.
11. N. Jin and S. Jordan, On the feasibility of dynamic congestion-based pricing in differentiated services networks, *IEEE/ACM Transactions on Networking*, 16(5): 1001–1014, 2008.
12. F. P. Kelly, A. K. Maulluo, and D. K. H. Tan, Rate control in communication networks: Shadow prices, proportional fairness and stability, *Journal of the Operational Research Society*, 49(3): 237–252, 1998.
13. P. Klemperer, Auction theory: A guide to the literature, *Journal of Economic Surveys*, 13(3): 227–286, July 1999.
14. V. Krishna, *Auction Theory*, Academic Press, San Diego, CA, 2002.
15. H. Lin, M. Chatterjee, S. K. Das, and K. Basu, ARC: An integrated admission and rate control framework for CDMA data networks based on non-cooperative games, *Proceedings of the 9th Annual International Conference on Mobile Computing and Networking* (*MobiCom*), San Diego, CA, 2003, pp. 326–338.
16. I. K. MacKic-Mason and H. R. Varian, Pricing congestible network resources, *IEEE Journal on Selected Areas in Communications*, 13 (7): 1141–1149, 1995.
17. P. Maille and B. Tuffin, Multibid auctions for bandwidth allocation in communication networks, *INFOCOM*, Hong Kong, China, 2004, pp. 54–65.
18. P. Marbach, Analysis of a static pricing scheme for priority services, *IEEE/ACM Transactions on Networking*, 12(2): 312–325, April 2004.
19. J. Mitola and G. Q. Maguire Jr., Cognitive radio: Making software radios more personal, *IEEE Personal Communications*, 6(4): 13–18, August 1999.
20. J. F. Nash, Equilibrium points in *N*-person games, *Proceedings of the National Academy of Sciences of the United States of America*, 36: 48–49, 1950.
21. I. Ch. Paschalidis and L. Yong, Pricing in multiservice loss networks: Static pricing, asymptotic optimality and demand substitution effects, *IEEE/ACM Transactions on Networking*, 10(3): 125–438, June 2002.

22. D. L. Reiley, Auctions on the Internet: What's being auctioned, and how?, *Journal of Industrial Economics*, 48(3): 227–252, September 2000.
23. A.-R. Sadeghi, M. Schunter, and S. Steinbrecher, Private auctions with multiple rounds and multiple items, *Proceedings of the 13th International Workshop on Database and Expert Systems Applications*, Washington, DC, 2002, pp. 423–427.
24. P. J. Seok and K. S. Rye, Developing MVNO market scenarios and strategies through a scenario planning approach, *7th International Conference on Advanced Communication Technology* (*ICACT*), Dublin, Ireland, Vol. 1, 2005, pp. 137–142.
25. N. Semret, Market mechanisms for network resource sharing, PhD dissertation, Columbia University, New York, 1999.
26. S. Sengupta and M. Chatterjee, Synchronous and asynchronous auction models for dynamic spectrum allocation, *International Conference on Distributed Computing and Networking* (*ICDCN*), IIT Guwahati, India, 2006, pp. 558–569.
27. S. Sengupta and M. Chatterjee, Designing auction mechanisms for dynamic spectrum access, *ACM/Springer Mobile Networks and Applications* (*MONET*), Special issue on *Cognitive Radio Oriented Wireless Networks and Communications*, 13(5): 498–515, 2008.
28. H. R. Varian, *Microeconomic Analysis*, 3rd edn., W.W. Norton and Company, Inc., New York, 1992.
29. W. Vickrey, Couterspeculation, auctions, and competitive sealed tenders, *Journal of Finance*, 16(1): 8–37, March 1961.
30. H. Wang, H. Xie, L. Qiu, A. Silberschatz, and Y. R. Yang, Optimal ISP subscription for Internet multihoming: Algorithm design and implication analysis, *Proceedings of IEEE INFOCOM*, Seattle, WA, Vol. 4, 2005, pp. 2360–2371.
31. W. Wang and B. Li, Market-driven bandwidth allocation in selfish overlay networks, *Proceedings of IEEE INFOCOM*, Miami, FL, Vol. 4, 2005, pp. 2578–2589.
32. W. Webb and P. Marks, Pricing the ether [radio spectrum pricing], *IEE Review*, 42(2): 57–60, 1996.
33. R. Wilson, Auctions of shares, *Quarterly Journal of Economics*, 93(4): 675–689, 1979.
34. A. D. Wyner, The wire-tap channel, *Bell Systems Technical Journal*, 54(8): 1355–1387, 1995.
35. Y. Xing, C. N. Mathur, M. A. Haleem, R. Chandramouli, and K. P. Subbalakshmi, Priority based dynamic spectrum access with QoS and interference temperature constraints, *IEEE International Conference on Communications* (*ICC'2006*), Istanbul, Turkey, June, 2006.
36. http://www.3gpp.org.
37. http://en.wikipedia.org/wiki/MVNO.
38. http://www.ebay.com/.
39. ET Docket-322, Notice of proposed rule making and order FCC, 2003.
40. http://www.sharedspectrum.com/inc/content/measurements/nsf/NYC_report.pdf.
41. http://www.paulweiss.com/PublicationAttachment/358fe5ec-442b-4528-982c-46a668184413/CTD5-11-07W.pdf.
42. http://www.wimaxforum.org.

APPLICATIONS AND SYSTEMS

Chapter 12

Cognitive Radio for Pervasive Healthcare

Phond Phunchongharn, Ekram Hossain, and Sergio Camorlinga

Contents

This chapter provides a big picture of the applications of cognitive radio technology in pervasive healthcare environments. First, a summary of the different wireless technologies and their applications in healthcare environment are discussed. Then, the key technical requirements in electronic healthcare applications are outlined, and the different types of electronic medical devices used in a healthcare environment are described. The impact of wireless transmissions on both active and passive medical devices and the corresponding electromagnetic interference (EMI) and electromagnetic compatibility (EMC) issues are discussed. It is shown how a cognitive radio can avoid interference with medical devices in different healthcare scenarios. Then, the requirements, constraints, and open issues in the development of cognitive radio technology for healthcare applications and services are discussed. Finally, to avoid EMI to medical devices, an EMI-aware cognitive radio transmission scheme is proposed for wireless access for eHealth applications.

12.1 Introduction

eHealth (electronic health) is an emerging medical service paradigm that employs information processing and communications to enhance traditional medical services. Among key enabling technologies, wireless communications provide the medical services with mobility and service availability. Examples of eHealth applications that utilize wireless technology include hospital information systems, remote patient monitoring, and telemedicine [1,2]. Despite its beneficial applications, an extensive use of wireless communications can interfere with electromagnetic interference (EMI)–sensitive medical devices. The interference can cause malfunctioning of those medical devices and potentially harm the patients who are using those medical devices.

In this chapter, we first overview and give key motivations for the introduction of wireless communications in hospitals in Section 12.2. In Section 12.3, we provide a survey on eHealth applications and their requirements and then a few examples of medical devices and their classifications. In Section 12.4, we discuss two of the most important issues of wireless eHealth applications: EMI and EMC. In Section 12.5, we present a cognitive radio system for eHealth applications. In Section 12.6, we discuss the problems and challenges in developing cognitive radio technology for eHealth applications and services in a hospital environment. Finally, we propose an EMI-aware cognitive radio transmission scheme to deal with EMI problem in healthcare environment.

12.2 Potential Wireless Devices and Technologies for Pervasive Healthcare

12.2.1 Wireless Devices for Pervasive Healthcare

- Cellular (mobile) telephones are the most widely used wireless devices in today's communications, and are expected to play an important role in eHealth applications.
- Personal digital assistants (PDAs) are portable devices that possess high computation power. PDAs can be used as calculators, document processors, audio and video players, and gaming consoles. Most of the recent PDAs are also equipped with wired and wireless connectivity. A PDA can be used as a mobile phone to host voice communication sessions or as a computer to access the Internet.
- Radio-frequency identification (RFID) [3,4] technology can be used to identify, track, and store item information. An RFID system consists of three parts: RFID tags, RFID readers, and a data management system. An RFID tag contains an identification code, while an RFID reader is an electronic device that can be used to retrieve identification codes. A data management system stores application information associated with the identification codes. In general, an RFID reader connects to the data management system to retrieve information associated with RFID tags.

  RFID is a radio wave innovation operating on low frequency (30–500 kHz), high frequency (13.56 MHz), or ultrahigh frequency (850–950 MHz, and 2.4–2.5 GHz) bands. With low frequency, the system cost is low, but it has a short operating range. On the other hand, the system that operates in high frequency incurs a higher cost, but it has a longer reading range.

12.2.2 Wireless Networking Technologies and Services for Pervasive Healthcare

- Wireless wide area network (WWAN) technology such as a cellular technology [5,6] provides wireless services over large geographical areas. Each base station (cell site) covers approximately 800 m in urban areas and 8 km in rural areas.
- Wireless local area networks (WLANs) is a class of wireless networks that has an operating range of 30–50 m for indoor and up to 900 m for outdoor.

Currently, most of the WLAN devices communicate using the IEEE 802.11 standard, which specifies physical and MAC layer protocols. As the operating range is relatively small, IEEE 802.11 [7] specifies low transmit power (<10 mW) to reduce interference and to save energy.

- Wireless personal area networks (WPANs) support short-range (within 10 m) connections, and operate at a very low power transmission. WPAN is implemented in most mobile phones to provide connectivity to neighboring devices such as laptops or hands-free headsets. Most of the WPAN devices (e.g., Bluetooth and ZigBee devices [9,10]) are based on the IEEE 802.15 standard [8].
- Wireless medical telemetry service (WMTS) is an important service to remotely monitor patients' vital signs (e.g., body temperature, heart rate, blood pressure, and respiratory rate among others) [11]. WMTS consists of two main components—wireless wearable sensors and central monitoring stations. These sensors monitor patients' vital signs and then transmit the data with low power (<1 mW) to a central monitoring station [12]. WMTS provides mobility to patients and permits remote monitoring of several patients at the same time.

  Currently, most medical telemetry devices operate as secondary users in 460–470 MHz, which the Federal Communications Commission (FCC) has specified for handheld devices (2 W or higher) and other mobile transmitters operated by police, ambulances, firefighters, emergency teams, taxis, and commercial trucks [13]. If there are more intense usage of primary services (e.g., handheld devices), the potential risk of interference to medical telemetry devices can be increased.

  To guard the telemetry from the RF interference, FCC established WMTS and also allocated certain frequencies and rules for this service. There are 14 MHz of spectrum bands (e.g., 608–614, 1395–1400, and 1429–1432 MHz) that have been allocated for use by licensed physicians, healthcare facilities, and certain trained and supervised technicians [14]. Therefore, the medical telemetry devices can operate without interference because they are operating in these channels as primary users. Furthermore, the WMTS specifies a frequency coordinator for WMTS transmitters and maintains a database to protect users from the frequency conflicts [15].
- Medical implant communications service (MICS) is similar to WMTS but medical implant devices are used inside the human body, for example, cardiac pacemakers and implanted defibrillators that transmit conditions of patients' heart with ultralow power for supporting of diagnostic or therapeutic functions. FCC has proposed the 402–405 MHz band for MICS [16]. This service is an unlicensed wireless radio service that permits users to employ medical implant devices without causing interference to other wireless devices.

12.3 Healthcare Environment

This section emphasizes key requirements in healthcare applications and the wireless technology features that could help meet the requirements. Also, we provide a classification of different medical devices used in a healthcare environment.

12.3.1 Challenges in a Clinical Environment

12.3.1.1 Human Error

In the United States, human errors in clinical environments account for 45,000–90,000 deaths per year and 770,000 injuries every year [17]. Statistics in [18] show that such human errors are caused by healthcare miscommunications. Wireless technology can be used to reduce the human errors in the information processing and gathering process. The doctors may prescribe via a PDA, and the prescription may be transmitted to the pharmacy through a wireless network [18,19].

12.3.1.2 Information Accessibility

Information accessibility helps expedite medical care processes and reduce human error mentioned earlier. Medical information can be classified into the following: (1) patient information, (2) technical information, (3) facility information. Patient information is the specific patient details such as allergies, blood type, and medical history. Technical information refers to medical details such as main effects, side effects, and efficacy. This information also includes knowledge about various diseases and how to deal with different symptoms. Facility information specifies resources available at the facility such as equipment and drug inventory, the list of registered specialists, and the availability of operating rooms. Wireless technology acts as a quick and convenient way to access this medical information. Physicians may use a wireless device to retrieve patient information at the point of care.

12.3.1.3 Service Accessibility

Wireless technology can help connect small clinics in remote or rural areas to a hospital. Operating at a small point of care, doctors may call for consultation from specialists in a hospital by using wireless connections.

12.3.1.4 Continuous Monitoring

Continuous monitoring is required for a special class of patients [1]. For example, patients suffering from chronic diseases (e.g., memory loss, heart disease, diabetes) have to be continuously monitored and taken care of in hospitals, which can be costly and inconvenient for some patients. Also, it could be fairly laborious for a caregiver to continuously monitor the patients. Wireless technology can be used for monitoring patients. Patients may have implanted wireless sensors or external sensors to monitor their physical conditions such as glucose level, blood pressure, blood temperature, or heart rate.

12.3.1.5 Time Efficiency

According to [18], a nurse walks 5 miles on average between the patient's bedside and nurse stations and spends 50 minutes per day to communicate with physicians over a phone. Also, medical staffs spend a large portion of their time trying to locate medical equipment. Wireless technology can help reduce unnecessary activities and improve time efficiency of healthcare services. Electronic prescription eliminates the ambiguity of handwritten prescriptions, and helps reduce the time that nurses require to confirm the prescription with doctors. Furthermore, RFID-based tracking and inventory management systems help locate medical assets in real time. Time management in a hospital would be more efficient with the introduction of wireless technology.

12.3.2 Examples of Healthcare Applications and Their Technical Requirements

This section shows a few healthcare application examples that use wireless technologies.

12.3.2.1 Hospital Information System

Hospital Information System (HIS) refers to an information system customized for hospital environments. The main functionalities of HIS are to collect medical data (including patient data, technical data, and facility data), to process and store data in a given format, and to present users (e.g., medical staff) with the data in a readable format. Wireless technology provides medical staffs with a quick and convenient access to HIS.

Doctors can use cellular phones or PDAs to retrieve patient information when they are on their way to the hospital, and can start treating the patient as soon as they reach the hospital. Containing facility information, HIS can help improve managing hospital operations such as inventory management and scheduling treatments too [1,2,17,19].

12.3.2.2 Patient Monitoring System

A patient monitoring system helps caregivers monitor patients who need continuous care [1,2]. Wireless implant and external sensors are used to monitor a patient's condition such as blood pressure, temperature, heart rate, or glucose level. These sensors can be programmed to send an alarm signal to the nearest point of care terminal upon the detection of a predefined condition such as high blood pressure.

12.3.2.3 Telemedicine

Telemedicine addresses the service accessibility issue. It provides healthcare delivery to patients in remote areas. For example, telemedicine applications include remote consultation, remote diagnosis, and patient information transfers. Healthcare professionals in a remote area can discuss with a specialist about patients' symptoms using WWAN services. The professionals in a remote area can also ask the specialist to see inside a patient's ear with a tele-otoscope, or to hear the patient's heartbeat with a tele-stethoscope and provide advice. When a patient in a remote area needs to be transferred to a hospital, patient information can be transmitted to the hospital in advance. The hospital can prepare a treatment plan when the patient is on the way and can start treating the patient upon arrival [20].

12.3.2.4 Wireless Medicine (Wireless Meds)

Wireless Meds incorporates bar coding and wireless technologies into healthcare services to reduce human errors due to miscommunications. Here, the doctor starts to issue a prescription via a PDA. The prescription is sent from the PDA to a pharmacy via WLAN. At the pharmacy, drug packages are bar-coded according to the prescription. The bar code contains drug information as well as patient's information. At the patient's bedside, nurses check whether the medicine bar codes match with the bar code associated with a patient before giving the medicine to the patient. This protocol assures that the patient receives the right drug with the right dose at the right time. However, if there is any mismatch with the bar code, the nurses may use a voice over WLAN (VoWLAN or VoWiFi) communication to consult with the doctor.

12.3.2.5 Tracking System

The main objective of a tracking system is to gain visibility over hospital resources such as patient information, equipment, and drug inventory. Tracking systems utilize two key technologies, RFID and WLAN. They operate as follows: an RFID tag or a WLAN transceiver is attached to the resource to be tracked. RFID readers such as WLAN access points are installed throughout the hospital. As the resource roams in a hospital, it passes through RFID readers or various WLAN access points. Tracking systems gather the location information collected by RFID readers or WLAN access points, and feed the information to the HIS. Then, users can obtain the real-time object location through HIS user interfaces. In a hospital, a tracking system can be used to locate equipments such as IV pump [18]. It can also be used for remote patient monitoring to track patient locations [20].

12.3.2.6 Intelligent Emergency Management System

An intelligent emergency management system is designed to deal with emergency events. It consists of two parts. The first part ensures that only one request is sent to the rescue units. The main idea is to collect location information from emergency reporting. Reporting entries with the same incident location tend to correspond to the same event. Only one request is sent to rescue units when multiple reporting entries are received. The second part uses road traffic information to compute the best driving route for rescue units [20].

12.3.2.7 Advanced Physical Rehabilitation System

Advanced physical rehabilitation systems use specially designed exercise routines and equipments to help patients regain physical ability. During a rehabilitation exercise routine, sensors are attached to various parts of patients to measure physical conditions such as heart rate. Traditionally, these sensors are connected to rehabilitation equipment using cables. Recently, wireless wearable sensors are introduced to replace traditional sensors [21].

Advanced physical rehabilitation systems utilize technologies such as wireless wearable sensors, PDAs, and WPAN. In the simplest form, patients can start exercise routines by following the instructions provided by specialists through PDAs. Wireless wearable sensors provide the system with greater flexibility. These sensors form a WPAN as soon as a patient gets close to rehabilitation equipment. Wearing wireless wearable sensors, patients no longer have to change the sensors when switching the equipment.

Wireless wearable sensors can also be incorporated in a so-called biofeedback system. Here, the measured physical condition is fed to the monitoring center. The monitoring center uses the collected information to adaptively recommend patients their next set of exercise routines. When the measured physical condition shows serious physical abnormality, the monitoring center may dispatch a health professional for immediate care. Also, when the exercise is prescheduled, the monitoring center may send out a message to alert the patient of a coming exercise session.

The eHealth applications mentioned here require different performance measures, address different challenges in a clinical environment, and utilize different wireless technologies. These features are summarized in Tables 12.1 and 12.2.

12.3.3 Electronic Medical Devices: Examples and Classifications

Different medical devices can be classified based on functions, physical properties, and locations, as shown in Table 12.3.

Table 12.1 Requirements of eHealth Applications

Applications	*Bandwidth*	*Latency*	*Packet Loss Probability*	*Reliability*	*Security*
Hospital information system	1–10 Mbps	<1 second	$<10^{-2}$	Moderate	Very high
Telemedicine, wireless meds, and intelligent emergency management system	10 kbps–1 Mbps	10–250 millisecond	$<10^{-4}$	Moderate	High
Patient monitoring and physical rehabilitation system	10–100 kbps	<300 millisecond	10^{-6}	Very high	High
Tracking system	≪1 kbps	<3–5 second	0	Very high	Moderate

Source: Soomro, A. and Cavalcanti, D., *IEEE Commun. Mag.*, 45, 114, 2007.

12.4 Issues Involved in Using Wireless Technology in a Hospital Environment

Although wireless technology provides many advantages for healthcare services, it can cause undesirable EMI at a hospital environment. This section presents the effect of EMI on the medical environment and EMC requirements in a healthcare environment.

12.4.1 Electromagnetic Interference

Electromagnetic waves are self-propagating waves that consist of electric and magnetic field components [24]. Table 12.4 shows various types of electromagnetic waves classified based on the frequency as well as the energy. Electromagnetic waves are classified into ionizing radiating waves and nonionizing radiating waves. The key difference between these two types is that ionizing radiating waves are characterized by high frequency and energy while nonionizing radiating waves are characterized by low frequency and energy. The radiation energy is given by (12.1), where E is defined as the radiation energy (eV), h is the fraction of the Planck constant over $2\pi(\sim 6.582 \times 10^{-16}\text{eV} \cdot \text{s})$, and f is the frequency in cycles per second (s^{-1}) [25]. Ionizing radiating waves are able to strip electrons off the molecule of the

Table 12.2 Applications of Wireless Technologies in eHealth Applications

eHealth Applications	*Medical Challenges*	*Wireless Devices and Technologies*
HIS	Human error, information accessibility	Cellular phone, PDA, WWAN, WLAN
Patient monitoring	Continuous monitoring	WLAN, WPAN, WMTS, MICS
Telemedicine	Service accessibility	WWAN, WLAN
Wireless meds	Human error, time efficiency	PDA, WLAN
Tracking system	Information accessibility, continuous monitoring, time efficiency	RFID, WLAN
Intelligent emergency management system	Information accessibility, time efficiency	Cellular phone, WLAN, WWAN
Physical rehabilitation system	Continuous monitoring, time efficiency	PDA, WLAN, WPAN, WMTS, MICS

exposed object.*

$$E = 2\pi hf. \tag{12.1}$$

Wireless communications use electromagnetic waves as an information carrier, and therefore, inevitably create EMI. EMI is an undesired electromagnetic wave, which can cause adverse effects on data transmission, biological systems, and medical devices [26]. For example, EMI causes power line voltage drops and interruptions, electrical fast transients (EFTs), electrostatic discharges, and radiated and conducted emission among others. Therefore, EMI is an important issue for wireless technology applications in healthcare environments.

12.4.1.1 Impact of EMI on Biological Systems

Biological effects are defined as the measurable response of a biological system to a stimulus or a change in the environment [24,27], which can be either harmful

* This is called the ionization process.

Table 12.3 Classification of Medical Devices

Medical Devices	*Classification Based on*		
	Functions	*Physical Properties*	*Locations*
Incubators	Life-supporting equipment	Passive	Neonatal care
Hearing aids	Life-supporting equipment	Passive	Home care
Pacemakers	Life-supporting equipment	Active or passive	Home care
Infusion pumps	Therapy devices	Passive	Ambulance, ICU, neonatal care
Foetal heart monitors	Diagnostic equipment	Passive	Examination room
Electrocardiograph (ECG) monitors	Diagnostic equipment	Active or passive	Emergency room, ICU, examination room, neonatal care
Anesthesia machines	Therapy devices	Passive	Operating room
Defibrillators	Life-supporting equipment	Passive	Ambulance, emergency room, operating room, ICU, neonatal care
Capnometers	Diagnostic equipment	Active or passive	Emergency room, ICU, examination room, neonatal care
Pulse oximeters (S_{O_2})	Diagnostic equipment	Active or passive	Emergency room, ICU, examination room, neonatal care

Table 12.3 (continued) Classification of Medical Devices

Medical Devices	*Classification Based on*		
	Functions	*Physical Properties*	*Locations*
Electroencephalography (EEG) monitors	Diagnostic equipment	Active or passive	Emergency room, ICU, examination room, neonatal care
Electromyography (EMG) monitors	Diagnostic equipment	Active or passive	Emergency room, ICU, examination room, neonatal care
Hematology analyzers	Diagnostic equipment	Active or passive	Emergency room, ICU, examination room, neonatal care
Holter monitors	Diagnostic equipment	Active or passive	Emergency room, ICU, examination room, neonatal care, home care
Telemetry monitors	Diagnostic equipment	Active	Emergency room, ICU, examination room, neonatal care, home care

Source: Railton, R. et al., Malfunction of medical equipment as a result of main borne interference, *Eighth International Conference on Electromagnetic Compatibility*, Edinburg, U.K., pp. 49–53, September 21–24, 1992. With permission.

or harmless. For example, increasing heart rate due to coffee intake is a harmless biological effect, while liver cirrhosis caused by chronic alcohol drinking is a harmful biological effect. Usually, harmful biological effects can be caused by ionizing EMI. Ionizing electromagnetic waves can cause a thermal effect, heated tissues, and increased body temperature. The eyes and the testicles are notably vulnerable to the thermal effect [24].

Table 12.4 Electromagnetic Spectrum

Electromagnetic Spectrum	*Frequency (Hz)*	*Energy (electron-volt, eV)*
Nonionizing radiation		
Power lines	$10–10^4$	$10^{-14}–10^{-9}$
Radio and television	$10^4–10^8$	$10^{-10}–10^{-7}$
Cellular radio	$10^8–10^9$	$10^{-6}–10^{-5}$
Microwave	$10^8–10^{12}$	$10^{-6}–10^{-3}$
Infrared	$10^{12}–10^{15}$	$10^{-3}–1$
Visible light	10^{15}	1–10
Ultraviolet	$10^{15}–10^{17}$	$1–10^3$
Ionizing radiation		
X-rays	$10^{17}–10^{20}$	$10^3–10^5$
Gamma rays	$10^{20}–10^{26}$	$10^5–10^{12}$

Source: Federal Communication Commission, Office of Engineering and Technology (OET), Questions and answers about the biological effects and potential hazards of radiofrequency radiation, http://www.fcc.gov/Bureaus/Engineering_Technology/Documents/bulletins/oet56/oet56e4.pdf, Available online, August 1999. With permission.

Most of the wireless communications use nonionizing radiation waves that have far less energy and cause small thermal effects to the human body. Although some researches showed that nonionizing EMI can lead to cancers and tumors, there is no significant evidence suggesting association between the radio frequency (RF)/microwave electromagnetic field exposure and some common brain tumors such as glioma and meningioma [28].

The potential risk of EMI is still a controversial health issue. The World Health Organization [27] reports that the low frequency exposure from mobile phones results in minor changes in brain activity, reaction time, and sleep patterns. Consequently, the U.S. Food and Drug Administration (FDA) agency suggests people to limit the duration of cell phone or other wireless device usage, and encourages the use of hands-free devices to increase the distance between the antenna and the user [29].

12.4.1.2 Electromagnetic Interference Caused to Medical Devices

EMI can cause medical equipment malfunctions such as display distortion, waveform distortion, howling, automatic restart, or automatic shut down. Depending on types of EMI and medical equipment, this malfunctioning could be reversible or irreversible. For example, EMI from mobile phones could stop an external cardiac pacemaker to stimulate pulses and syringe pump to generate alarms. While the pacemaker starts operating in its normal condition as soon as the EMI is reduced (i.e., when the mobile phones move away), the syringe pump does not. A clinician needs to reset the syringe pump, after it is exposed to EMI. The adverse effects from which the medical devices can and cannot return to their normal condition in the absence of EMI without human intervention are called reversible and irreversible malfunctions, respectively [30].

Several events and experiments showed that EMI can cause medical equipment malfunctions. For example, digital TV broadcasting systems can cause disruption to wireless heart-monitoring devices [31]. Small [32] reported experiments on EMI caused by cellular phones using the 900 and 1800 MHz on medical devices.

The following recommendations were made in [30]:

- Cell phones should not be present in the operating rooms, ICU (intensive care unit) rooms, and CCU (critical care unit) rooms.
- Cell phones should not be present within 1 m range of medical devices.
- Cell phones should be switched off in examination rooms and in-patient rooms.

12.4.2 Electromagnetic Compatibility

As EMI can cause medical device malfunctions and can lead to devastating impacts on healthcare services, active electromagnetic devices must be compatible with medical devices in an electromagnetic sense. According to [33–35], a device is said to be "electromagnetically compatible," if it can operate under its intended EM environment and does not introduce excessive EMI that may interfere with other devices. This section focuses on the EMC of medical devices and wireless communications devices.

12.4.2.1 Electromagnetic Compatibility for Medical Devices

The International Electrotechnical Commission (IEC) established two important standard series for medical electrical devices EMC: the IEC 60601-1 and the IEC 61000-4 standard series. IEC 60601-1 series specify general requirements for the safety of medical equipments. IEC 61000-4 series recommend testing and measurement techniques for EMC. The readers are referred to [36–42] for the testing and measurement techniques. The following discussions on IEC 60601-1-2 standard within the IEC 60601-1 series deals with the EMC requirements for medical electrical devices [43–46].

IEC 60601-1-2, which is sometimes referred to as IEC 601-1-2, defines the immunity level and compliance level for medical equipments. Immunity level is the maximum EM disturbance level in which medical devices can operate without performance degradation. Compliance level is the EM disturbance level that is below or equal to the immunity level. The standard defines seven types of EM disturbances for EMC as follows:

- *Electrostatic discharge*: Medical devices must withstand electrostatic charge transfer from/to another object with different electrostatic potential. The minimum electrostatic voltage that medical devices must be able to withstand and the EM environment guidance are shown in Table 12.5. The testing and measurement techniques are specified in IEC 61000-4-2 [36].
- *Radiated RF electromagnetic fields*: Although EM waves contain both electric and magnetic components, the electric component is more detrimental to medical devices. The requirement for this category is defined for the electric field (measured in volts per meter or V/m) only. Medical devices must operate normally in an anechoic chamber with an electric field generator. The electric field requirement for life-supporting and nonlife-supporting equipment and the EM environment guidance are given in Table 12.5. The testing and measurement techniques are specified in IEC 61000-4-3 [37].
- *Electrical Fast Transients (EFTs)*: When an induction load is connected or disconnected to a wire, an electric surge is generated. An inductive load can be an electrical device plugged into a line or be an input/output to a signal line. The minimum electric surge that medical devices must withstand and the EM environment guidance are shown in Table 12.5. The testing and measurement techniques are provided in IEC 61000-4-4 [38].
- *Surges*: EM pulses can cause fast and short duration electrical transients in power line voltage, also called a voltage spike. The minimum requirement of surges for AC power lines to ground and AC power lines to lines of medical devices and the EM environment guidance are given in Table 12.5. IEC 61000-4-5 [39] specifies testing and measurement techniques.
- *Conducted RF disturbances*: EM emission can generate undesired voltage on medical devices' external wires and cables. The minimum RF voltages that nonlife-supporting and life-supporting medical devices must be able to withstand inside and outside the ISM (industrial, scientific, and medical) band and the EM environment guidance are given in Table 12.5. The ISM band frequencies are 6.765–6.795, 13.553–13.567, 26.957–27.283, and 40.66–40.70 MHz. IEC 61000-4-6 [40] specifies the testing and measurement techniques.
- *Voltage dips* refer to short interruptions and voltage variations on power supply input lines. Voltage dips are generated by abrupt increase in load or source impedances in power lines. The immunity requirement measured in

Table 12.5 EMC Requirements for Medical Devices and Standards for Test Methods and Equipments

Immunity Requirements	*Immunity Level*	*Electromagnetic Environment Guidance*
Electrostatic discharge (ESD)		The devices should operate on wood, concrete, or ceramic tile floors. The relative humidity required is at least 30 percent when the floors are covered with synthetic material.
Nonconductive parts	±8 kV	
Conductive parts	±6 kV	
Radiated RF electromagnetic fields (at 80 MHz– 2.5 GHz)		RF communication device should be used within the separation distance (*d*), which can be calculated as follows:
Nonlife supporting		
80–800 MHz	3 V/m	$[3.5/E]^{a}\sqrt{P}^{b}$ m
800 MHz–2.5 GHz	3 V/m	$[7/E]\sqrt{P}$ m
Life supporting		
80–800 MHz	3 V/m	$[12/E]\sqrt{P}$ m
800 MHz–2.5 GHz	10 V/m	$[23/E]\sqrt{P}$ m
Electrical fast transients		The device should operate with main power quality of a typical commercial or hospital environment.
Power lines	±2 kV	
Signal lines	±1 kV	
Surges		The device should be operated with main power quality of a typical commercial or hospital environment.
AC power lines to ground	±2 kV	
AC power lines to lines	±1 kV	
Conducted RF disturbance (at 150 kHz–80 MHz)		RF communication device should be used within the separation distance (*d*), which can be calculated as follows:
Nonlife supporting	3 VRMS	$[3.5/V]^{c}\sqrt{P}$ m
Life supporting		
Outside ISM band	3 VRMS	$[3.5/V]\sqrt{P}$ m
Inside ISM band	10 VRMS	$[12/V]\sqrt{P}$ m

(continued)

Table 12.5 (continued) EMC Requirements for Medical Devices and Standards for Test Methods and Equipments

Immunity Requirements	*Immunity Level*	*Electromagnetic Environment Guidance*
Voltage dips (percent dip in U_T) 0.5 cycles 5 cycles 25 cycles 5 second	 >95 percent 60 percent 30 percent >95 percent	The device should be operated with main power quality of a typical commercial or hospital environment. An uninterruptible power supply or a battery is required when the device is required to continuously operate during power main interruption.
Power frequency magnetic field	3 A/m	The device should operate in power frequency magnetic field, which is at characteristic level of a location in a typical commercial or hospital environment.

Source: National Standard of Canada CAN/CSA—C22.2 No. 60601-1-2:03 (IEC 60601-1-2:2001), Medical electrical equipment—Part 1-2: General requirements for safety collateral standard: Electromagnetic compatibility—Requirements and tests, 2003. With permission.

[a] *E* is the actual radiated RF immunity level of the medical device (V/m)
[b] *P* is the maximum output power of the transmitter (W).
[c] *V* is the actual conducted RF immunity level of the medical device.

percentages of voltage dips over the AC main voltage prior to the application of the test level (U_T) and the EM environment guidance are shown in Table 12.5. This requirement is for all life-supporting with rated input power of 1 kVA or less and nonlife-supporting equipments with rated power greater than 1 kVA but the rated input current less than or equal 16 A per phase. Testing and measurement techniques are specified in IEC 61000-4-11 [41].

- *Power frequency magnetic field*: Electronic equipment often leads to magnetic fields at AC main frequencies and then cause problems for medical devices using CRT displays and Hall effect sensors. Table 12.5 shows the minimum magnetic field (measured in amperes per meter or A/m) that medical devices must be able to withstand and the EM environment guidance. IEC 61000-4-8 [42] specifies testing and measurement techniques.

12.4.2.2 Electromagnetic Compatibility for Wireless Transmitters

FCC proposed a guideline of RF energy exposure for wireless transmitters as shown in Table 12.6. The guideline limits the maximum exposure in terms of electric and magnetic field strength and power density at different frequency ranges from 300 kHz to 100 GHz. Electric and magnetic fields are more meaningful for the lower frequencies. On the other hand, the limits of higher frequencies are determined in terms of power densities. There are two types of RF exposures: controlled and uncontrolled. Controlled or occupational exposure limits are applied in the cases when people around the transmitter are fully aware of the potential of its exposure due to their employment and can protect themselves from the exposure. On the other hand, uncontrolled or general population exposure is applied in the cases when people (e.g., general public) around the transmitter are unaware of the potential of its exposure and cannot handle control over their exposure [24].

Tables 12.5 and 12.6 show that at the same frequency, a wireless transmitter can cause the maximum electric field much greater than the radiated RF immunity standard of medical devices. For example, at 80–300 MHz, the wireless transmitters can generate a maximum electric field of 27.5 and 64.1 V/m for uncontrolled and controlled exposures, respectively, while the minimum radiated RF immunity requirement of the IEC 60601-1-2 standard is only 3 V/m for both nonsupporting and supporting life equipment. These two standards are not conforming with each other. Furthermore, the IEC 60601-1-2 standard does not apply to the older versions of medical equipments, and the wireless transceivers can produce electric field strengths in the order of hundreds of volts per meter when it moves closer to the medical equipments. Therefore, the medical devices may fail due to RF interference.

12.5 Cognitive Radio for Wireless Communications in a Hospital Environment

In the conventional spectrum management, the spectrums are statically allocated to each licensed user. Currently, when the number of licensed users increase, it becomes difficult to find a vacant channel for new or existing services because most of the spectrum is already occupied. However, most of the licensed spectrum is rarely continuously used all the time and in space [47,48]. Cognitive radio is a novel paradigm to utilize the radio electromagnetic spectrum in an efficient way. This paradigm allows unlicensed users to opportunistically exploit frequency bands that are not heavily occupied by licensed users (i.e., white spaces or spectrum holes). This section describes how cognitive radio utilizes the spectrum in an effective way and then how to apply cognitive radio in a hospital environment to alleviate the interference between wireless devices and medical devices.

Table 12.6 Limits for the Maximum RF Energy Exposure

	Controlled Exposure				*Uncontrolled Exposure*			
Frequency (MHz)	*Electric Field (V/m)*	*Magnetic Field (A/m)*	*Power Density (mW/cm²)*	*Averaging Time (minutes)*	*Electric Field (V/m)*	*Magnetic Field (A/m)*	*Power Density (mW/cm²)*	*Averaging Time (minutes)*
0.3–1.34	614	1.63	100	6	614	1.63	100	30
1.34–3.04	614	1.63	100	6	$824/f$	$2.19/f$	$180/f^2$	30
3.0–30	$1842/f$	$4.89/f$	$900/f^2$	6	$824/f$	$2.19/f$	$180/f^2$	30
30–300	61.4	0.163	1.0	6	27.5	0.073	0.2	30
300–1,500	—	—	$f/300$	6	—	—	$f/1,500$	30
1,500–100,000	—	—	5	6	—	—	1.0	30

Source: Federal Communication Commission, Office of Engineering and Technology (OET), Questions and answers about the biological effects and potential hazards of radio frequency radiation, http://www.fcc.gov/Bureaus/Engineering_Technology/Documents/bulletins/oet56/oet56e4.pdf, Available online, August 1999. With permission.

Note: f denotes frequency (MHz).

12.5.1 How Does Cognitive Radio Utilize Electromagnetic Spectrum in an Effective Way?

In a cognitive radio system, there are two types of users: primary and secondary users. Primary users or licensed users are users who have legacy rights to use the spectrum. On the other hand, secondary users or unlicensed users have lower priority and can opportunistically use the frequency bands without any interference to primary users. Therefore, cognitive capabilities are required in secondary users [49]. In the context of a medical environment, a primary user/device could be an active medical device using RF transmissions, or a wireless device used by physicians/nurses for critical healthcare applications. A secondary user/device could be a wireless device used for normal data transfer applications (e.g., web browsing). Cognitive radio–based transmission process involves the following four steps: spectrum sensing, adaptive learning, spectrum decision, and transmission parameter setting (Figure 12.1).

12.5.1.1 Spectrum Sensing

A cognitive radio should have the ability to measure, sense, and be aware of the characteristics of the radio channel environment such as the availability of spectrum, power, interference and noise temperature, user requirement and application, and other operating restrictions [49]. In this step, the white spaces are found and the channel capacity of each space is determined. In addition, spectrum sensing should

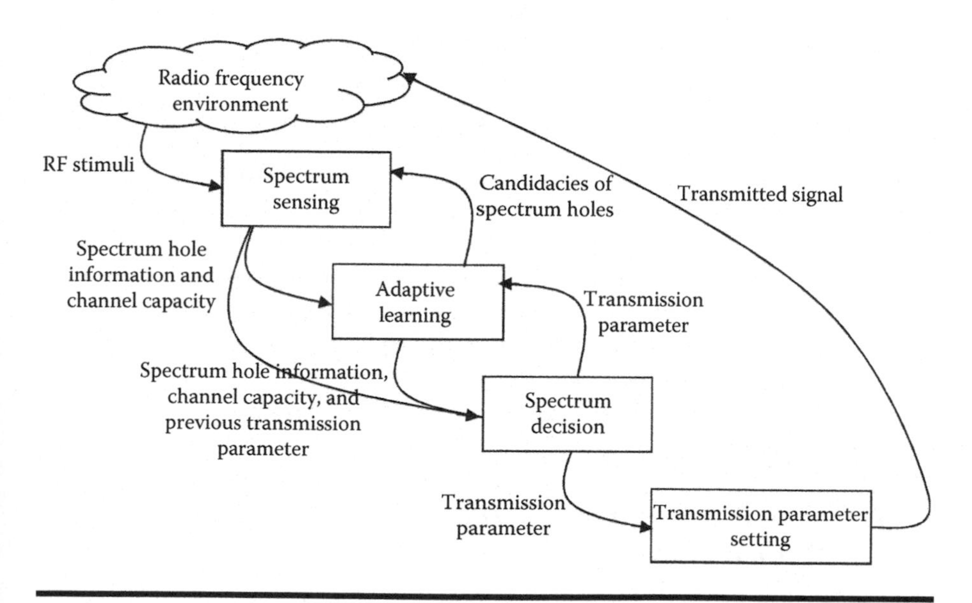

Figure 12.1 The cognitive radio cycle.

be able to explore spectrum opportunities in multiple dimensions (e.g., time, space, frequency, code, angle, and signal wave form).

12.5.1.2 Adaptive Learning

This step is used to learn the spectrum usage characteristics of primary users in multiple dimensions based on statistical analysis. Not only the primary users' usage but the secondary users' behaviors are also learned by this module. The previous and present sensing information and spectrum decision are used to analyze the secondary users' characteristics and requirements (e.g., the required data rate, the transmission mode, the bandwidth of transmission, the acceptable bit error rate, the data size, and the operating period). With this information, the spectrum decision can predict the future occupancy of the channels and reduce the probability of interference to primary users. In addition, a list of candidates is generated for the next sensing. The sensing module can thus save time to find possible spectrum holes and the appropriate white space and transmission parameters for the secondary users.

12.5.1.3 Spectrum Decision

Based on the information of spectrum sensing and adaptive learning module, a cognitive device can determine users' characteristics (e.g., the data rate, the transmission mode, and the transmission's bandwidth) and predict future white spaces. Then, an appropriate channel is selected according to the channel capacity of the spectrum holes and secondary users' requirements. Furthermore, the transmission parameters (e.g., transmit power and modulation technique) are determined for the cognitive device as well.

12.5.1.4 Transmission Parameter Setting

This module physically tunes the parameters of secondary transmitters following the spectrum decision module's transmission parameters before transmitting data in the selected channel. After the appropriate spectrum band is chosen and the transmission parameters are determined, the cognitive radio can communicate using the spectrum band. Nevertheless, the radio environment is dynamic. It can change all the time and over the space (e.g., primary user appearance, user movement, or channel variation). Therefore, a cognitive radio should keep track of the channel environment. When the currently used spectrum band becomes unavailable, the cognitive devices should seamlessly handoff to another spectrum band [50].

12.5.2 How Can Cognitive Radio Avoid Interference with Medical Devices?

As mentioned in Section 12.4, the EMI does not cause any hazardous biological effects to human beings; however, it has adverse effects on medical devices. Wireless devices can cause the potential risk of generating RF disturbance or radiated RF electromagnetic fields to electronic medical devices when the wireless devices come closer to medical devices below a certain distance. This distance depends on the maximum output power of the transmitter and maximum conducted and radiated RF immunity level of the medical device. For example, from Table 12.5, a 2 W mobile phone should be apart from a nonlife-supporting equipment of 3 V/m of maximum radiated RF immunity by at least 1.65 m.*

In addition, for wireless medical devices, other wireless devices also cause EMI problems if both types of devices operate on the same frequency at the same time in the same area. Hence, other wireless devices can cause wireless medical devices to malfunction even though they do not operate within a very close proximity as previously described. For example, as has been mentioned in Section 12.2 for WMTS, most of the current medical telemetry devices operate in the 460–470 MHz band that FCC specifies as the band to be used by 2 W or higher handheld and other mobile transmitters. Therefore, the handheld devices can cause undesired interference to wireless telemetry systems.

To avoid this problem, cognitive radio techniques can be used in a healthcare environment. In a hospital environment, the medical devices can be treated as primary or licensed users while other wireless devices can be treated as secondary or unlicensed users. Therefore, other wireless devices would need to have cognitive capabilities. The objective of such a cognitive radio system would be to control these wireless devices to transmit their data with the right power at the right time in the right area to avoid interference with medical devices and to effectively utilize the RF resources.

The cognitive radio systems first sense the environment before transmitting any data. Then, they find the opportunities to transmit their data without any interference with medical devices by adaptively tuning transmission parameters (e.g., transmit power and modulation technique) depending on the characteristics of the environment (e.g., locations of medical devices, RF immunity level of medical devices, and channel capacity). At the same time, the cognitive radio system dynamically learns the behaviors of the medical devices and other wireless devices (e.g., to predict spectrum occupancy and probability of interference with the medical devices in their vicinity) to increase the spectrum utilization and the safety of the medical devices.

12.5.3 Cognitive Radio for eHealth Applications and Services in a Hospital Environment

To describe how cognitive radios avoid interference with medical devices from other wireless devices, a cognitive radio system is proposed for a hospital environment. Then, three scenarios of how the system works in different healthcare environments are introduced.

* The distance is calculated by $[3.5/E]\sqrt{P}$, where $E = 3\,\text{V/m}$ and $P = 2\,\text{W}$.

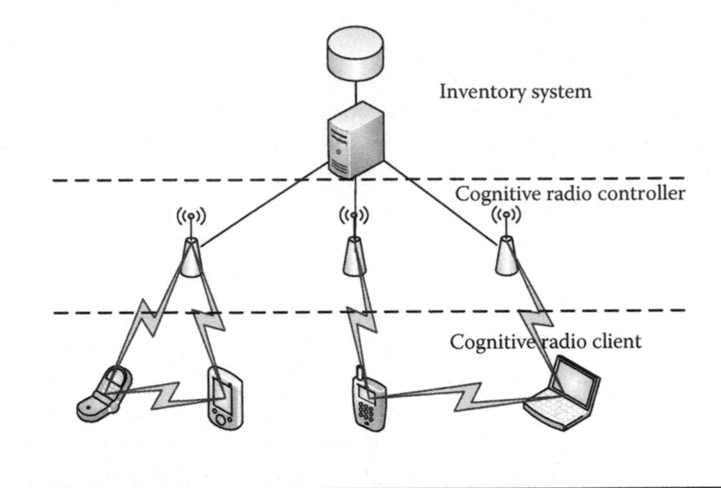

Figure 12.2 The cognitive radio system architecture.

The cognitive radio system consists of three main parts—the inventory system, the cognitive radio controller, and the cognitive radio client, as shown in Figure 12.2. The inventory system collects the information about all medical devices in the hospital (e.g., location, the RF immunity, and the transmission power). The locations of cognitive radio controllers are also included in the system. This information should be always updated.

When a cognitive client has data to transmit, the client will connect to the nearest cognitive radio controller using a weak signal (e.g., spread spectrum signal) that is not harmful to other medical devices in its vicinity. Then, the cognitive radio controller will sense the channel environment and retrieve the information of the medical devices from the inventory system to compute the appropriate transmission parameters (i.e., operating frequency, and transmission power) for the cognitive radio clients. Therefore, the cognitive radio clients can physically tune their transmission parameters to avoid harmful interferences to the medical devices.

The medical devices can be categorized into two main groups—passive and active devices (Figure 12.3). Each group has different ways to avoid interference as described below.

- The passive devices can have some EMI from other wireless devices when the wireless devices operate with excessive power and cause RF disturbances more than the RF immunity level of the medical devices. To determine the appropriate transmission power, cognitive radio controllers need to know the RF immunity levels of the medical devices, the operating frequency of the wireless devices, and the distance between the wireless devices and medical devices. The controller can sense the operating frequency and the location of

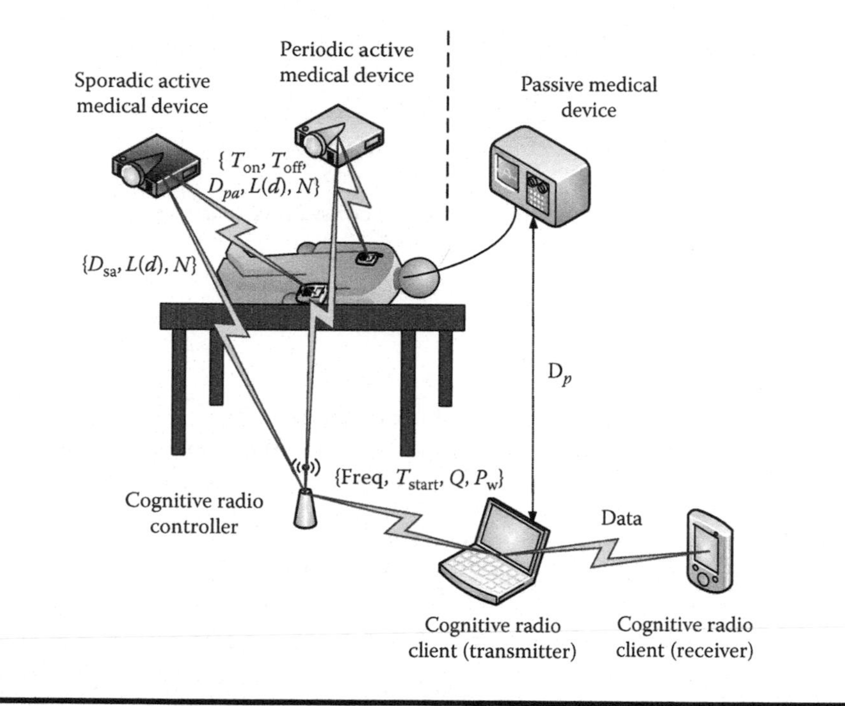

Figure 12.3 Cognitive radio environment in a hospital.

the client and retrieve the RF immunity level (V) and real-time location of the medical devices from the inventory system.

We assume that there are X passive medical devices in the transmission range from the client to the controller. Let P_w denote the vector of transmission power that will not cause any malfunction for all X medical devices. The controller can compute the optimal power for medical device i ($P_{w,i}$) as follows:

$$P_{w,i} = \left(\frac{D_{p,i}V_i}{k_i}\right)^2 \qquad (12.2)$$

where

k_i and V_i are the coefficient based on the operating frequency of the wireless device and the RF immunity level of device i

$D_{p,i}$ represents the distance between the electrical passive medical device i and the cognitive radio client

The cognitive controller can retrieve k_i and V_i of each medical device i from the inventory system and then estimate the values of $D_{p,i}$. When the controller obtains P_w, the maximum allowable transmission power (P_{max}) is given as

$$P_{max} = \min(P_w) \tag{12.3}$$

Therefore, we can define the power constraints to satisfy the interference constraints as follows:

$$0 \le P_t \le P_{max} \tag{12.4}$$

where

P_t is the transmission power of the cognitive client

P_{max} is the maximum transmission power of the cognitive client that may not cause harmful interference to medical devices

- An active wireless link can cause interference to another active link even if they do not operate in close proximity when both devices use the same channel at the same time in the same area. The presence of active devices can be observed by a sensing unit in the cognitive radio controller. By sensing, the controller can estimate the distance between the active medical device and the cognitive client (D_a), the total path-loss including shadowing and multipath fading effect ($L(d)$), and the background noise power (N). At the same time, the controller can retrieve the transmission power (P_a), the distance between the transmitter and the receiver (R), and the threshold of the signal-to-interference ratio (γ) of the medical device from the inventory system. The controller can then calculate the appropriate transmit power for the cognitive client (P_t) as follows:

$$P_t = \frac{\left(\frac{P_a L(R)}{\gamma}\right) - N}{L(D_a)}. \tag{12.5}$$

Furthermore, active medical devices can be divided into two subcategories based on the regularity of recurrence:

1. *Periodically active medical devices*: These devices have certain patterns of operating periods. Therefore, the controller can exactly determine the appropriate time to start the transmission (T_{start}) and the available periods of the spectrum (Q).
2. *Sporadically active medical devices*: These devices do not operate periodically. The controller might not be able to determine the exact value of T_{start} and Q. Therefore, the controller needs to periodically sense the channel to assure that the cognitive client will not interfere with active medical devices.

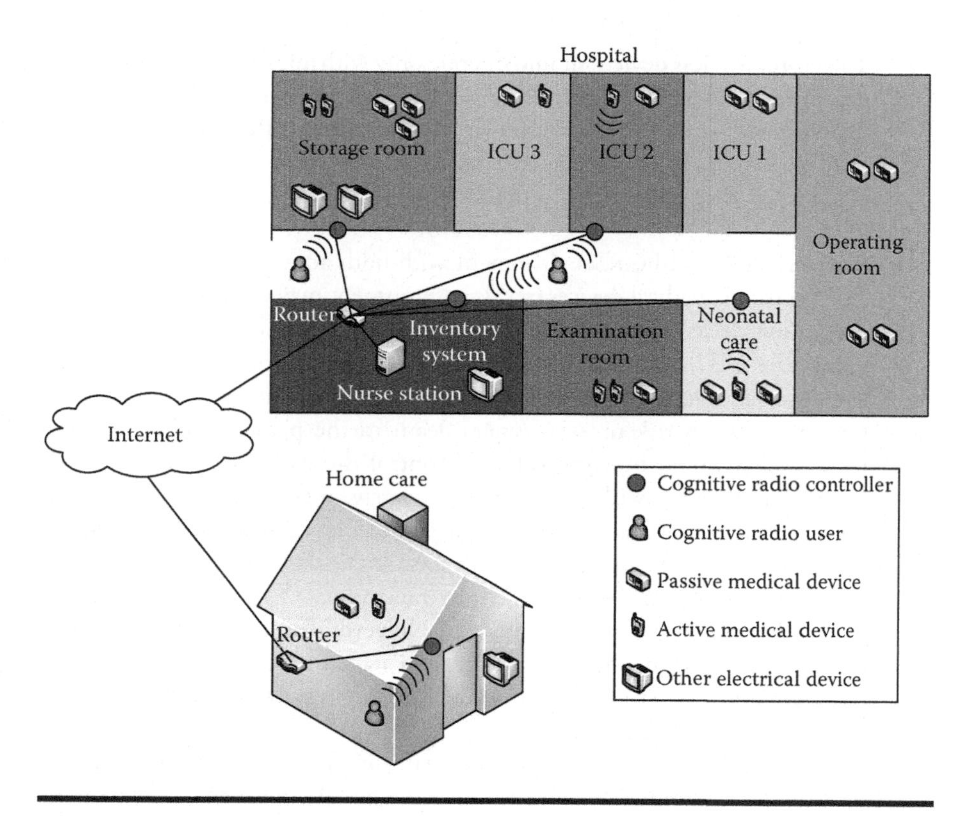

Figure 12.4 Different wards in a medical environment.

For eHealth applications in a hospital environment, cognitive radio techniques can be applied to avoid interference between medical equipments and other wireless devices. Therefore, patients can be assured that they will experience safe eHealth services. As mentioned in Section 12.3.3, there are different locations like hospital wards and home places among others where healthcare can be delivered (Figure 12.4). Each location has different medical devices for different purposes.

The healthcare environment can be categorized into three groups based on the physical properties of medical devices at each location. Each group has different methods of handling different medical environments as follows:

- (*Scenario 1*) A healthcare environment with only passive devices: All electronic medical devices in this environment are passive. An example of wards in this group is the operating room (OR). Using wireless devices with high transmission power nearby the operating room can cause malfunctioning of these medical devices due to EMI and cause harm to the patients in the OR.

Therefore, wireless devices should operate only with appropriate transmission power. In this case, the cognitive radio controller around the operating room should specify proper transmission power for the cognitive nodes, or alert them to switch off or change the status of the radio transceiver to sleep mode when they come close to the medical devices that are in the operating room.

- (*Scenario 2*) A healthcare environment with both active and passive devices: The electronic medical devices in this environment may consist of both active and passive devices. Examples of such an environment include the emergency room, the ICU, the examination room, and the ward for neonatal care among others. Malfunctioning of radio equipments in these rooms poses significant health risk or even causes death to the patients. The cognitive radio system would be responsible to control the wireless devices to avoid harmful interference with both active and passive medical devices. When the cognitive radio user has data to be sent, it will first connect to the cognitive controller to find the opportunity to transmit the data. The cognitive radio controller will allow a cognitive client to transmit data when it can be ensured that no harmful interference will occur. However, sometimes, the cognitive radio devices need to stop transmission and switch to sleep mode when they come close to the sensitive medical devices or when active medical devices appear in the operating frequency.
- (*Scenario 3*) A healthcare environment with other kinds of electrical equipments: This environment includes both active and passive medical devices and other electrical equipments (e.g., TV, radio, and computer monitor). Homecare is an example of such an environment. In a homecare environment, EMI to the medical devices may be caused due to other electronic devices (e.g., home appliances). In this case, the cognitive controller should consider the activity of these devices to determine the appropriate transmission power for cognitive radios. Besides, the cognitive controller may volunteer to notify the radio devices in the environment if it detects any harmful interference caused to the critical medical devices.

12.6 Challenges Related to Development of Cognitive Radio Technology for eHealth Applications

As explained in Section 12.5, a cognitive radio is designed to be aware of its surrounding environment, learn and adjust to its operating parameters depending on the changes of the environment for effective EMI alleviation, highly reliable communication, and efficient spectrum utilization. The following section focuses on the technical requirements and open issues related to the implementation of cognitive radio technology for eHealth applications.

12.6.1 Technical Requirements

To develop cognitive radio technology for eHealth applications, there exist six main requirements to protect electronic medical devices from potential risk of interference from other wireless communications devices.

12.6.1.1 Sensing Regularity

Cognitive radios need to be aware of the presence of medical devices in its vicinity before and during transmission. Therefore, cognitive controllers should sense its surrounding environment once every sensing period. The sensing period is the maximum time duration for which the primary user will not use the channel. Therefore, the cognitive users may not cause any harmful interference to the medical devices and can be unaware of the presence of medical devices during the sensing period. The sensing period should depend on the primary user's type and behavior.

12.6.1.2 Sensing Range

When a cognitive radio client transmits data, it can cause interference to other users within its transmission range, *D* as shown in Figure 12.5. To protect the primary users from the interference, the cognitive controller should sense the presence of the medical devices at least within distance *D*.

12.6.1.3 Processing Time

The cognitive radio controllers should be able to calculate the proper transmission parameters for the clients before and during transmission. Before transmitting data, the controllers need to carefully determine appropriate transmission parameters to achieve high transmission quality without any harmful interference to

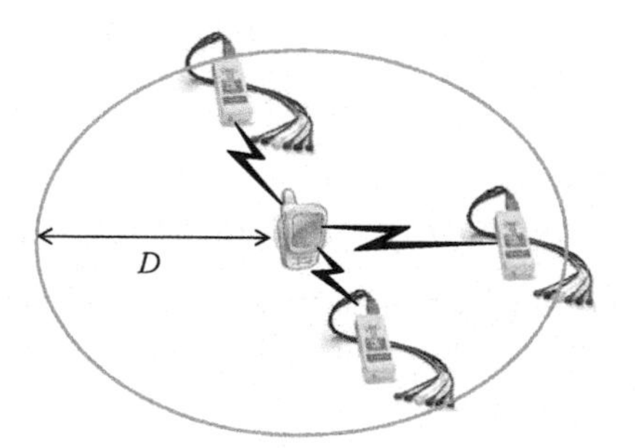

Figure 12.5 Spectrum sensing range.

medical devices. Intuitively, when the efficiency of spectrum decision increases, the complexity of computation and processing time also increases. If the duration of the computation of spectrum access decision is too long, the quality of service (QoS) can decrease. Moreover, the characteristics of surrounding environments change over time. Therefore, the cognitive radio should quickly sense the channel and make an effective decision before the channel state changes.

During transmission, medical devices can appear in the transmission range of the cognitive radio client or in the operating frequency anytime. To avoid unexpected interference, the controllers should be able to instantly detect the presence of medical devices and inform the clients to switch off or handoff from the band.

12.6.1.4 Adaptive Learning

Due to the dynamic nature of the medical environment, a cognitive radio controller should adaptively learn the changes in the environment to efficiently predict the opportunities for cognitive radio transmissions and save processing time.

12.6.1.5 Interference Penalties

For coexistence of wireless devices and medical devices, cognitive radio techniques should be designed with provisioning for interference penalty. A cognitive radio controller can predict that the channel will be available for a certain period and may allow a client to transmit data during the period. Nonetheless, if an active medical device reappears in the channel at the same time that the cognitive client utilizes the spectrum band, the controller should immediately notify the client to stop the transmission. As the client can cause some interference to the active device, it should have some penalties due to causing interferences to the medical devices. The penalty can be exercised by prohibiting the cognitive devices from using the available channels for a while or reducing the transmission rate of the client. The penalty should depend on the extent of interferences caused to the medical devices. If the interference is extremely severe, the client could be prohibited from using the channel anymore in the future.

12.6.1.6 Real-Time Positioning of Electronic Medical Devices

The real-time locations of medical equipments, especially passive medical devices, should be known by a cognitive radio controller. Therefore, this information should be always updated in the inventory system. Location-tracking systems (e.g., based on RFID) can be used to identify the locations of medical devices. Therefore, the controllers can alert the clients not to come too close to the electronic medical equipments.

12.6.2 Technical Challenges

The technical challenges and open issues in cognitive radio development for pervasive healthcare applications can be categorized into three main groups as discussed in the following text.

12.6.2.1 Challenges for General System Design

- *Spectrum handoff*: It refers to the process by which a cognitive radio client changes its operating frequency due to the appearance of medical devices or the movement of clients. Spectrum handoff can cause QoS degradation because the client needs to stop transmission or change to a new spectrum band. To improve the performance, the cognitive devices should use an efficient handoff strategy. When there are many available bands for a client, the cognitive radio controller should be able to select the best band based on the client's application requirements. Moreover, the controller should provide a connection management mechanism to reduce delays and losses during spectrum handoff [50].
- *No regular behavioral pattern of medical devices*: To save time to find available spectrum band, the cognitive radio should be able to predict the spectrum availability in the future based on previous activities of the medical devices. However, some devices may not have any regular behavioral pattern. In this case, the cognitive radio controllers may not be able to predict the channel availability accurately.
- *Noncognitive radio users*: In a practical scenario, both cognitive and noncognitive devices will exist in a hospital environment. While the cognitive radio devices try to avoid causing interference with medical equipments in the hospital, noncognitive radio devices are unaware of any interference caused to the medical devices. Therefore, the cognitive radio devices should be able to detect the interference caused by the non-cognitive devices to the medical devices and also try to mitigate the interference by alerting the relevant devices.
- *Multiple cognitive radio clients*: In the presence of multiple cognitive radio users, the cognitive radio controller should ensure fairness among the users as well as efficient use of the spectrum opportunities.

12.6.2.2 Challenges for Passive Medical Devices

- *On–off detection of passive devices*: The on–off status of a passive device is difficult to detect because this device does not transmit any signal during its operation. If the passive device is off, a cognitive client can exploit a spectrum opportunity to transmit its data even though it is close to the medical device. Therefore, the cognitive radio system should provide an effective way to detect the on–off status of passive devices without any potential risk.

12.6.2.3 Challenges for Active Medical Devices

- *Channel fading and shadowing*: The controllers may not be able to detect the presence of active medical devices in a severe fading and shadowing channel, even though the active device transmits data to the receiver [47]. This situation leads to the hidden primary user problem [49]. As a result, a cognitive radio node can cause harmful interference to the medical devices. The cognitive radio controllers should have the capability to differentiate between a faded or shadowed active device and white space.

  Cooperative sensing can be applied to solve this hidden primary user problem. In cooperative sensing, local channel measurements of a cognitive radio controller are shared with the measurements of other controllers and then the channel occupancy status can be determined in a more accurate fashion.
- *Adjacent channel interference*: This is caused by high transmission power in a radio channel. The telemetry devices in the WMTS and MICS bands may experience interference from devices in adjacent bands. Therefore, the cognitive controllers should consider this problem when determining transmission parameters for cognitive radio devices.
- *Multiple spectrum band exploitation*: To improve the QoS of cognitive radio transmission, the cognitive radio controllers can allow clients to transmit data over multiple spectrum bands at the same time [50]. For instance, if an active medical device reappears in a spectrum band, the secondary user needs to withdraw from this spectrum. Nonetheless, the client still maintains the communication because the client can use some other spectrum bands. Although this method will result in a more graceful quality enhancement compared to the single band case, the decisions on selecting the appropriate number of spectrum bands and the set of suitable bands would be challenging in a cognitive radio system.
- *Spread spectrum system*: With spread spectrum transmission (i.e., direct-sequence spread spectrum and frequency-hopping spread spectrum), the transmission power of active devices is difficult to detect because the power is diffused over a broader spectrum even though the actual required bandwidth is narrower [49]. One way to partially avoid this problem is to let the secondary users know the hopping or the pseudo-noise sequence of the primary users so that they can perfectly synchronize to a primary user's signal.

12.7 An EMI-Aware Cognitive Radio Transmission Scheme

In the system model shown in Figure 12.3, every user (including the primary users) transmits its data through the cognitive controller. The cognitive controller will periodically broadcast the spectrum usage of primary users to cognitive clients.

Then, each cognitive client has to make its own decision on spectrum access by itself based on this information and channel quality. We assume that there are N channels in the system and two of them will be selected to be temporarily used as broadcasting and logical common control channels, respectively. A broadcasting channel is used to broadcast the selected control channel that is used to identify transmission parameters (e.g., transmit power and channel) for each cognitive user via our EMI aware RTS-CTS mechanism (described in the following paragraphs). The cognitive controller will adaptively select the channel that has the highest probability of availability in that time to be used as the logical common control channel. The controller periodically broadcasts the logical common control channel every time period, called the broadcasting period. In contrast, the broadcasting channel should be used to synchronize a new user to the current cognitive network.

Figure 12.6 describes our spectrum access mechanism. First, a cognitive client waits for the common control channel broadcast from the controller in the pre-assigned channel. When the client has data to be transmitted, it first initiates an EMI-aware RTS-CTS mechanism on the selected common control channel. The controller then calculates the maximum allowable transmission power and assigns an appropriate channel to the client based on channel availability and quality and QoS requirements of the client. In each broadcasting period, the logical common channel is generally reserved for only EMI-aware RTS-CTS mechanism but it can

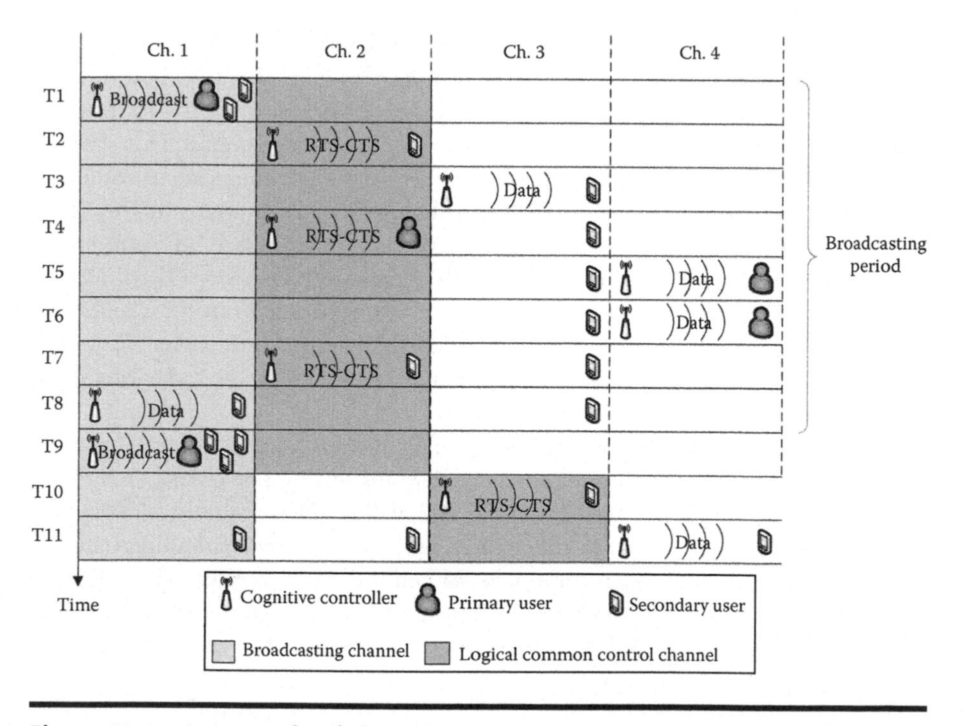

Figure 12.6 An example of channel access mechanism.

be assigned for only primary users in the case that the channel is the best for the primary user. If the best channel for the primary user is occupied by a secondary user, the cognitive controller will send a message to the secondary user to vacate from the channel and preempt the secondary user to a queue and then let the primary user to transmit in the channel. The controller serves the primary user until its transmission is complete. Once the primary user's transmission succeeds, the secondary user will be released from the queue and continue its transmission in the channel. For secondary users, the controller serves each user in a TDMA fashion with certain duration and more frequently alternate to the common control channel to be aware of the presence of a primary user. Moreover, if a new logical common control channel is occupied by a secondary user, the controller will preempt the user in to a queue and then find an idle channel for the secondary user.

For accessing a channel including the logical common control channel, every node has to follow the carrier sense multiple access with collision avoidance (CSMA-CA) protocol to avoid interference with other users [51,52]. In CSMA, a node senses the channel for a period of time (i.e., distributed coordinate function inter-frame spacing: DIFS) before transmitting. If the channel is busy, the node has to wait for the channel to become idle (hold-back time) and then keep sensing the channel for an additional random time (i.e., back-off time) before starting the transmission. The back-off time is randomly chosen from the range 0-contention window (CW). This can reduce collision in a multiple access system.

The traditional CSMA/CA protocol with RTS-CTS-based handshaking does not consider the effect of EMI on passive medical devices in a hospital environment. We propose a new RTS-CTS mechanism, called EMI-aware RTS-CTS mechanism, for pervasive healthcare applications in a medical environment. This scheme incorporates two methods based on types of transmission (i.e., uplink and downlink).

In uplink transmission (Figure 12.7), the cognitive client transmits its data to the cognitive controller. First, the cognitive client transmits RTS (request-to-send) packet to the cognitive controller with a pre-assigned power (P_{ctrl}). When the cognitive controller successfully receives the RTS packet from the client, the controller needs to calculate the maximum allowable transmission power (P_{max}) for the client that may not cause any harmful interference to medical devices in its vicinity by solving (12.2) and (12.3).

If P_{max} is greater than zero, the controller transmits CTS (clear-to-send) message to the client with the maximum of appropriate transmission power (P_{max}) and the channel quality information. On the other hand, if P_{max} is less than or equal to zero, the controller may not allow the client to transmit its data by responding with an NCTS (negative of clear-to-send) message. When the client receives CTS message, it can transmit its data with the appropriate power (P_t) to the controller. If the controller receives all data and checks that all data is complete, it will respond with ACK (acknowledge) message to the client.

For downlink transmission (Figure 12.8), the cognitive controller delivers the data to the right client. The controller must contend for accessing the channel with

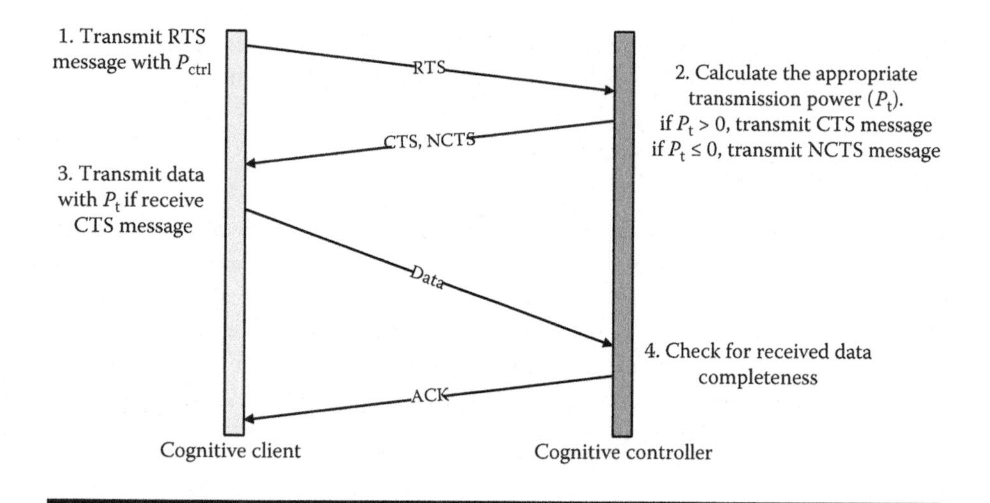

Figure 12.7 EMI-aware RTS-CTS scheme for uplink transmission.

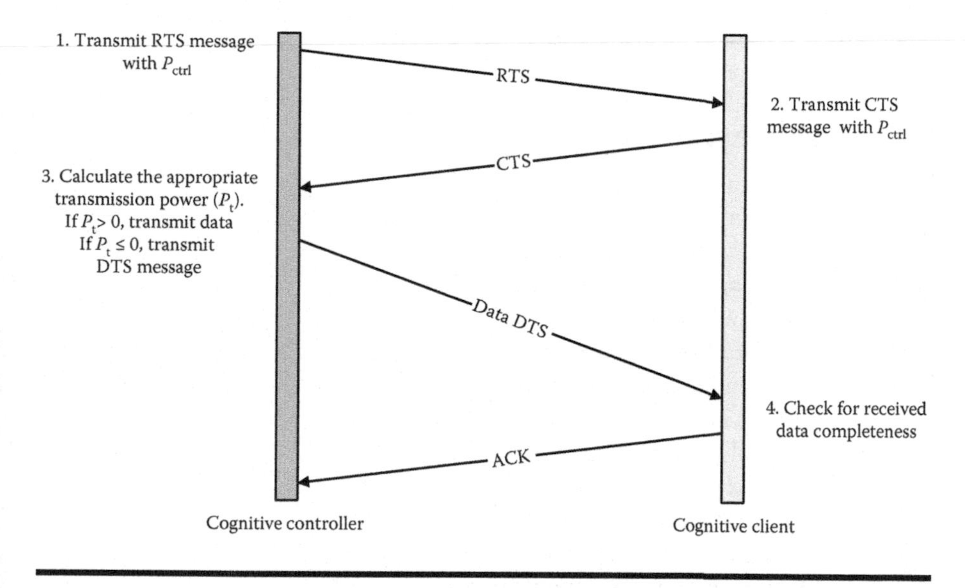

Figure 12.8 EMI-aware RTS-CTS scheme for downlink transmission.

CSMA-CA mechanism similar to the other clients. When the cognitive controller can access to the channel, it will perform the same steps as the client does by first transmitting RTS packet to the client. If the client receives the RTS packet, it will transmit CTS packet with P_{ctrl} to the controller. Once the controller receives the CTS, the controller can define the appropriate transmission power (P_t) from P_{max} by solving Equations 12.2 and 12.3. If P_{max} is greater than zero, the controller will transmit data with P_t. On the other side, if P_{max} is less than or equal to zero, the controller will transmit DTS (decline-to-send) message to the client.

12.8 Conclusion

We have outlined the requirements, constraints, and open issues in the development of cognitive radio technology for eHealth applications and services in a hospital environment. A cognitive radio-based network architecture for eHealth applications in a hospital environment has been presented. This architecture uses a channel access mechanism for secondary users that is aware of the EMI constraints for the biomedical devices. For performance analysis and optimization of the proposed system architecture, we refer the readers to [53].

Acknowledgment

This work was supported by TR*Labs* and in part by the Natural Sciences and Engineering Research Council (NSERC) of Canada.

List of Abbreviations

EFT	Electrical fast transient
eHealth	Electronic health
EMC	Electromagnetic compatibility
EMI	Electromagnetic interference
FCC	Federal Communications Commission
FDA	U.S. Food and Drug Administration
HIS	Hospital information system
IEC	International Electrotechnical Commission
MICS	Medical implant communications service
PDA	Personal digital assistant
RFID	Radio-frequency identification
WLAN	Wireless local area network
WPAN	Wireless personal area network
WWAN	Wireless wide area network

References

1. L. Czekierda, J. Danda, K. Loziak, M. Sikora, et.al, *Information Technology Solutions for Healthcare*, Springer, Berlin, pp. 85–109, October 29, 2007.
2. U. Varshney, *Pervasive Healthcare and Wireless Health Monitoring*, Springer Science + Business Media, the Netherlands, July 12, 2007.

3. Department of Commerce of Washington D.C., Radio frequency identification: Opportunities and challenges in implementation, April 2005.
4. Ministry of Economic Development of New Zeland, Radio frequency identification devices, http://data.rsm.govt.nz/planning/srd/discussion/discussion-06.html, Available online, August 2004.
5. W. C. Y. Lee, *Wireless & Cellular Telecommunications*, 3rd edn., The McGraw-Hill Companies, New York, pp. 1–5, 110–130, 175–181, 187–196, 2006.
6. W. C. Y. Lee, *Mobile Communications Engineering*, 2nd edn., The McGraw-Hill Companies, New York, pp. 7–11, 1997.
7. IEEE 802.11, Wireless LAN medium access control (MAC) and physical layer (PHY) specification, 2007.
8. IEEE 802.15, IEEE 802.15 working group for WPAN, http://ieee802.org/15/, Available online.
9. Bluetooth SIG, Bluetooth specification, http://www.bluetooth.com, Available online.
10. IEEE 802.15.4, Wireless medium access control (MAC) and physical layer (PHY) specifications for low-rate wireless personal area networks (LR-WPANs), 2006.
11. Wireless medical telemetry, http://wireless.fcc.gov/services/index.htm?job=service_home&id=wireless_medical_telemetry, Available online, October 3, 2003.
12. Telemetry/Telecontrol equipment, http://www.ofta.gov.hk/en/tcc/group_12.pdf, Available online.
13. Health Canada, Recent changes to US FCC rules in the 460–470 MHz band may affect Canadian wireless medical telemetry systems located near the US border, http://www.hc-sc.gc.ca/dhp-mps/alt_formats/hpfb-dgpsa/pdf/medeff/telemetry-telemedecine_system_2_nth-aah_e.pdf, Available online.
14. FCC establishes new wireless medical telemetry service, http://www.fcc.gov/Bureaus/Engineering_Technology/News_Releases/2000/nret0009.html, Available online, June 8, 2000.
15. Guidance on wireless medical telemetry risks and recommendations, http://www.fda.gov/cdrh/comp/guidance/1173.html, Available online, October 20, 2000.
16. Medical implant communications, http://wireless.fcc.gov/services/index.htm?job=service_home&id=medical_implant, Available online, June 1, 2003.
17. A. F. Graves, B. Wallace, S. Periyalwar, and C. Riccardi, Clinical grade—A foundation for healthcare communications networks, *Proceedings 5th International Workshop on Design of Reliable Communication Networks, 2005 (DRCN 2005)*, Italy, October 16–19, 2005.
18. IBM, Wireless healthcare delivery: Adapting to tomorrow's needs with mobile processes, www-304.ibm.com/jct09002c/university/scholars/skills/ssme/VSRBFinal1-16.pdf, Available online.
19. B. W. Podaima and R. D. McLeod, Point of care engineering and technology, *CMBEC29*, Vancouver, BC, Canada, June 1–3, 2006.
20. U. Varshney, Pervasive healthcare, http://www.cis.gsu.edu/uvarshne/papers/PervH.pdf, Available online, December 2003.
21. E. Jovanov, A. Milenkovic, C. Otto, and P. C. de Groen, A wireless body area network of intelligent motion sensors for computer assisted physical rehabilitation, *Journal of NeuroEngineering and Rehabilitation*, 2(1): 6, 2005.

22. A. Soomro and D. Cavalcanti, Opportunities and challenges in using WPAN and WLAN technologies in medical environments, *IEEE Communications Magazine*, 45(2): 114–122, February 2007.
23. R. Railton, G. D. Currie, G. A. Corner, and A. L. Evans, Malfunction of medical equipment as a result of main borne interference, *Eigth International Conference on Electromagnetic Compatibility*, Edinburg, U.K., September 21–24, 1992, pp. 49–53.
24. Federal Communication Commission, Office of Engineering and Technology (OET), Questions and answers about the biological effects and potential hazards of radiofrequency radiation, http://www.fcc.gov/Bureaus/Engineering_Technology/Documents/bulletins/oet56/oet56e4.pdf, Available online, August 1999.
25. L. Marchildon, *Quantum Mechanics*, Springer, Germany, 2002, pp. 4–6.
26. Center for Devices and Radiological Health, Draft guidance for industry and FDA staff-radio-frequency wireless technology in medical devices, http://www.fda.gov/cdrh/osel/guidance/1618.html, Available online, January 3, 2007.
27. World Health Organization (WHO), Establishing a dialogue on risks from electromagnetic fields, http://www.who.int/peh-emf/publications/EMF_Risk_ALL.pdf, Available online, October 2002.
28. G. Berg et al., Occupational exposure to radio frequency/microwave radiation and the risk of brain tumors: Interphone Study Group, Germany, *American Journal of Epidemiology*, Advance Access, July 27, 2006.
29. Questions & Answers, http://www.fda.gov/cellphones/qa.html, Available online.
30. H. Furuhata, Electromagnetic interferences of electric medical equipment from hand-held radiocommunication equipment, *1999 International Symposium on Electromagnetic Compatibility*, Tokyo, Japan, 1999, pp. 468–471.
31. T. Nguyen, A real and present wireless danger, http://www.shvoong.com/exact-sciences/504860-real-present-wireless-danger, Available online, April 14, 2007.
32. D. Small, Mobile phones should not be used in clinical areas or within a metre of medical equipment in hospitals, *Evidence-Based Healthcare and Public health*, 9(2): 114–116, December 2005.
33. Electromagnetic compatibility—EMC, http://www.fda.gov/cdrh/emc/, Available online.
34. R. Sitzmann, Electromagnetic compatibility in medical engineering, http://health.siemens.com/medroot/en/news/electro/issues/pdf/heft_2_98_e/09sitzma.pdf, Available online, 1998.
35. National Standard of Canada CAN/CSA - C22.2 No. 60601-1-2:03 (IEC 60601-1-2:2001) Medical electrical equipment—Part 1-2: General requirements for safety—collateral standard: Electromagnetic compatibility—Requirements and tests, 2003.
36. IEC 61000-4-2: Electromagnetic compatibility (EMC)—Part 4-2: Testing and measurement techniques—Electrostatic discharge immunity test, 2001.
37. IEC 61000-4-3: Electromagnetic compatibility (EMC)—Part 4-3: Testing and measurement techniques—Radiated, radio-frequency, electromagnetic field immunity test, 2008.
38. IEC 61000-4-4: Electromagnetic compatibility (EMC)—Part 4-4: Testing and measurement techniques—Electrical fast transient/burst immunity test, 2004.

39. IEC 61000-4-5: Electromagnetic compatibility (EMC)—Part 4-5: Testing and measurement techniques—Surge immunity test, 2005.
40. IEC 61000-4-6: Electromagnetic compatibility (EMC)—Part 4-6: Testing and measurement techniques—Immunity to conducted disturbances, induced by radio-frequency fields, 2006.
41. IEC 61000-4-11: Electromagnetic compatibility (EMC)—Part 4-11: Testing and measurement techniques—Voltage dips, short interruptions and voltage variations immunity tests, 2004.
42. IEC 61000-4-8: Electromagnetic compatibility (EMC)—Part 4-8: Testing and measurement techniques—Power frequency magnetic field immunity test, 2001.
43. D. Modi, IEC 601-1-2 and its impact on medical device manufacturers, *Proceedings in 19th International Conference—IEEE/EMBS*, Chicago, IL, October 30–November 2, 1997.
44. B. H. Bakker, EMC standard for medical electrical equipment, *The Institution of Electrical Engineers*, London, U.K., 1993.
45. S. Baisakhiya, R. Ganeasn, and S. K. Das, IEC 6061-1-2,2001: New EMC requirements for medical equipment, *Proceedings of INCEMIC 2003. 8th International Conference on Electromagnetic Interference and Compatibility*, Chennai, India, December 18–19, 2003, pp. 409–414.
46. Radiofrequency interference with medical devices, *IEEE Engineering in Medicine and Biology Magazine* 17(3): 111–114, 1998.
47. A. Ghasemi and E. S. Sousa, Spectrum sensing in cognitive radio networks: Requirements, challenges, and design tradeoffs, *IEEE Communications Magazine*, 46(4): 32–39, April 2008.
48. S. Haykin, Cognitive radio: Brain empowered wireless communications, *IEEE Journal on Selected Areas in Communications*, 23(2): 201–220, February 2005.
49. H. Arslan and T. Yucek, *Spectrum Sensing for Cognitive Radio Application*, Springer, Berlin, 2007.
50. I. F. Akyildiz, W. Y. Lee, M. C. Vuran, and S. Mohanty, NeXt generation/dynamic spectrum access/cognitive radio wireless networks: A survey, *Computer Networks: The International Journal of Computer and Telecommunications Networking*, 50(13): 2127–2159, September 2006.
51. T. Alexander, *Optimizing and Testing WLANs*, Elsevier Inc., Burlington, MA, 2007, pp. 1–26.
52. P. Chandra and D. Lide, *Wi-Fi Telephony*, Elsevier Inc., Burlington, MA, 2007, pp. 79–108.
53. P. Phunchngharn, D. Niyato, E. Hossain, and S. Camorlinga, An EMI-aware prioritized wireless access scheme for e-health applications in Hospital environments, submitted to the IEEE Transaction on Information Technology in BioMedicine.

Chapter 13

Network Selection in Cognitive Radio Networks

Yong Bai, Yifan Yu, and Lan Chen

Contents

With deregulation and technology evolution on cognitive radios (CR), mobile users may have the freedom to select and adapt to one preferred network from multiple radio access networks (RANs) that coexist in overlapping areas. To select a preferred serving RAN, the mobile terminal (MT) can first detect and sense the characteristics (attributes) of the network side and the user side. This chapter describes the relevant network attributes, user attributes, and the individual user's preference. For the convenience of mobile users, the disparity and conflict of these attributes need to be captured by network selection method for automatic decision making. MADM (multiple attribute decision making)-based methods are viewed as a promising approach to achieve this. This approach effectively aggregates the attributes and the individual user's preference. The procedure of the MADM-based network selection methods and the enhancements for the benefits of users and mobile operators are presented in this chapter. On the other hand, mobile operators can adjust the network attributes (e.g., price) to attract users to access and dynamically request a spectrum from a common spectrum pool to meet the varying service demands. The interaction of network selection of mobile users and spectrum allocation between RANs is analyzed by formulated service demand model and noncooperative competition model between RANs.

13.1 Introduction

In recent years, complementary deployment of both 3G wireless networks and IEEE 802.11-based WLANs is beginning to provide various voice and data services to mobile users. Each mobile user can select a cellular network or a WLAN for access manually or automatically based on his or her preferences on price and QoS. It is foreseen that more heterogeneous radio access networks (RANs) would be deployed in coexistence to provide various services. Hence, the mobile users have more alternatives of overlapping RANs for access to get the substitutable services (e.g., VoIP). Furthermore, with deregulation on network and spectrum management, mobile operators may open the opportunities for mobile users to access the network in a more flexible manner such as on per-service basis, and the mobile operators can dynamically request spectrum resources to meet varying service demands with shared spectrum usage between RANs.

This paradigm change brings forth more challenges in the design of radio resource management schemes both for mobile users and for RANs. The mobile users would be equipped with mobile terminals (MTs) (or called user equipment—UE) that have multi-mode radios, reconfigurability, or cognitive radio (CR) capability. From the discovered multiple overlapped RANs, each mobile user is able to select one network as his or her preferred serving RAN for initial network access (i.e., from idle mode to active mode) or vertical handoff. It is desirable that the decision of network selection can be made automatically for the convenience of mobile users. To make such an automatic decision, the MT first detects and senses relevant characteristics (attributes) from the network side and the user side. Then, mobile users face many attributes associated with networks (e.g., bandwidth, coverage, handoff latency), users (e.g., budget limit, residual battery power), and applications (e.g., QoS profile). These attributes may be disparate and even conflicting with each other. Hence, the mobile users need proper methods to capture these conflicts and trade off their preferences for selecting the serving RAN.

On the other hand, each RAN can adjust network attributes such as charged price and available spectrum to attract or repel mobile users to access. The network selection/reselection by mobile users leads to changeable network loadings, and the spectrum resource needs to be allocated correspondingly to meet the varying bandwidth requirements. Therefore, we need to analyze the interactions of network selection of mobile users and spectrum allocation between RANs.

The rest of this chapter is organized as follows. Section 13.2 describes relevant attributes for network selection. Section 13.3 presents MADM (multiple attribute decision making)-based network selection method and its enhancements. In Section 13.4, the impact of network selection on the spectrum allocation between RANs is investigated. Section 13.5 concludes this chapter.

13.2 Attributes for Network Selection

To select the most desirable network from the multiple RAN alternatives, the mobile users can utilize the objective attributes observed from both the network side and the user side. The attributes of the network side can be obtained by sensing from CR terminal or by detecting the network control signaling. The attributes of the user side can be obtained by detection (e.g., speed) and cross-layer signaling (e.g., QoS parameters). In addition, each mobile user can have his or her subjective judgments on the relative importance of attributes by expressing preferences on them.

As shown in Figure 13.1, the attribute information (network attributes, user attributes, and attribute preference) acts as the input to the network selection method to decide the order of selected networks. In the following paragraphs, we

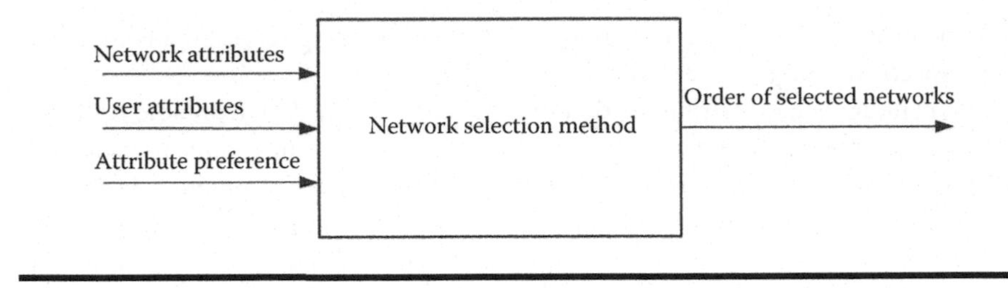

Figure 13.1 Network selection framework.

describe the network attributes, user attributes, and attribute preferences that are possibly involved in, but by no means exhaustive for, network selection.

13.2.1 Network Attributes

- Radio condition: The radio condition, for example, path loss, received reference symbol power, or received reference symbol Es/I_0, is collected by measurements from MTs.
- Price: The charged price can be presented per service or per user depending on the pricing policy. The charged price for a service can be different with respect to the classification of high-end/low-end users and primary/secondary users. The charged price can be dynamically advertised from the network and act as an incentive to attract users.
- Bandwidth: The bandwidth is the bandwidth offered by one RAN for a specific service.
- Coverage: The cell coverage of RAN can be deduced from the carrier frequency and the crossing rate of cells.
- Handoff latency: Handoff latency includes the latency spending on establishing connection with new target RAN and disconnecting with current serving network.

13.2.2 User Attributes

- QoS profile: QoS profile is a set of QoS parameters specified for a specific class of services, such as maximum/minimum data rates, packet loss rate, delay, etc.
- Willing-to-pay price: The mobile users may have their budget constraint for one service request. The budget constraint can be represented by a willing-to-pay price for the service of each user. When advertised prices of RANs exceed the highest willing-to-pay prices, the mobile users may decline their service requests.
- Residual battery power: The residual battery power is the measured left power that can be used on the MT.

- Power consumption: The power consumption is the battery power to support certain amount of data delivery on a specific RAN interface. It can be measured in milliamperes, the unit of the current drain on a service.
- Speed: The speed is the moving speed of mobile users. It can be detected by MT aided by the network.

13.2.3 Attribute Preference

The mobile users can use different formats to express their the preferences on the attributes. First, they may not give their preferences on attributes at all. Second, they can have their subjective judgments on the relative importance of the attributes and express their preferences in ordinal form by specifying the rank order of attributes. Third, they can give more precise information by expressing their preferences in quantitative cardinal forms such as relative importance ratio or weights on the attributes.

13.3 Network Selection Methods

After obtaining the relevant attributes, the next step for a mobile user is to capture the multiple conflicting attributes and make a proper decision on network selection.

Some researchers directly design functions over the attributes and weights assigned on the attributes [1,2]. As presented in [1], a cost function of a network is formulated in terms of the bandwidth, power consumption, and price. The network with the lowest value of cost function is selected. In [2], a consumer surplus function is defined as the difference of utility function and cost, and the consumer surplus function is used for the network selection.

This problem of network selection can also be tackled with the MADM technique [3,4]. MADM is a process to make preference decisions over the available alternatives that are characterized by multiple (usually conflicting) attributes. In MADM, the attributes of alternatives are constructed as the decision matrix. The attributes in the decision matrix are aggregated with relative importance weights to yield a ranking order of alternatives for decision making.

The MADM-based method is a promising solution due to its scalability on the number of attributes. Furthermore, it is a more generalized approach than the formulated function approach. Actually, the approach that explicitly defines functions over the attributes and weights can be viewed as that in which the MADM technique is implicitly utilized.

In the following paragraphs, we describe the procedure of MADM-based network selection, and we present some enhancements on the classical MADM method as proposed in [3]. The simulation illustration of the methods is given, and the implementation consideration is also discussed.

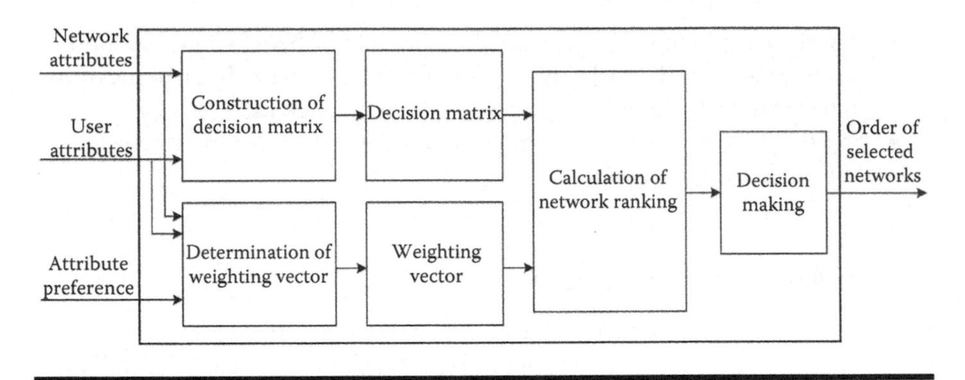

Figure 13.2 Procedure of MADM-based network selection.

13.3.1 MADM-Based Network Selection

The procedure of MADM-based network selection is shown in Figure 13.2. The main steps engaged in the MADM-based network selection are the construction of decision matrix, the determination of weighting vector, and the calculation of network ranking.

13.3.1.1 Construction of Decision Matrix

Let $S = \{S_1, S_2, \ldots, S_m\}$ be a discrete set of alternatives, $R = \{R_1, R_2, \ldots, R_n\}$ be a set of attributes, and $A = [a_{ij}]_{m\times n}$ be the decision matrix, where a_{ij} is a numerical attribute value of R_j at S_i ($i = 1, 2, \ldots, m$; $j = 1, 2, \ldots, n$).

The attributes used in MADM can be classified to benefit attributes and cost attributes from the user's perspective. The benefit attributes (e.g., bandwidth) are larger-the-better; the cost attributes (e.g., price, power consumption) are smaller-the-better.

The attributes used in MADM may have different dimensions. To measure all attributes in dimensionless units and to facilitate inter-attribute comparisons, the attribute values (i.e., the decision matrix A) need to be normalized. There are different ways of normalizing the attribute values.

One set of functions to normalize the attributes are

$$b_{ij} = \frac{a_{ij}}{a_j^{\max}} \quad i = 1, \ldots, m,\ j = 1, \ldots, n, \text{ for benefit attributes,} \tag{13.1}$$

$$b_{ij} = 1 - \frac{a_{ij}}{a_j^{\max}} \quad i = 1, \ldots, m,\ j = 1, \ldots, n, \text{ for cost attributes,} \tag{13.2}$$

where $a_j^{\max} = \max\{a_{1j}, a_{2j}, \ldots, a_{mj}\}$, $j = 1, \ldots, n$.

The attributes can also be normalized by the following set of functions

$$b_{ij} = \frac{a_{ij} - a_j^{\min}}{a_j^{\max} - a_j^{\min}}, \quad i = 1, \ldots, m, \ j = 1, \ldots, n, \text{ for benefit attributes,} \tag{13.3}$$

$$b_{ij} = \frac{a_j^{\max} - a_{ij}}{a_j^{\max} - a_j^{\min}}, \quad i = 1, \ldots, m, \ j = 1, \ldots, n, \text{ for cost attributes,} \tag{13.4}$$

where $a_j^{\max} = \max\{a_{1j}, a_{2j}, \ldots, a_{mj}\}$, and $a_j^{\min} = \min\{a_{1j}, a_{2j}, \ldots, a_{mj}\}$, $j = 1, \ldots, n$.

Another set of functions that can be used to normalize the attributes are

$$b_{ij} = \frac{a_{ij}}{\sqrt{\sum_{i=1}^{m} a_{ij}^2}}, \quad i = 1, \ldots, m, \ j = 1, \ldots, n. \tag{13.5}$$

13.3.1.2 Determination of Weighting Vector

Let w_j $(j = 1, 2, \ldots, n)$ be the relative importance weight on attribute R_j, where $\sum_{j=1}^{n} w_j = 1$ and $w_j \geq 0$ $(j = 1, 2, \ldots, n)$. $(w_1, w_2, \ldots, w_n)^T$ is the weighting vector. Weights on the attributes indicate the decision maker's judgments on the relative importance on the attributes. The weights can be elicited from the preference information given by the mobile users. As discussed in Section 13.2, the mobile users may not give their preferences on attributes, or they can express their preferences in ordinal or cardinal forms. According to the formats of available preference information, the following attribute-weighting methods can be employed correspondingly [5].

- Equal weighting: Under the assumption of no information about weights, the equal weights vector is given by

$$w_j = \frac{1}{n}, \quad j = 1, \ldots, n. \tag{13.6}$$

  Equal weights depend only on the number of attributes, so they preserve only categorical information from a mobile user's judgments.
- Rank-order centroid (ROC) weighting: If the rank order of attributes is known, but no other quantitative information about them, then we may assume that $w_{(1)} \geq w_{(2)} \geq \cdots \geq w_{(n)} \geq 0$, where (j) is the rank position of attribute j, and where $0 \leq w_{(j)} \leq 1$ and $\sum_j w_{(j)} = 1$. The weights for the attributes can be calculated using

$$w_{(j)} = \frac{1}{n} \sum_{k=j}^{n} \frac{1}{k}, \quad j = 1, \ldots, n. \tag{13.7}$$

These weights are called ROC weights because they reflect the centroid (center of mass) of the simplex defined by the ranking of the attributes. For example, with $n = 3$, $w_{(1)} = \frac{1}{3}\left(1 + \frac{1}{2} + \frac{1}{3}\right) = \frac{11}{18}$, $w_{(2)} = \frac{1}{3}\left(\frac{1}{2} + \frac{1}{3}\right) = \frac{5}{18}$, and $w_{(3)} = \frac{1}{3}\left(\frac{1}{3}\right) = \frac{2}{18}$.

- Rank-sum weighting: Rank-sum weights also use only the attribute ordering information as follows:

$$w_{(j)} = \frac{n+1-j}{\sum_{k=1}^{n} k} = \frac{2(n+1-j)}{n(n+1)}, \quad j = 1, \ldots, n \tag{13.8}$$

  where (j) is the rank position of attribute j.

  The rank-sum weights are flatter than ROC weights. For instance, with three attributes, the ROC weighting vector is $(0.611, 0.278, 0.111)$ and the rank-sum weighting vector is $(0.500, 0.333, 0.167)$.

- Ratio-scale weighting: When the mobile user gives both ordering information of the attributes and provides quantitative ratio-scale information $(r_1, \ldots, r_n)$ regarding the relative importance of attributes. The normalized weights are given by

$$w_j = \frac{r_j}{\sum_{k=1}^{n} r_j}, \quad j = 1, \ldots, n. \tag{13.9}$$

- Other weighting methods: To derive the weights, AHP (analytic hierarchy process) is discussed in [6].

These weight methods are used to obtain the weights from the subjective preference information from the mobile users. Actually, the objective preference information can be derived from the decision matrix A, and it can be integrated with the subjective preference information [7].

13.3.1.3 Calculation of Network Ranking

The problem concerned now is to rank the alternatives or to select the most desirable one based on the decision matrix and the weighting vector on the attributes. There are several computing techniques to calculate the ranking value of the alternatives based on the decision matrix and the importance weights. They are respectively referred to as SAW (simple additive weighting), MEW (multiplicative exponent weighting), TOPSIS (technique for order preference by similarity to ideal solution), and GRA (grey relational analysis). In computing the ranking value, the SAW and TOPSIS methods are used in [3,4] and the GRA method is used in [6]. The comparison of them is given for vertical handoff decision in [8].

In SAW, the overall ranking value of a candidate network is determined by the weighted sum of all the attribute values. The ranking value of each candidate network i is obtained by adding the normalized contributions from each attribute a_{ij} multiplied by the importance weight w_j assigned on the attribute j, that is,

$$d_i = \sum_{j=1}^{n} w_j b_{ij}. \tag{13.10}$$

Hence, the criterion to select the best alternative S_i^* is to choose the alternative with the largest ranking value, that is,

$$S_i^* = \underset{1 \le i \le m}{\operatorname{argmax}}\, d_i. \tag{13.11}$$

In MEW, the overall ranking value of a candidate network is determined by the weighted product of all the attribute values by

$$d_i = \prod_{j=1}^{n} a_{ij}^{w_j}. \tag{13.12}$$

Note that in (13.13), w_j is a positive power for benefit attribute $a_{ij}^{w_j}$, and a negative power for cost metrics $a_{ij}^{-w_j}$. As the ranking value of a network obtained by MEW does not have an upper bound [9], it is convenient to compare each network with the ranking value of the positive ideal network. This network is defined as the network with the best values in each attribute. For a benefit attribute, the best value is the largest. For a cost attribute, the best value is the lowest.

The value ratio between network i and the positive ideal is calculated by

$$\eta_i = \frac{\prod_{j=1}^{n} a_{ij}^{w_j}}{\prod_{j=1}^{n} \left(a_{ij}^*\right)^{w_j}} \tag{13.13}$$

where a_{ij}^* denotes the best value for each attribute.

The selected candidate network is the one with the largest ratio, that is,

$$S_i^* = \underset{1 \le i \le m}{\operatorname{argmax}}\, \eta_i. \tag{13.14}$$

TOPSIS first normalizes the attributes using (13.5) and then calculates the weighted normalized value. With the weighted normalized values, it finds the ideal solution and the non ideal solution. The ideal solution is obtained by using the best values for each attribute, and the non ideal solution is obtained by using the worst values for each attribute. Let c_i denote the relative closeness (or similarity) of the candidate network i to the ideal solution. The selected candidate network is the one that is the closest to the ideal solution, that is,

$$S_i^* = \underset{1 \le i \le m}{\text{argmax}}\ c_i. \tag{13.15}$$

GRA normalizes the decision matrix and composes a reference series S_0 from ideal attributes, then GRA calculates the GRC (gray relational coefficient), $\Gamma_{0,i}$, as the score used to describe the similarity between each candidate network and the ideal network. The selected network is the one that has the highest GRC, that is,

$$S_i^* = \underset{1 \le i \le m}{\text{argmax}}\ \Gamma_{0,i}. \tag{13.16}$$

13.3.2 Enhancements for MADM-Based Network Selection

The classical MADM-based methods provide an effective approach for network selection, but it has a few drawbacks to be employed by the mobile users and network operators. First, the price of services is usually charged based on the data volume delivered to the mobile users. For real-time services such as rate-adaptive video streaming, the mobile users have to pay more for the higher bandwidth, but the extra payment may only yield limited improvement on perception experience. In other words, the classical MADM-based methods may lead the user to select a network with higher bandwidth, but the marginal QoS improvement is unproportionate to his or her payment. Second, the classical MADM-based methods do not address the balance of user distribution among networks and the ping pong effect caused by frequent network switching. For example, several mobile users reside in the overlapping area of two networks A and B. If the network A has the higher ranking value than network B, all mobile users select network A, and the load of network A increases. Then, the users may suffer the degraded QoS in network A, and they may find that the ranking value of network A becomes lower than that of network B. Therefore, all the users switch to network B instead. Similarly, after accessing the network B, the user may find that the ranking value of network B falls down and that of network A rises up, and they have to switch back to network A. Thus, the users perform the networks reselection frequently, which leads to the ping pong effect and unbalanced user distribution between networks.

To address these issues, enhancements for MADM-based network selection method are proposed in [3]. First, the network attributes are proposed to be categorized into two categories. One category includes the attributes that are QoS-related and the mobile user needs to pay for; the other category includes the attributes that are either QoS-related but the mobile user does not need to pay for or not QoS-related. For instance, the available bandwidth of the network can be viewed as an attribute in the first category, and the power consumption can be viewed as an attribute in the second category. In the proposed scheme, the network attributes in

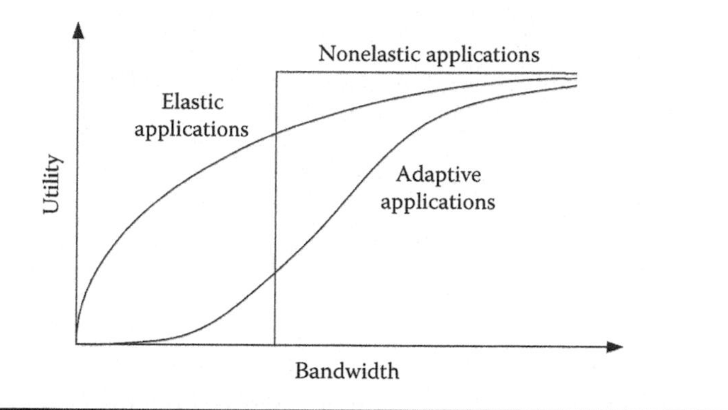

Figure 13.3 Utility of applications.

the first category is transformed to the utility function of the application, and the computed utility value is used in the decision matrix for further ranking calculation.

The utility functions can be designed corresponding to the classes of applications such as non elastic application (e.g., VoIP), elastic application (e.g., FTP, e-mail), and adaptive application (e.g., video streaming). Different formulas of utility functions have been studied to reflect the characteristics of applications, but the shapes of the functions designed for a class of applications are similar and shown in Figure 13.3.

Another enhancement is proposed in [3] to avoid the imbalance of network loading. In the decision making phase, it is proposed that the target network is selected randomly according to a probability determined by ranking values of the candidate networks. In other words, unlike the classical MADM-based method where the network with the maximum ranking value is selected, the proposed method possibly selects the network whose ranking value is not the maximum. To avoid choosing the networks that are under extremely bad conditions, only the networks whose utility value for an application is above a threshold are selected by the users. Let us denote the probability of the network i to be selected as P_i. P_i can be given by

$$P_i = \begin{cases} \dfrac{d_i}{\sum_{k \in \mathbf{S}} d_k} & i \in \mathbf{S} \\ 0 & i \notin \mathbf{S} \end{cases} \tag{13.17}$$

where

$\mathbf{S} = \{i \mid u_i \geq u_{\text{th}}, 1 \leq i \leq m\}$

m is the number of candidate networks

u_i is the utility value of network i

u_{th} is the minimum acceptable utility for a service from the view of user

To further suppress the Ping Pong effect between the networks, Yu et al. [3] additionally introduce the following network switching thresholds in the decision making stage,

- Criterion I: $u_c \leq u_{th}$ and $d(t) > d_c$
- Criterion II: $\frac{d(t)}{d_c} \geq 1 + \alpha$
- Criterion III: $u_c \leq u_{th}$ and $\frac{d(t)}{d_c} \geq 1 + \alpha$

where

u_c is the utility of the current serving network
u_{th} is the minimum acceptable utility for a service from the view of user
d_c is the ranking of the current serving network
$d(t)$ is the ranking of the targeted network to be accessed
α is a constant and $\alpha \in [0, 1]$

Criterion I focuses on keeping the continuity of the service by avoiding too aggressive network handoff. It is more suitable for the real-time applications. Criterion II cares more about selecting the suitable RAN in terms of ranking value than maintaining the continuity of service. It is more suitable for the non-real-time applications. Criterion III is largely the combination of Criterion I and Criterion II, which provides more conservative control on network handoff.

13.3.3 Simulation Illustration

In this section, we give simulation illustration to show the effectiveness of MADM-based method and its enhancements on the network selection. In the simulated scenario, there are two RANs alternatives available for mobile users to select. The attributes for the simulated scenario are listed in Table 13.1. Three attributes, bandwidth, price, power consumption, are considered in the MADM method. In the construction of the decision matrix, the bandwidth is viewed as the benefit attribute and is normalized using (13.1); price and power consumption are viewed as the cost attributes and are normalized using (13.2).

Suppose that the preference information given by the mobile users on the attributes is in the ordinal form. Two types of mobile users, high-end users and

Table 13.1 Simulation Parameters

Attributes	*RAN A*	*RAN B*
Bandwidth (kbps)	50	20
Price (cents/Mb)	1.0	0.5
Power consumption (mA)	100	50

Table 13.2 Network Selection Results

	Classical MADM		*Enhanced MADM*	
Application	*High-End*	*Low-End*	*High-End*	*Low-End*
Elastic	A	B	B	B
Adaptive	A	B	A	B

low-end users, are considered. The low-end users care more about the price of the service than the QoS offered by the network. The high-end users care more about the QoS experience than the price. The high-end users specify the relative importance order of attributes as bandwidth $\succ$ power consumption $\succ$ price; the low-end users specify the relative importance order of attributes as price $\succ$ bandwidth $\succ$ power consumption. By using the ROC weighting method, the weighting vector for the high-end users is $(0.611, 0.111, 0.278)$, and the weighting vector for the low-end users is $(0.278, 0.611, 0.111)$, for attributes (bandwidth, price, power consumption).

In this case, two types of services, elastic service and adaptive service, are considered. The utility function for elastic application is assumed to be $u = 1 - e^{-\frac{B}{\phi}}$, $\phi = 5$, and the utility function for adaptive application is assumed to be $u = 1 - e^{-\frac{B^2}{B+\mu}}$, $\mu = 450$. The bandwidth values of RANs are used in the classical MADM method, and the transformed utility values are used in the enhanced MADM method.

In the calculation of network ranking values, SAW is employed. Based on the calculated ranking values of RANs, the network selection results are listed in Table 13.2 and the results of utility and payment are shown in Figure 13.4 for respective types of users and applications with classical and enhanced MADM methods.

As shown in Figure 13.4a, the high-end users always get high utility for both elastic and adaptive applications. On the other hand, the low-end users always get low payment for both elastic and adaptive applications as shown in Figure 13.4b (assuming 10 Mb delivered data). It is also seen that the high-end user selects RAN A with the classical MADM method, and selects RAN B with the enhanced MADM method for the elastic application. The utility perceived by the high-end users with enhanced MADM only degrades about 0.2 percent. However, the total payment of the high-end users is saved about 50 percent. With the enhanced MADM method, the high-end users can make proper network selection to obtain satisfied QoS with payment saving.

Next, we study the effect of the user distribution with different network selection methods. Three network selection methods are considered including the classical MADM, the enhanced MADM with network handoff threshold, and the

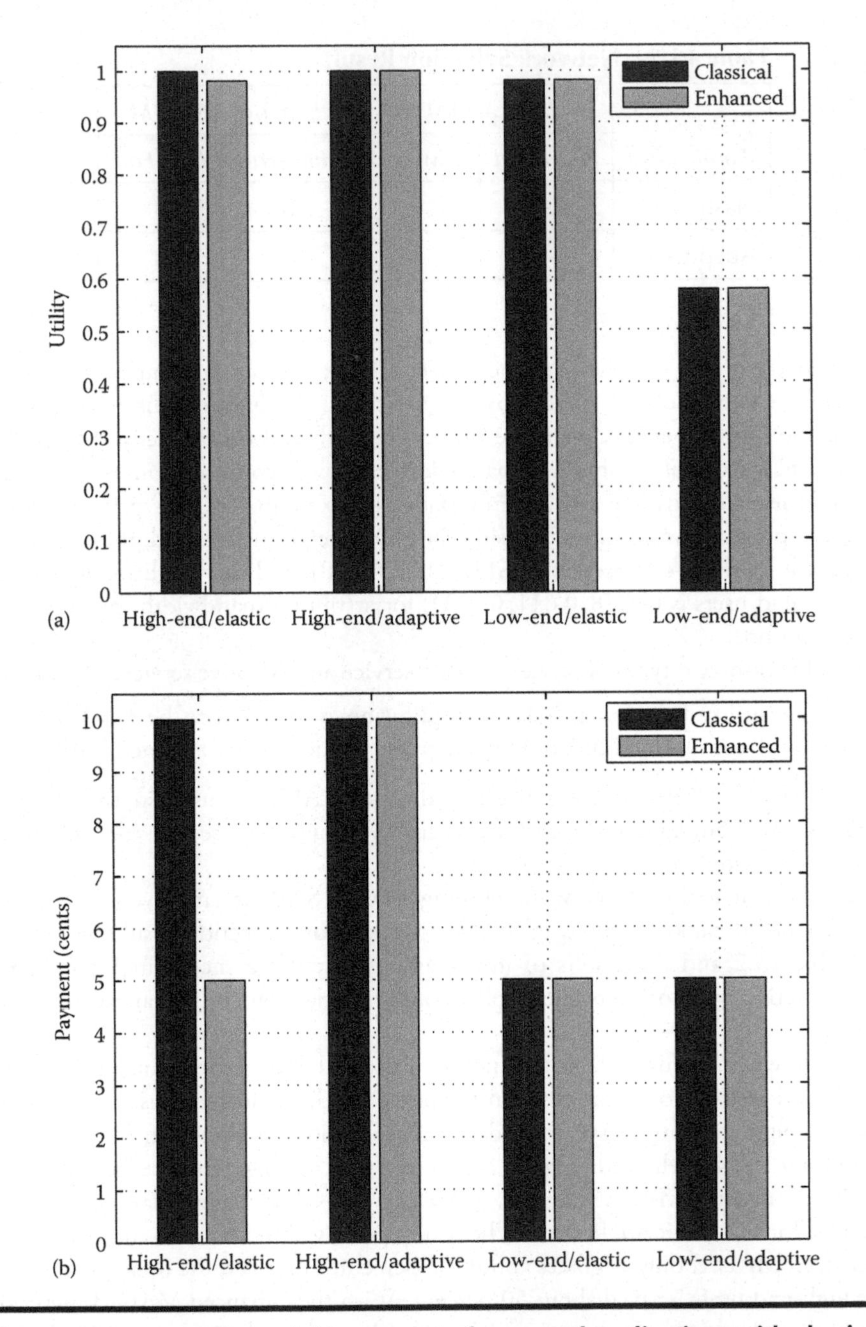

Figure 13.4 (a) Utility and (b) payment of users and applications with classical and enhanced MADM methods.

random-waiting method. In the random-waiting method, a random period is specified for mobile users before handoffs. Only if the network is consistently better than the current one in use for the stability period does the mobile user perform handoff.

To evaluate the distribution of users among the networks, the parameter of balance index (BI) is defined as

$$\text{BI} = \prod_{i=1}^{M} \frac{MK_i}{N} \tag{13.18}$$

where

N is the number of users
M is the number of networks
K_i is the average number of users in network i

The balance index is closer to 1 with more balanced user distribution between RANs.

To illustrate the effectiveness of the second enhancement for MADM-based method, Yu et al. [3] employ the handoff threshold and take into account the attribute of handoff latency. Figure 13.5 gives a trace of the balance index and the number of network reselections with different methods. It is observed that the user distribution is more balanced and the number of network reselections is reduced with the enhanced MADM compared with both the classical MADM and the random-waiting method.

13.3.4 Implementation Consideration

In the earlier discussion, the network selection method is assumed to be performed at MT carried by the mobile users. The network selection method can also be performed at the network side. There are pros and cons for these two approaches.

When the network selection is performed at MT, more intelligence needs to be added on the MT to enable it to sense the attributes, make calculation, and decision making. With this approach, the processing overhead at network side is low, but it may involve air interface signaling overhead to broadcast network attributes via control channel or beacons.

When the network selection method is performed in the network side, the MT has low processing overhead, but it still needs to do the measurements of network and user attributes, and report the attributes and preferences to the network via air interface signaling. In the network side, one central management unit for all involved RANs is needed to perform the network selection method for mobile users. The network selection result is notified to the mobile users afterward via the air interface of a specific RAN. With this approach, increased number and more accurate information of network attributes can be engaged in the MADM method, and more advanced load-balancing method can be employed by the network. However, the

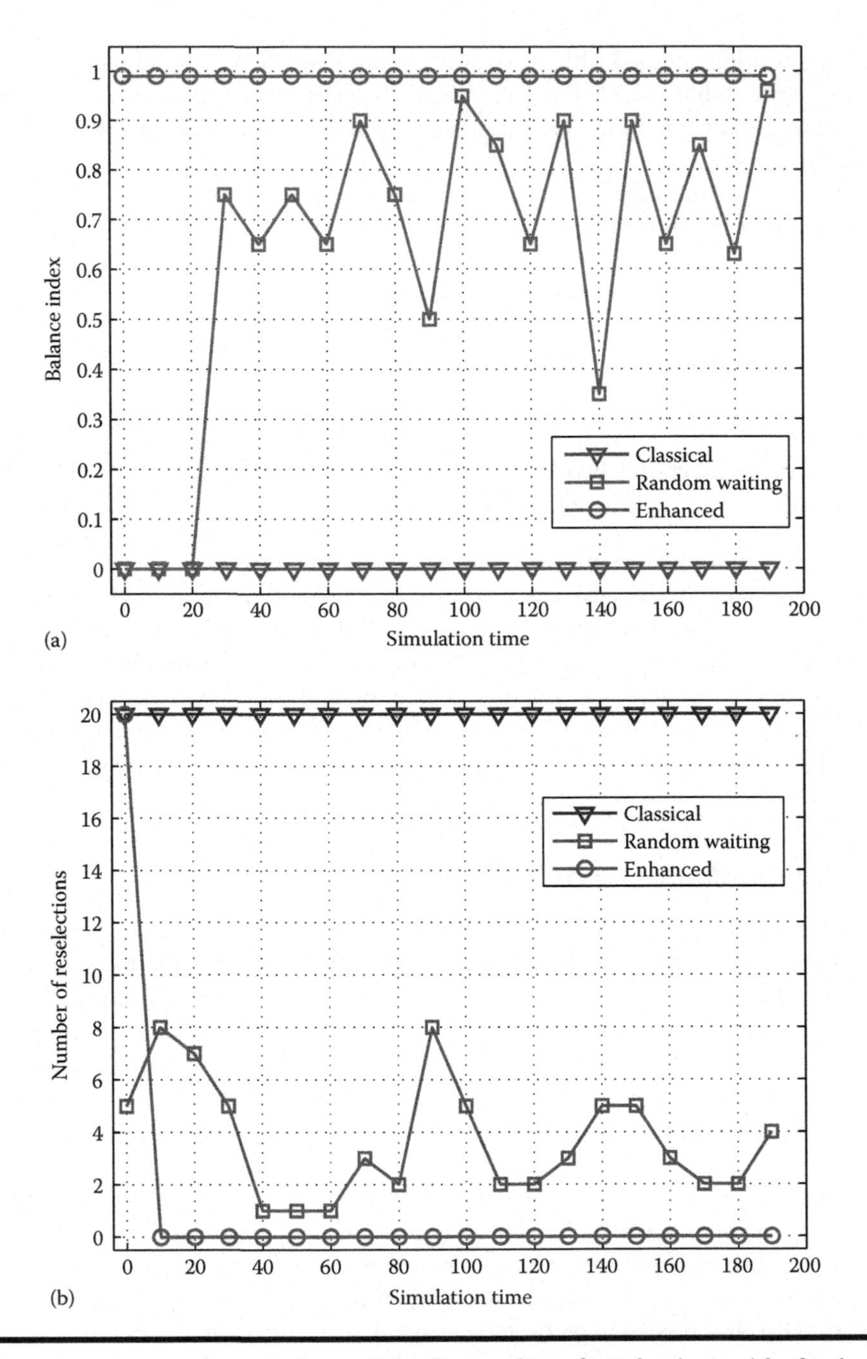

Figure 13.5 (a) Balance index and (b) the number of reselections with classical, random-waiting, and enhanced MADM methods.

processing overhead becomes higher in the network, and scalability in terms of the number of mobile users may become an issue.

13.4 Impact of Network Selection on Spectrum Allocation

To improve overall spectrum utilization in cognitive wireless networks, the RAN operators can dynamically acquire a spectrum by leasing or bidding from a common spectrum pool organized by a spectrum owner/broker. In such a spectrum market, the RAN operators need to decide the amount of spectrum to be purchased with intent to support more users for profit maximization. To make such a decision, a RAN operator needs to consider the cost of spectrum allocation that possibly be measured in the price of spectrum units (chunks).

Furthermore, the network operators face a common user pool when the RANs are deployed in the geographically overlapped areas and provide substitutable services (e.g., VoIP). The mobile users have the freedom to select and adapt to one RAN as his or her preferred serving RAN. Hence the RAN operators attempt to attract more users to fully exploit its purchased spectrum and obtain as much profit as possible. As price is an important economic incentive to users, it is often considered as an effective mechanism for traffic management. The RANs can adjust and advertise their prices to potential users. Hence, the RAN operators compete with each other in acquiring the spectrum from a common spectrum pool and also in attracting mobile users. The considered scenario is shown in Figure 13.6.

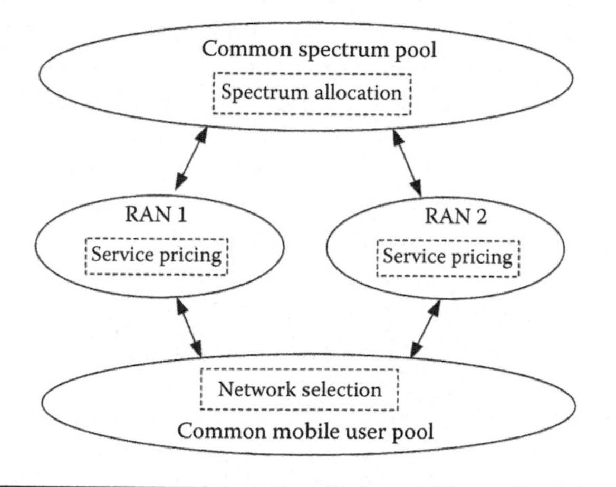

Figure 13.6 Interactions between spectrum allocation, service pricing, and network selection.

The strategic competition of spectrum allocation starts to gain the interest of researchers [10–12]. Their study mainly focuses on the auction and bidding strategies for dynamic spectrum management. Nevertheless, fewer works investigate both spectrum allocation and service pricing together. Sengupta et al. [13] cope with this topic by forming a noncooperative game between one network operator with users to decide service price and offered bandwidth, and then forming an auction model between multiple network operators for spectrum bidding. Thus, it does not link the spectrum allocation directly with the service pricing.

To improve the understanding of the interactions between spectrum allocation, service pricing, and network selection, we formulate a service demand model at the demand (user) side and a noncooperative game model of spectrum allocation between RANs at the supply (operator) side [14]. Then we derive the Nash equilibrium and its existence conditions of the competition game. Furthermore, we discuss the effects of influential factors (spectrum trading cost, spectrum efficiencies, service charge, willing-to-pay prices of users) on the competition results.

13.4.1 Service Demand from Mobile Users

When selecting the serving RAN, the mobile users may face multiple attributes associated with users, networks, and applications, and these attributes can be classified to benefit attributes and cost attributes from the users' perspective. When all the benefit attributes (e.g., bandwidth) offered by RANs to the demanded service are identical, the differentiation of services lies in the cost attribute, that is, price. Consequently, the mobile users may only consider the price charged for when selecting the serving RAN. By recognizing this choice strategy of users, each RAN is then motivated to adjust its advertised price to attract users for access.

The quantity of service demanded for RANs depends on the number of users with service requests and the network selection results of them. After collecting the price values advertised on the substitutable services from RANs, the mobile users first consider the RAN with the lowest price. In addition to comparing the price values of RANs for network access, a mobile user may decline his or her service requests when all the advertised prices of RANs exceed his or her willing-to-pay price. Hence, the willing-to-pay prices of users affect the service demand.

We assume that the amounts that users are willing to pay for are independent, identically distributed nonnegative random variables. Let $F(x)$ denote the proportion of users willing to pay a price of at most x. We call $F(x)$ the willingness-to-pay distribution with density $f(x)$. The probability that a user is willing to join the system at an advertised price x is $1 - F(x)$. We assume that x is subject to a uniform distribution on the interval $[a, b]$, that is, $f(x) = 1/b - a$ where $b > a > 0$.

Let N be the total (maximum) number of users that have service requests in the overlapped service area for two RANs. N is the number of the potential users known and to be attracted by two RANs. N can be estimated on account of the traffic activity of a common user pool. Assume that each potential user of N users only initiates one service request with the same QoS (e.g., data rate) requirements. Hence, the number of users and the number of services can be used interchangeably hereafter.

Let p_i be the price charged by RAN i, and let $D(p_i)$ be the quantity of service demanded for RAN i, and we further denote $D(p_i)$ by q_i. Next, we describe the service demand in relation to the service prices p_1 and p_2 advertised by two RANs. If $p_i > p_j$, RAN i's service demand q_i is zero where RAN j is the other RAN ($j = 2$ if $i = 1$, and $j = 1$ if $i = 2$). As RAN j has lower price charged for the users, none of the users select RAN i as their serving RAN.

If $p_i < p_j$, RAN i's service demand q_i can be represented by

$$q_i = \begin{cases} N & \text{if } p_i \leq a \\ N(1 - F(p_i)) = N\dfrac{b - p_i}{b - a} & \text{if } a < p_i \leq b \\ 0 & \text{if } p_i > b \end{cases} \tag{13.19}$$

Figure 13.7a gives an illustration of q_i for this case.

If $p_i = p_j$, the users can be split up to two RANs or assigned to one RAN according to rules when the RANs charge identical prices for the users.

Let $P(q)$ denote the inverse demand function $D^{-1}(p)$, which is the inverse function of $D(p)$. $P(q)$ gives us the price corresponding to a quantity q. Figure 13.7b illustrates the inverse demand function $P(q)$.

Let p be the price at which the service demand is exactly equal to q, that is, $D(p) = q$. Then by inversing the demand function $D(p)$ of (13.19), p can be derived as

$$p = P(q) = b - \frac{q}{N}(b - a). \tag{13.20}$$

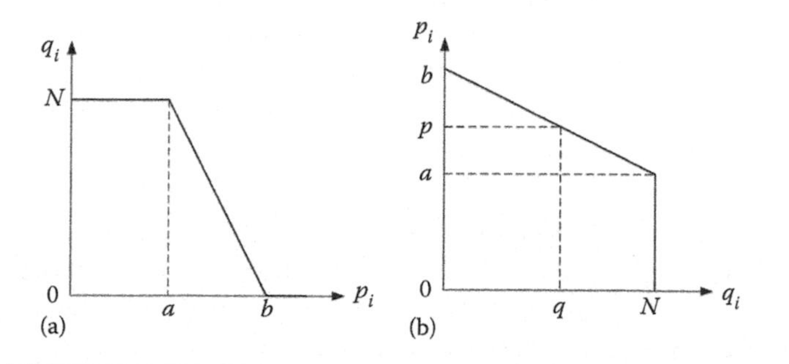

Figure 13.7 (a) The service demand ($p_i < p_j$) and (b) inverse service demand of RAN i.

13.4.2 Competition Model between RANs for Spectrum Allocation

Under the rule of fixed spectrum allocation, the capacities of RANs are restricted exogenously. However, with a flexible spectrum allocation, there arises a question that how much amounts of spectrum (or equivalently, what levels of capacity) the competing RANs choose. To address this question, we need a model of endogenous capacity choice. Fortunately, we can be enlightened by the "Cournot competition" model in economics [15,16].

"Cournot competition" (named after its creator—Antoine Augustin Cournot) refers to a model of oligopoly in which two or more firms compete by simultaneously choosing product quantities. Cournot competition can be formed as a noncooperative game in which each of $m \geq 2$ players (firms), $i = 1, 2, \ldots, m$, simultaneously chooses a quantity q_i for homogeneous product. Firms attempt to maximize their own payoffs (profits), assuming their rival's quantity is given.

In this case, RAN i's profit π_i can be represented as

$$\pi_i(q_1, q_2) = q_i P(q_1 + q_2) - C_i(q_i) \ (i = 1, 2). \tag{13.21}$$

where the price in the market is $P(q_1 + q_2)$ at which market demand equals the sum capacity $q_1 + q_2$.

A Cournot equilibrium is a Nash equilibrium of this game; that is, a pair of quantities $\left(q_1^*, q_2^*\right)$ such that, for player 1, $\pi\left(q_1^*, q_2^*\right) \geq \pi\left(q_1, q_2^*\right)$ for all q_1, and for player 2, $\pi\left(q_1^*, q_2^*\right) \geq \pi\left(q_1^*, q_2\right)$ for all q_2. At the point of equilibrium, neither firm can increase profits by unilaterally changing quantity.

The strategies of RANs to be decided are the amounts of spectrum units to be allocated, which are denoted by f_1 and f_2 for RAN 1 and RAN 2, respectively. We denote the cost of RAN i in purchasing the spectrum units f_i by $C(f_i)$.

The amounts of spectrum allocated can be transformed to the equivalent capacities of RANs, which can be measured in the number of services that can be supplied. Let k_1 and k_2 denote the capacities of RAN 1 and RAN 2, respectively. The transformation depends on the spectrum efficiency of respective RAN. Let α_1 and α_2 denote the spectrum efficiency factors of RANs, which are measured in the number of supported services per spectrum unit. Therefore, we have

$$f_i = \frac{k_i}{\alpha_i}, \text{ or equivalently } k_i = \alpha_i f_i. \tag{13.22}$$

The total amount of allocated spectrum of two RANs is $f_1 + f_2$, and the corresponding total sum capacity of two RANs expected on the service market is $k_1 + k_2 = \alpha_1 f_1 + \alpha_2 f_2$. Then by using (13.20) the price in the service market is $P(k_1 + k_2)$, and $P(k_1 + k_2)$ can be derived as

$$P(k_1 + k_2) = b - \frac{k_1 + k_2}{N}(b - a) = b - \frac{\alpha_1 f_1 + \alpha_2 f_2}{N}(b - a). \tag{13.23}$$

It is seen from (13.23) that $P(k_1 + k_2)$ can be represented as a function of f_1 and f_2. In the following discussion, $P(f_1 + f_2)$ is used as an equivalent to $P(k_1 + k_2)$.

By using (13.21), the profit of RAN i is represented as

$$\pi_i(f_1, f_2) = \alpha_i f_i P(f_1 + f_2) - C_i(f_i) \ (i = 1, 2). \tag{13.24}$$

In the formulated game, a Nash equilibrium is a pair $\left(f_1^*, f_2^*\right)$ of allocated spectrum units such that

$$f_1^* \in \text{argmax}\ \pi_1\left(f_1, f_2^*\right) = \alpha_1 f_1 P\left(f_1 + f_2^*\right) - C_1(f_1) \tag{13.25}$$

$$f_2^* \in \text{argmax}\ \pi_2\left(f_1^*, f_2\right) = \alpha_2 f_2 P\left(f_1^* + f_2\right) - C_2(f_2) \tag{13.26}$$

where $f_i^* \in \text{argmax}\ \pi_i\left(f_i, f_j^*\right)$ means that f_i^* should be in the solution set of $\text{argmax}\ \pi_i\left(f_i, f_j^*\right)$.

To derive the Nash equilibrium of the game, we first derive the optimal (or profit maximizing) action of each RAN for each possible output of its rival. This gives us the best response function (also known as reaction function) of each RAN. Let $B_1(f_2)$ and $B_2(f_1)$ denote the best response function of the RAN 1 and the RAN 2, respectively. A Nash equilibrium is a pair $\left(f_1^*, f_2^*\right)$ of spectrum units such that f_1^* is a best response to f_2^*, and f_2^* is a best response to f_1^*. That is, f_1^* is in $B_1\left(f_2^*\right)$ and f_2^* is in $B_2\left(f_1^*\right)$.

By letting $\partial\pi_1/\partial f_1 = 0$, we have

$$\frac{\partial\pi_1}{\partial f_1} = \alpha_1 P(f_1 + f_2) + \alpha_1 f_1 P'(f_1 + f_2) - C_1'(f_1) = 0 \tag{13.27}$$

$$\frac{\partial\pi_2}{\partial f_2} = \alpha_2 P(f_1 + f_2) + \alpha_2 f_2 P'(f_1 + f_2) - C_2'(f_2) = 0. \tag{13.28}$$

$B_1\left(f_2^*\right)$ and $B_2\left(f_1^*\right)$ are defined based on (13.27) and (13.28), respectively.

Let c be the cost of a spectrum unit, we have $C_1(f_1) = f_1 c = k_1 c/\alpha_1$, and $C_2(f_2) = f_2 c = k_2 c/\alpha_2$.

Now the equations of (13.27) and (13.28) become

$$\frac{\partial\pi_1}{\partial f_1} = b - \frac{2\alpha_1 f_1 + \alpha_2 f_2}{N}(b - a) - \frac{c}{\alpha_1} = 0 \tag{13.29}$$

$$\frac{\partial\pi_2}{\partial f_2} = b - \frac{\alpha_1 f_1 + 2\alpha_2 f_2}{N}(b - a) - \frac{c}{\alpha_2} = 0. \tag{13.30}$$

Then the best response functions of RANs can be obtained as

$$f_1^* = B_1(f_2) = \frac{1}{2\alpha_1}\left(\frac{b - \frac{c}{\alpha_1}}{b-a}N - \alpha_2 f_2\right) \tag{13.31}$$

$$f_2^* = B_2(f_1) = \frac{1}{2\alpha_2}\left(\frac{b - \frac{c}{\alpha_2}}{b-a}N - \alpha_1 f_1\right) \tag{13.32}$$

based on (13.29) and (13.30). $B_1(f_2)$ and $B_2(f_1)$ are illustrated in Figure 13.8.

Solving the two best response functions of (13.31) and (13.32), we get a single Nash equilibrium of the pair of allocated spectrum units of two RANs as

$$(f_1^*, f_2^*) = \left(\frac{N\left(b - \frac{2c}{\alpha_1} + \frac{c}{\alpha_2}\right)}{3\alpha_1(b-a)}, \frac{N(b - \frac{2c}{\alpha_2} + \frac{c}{\alpha_1})}{3\alpha_2(b-a)}\right). \tag{13.33}$$

The point of Nash equilibrium is the intersection of two best response functions as shown in Figure 13.8. Then at the point of Nash equilibrium, the pair of capacities of two RANs is

$$(k_1^*, k_2^*) = \left(\frac{N\left(b - \frac{2c}{\alpha_1} + \frac{c}{\alpha_2}\right)}{3(b-a)}, \frac{N\left(b - \frac{2c}{\alpha_2} + \frac{c}{\alpha_1}\right)}{3(b-a)}\right) \tag{13.34}$$

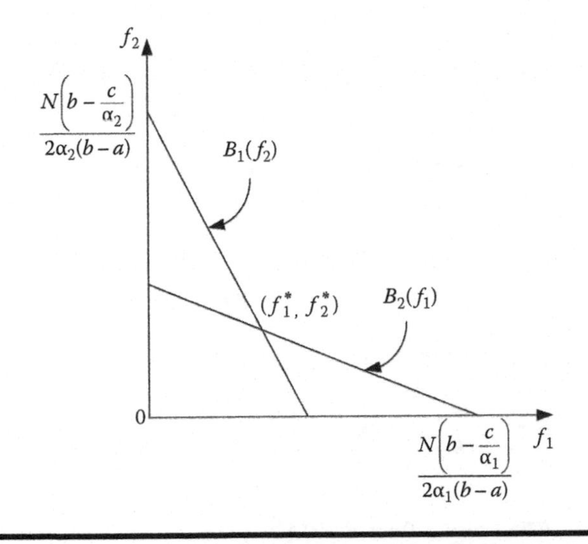

Figure 13.8 Best response functions and their intersection (Nash equilibrium).

by using (13.22). The aggregate capacity on the service market is $k_1^* + k_2^*$. Then by using (13.23) the price that each RAN expects to charge on a service request is

$$p_1^* = p_2^* = P\left(k_1^* + k_2^*\right) = \frac{1}{3}\left(b + \frac{c}{\alpha_1} + \frac{c}{\alpha_2}\right). \tag{13.35}$$

Next, by using (13.24), the profits expected by two RANs can be written as

$$\left(\pi_1^*, \pi_2^*\right) = \left(\frac{N\left(b - \frac{2c}{\alpha_1} + \frac{c}{\alpha_2}\right)^2}{9(b-a)}, \frac{N\left(b - \frac{2c}{\alpha_2} + \frac{c}{\alpha_1}\right)^2}{9(b-a)}\right). \tag{13.36}$$

Note that to guarantee the existence of the Nash equilibrium at $\left(k_1^*, k_2^*\right)$, we further need $k_1^* \geq 0$ and $k_2^* \geq 0$.

From $k_1^* \geq 0$, we have $\frac{b}{c} \geq \frac{2}{\alpha_1} - \frac{1}{\alpha_2}$; from $k_2^* \geq 0$, we have $\frac{b}{c} \geq \frac{2}{\alpha_2} - \frac{1}{\alpha_1}$. Therefore, we need the condition

$$\frac{b}{c} \geq \max\left(\frac{2}{\alpha_1} - \frac{1}{\alpha_2}, \frac{2}{\alpha_2} - \frac{1}{\alpha_1}\right), \tag{13.37}$$

to be satisfied or, equivalently, we need

$$\frac{c}{b} \leq \min\left(\frac{1}{\frac{2}{\alpha_1} - \frac{1}{\alpha_2}}, \frac{1}{\frac{2}{\alpha_2} - \frac{1}{\alpha_1}}\right), \tag{13.38}$$

or further equivalently, we need

$$\alpha_1 \geq \frac{3}{2\frac{b}{c} + \frac{c}{b}} \quad \text{and} \quad \alpha_2 \geq \frac{3}{\frac{b}{c} + 2\frac{c}{b}}. \tag{13.39}$$

It implies by (13.37) that the willing-to-pay price of user b should be greater than a certain value with a fixed spectrum unit cost c; it implies by (13.38) that the spectrum unit cost c should be less than a certain value with a fixed willing-to-pay price of users b; it also implies by (13.39) that the network efficiency factors of RANs should be large enough with fixed b and c. Otherwise, there is still no Nash equilibrium for this game.

13.4.3 Numerical Results and Discussion

To interpret the economic meanings of our analysis, in this section we present the numerical results of capacities and earned profits of RANs at the Nash equilibrium. Based on the exemplary results, we want to gain more insights of the effects of influential factors of the involved parties on the competition. The influential factors considered are the cost of spectrum unit asked by the spectrum broker, the spectrum efficiency of RAN operators, and the willing-to-pay price of users.

13.4.3.1 Capacities of RANs

Let η_k denote the ratio of capacities of RAN 1 and RAN 2 $\frac{k_1}{k_2}$. At the equilibrium, by using (13.34), η_k can be represented as

$$\eta_k = \frac{b - \frac{2c}{\alpha_1} + \frac{c}{\alpha_2}}{b - \frac{2c}{\alpha_2} + \frac{c}{\alpha_1}} = \frac{\eta_p - \frac{2}{\alpha_1} + \frac{1}{\alpha_2}}{\eta_p - \frac{2}{\alpha_2} + \frac{1}{\alpha_1}}$$

where $\eta_p = \frac{b}{c}$, which is the ratio of the willing-to-pay price of users b and the spectrum unit cost c. To satisfy $k_1 > k_2$, it can be derived that we just need $\alpha_1 > \alpha_2$. Therefore, the RAN with higher spectrum efficiency always yields a larger capacity in service provisioning.

At the equilibrium, the sum capacity of two RANs is represented as

$$k_1 + k_2 = \frac{N}{3(b-a)}\left(2b - \frac{c}{\alpha_1} - \frac{c}{\alpha_2}\right).$$

It is necessary to know when all users can be served by the two RANs. To satisfy $k_1 + k_2 \geq N$, it is derived that the following condition needs to be satisfied:

$$3a - b \geq \frac{c}{\alpha_1} + \frac{c}{\alpha_2}. \tag{13.40}$$

The condition of (13.40) is equivalent to

$$\frac{1}{\alpha_1} + \frac{1}{\alpha_2} \leq \frac{3a-b}{c} \tag{13.41}$$

and is further equivalent to

$$c < \frac{3a-b}{\frac{1}{\alpha_1} + \frac{1}{\alpha_2}}. \tag{13.42}$$

It implies by (13.40) that high concentration on willing-to-pay prices of users make the RANs to provide a larger sum capacity. It implies by (13.41) that the RANs tend to provide larger sum capacity if they have higher spectrum efficiencies. Furthermore, it implies by (13.42) that the RANs tend to provide larger sum capacities if spectrum broker asks lower spectrum price.

To give a numerical illustration of the capacities, Figures 13.9 and 13.10 show the capacity ratio η_k and sum capacity $k_1 + k_2$, respectively. In Figures 13.9 and 13.10, we set $\alpha_2 = 1$, and vary α_1 and η_p to see how the capacity ratio and the sum capacity change with them. It is seen that for a given η_p, the capacity ratio η_k increases as α_1 increases. Therefore, the larger the difference of spectrum efficiencies, the larger the difference of capacities. We set the total number of users

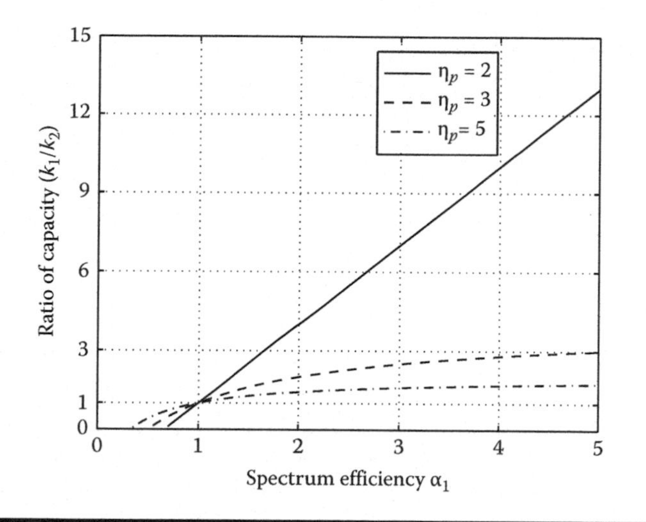

Figure 13.9 Capacity ratio η_k versus spectrum efficiency factors ($\alpha_2 = 1$, α_1 varies).

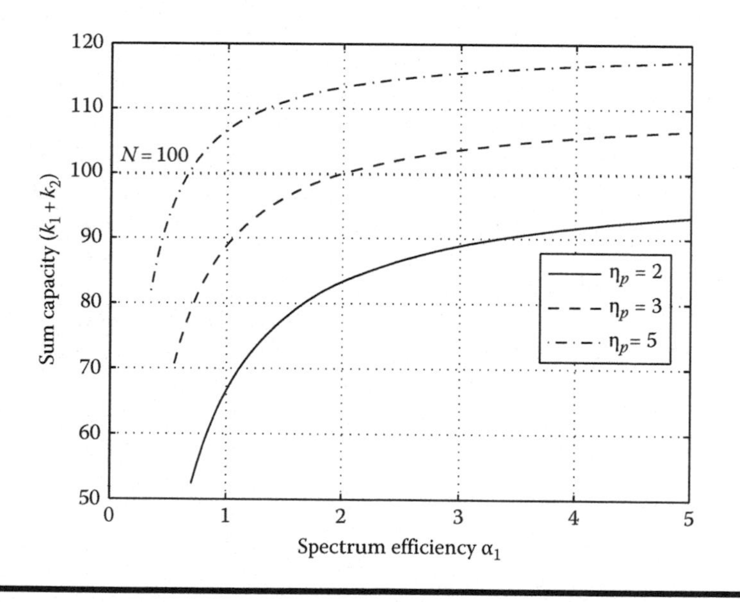

Figure 13.10 Sum capacity versus spectrum efficiency factors ($\alpha_2 = 1$, α_1 varies).

$N = 100$ in Figure 13.10. It can be observed that for a given α_1, the sum capacity $k_1 + k_2$ becomes larger as η_p increases, and it can exceed the number of total users (i.e., $k_1 + k_2 > 100$) under satisfied conditions. Nevertheless, the capacity ratio η_k becomes smaller as η_p increases. In other words, the difference of capacities of two RANs reduces with the increase of η_p.

13.4.3.2 Profits of RANs

Let η_π denote the ratio of profits of RAN 1 and RAN 2 $\frac{\pi_1}{\pi_2}$. At the equilibrium, by using (13.36), η_π can be represented as

$$\eta_\pi = \frac{\left(b - \frac{2c}{\alpha_1} + \frac{c}{\alpha_2}\right)^2}{\left(b - \frac{2c}{\alpha_2} + \frac{c}{\alpha_1}\right)^2} = \frac{\left(\eta_p - \frac{2}{\alpha_1} + \frac{1}{\alpha_2}\right)^2}{\left(\eta_p - \frac{2}{\alpha_2} + \frac{1}{\alpha_1}\right)^2}. \tag{13.43}$$

To satisfy $\pi_1 > \pi_2$, it can also be derived that we just need $\alpha_1 > \alpha_2$. Therefore, a higher spectrum efficiency always yields a higher profit for a RAN.

Figures 13.11 and 13.12 show the profit ratio η_π and sum profit $\pi_1 + \pi_2$, respectively. In Figures 13.11 and 13.12, we set $\alpha_2 = 1$, and vary α_1 and η_p to see how the profit ratio and the sum profit change with them. It is seen that for a given η_p, the profit ratio η_π and the sum profit $\pi_1 + \pi_2$ increase as α_1 increases. Thus, the RAN with higher spectrum efficiency obtains higher profit. The larger the difference of spectrum efficiency, the more significant the profit earned. On the other hand, the effect of η_p on the sum profit is not straightforward. It is observed in Figure 13.12 that for a given α_1, the sum profit $\pi_1 + \pi_2$ does not always increase with the increase of η_p. Actually, it is relevant to the specific values of α_1 and η_p.

Next, we investigate how the earned profits of RANs change with the cost of the spectrum unit c. Figure 13.13 shows the profits and the sum profit for $\alpha_1 = 1$ and $\alpha_1 = 5$ with the cost of spectrum unit. It is seen that the profits and sum profit of RANs drop as c increases if $\alpha_1 = 1$ and $\alpha_2 = 1$. However, the sum profit of

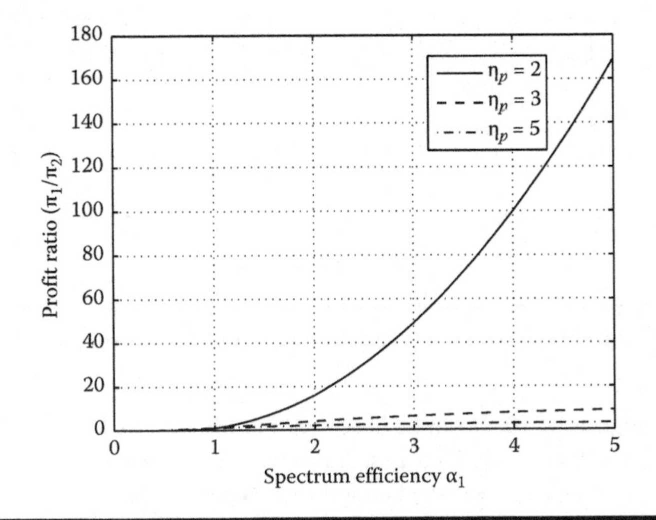

Figure 13.11 Profit ratio versus spectrum efficiency factors ($\alpha_2 = 1$, α_1 varies).

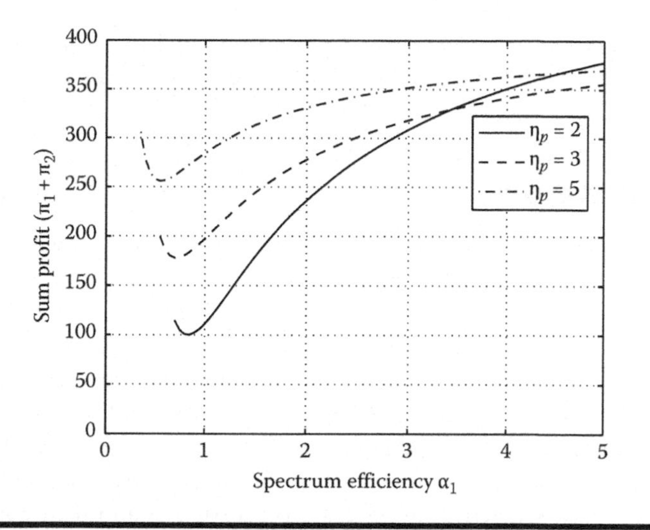

Figure 13.12 Sum profit versus spectrum efficiency factors ($\alpha_2 = 1$, α_1 varies).

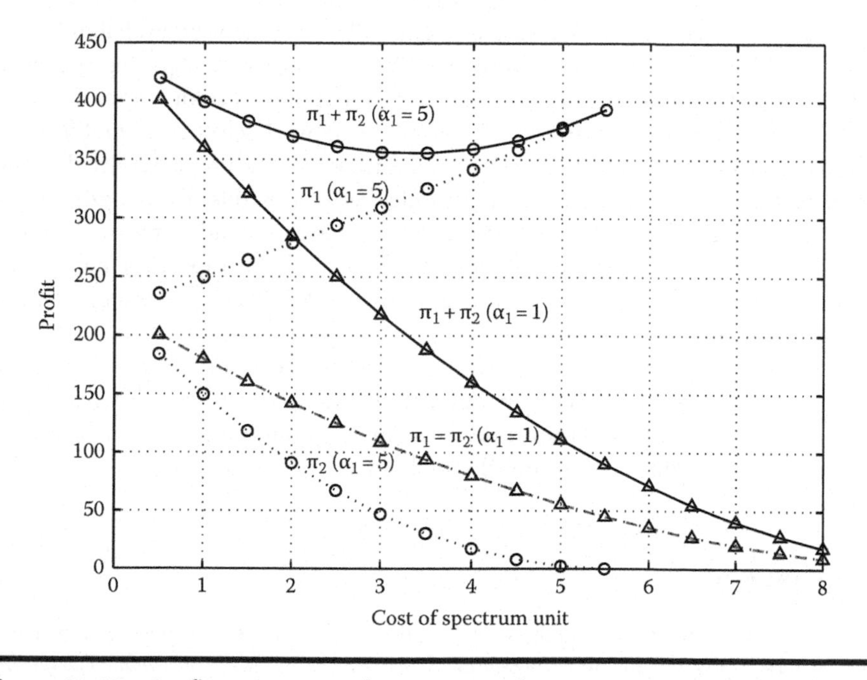

Figure 13.13 Profit versus cost of spectrum unit c ($\alpha_2 = 1$, $\alpha_1 = 1$ or 5).

RANs may not drop as c increases if $\alpha_1 = 5$ and $\alpha_2 = 1$. Actually, only the profit of RAN 2, the RAN with lower spectrum efficiency, always drops.

13.5 Conclusions

In this chapter, we consider the wireless networking environment where the mobile users have the freedom to select and adapt to one preferred serving RAN. We examine the relevant attributes for network selection including network attributes, user attributes, and user's preference. These attributes are disparate and usually conflicting with each other. Based on the attributes and preference information sensed and collected by the mobile user, automatic network selection method is needed to capture the conflicts and select the desirable RAN for a mobile user.

MADM-based methods are viewed as a promising approach to tackle this problem. This approach is effective to aggregate the attributes and their corresponding relative importance weights. Furthermore, it is scalable in terms of the number of relevant attributes, and provides a more generalized framework than other approaches. The procedure of the MADM-based network selection method is described in detail, and two enhancements for the MADM-based methods are presented for the benefit of mobile users (i.e., payment saving) and for mobile operators (i.e., load balancing and the alleviation of ping pong effect).

With deregulation and technology evolution, more intelligent engines for network selection, service pricing, and spectrum allocation may be available in CR networks. Multiple RANs would compete with each other in acquiring spectrum from a common spectrum pool and attracting mobile users from a common user pool. To investigate the interrelations between the network selection of mobile users and the spectrum allocation of wireless operators, we formulate the service demand model and noncooperative competition model between RANs. By analyzing these models, we obtain the Nash equilibrium and its existence conditions for the competition game. We further present the numerical results of capacities and earned profits of RANs at the Nash equilibrium and discuss the effects of influential factors on the competition between RANs.

We expect that our results can provide insights and hints for policy makers, wireless operators, and equipment manufacturers to make decisions on network and spectrum management and create a desirable mobile communication environment concerning the benefits of all involved stakeholders.

References

1. H. J. Wang, R. H. Katz, and J. Giese, Policy-enabled handoffs across heterogeneous wireless networks, in *Proceedings of the Second IEEE Workshops on Mobile Computing and Applications* (*WMCSA*), New Orleans, LA, 1999, pp. 51–60.

2. O. Ormond, J. Murphy, and G.-M. Muntean, Utility-based intelligent network selection in beyond 3G systems, in *Proceedings of IEEE International Conference on Communications (ICC)*, Istanbul, Turkey, 2009, pp. 1831–1836.
3. Y. Yu, Y. Bai, and L. Chen, Utility-dependent network selection using MADM in heterogeneous wireless networks, in *Proceedings of IEEE International Symposium on Personal, Indoor and Mobile Radio Communications (PIMRC)*, Athens, Greece, 2007, pp. 1079–1083.
4. W. Zhang, Handover decision using fuzzy MADM in heterogeneous networks, in *Proceedings of Wireless Communications and Networking Conference (WCNC)*, Atlanta, GA, 2004, pp. 653–658.
5. J. Jia, G. W. Fischer, and J. S. Dyer, Attribute weighting methods and decision quality in the presence of response error: A simulation study, *Journal of Behavioral Decision Making*, 11(2): 85–105, 1998.
6. Q. Song and A. Jamalipour, Network selection in an integrated wireless LAN and UMTS environment using mathematical modeling and computing techniques, *IEEE Wireless Communications Magazine*, 12(3): 42–48, June 2005.
7. J. Ma, Z. P. Fan, and L. H. Huang, A subjective and objective integrated approach to determine attribute weights, *European Journal of Operational Research*, 112(2): 397–404, 1999.
8. E. Stevens-Navarro and V. W. S. Wong, Comparison between vertical handoff decision algorithms for heterogeneous wireless networks, in *Proceedings of the VTC*, Melbourne, Australia, 2006, pp. 947–951.
9. K. Yoon and C. L. Hwang, *Multiple Attribute Decision Making: An Introduction*. Sage, Thousand Oaks, CA, 1995.
10. W. Webb, The role of economic techniques in spectrum management, *IEEE Communications Magazine*, 36(3): 102–107, 1998.
11. P. Leaves, K. Moessner, and R. Tafazolli, Dynamic spectrum allocation in composite reconfigurable wireless networks, *IEEE Communications Magazine*, 42(5): 72–81, May 2004.
12. M. Pan, R. Liu, J. Chen, Z. Feng, Y. Wang, and P. Zhang, Dynamic spectrum access and joint radio resource management combining for resource allocation in cooperative networks, in *Proceedings of Wireless Communications and Networking Conference (WCNC)*, Hong Kong, China, 2007.
13. S. Sengupta, M. Chatterjee, and S. Ganguly, An economic framework for spectrum allocation and service pricing with competitive wireless service providers, in *Proceedings of Dynamic Spectrum Access Networks (DySPAN)*, Dublin, Ireland, 2007, pp. 89–98.
14. Y. Bai and L. Chen, Flexible spectrum allocation methods for wireless network providers, to appear in *The Mediterranean Journal of Computers and Networks*, Special Issue on *Recent Advances in Heterogeneous Cognitive Wireless Networks*.
15. M. J. Osborne, *An Introduction to Game Theory*. Oxford University Press, Inc., New York, 2004.
16. R. Gibbons, *A Primer on Game Theory*. Prentice Hall, New York, 1992.

Chapter 14

Cognitive Radio Networks: An Assessment Framework

Mikhail Smirnov, Jens Tiemann, and Klaus Nolte

Contents

Cognitive radio (CR) networks are believed to be self-managed. Do we know how to engineer such systems? And, if such engineering is done, do we know how to assess the properties of a self-managed system? In addition to performance evaluation and testing to guarantee for the consistency, safety, purposefulness, security, and the efficiency of the operation of such systems engineers require assessment—systematic evaluation (of nonconventional to previously engineered systems) of properties such as cognition, learning, aware sensing, and perception. Technology-wise we analyze the problem area of flexible spectrum management (FSM) and suggest that both (self-) management and assessment problems can be solved within the same framework—situation-aware behavioral composition—the key to our policy approach. We describe a model-driven assessment framework: the driving model in the cognition cycle is the enabler of robust control with agile policies—here the "subject–object" policy-based self-management framework is introduced; this framework is interpreted from the control theoretical viewpoint. We then compare assessment with the performance evaluation of cognitive networking, and conclude that the evaluation of the process correctness is the main difference between the two. We derive primary assessment metrics and build some derivatives that, using the specific objectivity framework of G. Rasch, allow us to define the assessment of process correctness in a way that does not depend on a particular cognition algorithm. We report the ongoing research and show some early results, followed by the open issues section.

14.1 Introduction

Today, wireless mobile networking is largely facilitated by the organic growth of coverage; as stated in [1] with 20 million installed access points, the coverage continues to grow by tens of millions of access points per year. Although coverage contributes directly to the empowerment of end users (including appliances and facilities), it, at the same time, demands even greater acceleration in the growth of end-user services and service infrastructures. As it is widely agreed, the natural and mostly required way to achieve the needed growth in services (strongly supported by the multi-radio mobile terminals appearing on the market) is along the path of convergence of service infrastructures over heterogeneous radio access technologies (RAT), which, in wireless access domain, translates into the need for flexible spectrum management (FSM), in which cognitive radio (CR) systems and networks will operate opportunistically yet efficiently.

According to the studies reported in [2], practically deployable FSM techniques should address the two sets of requirements—technical (including radio resource management, reconfiguration, and efficiency) and nontechnical (regulatory and business aspects). At the same time, thus converged (wireless) service infrastructures shall exhibit a novel property—they shall facilitate the secondary market [3]. In [4] this future secondary market is compared to the existing energy secondary market, where the exchange demand exists on a hourly basis, with the conclusion that "licensed spectrum commodity secondary market might have a near-real-time exchange demand." It is clear that conventional control and management techniques cannot be applied in these likely future scenarios; it is also commonly agreed (after the pioneering XG initiative of DARPA [5,11]) that needed control mechanisms should be devised within the policy domain.

Policy-based cognitive networking, however, appears to be much harder to design than policy-based management (PBM) of less complex systems. First, conventional PBM solutions are limited in scope: security policy is dealing only with the access control, while QoS policy is dealing only with the device configuration(s), while FSM policies should take into account both security and configuration concerns and regulatory and business ones. Second, conventional radio technologies were not designed with PBM in mind; hence their functional parts require modifications that are always painful. Third, in multi-vendor-, multi-layer-, multi-technology-converged networking environment (ecosystem), policies will originate from multiple sources and shall require policy multiplexing, which is currently an unsolved problem. Last but not least, as stated in [6] there is a need not only to "translate policy rules into radio behaviour controls" but also to "control operating rules based on policies and situations." These challenges are calling for a novel approach for policy handling that is commonly referred to as self-management.

Self-management is known to be the goal of autonomic computing [7]; a similar concept is recognized within the autonomic networking [8], where self-management is achieved by informed autonomic decisions, for example, as postulated by the AN's flavor—declarative networking [9,10] with the help of distributed continuous query processing (DCQP). The declarative networking approach is attractive for FSM for the following benefits: first, it is essentially distributed PBM and is multiplexing-friendly; second, DCQP is conceptually simple; third, it allows easy composition of components into a larger system (supported by data fusion); and, finally, DCQP is inherently capable of self-management (supported by database reflection technique). Besides these benefits the DCQP as a foundational technique for FSM can be justified by yet another feature—ability for self-retrospection, which is the key requirement of any assessment.

Despite its benefits the DCQP cannot be regarded as the ultimate answer for FSM because it is not readily capable of separating the two concerns—query processing and query routing (query plan), which are tightly coupled [10]. It is well recognized, however, that such separation is an important requirement for pragmatic FSM implementation; as stated earlier policy-based FSM will need to intervene with the implementation of the functionality of different RATs in a policy agile manner (the XG Working Group [11] rightfully demands that "policies are not embedded in the radio, but can be loaded on-the-fly"), which in turn translates either into a complete unbundling of functionalities from policy-based control or, at least, into their much looser than currently coupling.

We see self-management as the radical departure from the traditional "manager-managed object" paradigm; instead the policy-friendly paradigm of "subject–object" communication should become the cornerstone of cognitive networking. However, if future networks will be self-managed, how do we know about this? If future network devices will learn and act based on knowledge, how do we design such devices, and how do we test, evaluate, and assess their behavior?

In general, the assessment of cognitive systems that are capable of self-management is required to guarantee consistency, safety, purposefulness, security, and efficiency of their operation; assessment is an essential part of self-management and must become part of the design process. Some assessment mechanisms perhaps should be routinely performed at run-time using, when possible, already deployed mechanisms (e.g., measurement, sensing, and context management). Policy-based self-management in FSM is more challenging due to the required loose coupling between policy and functionality.

The rest of this inevitably cross-disciplinary chapter is structured as follows. First, we distinguish our policy approach from a body of related work and claim that our innovation is in the behavioral composition and situation awareness. We then describe our model-driven assessment framework: first, we show that the driving model in the cognition cycle is the enabler of robust control with agile policies—here our "subject–object" PBM framework is introduced; the semantic of this framework is interpreted from the control theoretical viewpoint; we then compare assessment with the performance evaluation of cognitive networking, and conclude that the evaluation of the process correctness is the main difference between the two. We derive primary assessment metrics and build some derivatives that, using the specific objectivity framework of G. Rasch, allow us to define the assessment of process correctness in a way that does not depend on a particular cognition algorithm. We report the ongoing research and show some early results, followed by the open issues section.

14.2 Agile Policy and Control Loops

DARPA XG program was perhaps the first large-scale effort to address the issue of PBM for FSM, and perhaps the first to realize that the required unbundling of functionality and control can be realized if the behavior of a network element is considered as a resource. The XG Working Group [5] argues that the novel architecture for FSM must meet the following requirement: "A core set of behaviours must be identified in such a manner that a viable architecture where only the core set needs to be considered for regulatory approval is possible" and defines the following top-level abstract behaviors (AB): sensing, identification, dissemination, allocation, and the use of opportunities. These abstract behaviors are considered to be similar to a secure kernel in conventional computing: kernel behavior is completely predictable, while the outer behavior can be any.

The XG Working Group [3] outlines important requirements for FSM policies, such as policy consistency, the adaptation of radios to policies, openness for new capabilities, and the stressed earlier separation of policy and behaviors. The latter requirement is essential for the understanding of the innovation of our work, hence we quote from [3]: "It is a goal of the XG program to separate the policies that govern an XG system from the behaviours of the system. This will ease accreditation of XG systems by enabling regulators to accredit radios based on the ability to interpret policies correctly to obtain desired behaviour, not (as today) by conformance of the implementation to specific pre-determined policies. These behaviours are discussed in the XG Abstract Behaviours RFC [XGAB]. In the absence of such separation, accreditation would need to be performed every time a policy was changed even in a minor way."

There are several issues with this requirement that sketch the accreditation (which is obviously relevant to the assessment we are working on); we try to address these issues in our assessment approach. First and fundamentally, it is not clear how a policy, when understood (after M. Sloman) as a rule defining a choice in the behavior of a system, can be separated from a behavior that hosts choices; our take on this is to consider concurrently temporal and spatial aspects of this separation. Second, the required ability to interpret policies correctly is proposed in [3] to be verified by the comparison of the resulting behavior with the desired behavior, practically meaning that the policy set as such is outside the accreditation. In our approach not only the resulting behavior but the set of given policies also has to be assessed; we put forward the compliance of both to the situation, which is defined in the following text. Finally, as XG is silent* about abstract behaviors it appears to us that the missing notion of the situation is critical. Our assessment framework considers permanent verification of a behavior based on the observation of the exhibited behavior, of given policies and of the situation; the latter includes the current system under assessment (SuA). In our approach (inspired by the control theory and by theoretical computer science as explained in the following text), the given (external to the SuA) policies are considered as peer entities to those functional behaviors that host choices controlled by policies. We consider the SuA's resulting behavior as the behavioral composition of these two types of peers. In doing so we agree with the statement in [12] that in the behavioral composition "the constraints can be represented like any another component behaviour," and further extend the approach to consider any policy as a functional behavior constraint.

In considering spatial and temporal separation of policies from behaviors, we also follow the trend known in wireless communication as spatial multiplexing; for example, talking about the need to exploit the waveform diversity Ramanathan [13]

* The XGAB document is not available online.

states that a link is no longer just an input to control mechanisms (such as routing protocols), but also a parameter that can and needs to be controlled, which leads to a challenging multidimensional topology control problem, and opens an opportunity of distributed real-time optimal algorithms for custom-made topologies. Our behavioral composition of policy behaviors and functional behaviors is formulated as this topology control problem but in the control-theoretical sense.

14.3 Assessment Framework

We advocate a model-driven decision process that has to be implemented in each network element; the decision process can be characterized by the two strongly related aspects:

- Decision process selects one, either the best, or the only feasible, or selected by a tie breaking, etc., alternative branch out of potentially possible behaviors.
- Metric (preference) used in the selection process must be self-reflective, meaning that its value used in the selection process must be modified after the selection, both the selection and the modification should be done as prescribed by the driving model.

In these model-driven settings, the question of assessing a model itself can be of major importance; if a model is proved to react satisfactorily in all operational conditions (contexts) and is proved to generate correct calls for governance in all feasible exceptions, then one can use this model as the driving one.

We believe that FSM-driving models should be relatively simple mathematical models. Assume that such a driving model is found for each controller; the key question then is how these potentially different models generate information flow that helps to coordinate the controllers? The assumption that models are different is a working one and has to be very strongly justified; because such a justification is not readily available, one can assume that controllers can use the same model, provided that the purposeful behavior and the model behavior are similar.

Thus, we think that both FSM and its assessment are all about the comparison of behaviors in a metric space, which we plan to construct in future work following the spirit of composition rules defined in [14], though the abstract behavior types should be that instantiated as appropriate for FSM.

In general the complete assessment framework might be a hierarchy of components: at the highest level an ecosystem (multi-vendor-, multi-layer-, multi-technology-converged networking environment) assessment should be performed, based, for example, on fitness metric; this is followed by the assessment of common service attributes (fairness, robustness, versatility, and cost-efficiency) per service type; finally followed by the assessment of service-specific attributes per network element or feature. The complete framework and its components is our future work; we report the current approach in the following text.

14.3.1 Toward the Driving Model

We model autonomic and cognitive SuA as a quadruplet $E = \langle P, B, S, M \rangle$, where E is a current ecosystem, that is, existing composition; P is a current set of given policies that are treated as constraints to potential behaviors, B is a current set of potentially possible (functional) behaviors, S is a current system's knowledge about the environment (situation) as perceived by the SuA, and M is the driving model, in this case the model used for the FSM.

We define situation as system-perceived context, where context is the state of operational environment. From the operational environment viewpoint, the context has larger scope; however due to self-management capabilities of E its perception of context is relative to the purpose, which is a communication service under provisioning; hence a situation in our view has end-to-end properties and abstracts away those context information that are redundant to the service.

Following [14], from which we borrow largely the motivation for behavioral composition, we formalize both behaviors and policies as components of the two types, and consider E as a dynamic composition of these components. Without loss of generality we assume that each RAT is a policy domain, meaning that all behaviors of all entities using particular RAT are constrained by the same set of policies. This way, at any moment in time the ecosystem E is a result of the dynamic composition of the population of multi-radio cognitive mobile terminals with their potential behaviors composed with constraining policies per RAT. We demonstrated in [15] the model derivation by an example of a future FSM environment that follows the simplifying assumptions and the "wall-hole" protocol technique suggested in [16], also assuming the existence of a dedicated coordination channel (it can be also considered as a cognition channel similar to CPC introduced in [17]), to which secondary users are assumed to listen permanently; we show in [15] that it is the coordination channel that practically facilitates the behavioral composition. The spectrum is assumed to be divided into frequency bands, some having both primary and secondary users; it is the primary usage policy that shall constrain the behavior of secondaries.

14.3.2 Some Results in Policy

This section quickly reviews relevant assessment results in policy, which make a background for the next section, where a general framework for policy-based control is presented. We largely follow the classical approach of [18], and the new results, such as policy continuum, introduced in [19].

Moris Sloman defines what a policy is in an extremely illuminating manner: policy is a rule that defines a choice in the behavior of a system. Types of policy can be further defined based on the types of behavior. Authorization policy defines what actions a subject is permitted or is forbidden to perform on a set of target objects, while obligation policy defines what actions must or must not be performed on a set of target objects. Thus defined policies are not conflict free; the following types of conflicts are known: feasibility conflicts, required actions are not available on a target object; modality conflicts, two or more policies are mutually exclusive with regard to the same target object.

Both modality and feasibility conflicts can be local, detected on one target object, or global, detected on a set of objects, where the former are tractable during the design of the policy-based system, while the latter present a challenge both at the design and at the operation phases. M. Sloman gives a set of general recommendations that help to overcome policy conflicts: sub-domain policy overrides domain policy; negative authorization overrides positive authorization; recent policy overrides an older one; short-term policy overrides long-term policy. Although application-specific care of course must be taken in following these recommendations, the general logic of the approach can be adopted. This general logic is the one of an admission control for policies based on meta-policies.

Meta-policies are policies that are defined for roles rather than for particular instances that might implement these roles. A role is equivalent to a position within a role hierarchy; this position is defined by a set of authorization policies and by a set of privileges to set obligation policies. Similar to abstract data types (ADT) known in object-oriented programming a role can be formally defined by a role class with templates for the associated policies, with inheritance mechanism, and with cross-references that allow traceability between parent and child policies. Cognitive networking is highly demanding in handling policy agility, which practically makes a conflict-free operation an exception rather than a rule; policies representing business goals must decide in uncertainty and under multiplexing. Such high demands require the policy continuum [19]—the entire process of the refinement of all policies from business goals—to detect and to resolve conflicts during a cognitive operation. These definitions of policy nicely separate the two concerns—the design of a system behavior and the design of rules that might control these behaviors—and suggests how to design agile policies.

14.3.3 Robust Control with Agile Policies

This section provides a framework for the design of policy-controlled systems—the codesign of subjects and objects that allow the separation of concerns mentioned earlier; the framework also serves as a foundation for the assessment of policy-controlled systems.

Conventionally designed systems are not suitable for policy-based control as their behavior choices are designed within internal control loops and thus do not allow external rules to influence the control. To overcome the issue in early policy-based systems, the designers had to assume a certain policy middleware, usually termed a policy agent, that by assumption knew the policies of all subjects and the actions available on all objects; the policy agent in such designs served as a mediator and simultaneously as the policy admission controller between subjects and objects.

Table 14.1 Policy Semantics (Middle Ware Viewpoint)

Policy Types	*Semantics*
$S \rightarrow A+ \rightarrow T(a_i)$	Subject may request the *i*th action on target object
$S \rightarrow A- \rightarrow T(a_i)$	Subject may not request the *i*th action on target object
$S \rightarrow O+ \rightarrow T(a_i)$	Subject must request the *i*th action on target object
$S \rightarrow O- \rightarrow T(a_i)$	Subject must not request the *i*th action on target object

In the attempt to eliminate the policy agent and to cater for the design of systems that are natively controlled by policies, we advocate that the composition principle is useful. First, we concentrate on policy semantics as seen by a policy agent (a middleware viewpoint) and second we interpret the same policies from the viewpoint of a system, in which subjects are composed with objects.

In our policy notation, the semantics of all four possible types of policies that a policy agent hosts for all subjects and objects in a system are given in the Table 14.1.

From the theoretical computer science viewpoint, the semantics presented in Table 14.1 are those of a glue code based on ADT and are termed the rendezvous semantics, which can be explained with the following quote from [14]: "The method invocation semantics of object oriented message passing implies a rather tight semantic coupling between the caller and callee pairs of objects. By this semantics, if an object c sends a message $m(p)$ to another object e, then c is invoking the method m of e with the actual parameters p. For this to happen: (i) c must know (how to find) e; (ii) c must know the syntax and the semantics of the method m of e; (iii) e must (pretend to) perform the activated method m on parameters p, and return its result to c upon its completion (the pretense refers to when e delegates the actual execution of m to a third object); and c typically suspends between its sending of m and the receiving of its (perhaps null) result."

To explain the policy semantics from the composition viewpoint, we apply the zero knowledge reasoning to both types of entities that appear in a composition—to subjects and to objects. The zero knowledge breaks the rendezvous semantic; in particular, all but the first must requirements from the previous paragraph are relaxed. We substitute the first must know (how to find) by an assumption that all entities in a composed system are fully connected (or can discover such a connectivity) and can pass arbitrary messages to each other. These arbitrary messages will be those that each entity can form and send; without loss of generality we can assume that these messages are those that subjects and objects are configured to form and send under certain conditions. Of course, an action of passing a particular message triggered by some conditions can be seen as a policy; however, for the moment we intentionally forget about this. Instead, we imagine that entities pass messages (that they can pass) unconditionally and for a reasonably long period of time; we also

Table 14.2 Policy Semantics (Compositional Viewpoint)

Role	*Relation*	*Configured Policy*	*Discovered Policy*	*Configuration Purpose*
Object	Server	*A*	*O*	Safeguard
Subject	Client	*O*	*A*	Behavior

assume that all sent and received messages are time stamped and stored in respective received and sent buffers. Eventually each entity will pass all messages it can and will receive all possible messages in the entire composed system.

When this happens, we require each entity to start relating (matching) all sent and received messages to the entity's configuration conditions. During this hypothetical matching process, the subjects of the composed system will observe that all sent messages match the configuration, and some of the received messages also match the configuration; the objects will observe that some of the sent messages and some of the received messages match the configuration. This reflects the fundamental difference between the two roles—subjects and objects; in our mental exercise the subjects discover the obligation policies (all sent messages are configured as must request) and the objects discover the authorization policies (some sent messages are matching some incoming requests that may be answered), by which they were configured. Further analysis of received and sent buffers will allow subjects to infer per obligation policy the sets of objects that have positive and negative authorization for the respective requests; by the similar analysis objects may infer per authorization policy the sets of subjects that may and may not request respective actions. Table 14.2 summarizes the compositional semantics of policies.

We argue that in a correctly composed system the subjects act as clients, while the objects act as servers; the objects' configuration serves the purpose of a safeguard, while subjects' configurations are behavior definitions. This argument facilitates the design of systems in which subjects and objects can, in principle, be dynamically composed; yet for such systems to deliver controllable and predictable behavior we need certain facilitation from the control theory.

Recently published formulation of an unsolved control theoretical problem—decentralized control with communication between controllers [20]—clearly states that "control of traffic on the Internet is a concrete example" of that unsolved problem. The problem reads: to synthesize *R* controllers and a communication protocol for each directed tuple of controllers, such that when all the controllers use their received communications the control objective is met as well as possible. Notably, a communication protocol is called a communication constraint in [20]; it concludes that the basic underlying problem is "What information of a controller is so essential in regard to a control purpose that it has to be communicated to other controllers?" It also claims that the problem for discrete-event systems is likely undecidable in

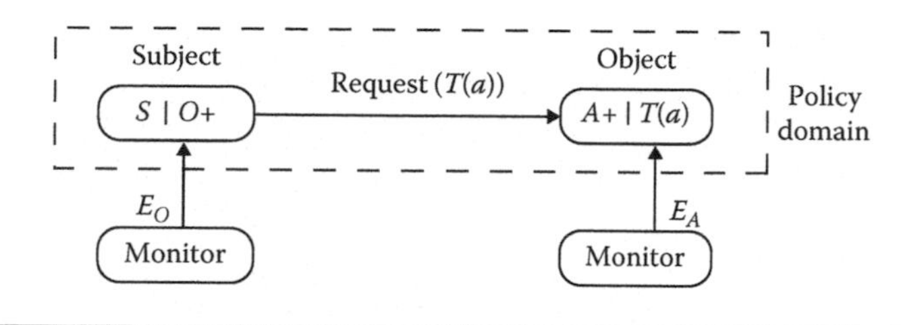

Figure 14.1 Dependencies within a conventional policy domain.

general; however, at the same time it mentions particular cases (CSMA/CD, IEEE 802.11), in which formulation and analysis of simple control laws and necessary conditions helped to achieve a solution. We claim that finding the right information is the key to both problems—decentralized control and its assessment. However, in the scenarios in which the coordination and its assessment need to be achieved between cognitive elements that coexist in different realms (different RATs) the problem gets even more difficult. To discover the essential, in control theoretical sense information, we look further at the co-design of subjects and objects.

Figure 14.1 outlines the dependencies within a single policy domain; subject S is configured with the positive obligation policy ($O+$ to conditionally request an action a on the target object T, that is, $T(a)$, the object T is configured to conditionally authorize this request. We now concentrate on policy conditions; on the subject side the condition triggers the obligation, that is, a request; on the object side the condition might reject or otherwise permitted the request. In the majority of practical cases, the obligation is triggered by an event (E_O) external to the subject; we introduce another role (monitor) to represent an entity that generates the event that triggers subject's obligation. On the object side the conditions of otherwise positive authorization can be those defined by safety, state dependencies, conflict resolution, platform- or device-dependencies, etc. Again, we introduce the monitor role to be responsible for generating the condition, or without loss of generality just an event (E_A).

A policy domain—a logical collection of subjects and objects that are configured with common policies—does not by default include the monitor role. In the design of conventional policy-based systems both types of events were assumed to be generated by a middle ware; thus the policy design was limited to the two roles—subjects and objects. We shall now demonstrate how to extend the policy domain concept to include the monitor role, and how the extended policy domain helps to achieve policy controllability.

Figure 14.2 shows the dependencies within the extended policy domain; not only subject and object roles are configured with respective policies, but the monitor role is also configured explicitly with the notification policy. Obviously, the notification policy can be classified as a positive obligation to notify; thus a set of configured notification policies for a particular instance of a monitor defines its monitoring behavior.

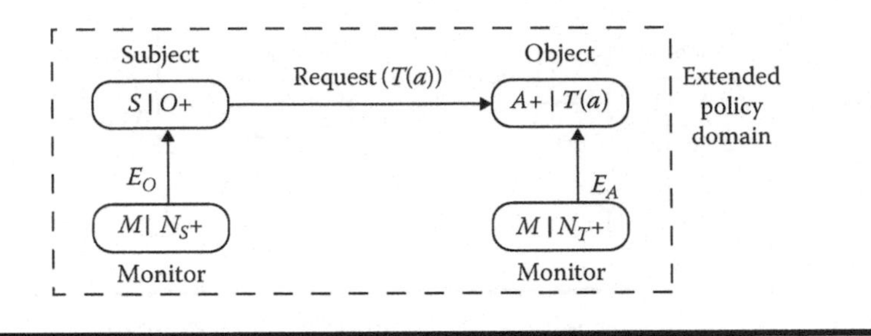

Figure 14.2 Dependencies within extended policy domain.

As it is clear from the previous discussion the instances of the role monitor that notify subjects and objects belong to essentially different realms. Subjects need to be notified on changes in the external environment as a whole (ecosystem, multi-RAT environment), to which changes the subjects need to react (adapt), while objects need to be notified on mainly internal events, derived from the states of resources of a single RAT. In the most generic settings, the target objects in our framework represent all types of resources that actually do some useful work; equally generic these resources can be assumed as shared between multiple requests. Thus, without loss of generality a role object has a property of a multiplexer, and its configured authorization policy acts as an admission control to a multiplexed resource. At this time, we need to look at the extended policy domain from the viewpoint of the control theory. From the control theoretical perspective, all three roles—subjects, objects, and monitors—are controllers, and all their pair-wise communications (request, action, event) contribute to the common control purpose, but very differently. This is best explained by the control-theoretical comparison of the four policies and their modalities made in Table 14.3.

Table 14.3 Policy Semantics (Control Viewpoint)

Policy Type	*Semantics*
$O+$	Request a behavior change (adaptation)
$O-$	Refrain from behavior change (stability, robustness)
$A+$	Admit behavior change request (multiplex the request)
$A-$	Deny behavior change request (for multiplexer fairness)
N_S+	Detect change/opportunity (external feedback)
N_S-	Refrain from notification (uncertainty, conflict)
N_T+	Detect change/opportunity (internal feedback)
N_T-	Refrain from notification (uncertainty, conflict)

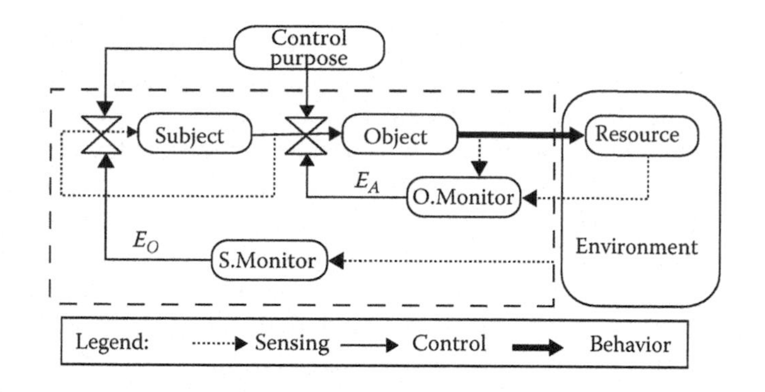

Figure 14.3 Control loops in extended policy domain.

Following this semantics we consider an extended policy domain as a distributed controller system with the three types of information exchange shown in Figure 14.3.

The thick line in Figure 14.3 denotes a flow of information that makes a distributed controller system's behavior externally visible; dotted lines represent sensing including the recording of the own behavior; normal lines depict policy control, including the possibility for the controller system to get the control purpose. Figure 14.3 distinguishes clearly between subjective (S.Monitor) and objective (O.Monitor) monitoring. We shall use this framework model now for the following: to exemplify cognition within the two control loops and to verify that the model is indeed correct for the assessment of any cognition algorithm; we also use it to derive the requirements for the assessment of CR systems.

14.3.4 Cognition and Assessment

Our distributed controller system—an interpretation of the extended policy domain—is essentially a composite system; thus its assessment can proceed following the "specifically objective" frame of reference introduced by G. Rasch. However, though this is sufficient, it is not the necessary condition to claim that this composition is exactly the correct one for the assessment of CR systems. This brings us to the issue of cognition in CR systems; without trying to define what cognition is, we instead try to define what is the purpose of cognition in CR, meaning, for the assessment, that if the purpose of cognition is met the assessment of underlying cognition can be high, and low otherwise.

It is important to compare now the assessment of CR networks with their performance evaluation. Not surprisingly, the chapter of [21] that deals with performance evaluation asks somewhat different, but very relevant and useful, questions. They are the following: What happens when we deploy CRs? Will they perform as we expect when faced with a realistic environment? And exactly what is realistic environment?

To answer these questions, the authors of the chapter the take a slightly different view on CR from other chapters; instead of using cognition cycle to explain the operation, the chapter considers the cognition cycle as a limited model of the outside world ("the environment, in which the cognitive radios are observing, learning and reacting"). The chapter says "A more realistic view of the operation of cognitive radios would depict an outside world whose state is jointly determined by the adaptations of several cognitive radios . . . cognitive radios react to an outside world determined by both 'dumb' and other cognitive radios"—the former is much more difficult to understand than the latter. Later the chapter states ". . . abstractly, this interaction can be viewed as a recursive interaction decision process in which each adaptation can spawn an infinite sequence of adaptations" [21].

This type of reasoning when applied to the assessment allows us to map this recursive interaction decision process onto our extended policy domain model—our distributed controller system is not seeking point correctness but rather process correctness [22] of the behavior. We broaden the coexistence issue of dumb and CRs by the assumption that CRs might be heterogeneous; this assumption is captured by allowing uncertainty as an outcome of the monitoring behavior. In full accordance with [21] our model of the external world used by a single CR for its adaptations is built and maintained by subjective and objective monitoring of adaptations—of own adaptations, recorded as a history of behaviors, and of adaptations of others—recorded after the sensing of multiplexed resources.

We define ideal cognition as the one that remembers all possible situations; assuming unlimited resources for the decision making the ideal cognition always helps to select the best match behavior based on the past experience, which includes sufficient training. Assessment applied to the ideal cognition should yield the highest value. Implicitly, our definition of ideal cognition also assumes the ideal decision process, which in these ideal settings can be considered as the problem of pattern recognition without noise—fairly easy task for any search algorithm under these ideal assumptions. In reality, we must assume that cognition (or the training of it) is not exhaustive, that decision process must act in uncertainty and within constraints, and that sensing is being performed in noisy environments. All these factors contribute to the uncertainty and can be captured by the confidence level, which is our characterization of the situation presented to the decision process. At the same time, the decision process itself can be characterized by its ability to produce more or less correct decisions, where the degree of correctness will be of course judged based on the delivered performance.

14.3.5 Objective Monitoring Metrics

This section demonstrates that a distributed controller system is capable of reflecting cognition in both control loops in the implementation agnostic manner; meaning

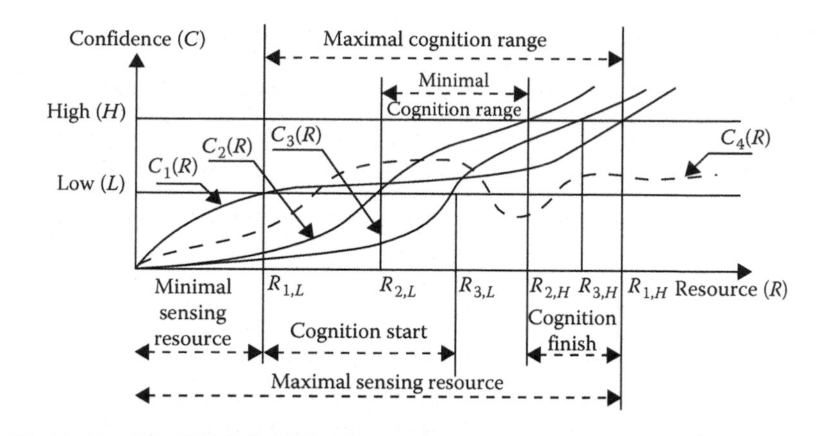

Figure 14.4 To the definition of primary cognition metrics.

that regardless of the cognition algorithm the distributed controller system can be used as a model of CR for the assessment of cognition algorithms.

Objective monitoring in our distributed controller system can be modeled as the sensing of a signal in a generally noisy environment. The quality of objective monitoring depends on the amount of spent resources, such as time and energy. In CR systems, spectrum sensing is the first task that is empowered by cognition; from the assessment viewpoint we are interested in metrics that will allow an object to enforce the policy triggered by the maximal likelihood (confidence) of sensed value to the target event (E_A). The purpose of cognition in the objective monitoring is to compute the confidence of the sensed event.

Let us consider the confidence (C) as a function of spent resource (R). Figure 14.4 introduces three confidence functions ($C_i(R)$ shown with solid lines) that monotonously increase with the amount of resources spent for both sensing and cognition. Confidence in detecting the target event might grow from zero to the value denoted as low (L), that is, $C_i(R_i, L) = L$, and eventually proceed to the value denoted as high (H), that is, $C_i(R_i, H) = H$. With the confidence above low our objective monitor enforces cognition that helps to unambiguously determine the target event by various means—radio-specific as well as radio-agnostic. The confidence level low can be characterized by the minimal amount of sensing resource that is needed to start learning. When our confidence function is between low and high the cognition cycle is applied perhaps as a supervised sensing. We define in Section 14.3.5 and show in Table 14.4 our primary cognition metrics for objective monitoring—likely typical areas and points of the confidence function.

We use these primary metrics to engineer some derivatives useful for the establishment of the needed for the assessment relation between cognition in objective monitoring and cognition in subjective monitoring. The first derivative shown in

Table 14.4 Primary Cognition Metrics (Objective Monitoring)

Acronym	*Full Name*	*Meaning and Computation*
MiSR	Minimal sensing resource	Required to trigger the cognition; the smallest of all experienced $C(R)$
MxSR	Maximal sensing resource	Allowed to be spent in the detection of the target event; configured upper bound
CSR	Cognition start range	The resource range between the minimal MiSR (the earliest cognition trigger) and the maximal MiSR (the latest cognition trigger); the difference between maximal and minimal values of MiSR for all $C(R)$
CFR	Cognition finish range	The resource range between the minimal MxSR (the earliest generation of the target event) and the maximal MxSR (the latest generation); the difference between maximal and minimal values of MxSR for all $C(R)$
MiCR	Minimal cognition range	The minimal resource range between MiSR and MxSR; difference for one function minimal for all $C(R)$
MxCR	Maximal cognition range	The maximal resource range between MiSR and MxSR; difference for one function maximal for all $C(R)$

Figure 14.5a is the cognition envelope; it shows the boundaries of confidence maximally obtained for particular resources spent on cognitive sensing. The shape of this envelope within [MiSR,MxSR] is almost arbitrary as the envelope is only an instrument for further deliberations. The next derivative is shown in Figure 14.5b; it is also instrumental and is termed cognition entropy. The cognition entropy is modeled after the binary entropy function known as Bernoulli trial because the goal of the objective monitoring is to detect with high confidence the presence or the absence of the target event. We deliberately apply two entropy functions separately for low and high confidence outcomes because—as it appears almost in all studies on cognition (see, e.g., the survey in [23] and [24])—the process of cognition is nonlinear in many respects.

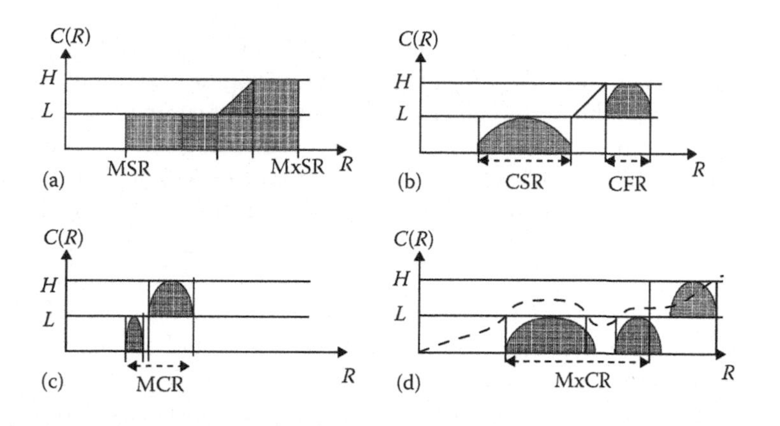

Figure 14.5 Secondary cognition metrics. (a) Cognition envelope, (b) cognition entropy (CR case), (c) cognition entropy (MiCR case), and (d) cognition dynamics.

The diagram in Figure 14.5b corresponds to the MxCR case of cognitive sensing, in which the detection of low confidence requires on average more resources than the detection of the high confidence. For a typical CR example, this can be a case of the wider spectrum sensing at the beginning of the cognition cycle, in which a hint of an opportunity is detected, and that is followed by more focused sensing around the found hint. Contrary to the MxCR case, the diagram labeled (c) in Figure 14.5 shows the MiCR case—an initial opportunity hint is on average sensed faster than the more detailed sensing that follows. As sensing in CR is a continuous activity, or at least spreads several consecutive sensing intervals, it is important to consider the dynamics of cognitive sensing.

The goal of cognitive sensing is to learn from experience, and the experience in our case is given by sensing parameters and by the achieved results in a number of consecutive intervals. As the confidence function $C_4(R)$ (dashed line in Figure 14.4) shows, the cognition process might loose the obtained confidence level; for example, an already detected opportunity hint might be broken by suddenly sensed collision in place of the hint. When the low confidence is lost after the CSR ends and not obtained again before the CFR starts, it is perhaps reasonable for objective monitoring to consider the target event as not detected in this cognition cycle and to start the next cognition cycle with the low confidence value as the outcome of the previous cycle. This reasoning—discrete cognition cycles but continuous learning—is shown in the diagram labeled (d) in Figure 14.5; one can consider that continuous learning is achieved by a sliding window of the size (in this case) of MxCR that relates the outcomes of the consecutive cognition cycles.

The goal of the assessment of cognitive sensing is to evaluate how well the competence gained in the process of learning is serving the control purpose. The control purpose in our distributed controller system is given differently to different controllers; better to say that control purpose is represented as different state data within different controllers. To the objective monitoring the control purpose state is given as a set of conditions (events) enforcing the authorization policies set on the object part of CR; to the object the control purpose state is given as a combination of authorization policies and behavior requests from the subject. When all three representations match in the policy domain (subject requests only those behaviors that are authorized by needed events detected in the sensed environment) and in the time domain (events are still valid, both authorization and obligation policies are still enforced) the control purpose is obviously met. However, CR systems must operate in uncertainty; following this reasoning we could define the two types of uncertainty—policy uncertainty, which in our system appears as the mismatch (conflict) between policies, and time uncertainty appearing as the lack of resources for complete cognition.

Instead, as our distributed controller system acts as a composition of subject and objects, we would like to apply the methodology inspired by the Rasch model in the attempt to eliminate perhaps multiple and peculiar differences of objective monitoring in different spectra. Each objective monitor has its individuality; all the individualities must be abstracted for a subject to make the best match to the control purpose. Following the cognition dynamics introduced in Figure 14.5d we shall consider a sequence of adjacent cognition cycles as a sequence of tests and apply to the outcomes of these tests the specific objectivity framework, which we briefly introduce next.

14.3.6 Specific Objectivity Framework

The term "specific objectivity" and the corresponding method were developed in the 1960s by Georg Rasch, a statistician from Copenhagen, while he was studying reading disabilities exhibited by various groups of humans. Rasch Model provides a framework for comparison of test outcomes (h) produced by different humans, as shown in (14.1),

$$h_n i = z_n \times E_i \tag{14.1}$$

where

z_n is the ability parameter of the nth human
E_i is the easiness parameter of the i-th test

We use the opportunity in this section to formulate as dense as possible, but closely following the logic of [25], the specific objectivity framework in its basic form.

The basic or determinate framework ($\mathbf{F}_d$) considers the two sets of elements—the set of objects (O) and the set of agents (A)—that both appear as factors in the comparison. Any object may enter into a well-defined contact with any agent, which may yield certain results; all results are said to form the set of results (R). Hence, the basic framework is given by (14.2) through (14.4).

$$\mathbf{F}_d = \langle \mathbf{O}, \mathbf{A}, \mathbf{R} \rangle \tag{14.2}$$

$$R = r(O, A), r = \mathbf{O} \times \mathbf{A} \rightarrow \mathbf{R} \tag{14.3}$$

$$u(R_1, R_2) = u(r(O_1, A), r(O_2, A)) = v(O_1, O_2|A) \tag{14.4}$$

The framework ($\mathbf{F}_d$) is termed determinate because it is required that each contact determines the result uniquely as in (14.3), where r is a reaction function of both O from O and A from A. As O, A, and R can be qualitative concepts the reaction function r is a single-valued mapping, also shown in (14.3). Following (14.3) any two objects O_1 and O_2 can be compared with regard to their reactions to contacting agent A, the values of this comparing function $u(R_1, R_2)$ form a collection U, the values of which can be also qualitative. In reality, the values of U are statements about the objects O_1 and O_2 as it is shown in (14.4), which also uses the notation for the comparator v explicitly conditioned on the agent A used for the comparison, which is said to be a local comparison in $\mathbf{F}_d$.

For the global comparison in $\mathbf{F}_d$ the comparator v should not depend on the instance of A; if this happens then our pair-wise comparison of objects is said to be specifically objective. The meaning of the objective is twofold: first, the comparison of objects O_1 and O_2 does not depend on the instance of A and second, it does not depend on other objects from O. The meaning of the objective specifically denotes that the comparison is specific to the framework $\mathbf{F}_d$. As it is demonstrated in [25] and in multiple online publications, the global comparison is indeed possible—even in the bi-factorial framework—provided that all three sets in (14.2) can be characterized by scalar parameters, which explains the direct applicability of the Rasch model for the assessment of CR systems.

To illustrate the technique with real assessment data (though limited of course and not technical) we present here the results of the assessment of a group of 46 students of the Technical University of Berlin that in 2006–2008 took the course "Advanced Internet Services" (lecture course 0432 L 746 OKS). As part of the written examination of the course all students were given 16 questions with 4 alternative answers per question. From the Rasch model viewpoint, we interpret this experiment as a series of 16 tests performed by 46 subjects. The ability parameter of a subject was computed as the percentage of correctly answered questions; the easiness parameter of each test was computed as the percentage of humans that provided correct answers. Despite the fact that each question had four alternatives our interpretation is dichotomous because we count only correct–incorrect outcomes. To simplify the outcome of the analysis we have analyzed groups of subjects instead of all individuals; the subjects were grouped based on their equal abilities shown in the tests. Figure 14.6a shows the distribution of relative easiness of all tests, Figure 14.6b shows the distribution of relative abilities of groups of subjects; both scales are normalized.

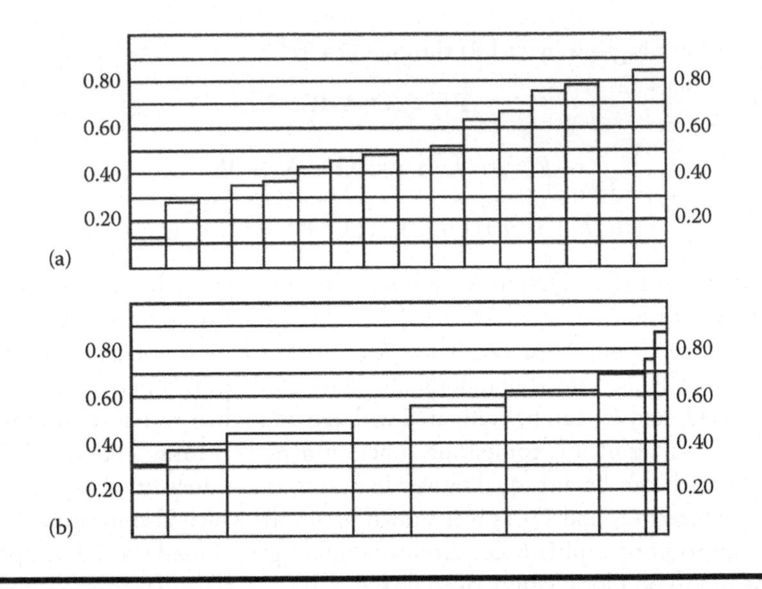

Figure 14.6 Normalized scales of factors. (a) Relative easiness of tests and (b) relative abilities of subjects.

Indeed in full conformance with the Rasch model the location of a particular subject on the common horizontal scale corresponds to the ability of that subject to answer correctly with 50 percent probability the question of the easiness that corresponds to the same common location on the scale; the easier the question, the higher is the probability of correct answer for the given subject, and vice versa. Interestingly, the effect of grouping shows that a group has a higher potential for correct answers than the same set of subjects taken individually. Next, we show how the specific objectivity framework helps to guide the assessment of CR systems modeled as a distributed controller system, which in turn is the model for our extended policy domain.

14.3.7 Assessment of Process Correctness

The assessment of process correctness is the assessment of how well the control purpose is met by the composed behavior of our distributed controller system that represents a single policy domain, in which all subjects and objects are configured with similar policies. From this viewpoint, we argue that the purpose of cognition can be formulated as continuous consensus building between pairs of controllers. We attempt now to demonstrate that regardless of the implemented cognition algorithm our distributed controller system is the correct model for the assessment of those algorithms. Following the cognition dynamics introduced in Figure 14.5d we shall consider a sequence of adjacent cognition cycles as a sequence of tests and apply the specific objectivity framework to their outcomes.

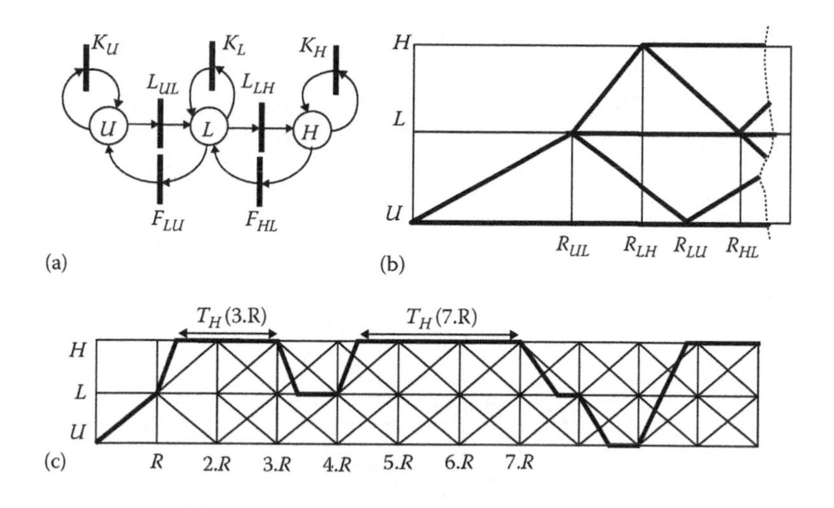

Figure 14.7 Cognition process (objective monitoring). (a) Major cognition phases, (b) resource constraints in the cognition dynamics, and (c) expected cognition process.

Cognition dynamics in our settings is constrained by the amount of resources that can be used for the objective sensing and by the allowed sequences of major cognition phases. Major cognition phases are as follows: at bootstrapping the level of confidence in detecting the target event is uncertain (U); cognition then can proceed to the low (L) level, and from low to high (H), or eventually from high to low, and from low to uncertain as shown by the Petri net in Figure 14.7a. The seven transitions shown by the Petri net are the only ones allowed in our model; this is the constraint on possible sequences of cognition phases; the transitions between phases can be clustered as learning transitions (L_{UL}, L_{LH}), forgetting transitions (F_{LU}, F_{HL}), and keeping the confidence-level transitions (K_U, K_L, K_H).

For each place in the Petri net the following holds: if neither learning nor forgetting transition happens within the resource constraints, then the keeping transition fires, which marks the start of a new cognition cycle with the previous confidence level, otherwise (if either learning or forgetting happens) the new cognition cycle starts with the newly obtained level of confidence. These dynamics can be observed with the initial fragment of the reachability tree of the Petri net shown in Figure 14.7b: horizontal lines labeled U, L, and H represent the confidence levels; the resource consumption is assumed increasing (not necessarily linearly; the linear form is used for simplicity) from left to right; we draw a thick line starting at one confidence level and reaching the neighboring confidence level to depict the transition between the two levels. The reachability tree fragment is shown for the initial marking $(U, L, H) = (1, 0, 0)$, the tree continues infinitely as a regular pattern with two arbitrary transitions from the phases U and H, and with three arbitrary transitions from the phase L.

We use Figure 14.7b to depict the constraints on the four transitions and all but the first one are shown as happening in some arbitrary order, which is yet constrained by the inequalities (14.5).

$$\begin{aligned} R_{UL} &\leq \frac{\mathrm{CSR}}{2} + \mathrm{MiSR} \\ R_{LH} - R_{UL} &\leq \frac{\mathrm{CFR}}{2} \\ R_{LU} - R_{UL} &\leq \frac{\mathrm{CSR}}{2} \\ R_{HL} - R_{LH} &\leq \frac{\mathrm{CFR}}{2} \end{aligned} \tag{14.5}$$

These inequalities simply mean that objective monitoring has its unique internal structure fully defined by cognition phases and parameters. This structure can itself be exploited within the cognition to improve its results; however, it is mostly important in the assessment of CR systems. We choose deliberately the resource constraints for learning and forgetting transitions to be 50 percent of respective average resource ranges because this corresponds to the maximum of cognition entropy for the MxSR case in Figure 14.5b), meaning that the worst-case ability parameter of an objective monitor (z_n in the expression (14.1)) is set to $z_n = 0.5$.

CR systems are expected to adapt to the changes in the environment, which requires the adaptation of the cognition itself. When applied to the objective monitoring this means that the reachibility tree of the Petri net model—and each individual branch of it—will be more and more departing from the regular pattern shown in Figure 14.7c as the trellis of thin lines; the trellis is a product of arbitrary transition between confidence levels in a slotted (equal resource R for each cognition cycle) sensing. We expect that good CR sensing with the increase of knowledge will tend to keep the high confidence level for longer periods, it will be returning to uncertain level only in a few really critical cases, while transiting the low level relatively quickly in both directions. The expected (good) cognition process is shown as the thick line in Figure 14.7c; the process is shown to keep the high confidence level almost 70 percent of the first seven after bootstrapping cognition cycles, which can be defined as the weight of the high confidence phase in the dynamic cognition process following the usual computation of the ratio of cognition slots with high level to all elapsed slots, subject to appropriate ageing of "old" slots.

Obviously, out of the three cognition phases only two—high and low confidence in detecting a target event—are reasonable to weight, as discussed earlier, proportional to the obtained and yet valid knowledge. We believe that the two types of knowledge, respectively for low and for high confidence, need to be distinguished as they apply significantly different CR-sensing strategies; for example, following [24] the strategy in the low-confidence phase might require larger entropy.

It is reasonable to share the weights of cognition phases of objective monitor with the subjective monitor. As we cannot expect that CR systems will operate always in high confidence of all target events, our distributed controller system makes decisions in uncertainty; more precisely, the uncertainty characterizes the behavior of both monitors, while the subject and the object are making very certain decisions. The subject's decision is to select for the enforcement that obligation policy that minimizes the risk; the object's decision is to apply the right authorization policy to the requested action. Both the subject and the object might lot have enough certain information about events and conditions that prefix their respective policies; the decision in this case is to select the right—the one minimizing the risk—policy and to enforce it as if the policy is the fully qualified one. Again, in both cases, missing (uncertain) part of required information (i.e., events and conditions) is substituted by the prediction. An object uses cognition to predict the confidence level of the target event; a subject uses prediction to quantify the risk of the decision taken.

It is clear now that the efficiency requires that both types of predictions need to be made in consensus; even more than that—it appears that the consensus mechanism is helpful for both types of predictions. As a background motivation for choosing the consensus building as the right prediction technique let us recall Delphi method proved to be very successful exactly in the area of prediction; let us recall the phenomenon of wise crowds [26]. Even in our small experiment with the students we have observed that out of 64 possible choices in the test the students have mostly converged on 28 alternatives, which include 15 out of 16 correct choices.

Indeed, our distributed controller system is well suited for building a consensus between pairs of controllers because they represent a single policy domain. Objective and subjective monitoring, though performed in different realms, is the coordinated prediction of the best behavior in a given situation. Objective monitors perceive the spectrum situation as the confidence level of detection of target events; subjective monitor perceives the decision situation as a collection of said confidence levels. Cross-adjustments of parameters between pairs of controllers is the consensus that brings the CR device to the converged behavior choices.

14.4 Practical Assessment of Cognitive Networking

14.4.1 The E^3: Methodology and Approach

One of the tasks addressed by the EU-funded research* on end-to-end efficiency (E^3 project) in the "Network of the Future" area is behavioral specification and assessment, which includes scenarios, use cases, and the assessment framework. This

* End-to-End Efficiency (E^3) Integrated Project of the 7th Framework Programme of the European Commission, Grant Agreement 216248, URL: https://ict-e3.eu

task analyses the technical aspects of project scenarios and use cases, in correlation with business models. The main focus is on the OAM (operation administration and maintenance) for network operators to support upcoming multi-operator and multi-RAT scenarios and the use of flexible (dynamic) spectrum management. This includes network planning, configuration, and optimization on the operator side, but also the introduction of autonomic and cognitive functionalities in the mobile terminal. The target result is the description of network elements using *self-x* capabilities to increase the overall system performance. Based on the scenarios and use cases of this task and the introduction of self-x capabilities for the network elements, there are new measures needed to test and assess such a network system. This includes procedures to apply classical conformance tests and performance measurements and also the introduction of new metrics to assess the level of autonomicity and cognition of the whole system, parts of the system, or single components. The assessment framework defines metrics, procedures, and interfaces to test and compare the self-x capabilities of network (sub)systems according to common use cases.

This activity extends the work on the overall E^3 system on the aspect of test and assessment of functionality. The results will be used to evaluate and benchmark the implemented self-x algorithms from other parts of the project in practical demonstrations and at the same time extend the theoretical concepts of the new area of cognitive networking and autonomic communication, which is the assessment. Test and assessment procedures for E^3 systems extend the current methods in two directions. First, the process of adaptation itself needs to be assessed by the appropriate metrics for a full evaluation of new functions (e.g., cognitive/learning functions). Additionally for applying current test and performance measurement procedures, the adaptation part of the algorithms needs to be controlled to achieve comparable results. The approach for both methods is the identification and decoupling of the adaptive parts of the algorithms, in the design of the algorithms as well as for assessment purposes. The results of the assessment framework are metrics to describe the impact of adaptive algorithms inside the equipment and methods to control the influence of these adaptive algorithms in equipment performance test. This can be achieved by extending the existing mechanism that support the context or situation awareness of the network elements.

Assessment is considered as part of standardization activities of the project, which target ETSI, IEEE, and specifically for the assessment part the Assessment Working Group of the ACF [27]. The project adopts the CR system approach that is under development in the IEEE P1900.4 and contributes to its FSM environment model [28].

This system assumes that numerous heterogeneous wireless systems are operated by a (meta-) operator. It introduces the three building blocks shown in Figure 14.8, borrowed from [28]:

- *Network reconfiguration manager* to define resource selection constraints (policies) for user devices.

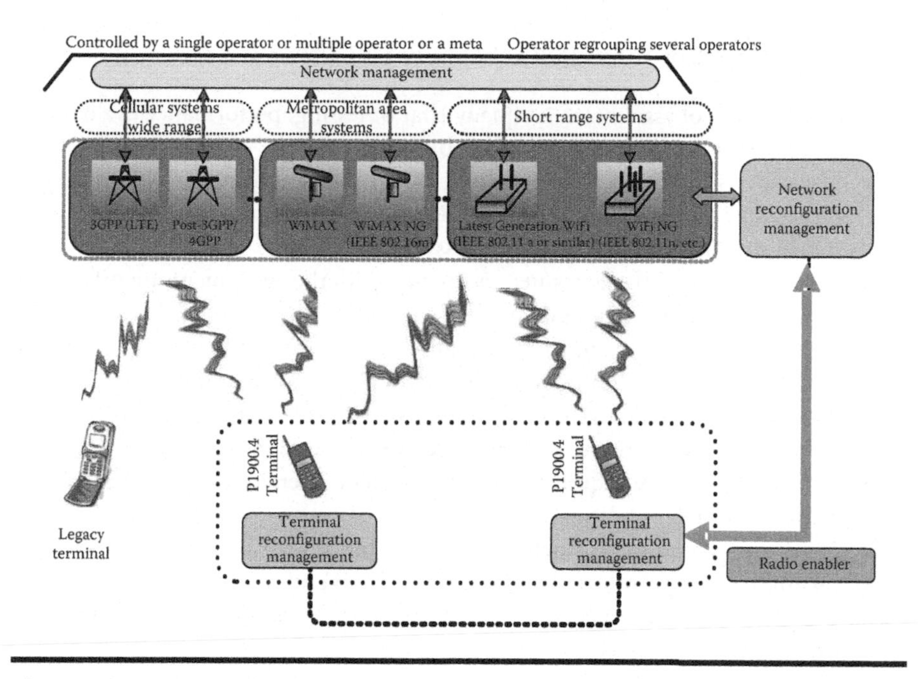

Figure 14.8 IEEE P1900.4 architecture.

- *Radio enabler of reconfiguration management* to communicate policies and context information to user devices.
- *Terminal reconfiguration manager* for distributed resource selection subject to policies by user devices. (Resources considered are frequency bands and RAT available.)

The two generic use cases dealing with FSM include spectrum allocation (dynamic coordination of spectrum resources among radio access networks triggered by traffic demands), and dynamic spectrum access (independently selected frequency, secondary entity in the spectrum).

14.4.2 Building a Consensus on Assessment

The assessment of cognitive and autonomic systems that under changing situations are expected to adapt both their behavior and their decision processes using the cognition (machine learning) algorithms in their control mechanisms is a novel field. Bearing this in mind we approached individual experts with diverse backgrounds and knowledge with a questionnaire seeking to build a consensus on understanding of and requirements for the assessment. The questionnaire asks eight questions about the assessment giving four alternatives per question. We report here the main findings of this study followed by the questionnaire and expert opinions reflected as relative weights per alternative.

A summary based on the majority of opinions is follows:

- The purpose of assessment is equally that of testing, performance evaluation, and the evaluation of cognition capabilities.
- The experts are sure that assessment is possible to specify at the system design phase; they are less sure that this specification can be standardized but nobody thinks that this specification is impossible.
- Responsibility for the assessment is for sure with the equipment manufacturer or with the certification body.
- Assessment placement in the design life cycle is for sure covering all phases but also can be a dedicated phase before deployment (before market entry).
- Assessment methodology is likely to be a mixture of methods with a tendency to be the extension of the performance evaluation methodology.
- Practical assessment requires special assessment interface but no other special instrumentation.
- Assessment process must include cognition model, network context handling capabilities, and complete logging of all assessment events.
- Security considerations of the assessment must be specified as concluded by the majority of experts.

The state of the art of the consensus is captured in Table 14.5; our next goal is to repeat the questionnaire at the second Delphi step, the result of which, as reported in many practical cases, will demonstrate the complete convergence of opinions.

We conclude that the assessment of cognitive networking is a novel field with its distinct methodology, cross-disciplinary in nature; further exploration in this field is required for the success of future networking; some of the research challenges that are already clear are reported in the next section.

14.5 Open Issues

There is an amazing similarity in the description of a use case and a policy; in particular both coincide in all but one field. This different field—the behavior description—is explicitly mentioned in the use case and is implicitly assumed to be enforced when the subsequent policy is enforced. Does this allow to fully assess a system based on use cases, at least theoretically? We envisage future cognitive networking in the environment populated by fully and partially qualified policies together with fully and partially qualified functions, in which environment they purposefully mate and mutate, compose (=evolve) and decompose, behave, etc., being assessed based on the complexity of exhibited behavior. This vision together with the reported work allow us to formulate some open issues that are the research challenges in the field of the assessment of CR networks.

The Rasch model gives a frame of reference for the design of a number of artificial situations that provide the foundation for the laboratory assessment of composite

Table 14.5 Expert Opinions on Assessment

Question (Statement Starts)	*Alternative Answers (Statement Ends)*	*Relative Weight*
The purpose of assessment is to...	test, to verify the conformance of implementation to the specification	0.43
	perform benchmarking of alternative implementations	0.14
	evaluate the cognition capability of a SuA (system under assessment)	0.43
	evaluate the process of SuA's dynamic adaptation (rather than testing a static system)	0.14
Definition of assessment in a system specification is...	impossible, as it will require to specify close to infinity number of use cases	0.00
	possible but must be limited to a reasonable number of reference assessment use cases	0.71
	impossible, as the assessment must comply with the National Regulation (certification)	0.00
	to be a reference to (future) internationally standardized harmonized standard	0.29
Responsibility for assessment...	is always with the manufacturer of an SuA, because what is assessed is embedded	0.57
	is defined by the reconfiguration responsibility cases (by a user, before and after market entry)	0.14
	is with the certification body	0.43
	is with the benchmarking body	0.29
In the design life cycle the assessment is...	covering all phases, from a concept down to acceptance testing	0.71
	only within validation and verification of the implementation	0.00
	a dedicated phase after deployment (after market entry)	0.00
	a dedicated phase before deployment (before market entry)	0.43

(continued)

Table 14.5 (continued) Expert Opinions on Assessment

Question (Statement Starts)	*Alternative Answers (Statement Ends)*	*Relative Weight*
Assessment methodology is...	an extension of testing methodology	0.29
	an extension of performance evaluation methodology	0.43
	unique (needs to be developed anew)	0.14
	a mixture of methods from various areas, including also machine learning, AI, etc.	0.29
Practical assessment requires from a system under assessment...	source code (complete implementation specification) as it must know possible internal states	0.00
	special assessment interface to allow special operations for the assessment	0.43
	no special instrumentation; the black box approach should be enough	0.57
	self-assessment capabilities	0.00
Assessment process must include...	a training phase for a SuA to learn/recognize changes in situation under assessment stimuli	0.43
	evaluation of a model used by a SuA for cognition (sensing, perceiving, etc.) purposes	0.43
	evaluation of SuA's ability to handle network context	0.29
	complete logging of all events during the assessment	0.29
Security considerations in assessment...	are not relevant	0.00
	must be specified	0.71
	can be added after the assessment	0.29
	require further study	0.29

systems. We have explored the model for a simple case while in theory it is applicable in multifactor settings as well, which opens a wide research opportunity.

In our assessment framework we intentionally try to abstract from the precise model of the cognition process; taking the cognition model into account provokes many questions. Does the cognition model present in all behavior types (e.g., sensing, predicting, usage, and dissemination)? Is it the same model? Learning should result in better performance, hence one needs a performance evaluation—with and without learning? Or, is performance irrelevant because it is defined by functional characteristics? Does the assessment want to make the cognition observable?

Due to its potential impact on the acceptability of new technologies the assessment of cognitive networking should enforce the interoperability by standardized metrics. These metrics are likely clustered into the three groups: first, related to the adaptation dynamics (robustness, sensitivity, agility, etc.); second, related to the community of network elements (trust by distributed validation, community management, bootstrapping into a community, etc.); and third, related to governance (call for governance, identification of context areas, etc.)—all being open research questions.

Assessment is about the comparison of behaviors in a metric space, which needs to be constructed perhaps following the spirit of composition rules defined in [14], though the abstract behavior types should be that specific for the cognitive networking; one also needs to design out of these different models an information flow that helps to coordinate controllers. In future FSM environments, one also needs to be able to assess the behavior of a dedicated coordination channel that practically facilitates the behavioral composition with unique internal structure fully defined by cognition phases and parameters.

Acknowledgments

This work is performed in project E^3, which has received research funding from the Community's Seventh Framework programme. This chapter reflects only the authors' views and the community is not liable for any use that may be made of the information contained therein. The contributions of colleagues from E^3 consortium are hereby acknowledged.

Abbreviation List

AB Abstract Behaviour
ACF Autonomic Communications Forum
ADT Abstract Behaviour Type
CFR Cognition Finish Range
CPC Cognitive Pilot Channel
CR Cognitive Radio

CSR	Cognition Start Range
CSMA/CD	Carrier Sense Multiple Access/Collision Detection
DCQP	Distributed Continuous Query Processing
FSM	Flexible Spectrum Management
MiCR	Minimal Cognition Range
MiSR	Minimal Sensing Resource
MxCR	Maximal Cognition Range
MxSR	Maximal Sensing Resource
PBM	Policy Based Management
QoS	Quality of Service
RAT	Radio Access Technology
SuA	System under Assessment

References

1. G. Karlsson, Extendable wireless local-area networks, in Perspectives Workshop: Autonomic Networking, Seminar N 06011,03.01.-06.01.06, Schloss Dagstuhl, http://kathrin.dagstuhl.de/06011/Materials2/
2. The E2R II flexible spectrum management (FSM) framework and cognitive pilot channel (CPC) concept technical and business analysis and recommendations, E2R II White Paper, November 2007, 52 p., on-line at http://e2r2.motlabs.com/whitepapers
3. XG Working Group, XG policy language framework, Request for comments, v.1.0, April 16, 2004, prepared by BBN technologies, Cambridge, MA, on-line at http://www.ir.bbn.com/projects/xmac/index.html
4. B. Lane, Cognitive radio, in *Cognitive Radio Technologies Proceeding (CRTP)* ET Docket No. 03-108, on-line at http://www.fcc.gov/oet/cognitiveradio/
5. XG Working Group, The XG architectural framework, Request for comments, v.1.0, prepared by BBN Technologies, Cambridge, MA on-line at http://www.ir.bbn.com/projects/xmac/index.html
6. P. Marshall, Beyond the outer limits, in Cognitive Radio Technologies Proceeding (CRTP) ET Docket No. 03-108, on-line at http://www.fcc.gov/oet/cognitiveradio/
7. Wikipedia contributors, Autonomic computing, Wikipedia, The Free Encyclopedia, http://en.wikipedia.org/w/index.php?title=AutonomicComputingandoldid=175412832 (accessed January 10, 2008).
8. Wikipedia contributors, Autonomic networking, Wikipedia, The Free Encyclopedia, http://en.wikipedia.org/w/index.php?title=AutonomicNetworkingandoldid=152692107 (accessed January 10, 2008).
9. T. Roscoe, Declarative networking, in Perspectives Workshop: Autonomic Networking, Seminar N 06011,03.01.-06.01.06, Schloss Dagstuhl, on-line at http://kathrin.dagstuhl.de/06011/Materials2
10. B. T. Loo, J. M. Hellerstein, I. Stoica, and R. Ramakrishnan. Declarative routing: Extensible routing with declarative queries, *Proceedings of ACM SIGCOMM*, Philadelphia, PA, August 2005, on-line at http://p2.berkeley.intel-research.net/p2pubs.php

11. XG Working Group, The XG vision, Request for comments, v.2.0, prepared by BBN Technologies, Cambridge, MA, on-line at http://www.ir.bbn.com/projects/xmac/index.html
12. M. Beauvois, Brenda: Towards a composition framework for non-orthogonal non-functional properties, in *Distributed Applications and Interoperable Systems*, LNCS, Vol. 2893/2003, pp. 29–40, on-line at http://www.springerlink.com/content/qda3mkrxhlaxmpee/
13. R. Ramanathan, Wireless research: Challenges and future directions, Industry Panel: NSF NETS Meeting, December 2007, on-line at http://www.ece.iit.edu/nsfwmpi/abstracts/Event*speakers*/Ramanathan$_n$*ets*-panel.ppt
14. F. Arbab, Abstract behavior types: A foundation model for components and their composition, in Formal methods for components and objects pragmatic aspects and applications, *Science of Computer Programming*, 55(1–3):3–52, 2005, ISSN:0167-6423
15. K. Nolte, M. Smirnov, J. Tiemann, D. Witaszek, End-To-End Efficiency assessment in autonomic and cognitive systems, in ICT-MobileSummit 2008 Conference Proceedings Poster Paper, P. Cunningham and M. Cunningham (Eds.), IIMC International Information Management Corporation, 2008, ISBN: 978-1-905824-08-3.
16. C. Santivanez, R. Ramanathan, C. Partridge, R. Krishnan, M. Condell, and S. Polit, Opportunistic spectrum access: Challenges, architecture, protocols, in ACM International Conference Proceeding Series; Vol. 220, *Proceedings of the 2nd Annual International Workshop on Wireless Internet*, Boston, MA, 13, 2006.
17. The E2R II flexible spectrum management (FSM) framework and cognitive pilot channel (CPC) concept technical and business analysis and recommendations, E2R II White Paper, November 2007, 52 p., on-line at http://e2r2.motlabs.com/whitepapers
18. M. Sloman and E. Lupu Policy specification for programmable networks, *Proc. IWAN99*, Berlin, Germany, URL http://www-dse.doc.ic.ac.uk/mss
19. J. Strassner, Autonomic administration: HAL 9000 meets gene roddenberry, keynote presentation at USENIX 2007.
20. J. H. van Schuppen, Problem 4.4 decentralised control with communication between controllers, in *Unsolved Problems in Mathematical systems and Control Theory*, V. D. Blondel, and A. Megretski (Eds.), Princeton University Press, Princeton, NJ, 2004.
21. J. O. Neel, J. H. Reed, and A. B. MacKenzie, Cognitive radio performance analysis, pp. 501–579, Chapter 15 in *Cognitive Radio Technology*, Bruce Fette (Ed.), Newnes/Elsevier, Burlington, MA, 2006, 622 p.
22. S. Dobson, Facilitating a well-founded approach to autonomic systems, in *Proceedings of the 5th IEEE Workshop on Engineering of Autonomic and Autonomous Systems* (*EASe 2008*), Belfast, Ireland, 2008.
23. D. Vernon, G. Metta, and G. Sandini, A survey of artificial cognitive systems: Implications for the autonomous development of mental capabilities in computational agents, *IEEE Transactions on Evolutionary Computation*, special issue on *Autonomous Mental Development*, 11(2): 151–180, 2007. On-line at http://www.vernon.eu/publications.htm
24. R. V. Belavkin and F. E. Ritter. The use of entropy for analysis and control of cognitive models. In F. Detje, D. Doerner, and H. Schaub, (Eds.), *Proceedings of the*

Fifth International Conference on Cognitive Modeling, pp. 21–26, Bamberg, Germany, 2003. http://citeseer.ist.psu.edu/belavkin03use.html
25. G. Rasch, On specific objectivity: An attempt at formalizing the request for generality and validity of scientific statements, 1977, on-line at http://www.rasch.org/memo18.htm
26. Wikipedia contributors, The wisdom of crowds, Wikipedia, The Free Encyclopedia, http://en.wikipedia.org/w/index.php?title=The*WisdomofCrowdsoldid*=218990817 (accessed June 15, 2008).
27. Assessment Working Group, ACF, URL: http://autonomic-communication-forum.org/node/64
28. SCC41 (IEEE P1900) Plenary Meeting, Working Group 4 Overview and Report, Dublin, Ireland, April 20, 2007.

Index

A

B

C

D

E

H

I

J

L

M

N

O

Q

R

S

T

U

V

W